ACTIVE-LEARNING WORKBOOK

Second Edition

Erin C. Amerman
Florida State College at Jacksonville

Editor-in-Chief: Serina Beauparlant
Courseware Portfolio Manager: Jennifer McGill Walker
Content Producers: Dapinder Dosanjh and Jessica Picone
Managing Producer: Nancy Tabor
Courseware Director, Content Development: Barbara Yien
Full-Service Vendor: Cenveo® Publisher Services
Main Text Cover Designer: Jeff Puda
Copy-editor: Joanne E. (Bonnie) Boehme
Manufacturing Buyer: Stacey Weinberger
Director of Product Marketing: Allison Rona
Senior A&P Specialist: Derek Perrigo

Cover Photo Credit: GettyImages/Ascent Xmedia

www.pearson.com

ISBN-13: 978-0-134-75750-6
ISBN-10: 0-134-75750-5

Contents

To the Professor

You may be wondering, "Why a workbook?" And that's a fair question to ask. With all of the technology available, why produce something as decidedly low-tech as a workbook?

The answer is simple: It works. Many studies have been published in the last few years that have shown that students learn best when they are *actively* engaged in the material. Unfortunately, reading a textbook is largely a passive process. Some students make it more active by highlighting or taking notes, but still the majority of textbook reading is passive. For this reason, students often don't retain what they have read. This isn't unique to students—we have all had that moment when we finished a paragraph or page and had to go back and re-read it because we can't remember anything that we just read. For us it's an annoyance, but for our students, it leads to feelings of frustration, a lot of wasted time, and poor performance on exams.

This companion workbook, *Active-Learning Workbook* (*ALW*), aims to relieve these feelings of frustration and help students make better use of their time. How? By making their textbook reading a more active process. Unlike most workbooks and study guides that help students review material after they have read a chapter, *ALW* is intended to be used *while* the student reads the chapter.

Each chapter in *ALW* is broken into modules that correspond precisely with the modules in the Amerman textbook. At the beginning of each module, the student is encouraged to practice the SQ3R method of textbook reading that they were taught in Chapter 1 of their textbook. Following key terms and a question survey are a variety of exercises that students are asked to complete while they read, including identification, building summary tables, coloring, drawing, describing, fill-in-the-blank, and more. In addition, most chapters have "Team Up" exercises that encourage students to work together in groups.

The wide variety of activities in *ALW* will not only help students retain and understand more of what they read, but it will also allow them to take charge of their learning. They can choose the activities that benefit them the most and even invent new activities. Assuming responsibility for how they learn will benefit them not just in this course, but in all aspects of their lives. **This workbook can be packaged as a print supplement with the companion textbook at no additional cost.** It is also available in downloadable PDF files in the Study Area of Mastering™ Anatomy & Physiology. You can also access the workbook chapters in editable Word files in the Instructor Resources in Mastering™ Anatomy & Physiology, along with the answer key to the exercises.

To the Student

Have you ever wondered why you can sometimes read an entire page in a textbook and not remember a single detail by the time you reach the end of the page? Well, there are two main problems that can cause this. Problem 1 is that the material is brand new to you, so you don't have any frame of reference to help your brain form memories and connections. Problem 2 is that reading is largely a passive process, which doesn't necessarily engage multiple parts of your brain, making the information less likely to be encoded in your long-term memory.

Another potential problem is lack of interest in the subject, but we all know that that isn't the case with A&P, right?

To help prevent this from happening and make the most out of your study time, we present you with the *Active-Learning Workbook*, a companion workbook to go with your text. This book isn't intended to be used as a separate study guide. Instead, you are meant to use it as you read through a chapter in the textbook. Here's how it works:

- Each module opens with a list of key terms for you to define *before* you read the module. These terms can be found defined in the glossary or in the module itself in boldface. After you have defined the terms, you are asked to survey and form questions about the module's content. Both of these activities help to solve Problem 1, because now the material and terms are more familiar to you.

- Following the survey are several different types of activities, including drawing, labeling, sequencing, tracing, and more. Each activity is intended to get you to actually *do* something while you're reading. So, for example, as you read about the parts of the cell, you label a cell diagram and write out the functions of its organelles. These exercises were designed to solve Problem 2, as they make the reading active, which engages more regions of your brain. The more parts of your brain that you engage in the learning process, the easier it is for your brain to encode that information in your memories.

So what are you waiting for? Turn the page and let's get started!

My A&P Course Reference Sheet

Following is a reference sheet that you can use to keep all of the relevant data about your A&P course and study partners handy and organized. Be sure to fill this out during your first week of classes, and update it as needed.

About the course:

Section number: ____________________

Professor: ____________________

Professor's phone number: ____________________

Professor's email/website: ____________________

Professor's office hours: ____________________

Class time and room: ____________________

Lab time and room: ____________________

Required materials to bring to class: ____________________

____

Required materials to bring to lab: ____________________

____

Teaching assistant's contact information: ____________________

Exams and quizzes:

Number of class exams/quizzes: ____________________

Format of exams/quizzes: ____________________

Dates of class exams/quizzes: ____________________

____

Number of lab exams/quizzes: ____________________

Format of lab exams/quizzes: ____________________

Dates of lab exams/quizzes: ____________________

____

Other projects/assignments: ____________________

____

Due dates for other projects/assignments:

____

Study resources:

Course website: ____________________

Mastering A&P Login: ____________________

Computer/learning center room and hours: ____________________

Open lab hours: ____________________

Other class-specific or college-specific resources: ____________________

____

Lab group/Study partners:

Lab group members' names and contact information:

- ____________________
- ____________________
- ____________________
- ____________________
- ____________________

Other study partners' names and contact information:

- ____________________
- ____________________
- ____________________
- ____________________
- ____________________

Scheduled study group meeting times:

- ____________________
- ____________________
- ____________________
- ____________________
- ____________________

1 Introduction to Anatomy and Physiology

In this chapter, we begin our discussion of the world of anatomy and physiology with a look at the language of A&P, how the body is organized, and some of the core principles that you will see repeatedly throughout the course. The chapter also addresses something that is likely already on your mind: how to succeed in your anatomy and physiology course.

What Do You Already Know?

Before we take a journey into the unknown, it is a good idea to inventory what you already know—or at least what you *think* you already know. Try to answer the following questions before proceeding to the next section. If you're unsure of the correct answers, give it your best attempt based on previous courses, previous chapters, or just your general knowledge.

- What is the smallest unit of life?

  ____

- What are the characteristics of living organisms?

  ____

- What is homeostasis?

  ____

Module 1.1: How to Succeed in Your Anatomy and Physiology Course

Module 1.1 in your text gives you tips on the many ways to successfully use this text and its associated resources and study for your quizzes and exams. By the end of the module, you should be able to do the following:

1. Describe how to determine your learning modality, read a textbook, budget your time, and study for quizzes and exams.
2. Explain how to make the best use of class and laboratory time.
3. Describe how to use this book and its associated materials.

Think About It: Learning Modalities

The following table features a list of learning modalities and strategies for studying that tend to suit each modality. Circle each strategy that has worked for you in the past or that you think would likely work for you now.

Learning Modalities and Study Strategies

Learning Modality	Study Strategies
Visual	• Make flash cards for terms or concepts • Color-code your notes, flash cards, and diagrams • Write notes for yourself in your own words to describe text and figures • Rewrite terms and facts that you need to remember repeatedly • Write mini-lectures and teach members of a group about a topic • Build your own tables and diagrams • Use coloring, tracing, and drawing exercises
Auditory	• Record your lectures if your professor gives you permission; listen to the recordings before you go to sleep and while you are in the car • Talk aloud as you read or make notes into a recording device so that you can play back the recording later • Work with a study partner or study group and discuss the concepts and facts together • Develop teaching activities in which you teach members of your study group about a specific topic (examples of such activities are available in this workbook)
Tactile	• Spend extra time in lab manipulating three-dimensional models and experiments • Do experiments to see how things work
Kinesthetic	• Watch software animations that model anatomical structures and physiological processes • Work with a partner or study group • Develop demonstrations that show key principles (examples of such activities are available in this workbook)

After reading through this table and circling different studying methods, take a look at the choices you selected. Is there a trend? Do you seem to favor one learning modality most? If so, what is your learning modality?

____

Practice It: Survey and Question

Your text discusses the SQ3R approach to textbook reading. Since there's no time like the present, why not start now? Go ahead and survey the last four modules of this chapter. Then, formulate at least two questions per module that you will be able to answer as you read and write them here.

Question 1: ________________________________

Answer: ________________________________

Question 2: ________________________________

Answer: ________________________________

Question 3: ________________________________

Answer: ________________________________

Question 4: ______________________________

Answer: ______________________________

Question 5: ______________________________

Answer: ______________________________

Question 6: ______________________________

Answer: ______________________________

Question 7: ______________________________

Answer: ______________________________

Question 8: ______________________________

Answer: ______________________________

Key Concept: How can you use this text and its associated resources to help you succeed?

____

____

Module 1.2: Overview of Anatomy and Physiology

Now we start getting to the "heart" of the matter (no pun intended) as we begin to explore the human body. When you finish this module, you should be able to do the following:

1. List the characteristics of life and the processes carried out by living organisms.
2. Describe the major structural levels of organization in the human body and explain how they relate to one another.
3. Define the types of anatomy and physiology.
4. Describe the organ systems of the human body and their major components.
5. Explain the major functions of each organ system.

Build Your Own Glossary

Below is a table with a list of key terms from Module 1.2. Before you read the module, use the glossary at the back of your book or look through the module to define the following terms.

Key Terms of Module 1.2

Term	Definition
Cell	
Chemical	

Term	Definition
Metabolism	
Excretion	
Tissue	
Organ	
Organ system	
Gross anatomy	
Microscopic anatomy	
Histology	
Cytology	

Key Concept: A virus is unable to carry out its own metabolic reactions or reproduction, and it relies on a host cell that it infects in order to perform these processes. Is a virus alive? Why or why not?

____

____

Describe It: How to Build a Human

Imagine you are teaching a group of high school students about the human body, and you have decided to give them a "recipe for building a human." Write a paragraph or draw a diagram to explain to your students how a human is "built," starting with the most basic level of organization and ending with a fully functional organism.

Build Your Own Summary Table: Organ Systems

As you read Module 1.2, build your own summary table about the 11 organ systems by filling in the information below.

Summary of Organ Systems

Organ System	Main Organs	Main Functions
Integumentary system		
Skeletal system		

Muscular system		
Nervous system		
Endocrine system		
Cardiovascular system		
Lymphatic system		
Respiratory system		
Digestive system		
Urinary system		
Reproductive system		

Module 1.3: The Language of Anatomy and Physiology

Learning a new subject often feels like studying a foreign language, and nowhere is that feeling more prevalent than in anatomy and physiology. This module in your text introduces you to the language of A&P and some of the most common terms you will encounter throughout this book. When you complete this module, you should be able to do the following:

1. Describe a person in anatomical position.
2. List and define the major directional terms used in anatomy.
3. Describe locations of body structures using regional and directional terminology.
4. List and describe the location of the major anatomical regions of the body.
5. Identify the various planes in which a body or body part might be dissected, and describe the appearance of a body sectioned along each of those planes.

Build Your Own Glossary

Below is a table with a list of key terms from Module 1.3. Before you read the module, use the glossary at the back of your book or look through the module to define the following terms.

Key Terms of Module 1.3

Term	Definition
Anatomical position	
Axial region	
Appendicular region	
Sagittal plane	
Frontal plane	
Transverse plane	

Key Concept: Why is anatomical position important?

____

____

Build Your Own Summary Table: Directional Terms

The best way to master the directional terms is simply to practice them over and over again. So first, define each pair of directional terms in the following table for your reference. Then, it's time to practice. Come up with 10 statements using directional terms, similar to those in Figure 1.6 in your text. For extra practice, work with a partner and critique and correct each other's statements to ensure that you both are on the right track.

Definitions of Directional Terms

Term	Definition	Term	Definition
Anterior		Posterior	
Superior		Inferior	
Medial		Lateral	
Proximal		Distal	
Superficial		Deep	

Practice

- ______________________
- ______________________
- ______________________
- ______________________
- ______________________
- ______________________
- ______________________
- ______________________
- ______________________
- ______________________

Key Concept: How do proximal and distal differ from superior and inferior?

____

____

Identify It: The Regions of the Body

Color and label Figure 1.1 with noun or adjectival terms for the regions of the body as you read Module 1.3.

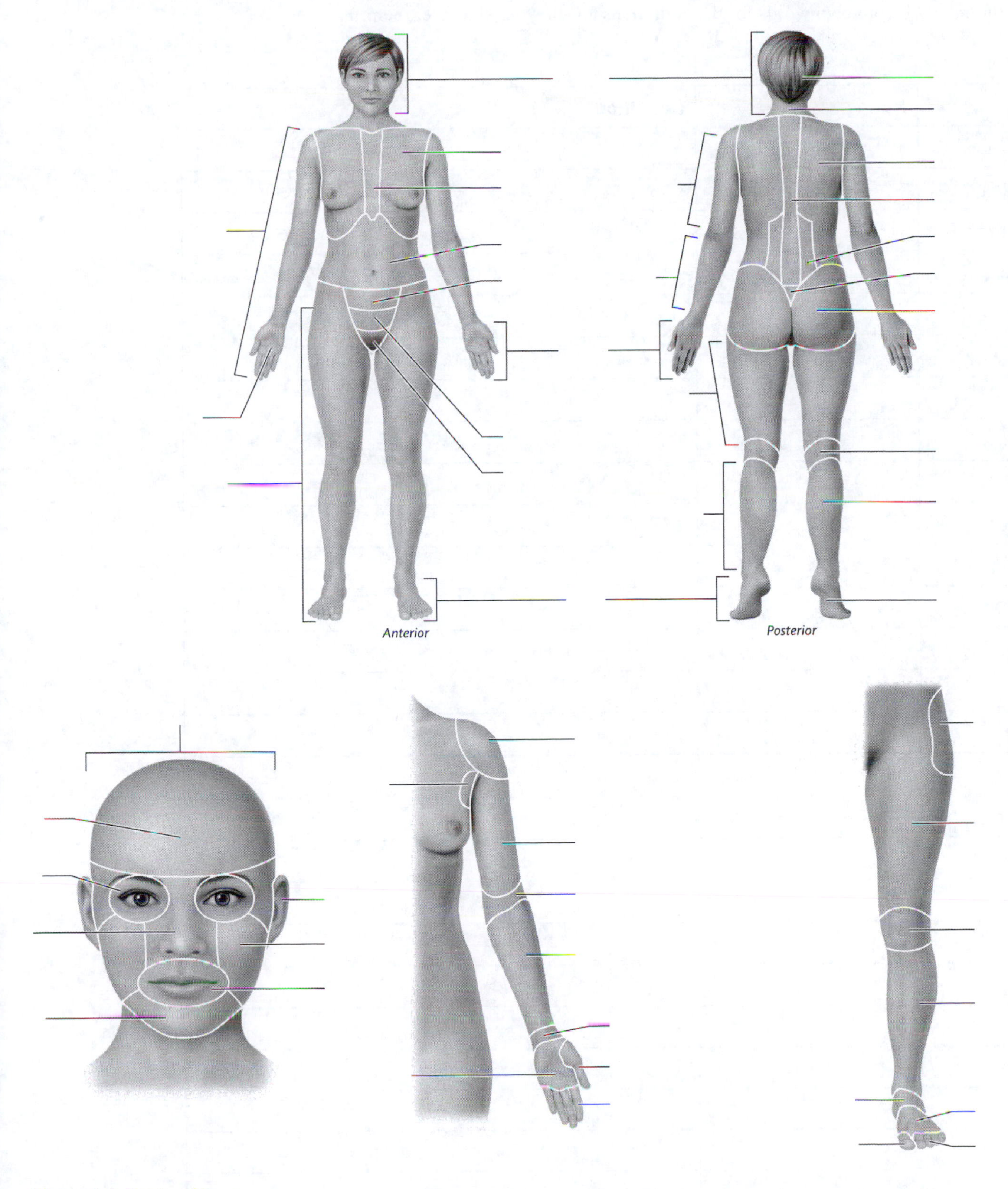

Figure 1.1 Regional terms.

Build Your Own Summary Table: Regional Terms

Now that you've labeled all the regions, you can build yourself a summary table of the regional terms. The table below contains each regional term in its adjectival format. Complete the table by filling in the noun form of the term and its definition. Use your own words for the definitions to help you remember them.

Regional Terms

Adjective	Noun	Definition
Abdominal region		
Acromial region		
Antebrachial region		
Antecubital region		
Aural (or otic) region		
Axillary region		
Brachial region		
Buccal region		
Carpal region		
Cephalic region		
Cervical region		
Coxal region		
Cranial region		
Crural region		
Digital region		
Dorsal region		
Femoral region		
Frontal region		
Gluteal region		
Hallucis region		
Inguinal region		
Lumbar region		
Manual region		
Mental region		
Metacarpal region		
Metatarsal region		

Nasal region		
Occipital region		
Ocular region		
Oral region		
Palmar region		
Patellar region		
Pelvic region		
Pedal region		
Plantar region		
Pollicis region		
Popliteal region		
Sternal region		
Sural region		
Tarsal region		
Thoracic region		
Vertebral region		

Draw It: Planes of Section

In the boxes provided, draw a face or some other body part, then draw the section indicated by the text underneath the box.

Transverse section

Parasagittal section

Frontal section

Midsagittal section

Key Concept: Why do we use sections in anatomy?

___

___

Module 1.4: The Organization of the Human Body

Module 1.4 in your text teaches you about the basic layout of the human body and its many fluid-filled cavities. At the end of this module, you should be able to do the following:

1. Describe the location of the body cavities, and identify the major organs found in each cavity.
2. Describe the location of the four quadrants and nine regions of the abdominopelvic cavity, and list the major organs located in each.
3. List and describe the serous membranes that line the body cavities.

Build Your Own Glossary

Following is a table with a list of key terms from Module 1.4. Before you read the module, use the glossary at the back of your book or look through the module to define the terms.

Key Terms of Module 1.4

Term	Definition
Posterior body cavity	
Cranial cavity	
Vertebral cavity	
Anterior body cavity	
Thoracic cavity	
Mediastinum	
Pleural cavity	
Pericardial cavity	
Abdominopelvic cavity	
Peritoneal cavity	
Serous membrane	

Identify It: Body Cavities

Color and label the body cavities illustrated in Figure 1.2.

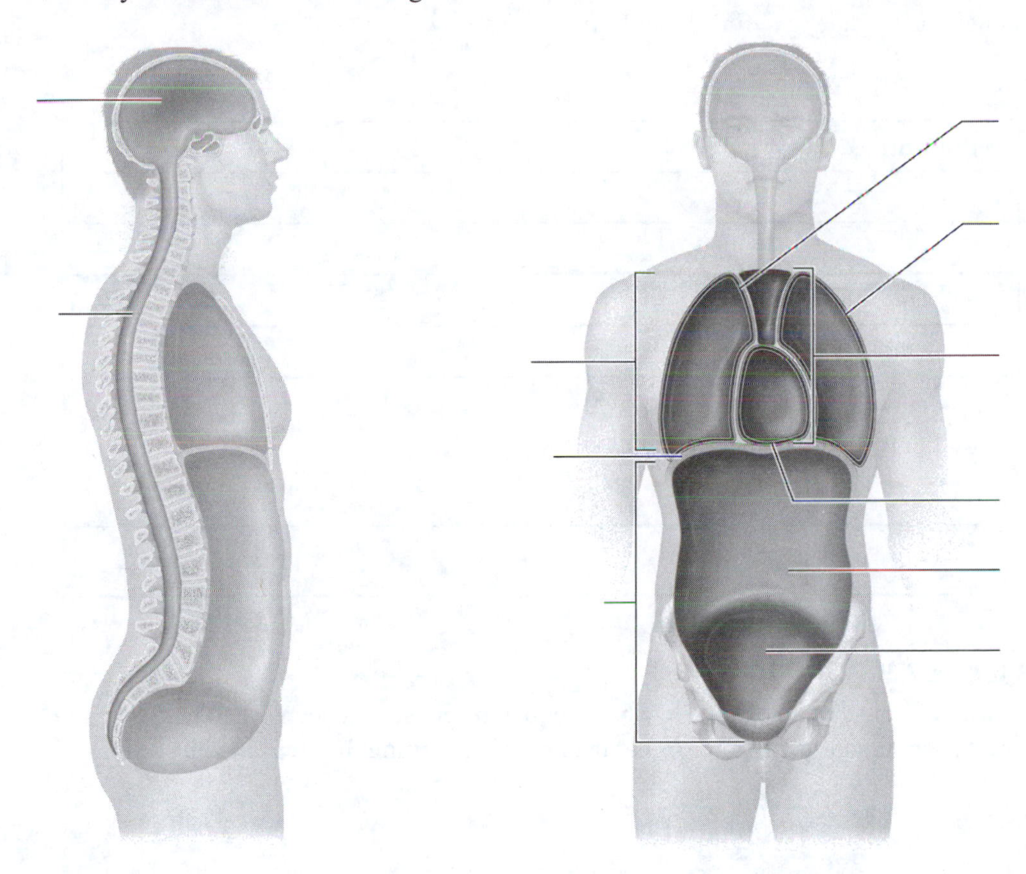

Figure 1.2 The posterior and anterior body cavities.

Key Concept: How does a serous membrane envelop an organ? What is its function?

____

____

Module 1.5: Core Principles in Anatomy and Physiology

There are certain themes and principles that we see repeatedly in A&P. We call these the "Core Principles," and Module 1.5 introduces you to them. At the end of this module, you should be able to do the following:

1. Describe the principle of homeostasis.
2. Describe the components of a feedback loop, and explain the function of each component.
3. Compare and contrast negative and positive feedback, and explain why negative feedback is the most commonly used mechanism to maintain homeostasis in the body.
4. Describe how structure and function are related.
5. Define the term gradient, and give examples of the types of gradients that drive processes in the body.
6. Describe how cells communicate with one another, and state why such communication is necessary in a multicellular organism.

Build Your Own Glossary

Below is a table with a list of key terms from Module 1.5. Before you read the module, use the glossary at the back of your book or look through the module to define the following terms.

Key Terms of Module 1.5

Term	Definition
Homeostasis	
Negative feedback loop	
Positive feedback loop	
Gradient	

Key Concept: What does it mean to maintain homeostasis in the body?

____

____

Describe It: A Negative Feedback Loop

Fill in the boxes in Figure 1.3 with the steps of a negative feedback loop in response to a decrease in body temperature. Try to describe each step in your own words rather than simply rewriting the steps in your text.

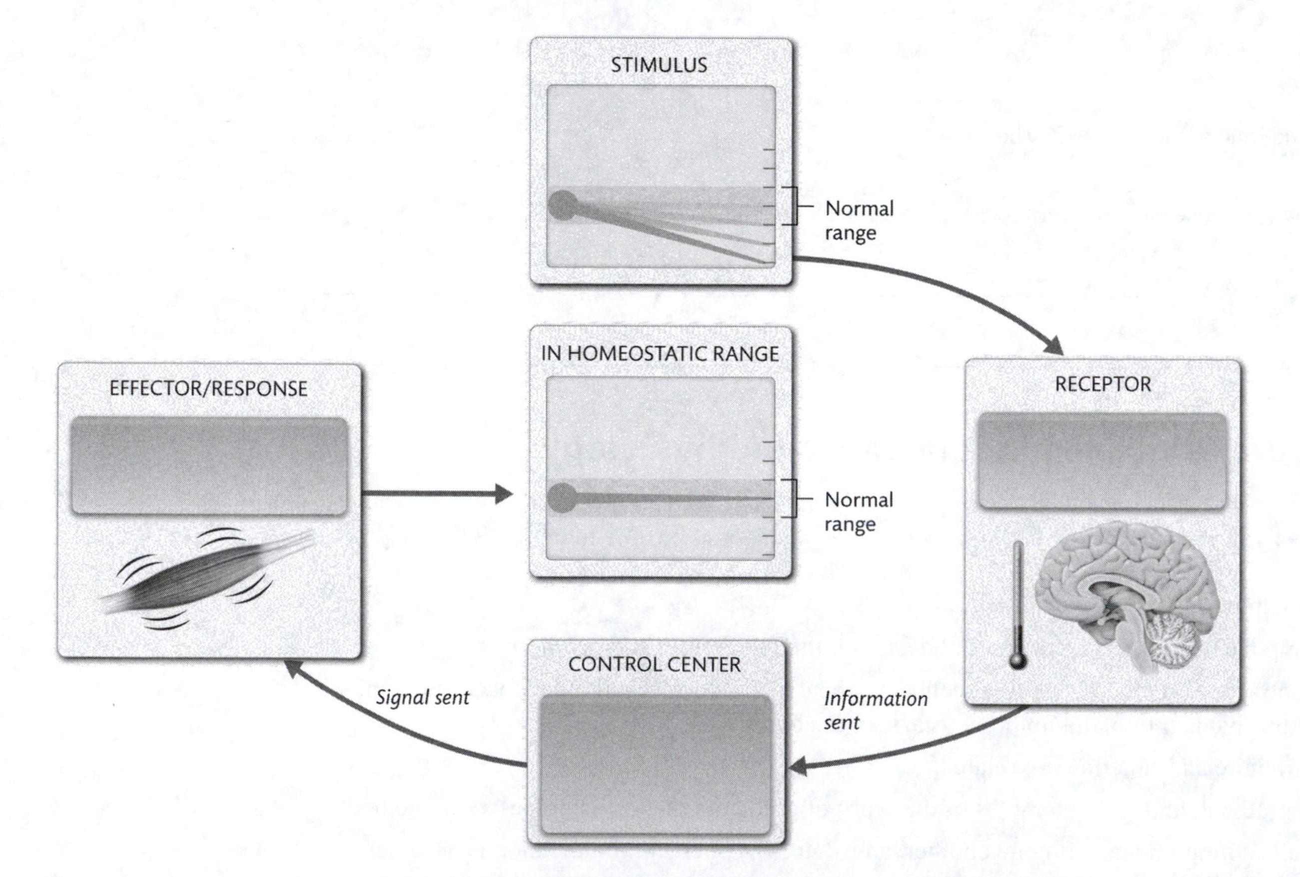

Figure 1.3 Comparison of how negative feedback mechanisms control room and body temperature.

Team Up

Make a handout to teach the differences between positive and negative feedback loops. You can use Figures 1.13 and 1.14 in your text as a guide, but the handout should be in your own words and with your own diagrams. At the end of the handout, write a few quiz questions. Once you have completed your handout, team up with one or more study partners and trade handouts. Study your partners' diagrams, and when you have finished, take the quiz at the end. When you and your group have finished taking all the quizzes, discuss the answers to determine places where you need additional study. After you've finished, combine the best elements of each handout to make one "master" diagram for the differences between positive and negative feedback loops.

Key Concept: How might a malfunctioning receptor affect the functioning of a negative feedback loop?

____

____

Key Concept: Think of some examples of commonly known organs (e.g., a bone, the brain, the trachea ["windpipe"], and the stomach). How would their functions be affected by a change in structure? For example, how would the stomach function if it were made of solid muscle instead of hollow?

____

____

Describe It: The Core Principles

Identify each of the four core principles represented by the illustrations in Figure 1.4 below. Then, in your own words, write a sentence or two describing the core principle and explaining why it is important to anatomy and physiology.

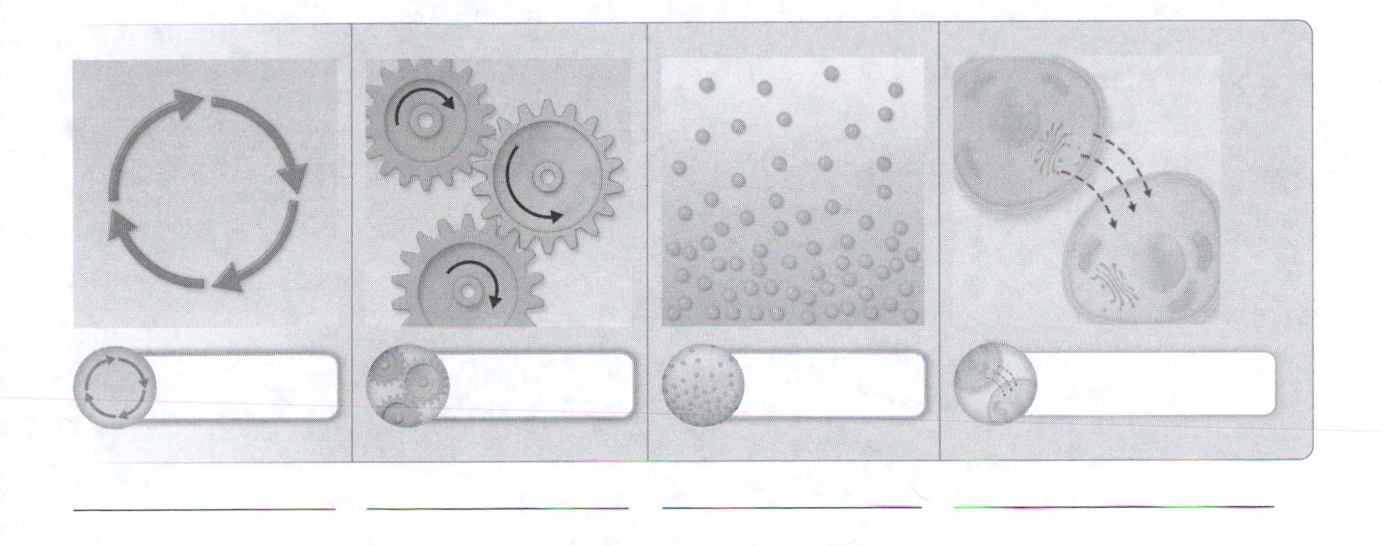

Figure 1.4 The core principle icons.

What Do You Know Now?

Let's now revisit the questions you answered at the beginning of this chapter. How have your answers changed now that you've worked through the material?

- What is the smallest unit of life?

  ____

- What are the characteristics of living organisms?

  ____

- What is homeostasis?

  ____

2 The Chemistry of Life

Our exploration of the human body starts at the smallest level of organization: the chemical level. This chapter introduces you to basic principles of chemistry, which form the foundation for both our anatomy and our physiology.

What Do You Already Know?

Before we take a journey into the unknown, it is a good idea to inventory what you already know—or at least what you *think* you already know. Try to answer the following questions before proceeding to the next section. If you're unsure of the correct answers, give it your best attempt based on previous courses, previous chapters, or just your general knowledge.

- What is the basic structure of an atom?

  ____

- Which elements make up the human body?

  ____

- What are the major macromolecules in the body?

  ____

Module 2.1: Atoms and Elements

Module 2.1 in your text introduces you to the chemical level of organization and the fundamental unit of chemicals, the atom. By the end of the module, you should be able to do the following:

1. Describe the charge, mass, and relative location of electrons, protons, and neutrons.
2. Compare and contrast the terms atoms and elements.
3. Identify the four major elements found in the human body.
4. Compare and contrast the terms atomic number, mass number, isotope, and radioisotope.
5. Explain how isotopes are produced.

Build Your Own Glossary

Below is a table listing key terms from Module 2.1. Before you read the module, use the glossary at the back of your book or look through the module to define the following terms.

Key Terms for Module 2.1

Term	Definition
Matter	
Chemistry	
Atom	
Proton	
Neutron	
Electron	
Electron shell	
Atomic number	
Element	
Mass number	
Isotope	

Survey It: Form Questions

Let's practice the SQ3R method of textbook reading that we introduced in Chapter 1. Before you read the module, survey it and form at least three questions for yourself. When you have finished reading the module, return to these questions and answer them.

Question 1: ______________________________

Answer: ______________________________

Question 2: ______________________________

Answer: ______________________________

Question 3: ______________________________

Answer: ______________________________

Draw It: The Atom

Draw the following atoms using a classic model of the atom. Remember to have the appropriate number of electrons in each electron shell.

Oxygen: 6 protons and 6 neutrons

Magnesium: 12 protons and 12 neutrons

Fluorine: 9 protons and 9 neutrons

Key Concept: What makes hydrogen unusual in its placement in the periodic table?

____

____

Key Concept: Why does knowing an atom's atomic number allow you to determine its number of electrons? Can you determine how many neutrons an atom has by its atomic number? Why or why not?

____

____

Module 2.2: Matter Combined: Mixtures and Chemical Bonds

Now we look at what happens when we combine two or more atoms and the ways in which we can combine them. When you finish this module, you should be able to do the following:

1. Describe the three different types of mixtures, and explain the difference between a solvent and a solute.
2. Explain how the number of electrons in an atom's valence shell determines its stability and ability to form chemical bonds.
3. Distinguish between the terms molecule, compound, and ion.
4. Explain how ionic, nonpolar covalent, and polar covalent bonds form, and discuss their key differences.
5. Describe how hydrogen bonds form and how they give water the property of surface tension.

Build Your Own Glossary

Below is a table listing key terms from Module 2.2. Before you read the module, use the glossary at the back of your book or look through the module to define the following terms.

Key Terms for Module 2.2

Term	Definition
Mixture	
Molecule	
Suspension	
Colloid	
Solution	
Chemical bond	
Valence electron	
Ionic bond	
Ion	
Nonpolar covalent bond	
Polar covalent bond	
Hydrogen bond	

Survey It: Form Questions

Before you read the module, survey it, and form at least three questions for yourself. When you have finished reading the module, return to these questions and answer them.

Question 1: ______________________________

Answer: ______________________________

Question 2: ______________________________

Answer: ______________________________

Question 3: ______________________________

Answer: ______________________________

Identify It: Mixtures

Identify each of the mixtures in Figure 2.1. Then, write the main properties of each type of mixture.

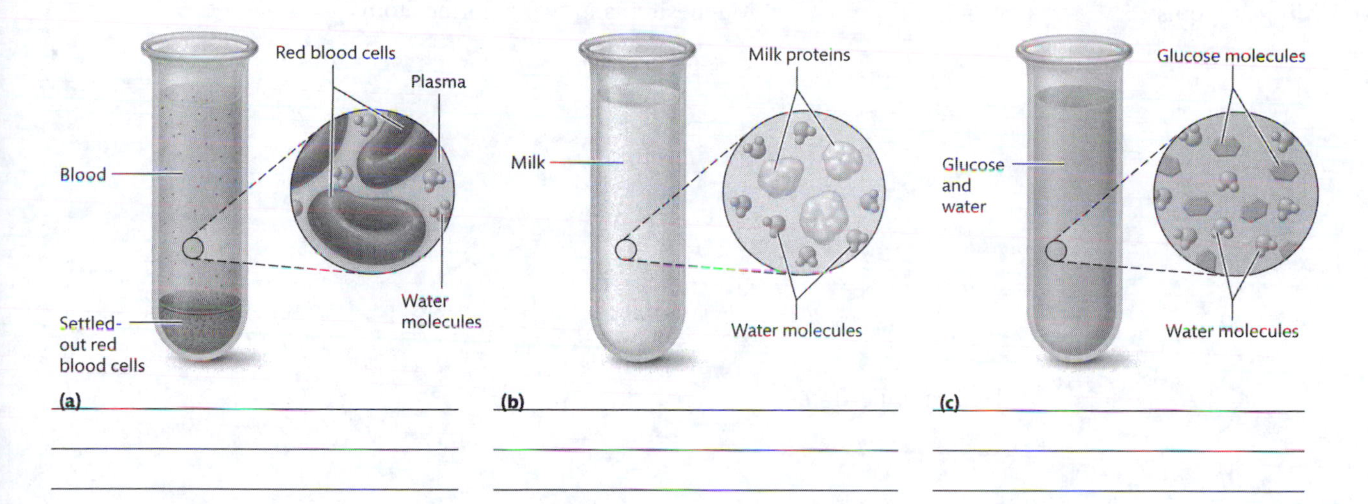

Figure 2.1 Types of mixtures.

Key Concept: What is the key difference between the components of a mixture and the components of a chemical bond?

Key Concept: Which electrons are important to consider in chemical bonding? Why are these electrons so important?

Draw It: Atoms to Ions

How many electrons would each of the following elements need to gain or lose to become stable? Draw the resulting particle, and indicate its charge. (Refer to your drawings in *Draw It: The Atom* on p. 17 in this workbook for help.)

Oxygen:

Magnesium:

Fluorine:

Draw It: Chemical Bonds

Draw and identify the bonds that would result if the following atoms formed chemical bonds:

Oxygen and two hydrogen atoms:

Magnesium and two chlorine atoms:

Two fluorine atoms:

Key Concept: What is the fundamental difference between an ionic and a covalent bond?

____

____

Team Up

Make a handout to teach the differences between ionic, polar covalent, and nonpolar covalent bonds. You can use the Concept Boost in your text on pages 40–41 as a guide, but the handout should be in your own words and with your own diagrams. At the end of the handout, write a few quiz questions. Once you have completed your handout, team up with one or more study partners and trade handouts. Study your partners' diagrams, and when you have finished, take the quiz at the end. When you and your group have finished taking all the quizzes, discuss the answers to determine places where you need additional study. After you've finished, combine the best elements of each handout to make one "master" diagram for the differences between ionic, polar covalent, and nonpolar covalent bonds.

Key Concept: How do hydrogen bonds differ from polar covalent bonds?

____

____

Module 2.3: Chemical Reactions

The number of chemical reactions that occur in the human body each second is staggering, totaling in the hundreds of trillions to quadrillions. This module introduces you to how chemical reactions occur and the many different factors that regulate them. When you complete this module, you should be able to do the following:

1. Explain what happens during a chemical reaction.
2. Describe the forms and types of energy, and apply the principles of energy to chemical bonds and endergonic and exergonic reactions.
3. Describe and explain the differences between the three types of chemical reactions.
4. Describe the factors that influence reaction rates.
5. Explain the properties, actions, and importance of enzymes.

Build Your Own Glossary

Below is a table listing key terms from Module 2.3. Before you read the module, use the glossary at the back of your book or look through the module to define the following terms.

Key Terms of Module 2.3

Term	Definition
Chemical reaction	
Reactant	
Product	
Potential energy	
Kinetic energy	
Endergonic reaction	
Exergonic reaction	
Catabolic reaction	
Exchange reaction	
Anabolic reaction	
Activation energy	
Enzyme	

Survey It: Form Questions

Before you read the module, survey it and form at least three questions for yourself. When you have finished reading the module, return to these questions and answer them.

Question 1: ______________________________

Answer: ______________________________

Question 2: ______________________________

Answer: ______________________________

Question 3: ______________________________

Answer: ______________________________

Complete It: Energy and the Cell

Fill in the blanks to complete the following paragraph that describes the types of energy and how it applies to the cell.

Energy is defined as the capacity to do ____________. Energy that is ready to be released is called ____________ ____________, whereas energy in motion is called ____________ ____________. Chemical energy is the energy inherent in ____________ ____________. The energy generated by the movement of charged particles is called ____________ ____________. Energy transferred from one body to another is known as an ____________ ____________.

Key Concept: You are in a lab trying to carry out a reaction. You find that if you do nothing, the reaction does not proceed. But if you add heat, the reaction proceeds to completion. Is this reaction endergonic or exergonic? Why?

____

____

Identify It: Chemical Reactions

Identify each of the following types of reactions:

$MgCl_2 + 2LiI \rightarrow MgI_2 + 2LiCl$

$C_6H_{12}O_6 + C_6H_{12}O_6 \rightarrow C_{12}H_{22}O_{11} + H_2O$

$NaOH \rightarrow Na^+ + OH^-$

Draw It: Reaction Rates

In the boxes provided, draw a curve on a graph to represent the type of reaction indicated below. Label your drawings with the following terms: transition state, activation energy, reactants, and products.

Endergonic reaction, no enzyme	**Exergonic reaction**	**Endergonic reaction with enzyme**

Key Concept: How important are enzymes to our biology? Explain.

____

____

Module 2.4: Inorganic Compounds: Water, Acids, Bases, and Salts

Module 2.4 in your text explores biochemistry, teaching you about the basic properties and physiological importance of many different inorganic compounds. At the end of this module, you should be able to do the following:

1. Discuss the physiologically important properties of water.
2. Explain why certain molecules and compounds are hydrophilic and why others are hydrophobic.
3. Describe the properties of acids and bases with respect to hydrogen ions.
4. Explain what the pH scale represents, and why a given value is acidic, neutral, or basic.
5. Explain the function of a buffer.
6. Define the terms salt and electrolyte, and give examples of physiological significance.

Build Your Own Glossary

Below is a table listing key terms from Module 2.4. Before you read the module, use the glossary at the back of your book or look through the module to define the following terms.

Key Terms of Module 2.4

Term	Definition
Inorganic compound	
Organic compound	
Water	

Term	Definition
Hydrophilic	
Hydrophobic	
Acid	
Base	
pH scale	
Buffer	
Salt	

Survey It: Form Questions

Before you read the module, survey it and form at least three questions for yourself. When you have finished reading the module, return to these questions and answer them.

Question 1: ______________________________

Answer: ______________________________

Question 2: ______________________________

Answer: ______________________________

Question 3: ______________________________

Answer: ______________________________

Describe It: The Properties of Water

Write a paragraph as if you were explaining to an audience of nonscientists why the properties of water make it a good solvent for living organisms.

Key Concept: What makes a nonpolar covalent molecule hydrophobic?

___

___

Practice It: Hydrophobic and Hydrophilic Molecules

Identify each of the following molecules as being ionic, polar covalent, or nonpolar covalent. Then, determine whether the molecule is likely to be hydrophilic or hydrophobic.

Molecule	Type of Molecule	Hydrophilic or Hydrophobic
I_2		
$C_5H_{10}O_5$		
CaF_2		
KCl		
$C_{16}H_{32}O_2$		
$C_2H_5NO_2$		

Think About It: Acids, Bases, and Buffers

You have a beaker filled with pure water that has a pH of 7. Predict what would happen in each of the following situations:

1. You add several milliliters of concentrated HCl:
   a. What happens to the number of hydrogen ions in the solution?
   b. What happens to its pH?
2. You add several milliliters of concentrated NaOH:
   a. What happens to the number of hydrogen ions in the solution?
   b. What happens to its pH?
3. You add a buffer to the water, then add several milliliters of NaOH:
   a. What happens to the number of hydrogen ions in the solution?
   b. What happens to its pH?
4. You add a buffer to the water, then add several milliliters of HCl:
   a. What happens to the number of hydrogen ions in the solution?
   b. What happens to its pH?

Key Concept: Why does the pH value change in the opposite direction of the hydrogen ion concentration?

____

____

Build Your Own Summary Table: Inorganic Molecules

As you read Module 2.4, build your own summary table about the different types of inorganic molecules by filling in the information in the boxes below.

Summary of the Inorganic Molecules of the Human Body

Type of Molecule	Chemical Structure	Main Functions
Water		
Acids		
Bases		
Buffers		
Salts		

Module 2.5: Organic Compounds: Carbohydrates, Lipids, Proteins, and Nucleotides

This module turns to the other facet of biochemistry: organic compounds, their structure, and their physiological roles. At the end of this module, you should be able to do the following:

1. Explain the relationship between monomers and polymers, and describe how they are formed and broken down by dehydration synthesis and hydrolysis reactions.
2. Compare and contrast the general molecular structures of carbohydrates, lipids, proteins, and nucleic acids, and identify their monomers and polymers.
3. Identify examples of carbohydrates, lipids, proteins, and nucleic acids, and discuss their functional and structural roles in the human body.
4. Describe the four levels of protein structure, and explain why protein shape is important for protein function.
5. Describe the reaction for ATP hydrolysis, and explain the role of ATP in the cell.

Build Your Own Glossary

Following is a table listing key terms from Module 2.5. Before you read the module, use the glossary at the back of your book or look through the module to define the following terms.

Key Terms of Module 2.5

Term	Definition
Monomer	
Polymer	
Dehydration synthesis	
Hydrolysis	
Carbohydrate	
Monosaccharide	
Polysaccharide	
Lipid	
Fatty acid	
Triglyceride	
Phospholipid	
Protein	
Amino acid	
Peptide	
Denaturation	
Nucleotide	
Adenosine triphosphate (ATP)	
Deoxyribonucleic acid (DNA)	
Ribonucleic acid (RNA)	

Survey It: Form Questions

Before you read the module, survey it and form at least three questions for yourself. When you have finished reading the module, return to these questions and answer them.

Question 1: ____________________

Answer: ____________________

Question 2: ____________________

Answer: ____________________

Question 3: ____________________

Answer: ____________________

Draw It: Formation and Breakdown of a Disaccharide

Let's start by building a disaccharide. In the space below, draw the molecular structure of two glucose molecules that then combine by dehydration synthesis to form a molecule of maltose. See Figure 2.14 in your text for the structure of glucose.

Now let's do the opposite and break down a disaccharide: Start by drawing a sucrose molecule (see Figure 2.15 in your text for its structure) and a hydrolysis reaction, including the resulting glucose and fructose molecules.

Key Concept: What is the main role of carbohydrates in the body? In what form are carbohydrates stored?

____

____

Key Concept: Why are lipids nonpolar but carbohydrates polar? What makes phospholipids unique?

____

____

Complete It: Proteins and Amino Acids

Fill in the blanks to complete the following paragraph that describes the types of organic molecules.

A peptide consists of ____________ ____________ joined by ____________ bonds. A long peptide is known as a ____________, and it is not until it is ____________ that it is considered a protein. A protein has ____________ possible levels of structure. A protein's primary structure is its ____________ ____________ sequence. In a protein's secondary structure, the patterns of ____________ ____________ and ____________ ____________ are common. A protein's ____________ structure refers to its three-dimensional shape, and its ____________ structure refers to the assembly of multiple polypeptide chains.

Key Concept: Why is a protein's structure important?

____

____

Identify It: Types of Organic Molecules

Identify each of the organic molecules in Figure 2.2 as a carbohydrate, lipid, amino acid/peptide, or nucleotide/nucleic acid base.

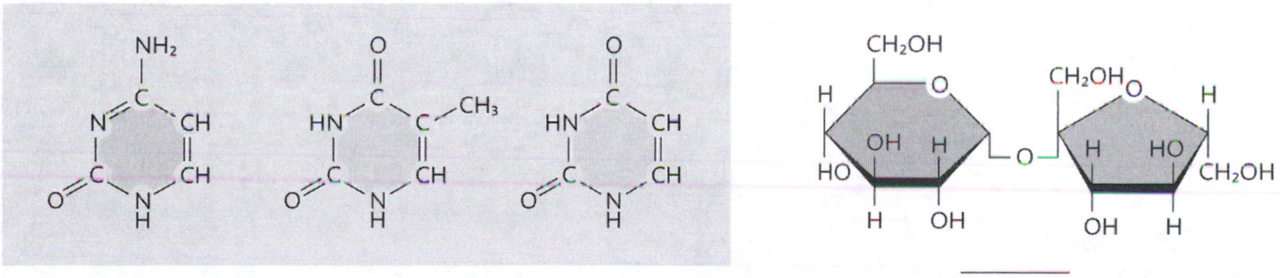

____

____

Figure 2.2 Organic molecules.

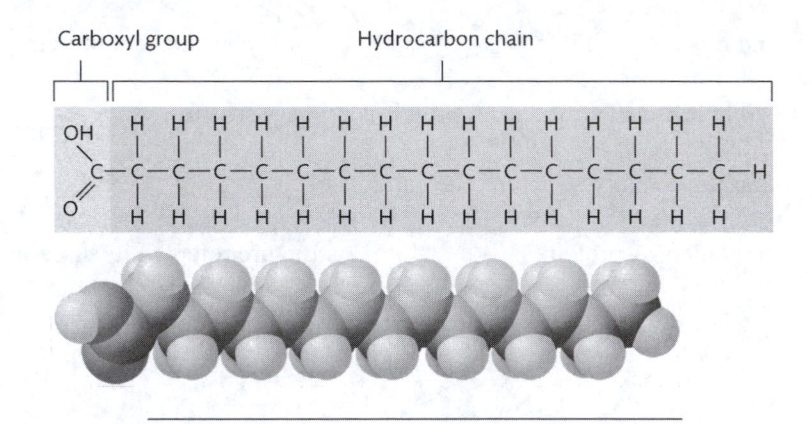

Figure 2.2 (*continued*)

Key Concept: What role does DNA play in the human body?

What Do You Know Now?

Let's now revisit the questions you answered in the beginning of this chapter. How have your answers changed now that you've worked through the material?

- What is the basic structure of an atom?

  ____

- Which elements make up the human body?

  ____

- What are the major macromolecules in the body?

  ____

3 The Cell

We now turn to how chemicals organize into the functional units called cells. This chapter introduces you to the structure and function of cells and their components, along with the manner in which cells produce proteins and reproduce themselves.

What Do You Already Know?

Try to answer the following questions before proceeding to the next section. If you're unsure of the correct answers, give it your best attempt based on previous courses, previous chapters, or just your general knowledge.

- What are the basic components of a cell?

  ____

- What is DNA, and what is its function?

  ____

- How does a cell in the human body reproduce itself? Can all cells reproduce in this manner?

  ____

Module 3.1: Introduction to Cells

Module 3.1 in your text introduces you to the cellular level of organization and the fundamental unit of life, the cell. By the end of the module, you should be able to do the following:

1. Identify the three main parts of a cell, and list the general functions of each.
2. Describe the location and components of intracellular and extracellular fluid.
3. Explain how cytoplasm and cytosol are different.
4. Define the term organelle, and describe the basic functions of organelles.

Build Your Own Glossary

Following is a table listing key terms from Module 3.1. Before you read the module, use the glossary at the back of your book or look through the module to define the following terms.

Key Terms for Module 3.1

Term	Definition
Cell metabolism	
Plasma membrane	
Cytoplasm	
Cytosol	
Organelle	
Cytoskeleton	
Nucleus	

Survey It: Form Questions

Let's practice the SQ3R method of textbook reading that we introduced in Chapter 1. Before you read the module, survey it and form at least two questions for yourself. When you have finished reading the module, return to these questions and answer them.

Question 1: ______________________________

Answer: ______________________________

Question 2: ______________________________

Answer: ______________________________

Key Concept: Why are enzymes so important for cell function? (*Hint*: Consider the basic processes carried out by cells.)

____

____

Practice Drawing: A Generalized Cell

In the space to the right, draw a basic cell like the one in Figure 3.1 in your text. Label your drawings with the following terms: plasma membrane, cytoplasm, cytosol, organelles, cytoskeleton, and nucleus.

Module 3.2: Structure of the Plasma Membrane

Now we look at the structure of our cellular "fence": the plasma membrane. When you finish this module, you should be able to do the following:

1. Describe how lipids are distributed in a plasma membrane, and explain their functions.
2. Describe how carbohydrates and proteins are distributed in a plasma membrane, and explain their functions.
3. Explain the overall structure of the plasma membrane according to the fluid mosaic model.

Build Your Own Glossary

Following is a table listing key terms from Module 3.2. Before you read the module, use the glossary at the back of your book or look through the module to define the following terms.

Key Terms for Module 3.2

Term	Definition
Phospholipid bilayer	
Fluid mosaic model	
Integral protein	
Peripheral protein	
Receptor	
Ligand	

Survey It: Form Questions

Before you read the module, survey it and form at least three questions for yourself. When you have finished reading the module, return to these questions and answer them.

Question 1: ______________________________

Answer: ______________________________

Question 2: ______________________________

Answer: ______________________________

Question 3: ______________________________

Answer: ______________________________

Key Concept: Why is the amphipathic nature of phospholipids important for the plasma membrane? What would happen if phospholipids were either all polar or all nonpolar?

____

____

Practice Labeling: Components of the Plasma Membrane

Identify and color-code each component of the plasma membrane in Figure 3.1. Then, list the main function(s) of each component.

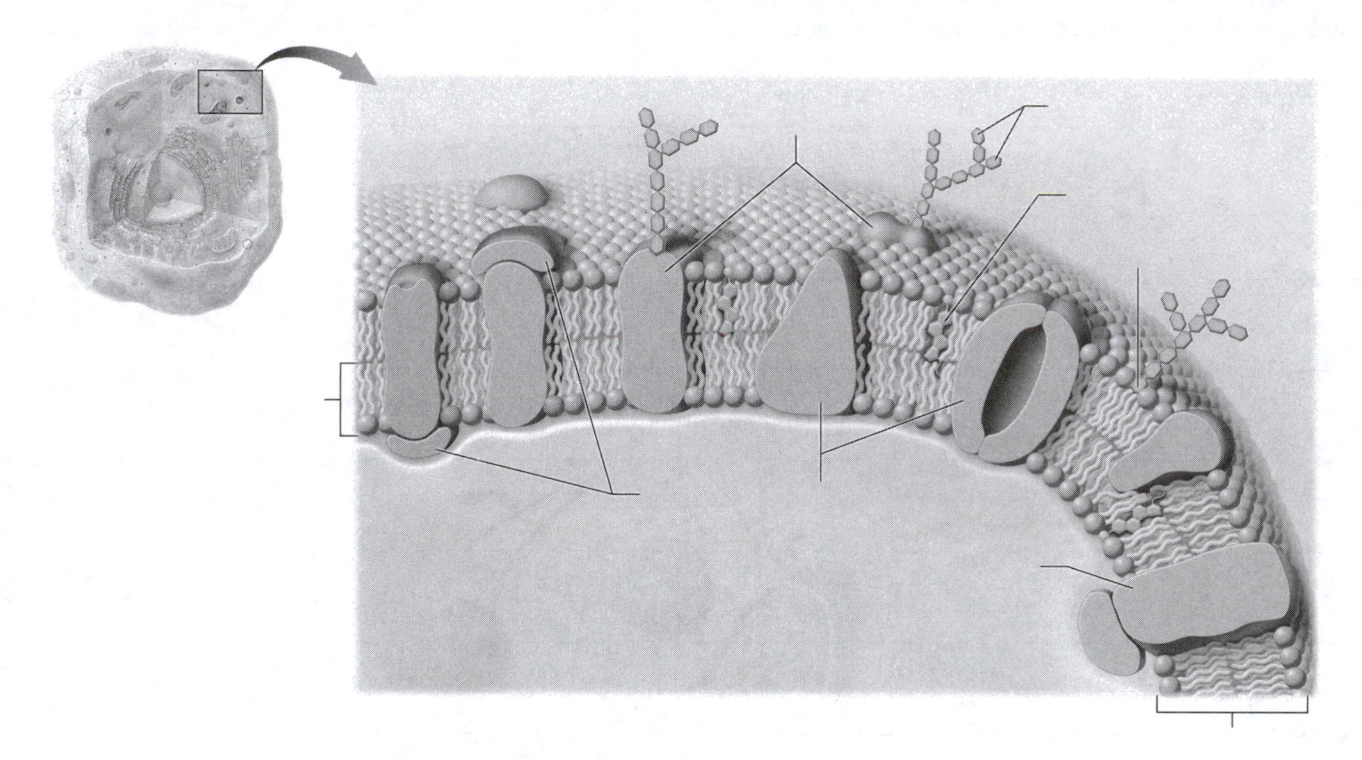

Figure 3.1 The plasma membrane.

Key Concept: What roles do proteins play in the plasma membrane?

____

____

Module 3.3: Transport across the Plasma Membrane

Substances must be able to move in and out of a cell through the plasma membrane in order for the cell to maintain homeostasis and perform its functions. This module explores how transport across the membrane occurs and the many different factors that regulate these processes. When you complete this module, you should be able to do the following:

1. Describe thc energy requirement for and the mechanism by which solute movement occurs in simple and facilitated diffusion.
2. Describe the process of osmosis and the direction of solvent movement.
3. Compare and contrast the effects of hypertonic, isotonic, and hypotonic conditions on cells.
4. Describe the energy requirement for and the mechanism by which solute movement occurs in primary and secondary active transport.
5. Compare and contrast the mechanism by which movement occurs and the types of molecules moved for the different types of vesicular transport.

Build Your Own Glossary

Below is a table listing key terms from Module 3.3. Before you read the module, use the glossary at the back of your book or look through the module to define the following terms.

Key Terms of Module 3.3

Term	Definition
Selectively permeable	
Concentration gradient	
Simple diffusion	
Facilitated diffusion	
Osmosis	
Tonicity	
Isotonic	
Hypertonic	

Term	Definition
Hypotonic	
Primary active transport	
Secondary active transport	
Membrane potential	
Phagocytosis	
Pinocytosis	
Receptor-mediated endocytosis	
Exocytosis	
Transcytosis	

Survey It: Form Questions

Before you read the module, survey it and form at least three questions for yourself. When you have finished reading the module, return to these questions and answer them.

Question 1: ______________________________

Answer: ______________________________

Question 2: ______________________________

Answer: ______________________________

Question 3: ______________________________

Answer: ______________________________

Key Concept: What makes a transport process passive?

____

____

Identify It: Types of Passive Transport

Identify the types of passive transport in the illustrations in Figure 3.2.

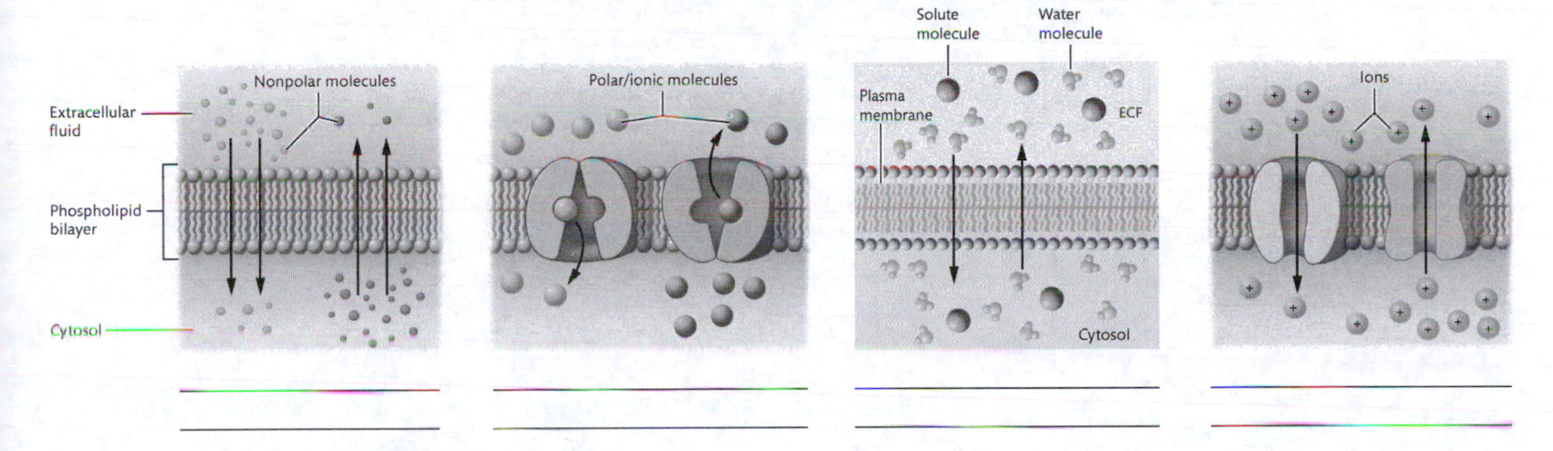

Figure 3.2 Types of passive membrane transport.

Key Concept: How do the forces that drive diffusion and osmosis differ?

____

____

Sequence the Events: Stages of the Na^+/K^+ Pump

Write in the steps of the Na^+/K^+ pump in Figure 3.3. You may use Figure 3.10 in your text as a reference, but write the steps in your own words. Be sure to color-code and label the Na^+ and K^+ so you can see which ions are moving in each step.

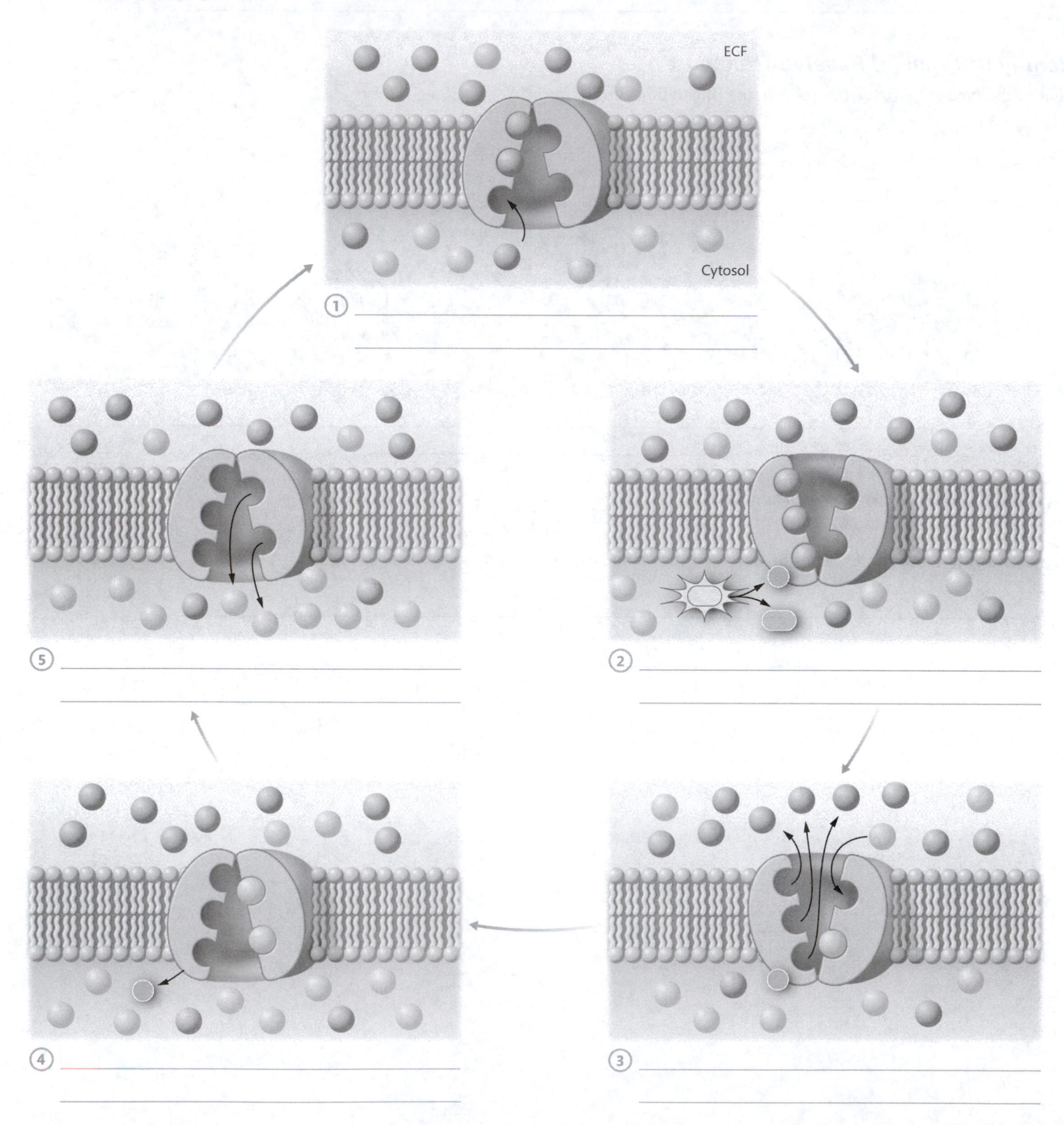

Figure 3.3 The Na^+/K^+ pump.

Key Concept: How is a concentration gradient established in secondary active transport? How does it drive the transport of another molecule?

____

____

Draw It: Vesicular Transport

In the boxes provided, draw what happens during each type of vesicular transport. Label your drawings with the following terms, where appropriate: plasma membrane, pseudopods, receptor, vesicle, ligand, and protein-coated pit.

Phagocytosis	**Pinocytosis**	**Receptor-mediated endocytosis**	**Exocytosis**

Complete It: Transport across Membranes

Fill in the blanks to complete the following paragraphs that describe the types of energy and how they apply to the cell.

A passive process is one that requires no external input of ____________. Diffusion is driven by the potential energy of the ____________ ____________, whereas the driving force in osmosis is called ____________ ____________. ____________ ____________ is a passive process that requires the use of a protein channel or carrier.

Active processes require the input of energy in the form of ____________. ____________ ____________ transport uses this energy source indirectly by creating an unequal distribution of a certain solute. Exocytosis is a type of transport that uses ____________ to move molecules ____________ the cell. Endocytosis uses these structures to transport molecules ____________ the cell.

Team Up

Make a handout to teach the differences between osmosis, the different types of diffusion, primary active transport, secondary active transport, and vesicular transport. You can use Table 3.1 in your text on page 86 as a guide, but the handout should be in your own words and with your own diagrams. At the end of the handout, write a few quiz questions. Once you have completed your handout, team up with one or more study partners and trade handouts. Study your partners' diagrams, and when you have finished, take the quiz at the end. When you and your group have finished taking all the quizzes, discuss the answers to determine places where you need additional study. After you've finished, combine the best elements of each handout to make one "master" diagram for the differences between the various types of membrane transport.

Module 3.4: Cytoplasmic Organelles

Module 3.4 in your text explores the cytoplasmic organelles, those tiny cellular "machines" that perform a variety of functions within the cell. At the end of this module, you should be able to do the following:

1. Describe the structure and function of each type of organelle.
2. Explain how the organelles of the endomembrane system interact.

Build Your Own Glossary

Below is a table listing key terms from Module 3.4. Before you read the module, use the glossary at the back of your book or look through the module to define the following terms.

Key Terms of Module 3.4

Term	Definition
Mitochondria	
Peroxisomes	
Ribosomes	
Endomembrane system	
Rough endoplasmic reticulum	
Smooth endoplasmic reticulum	
Golgi apparatus	
Lysosomes	

Survey It: Form Questions

Before you read the module, survey it and form at least three questions for yourself. When you have finished reading the module, return to these questions and answer them.

Question 1: ______________________________

Answer: ______________________________

Question 2: ______________________________

Answer: ______________________________

Question 3: ______________________________

Answer: ______________________________

Key Concept: What does it mean that organelles allow for compartmentalization? What advantage does compartmentalization give to a cell?

____

____

Practice Labeling: Parts of the Cell

Identify and color-code each component of the cell in Figure 3.4. Then, list the main function(s) of each component.

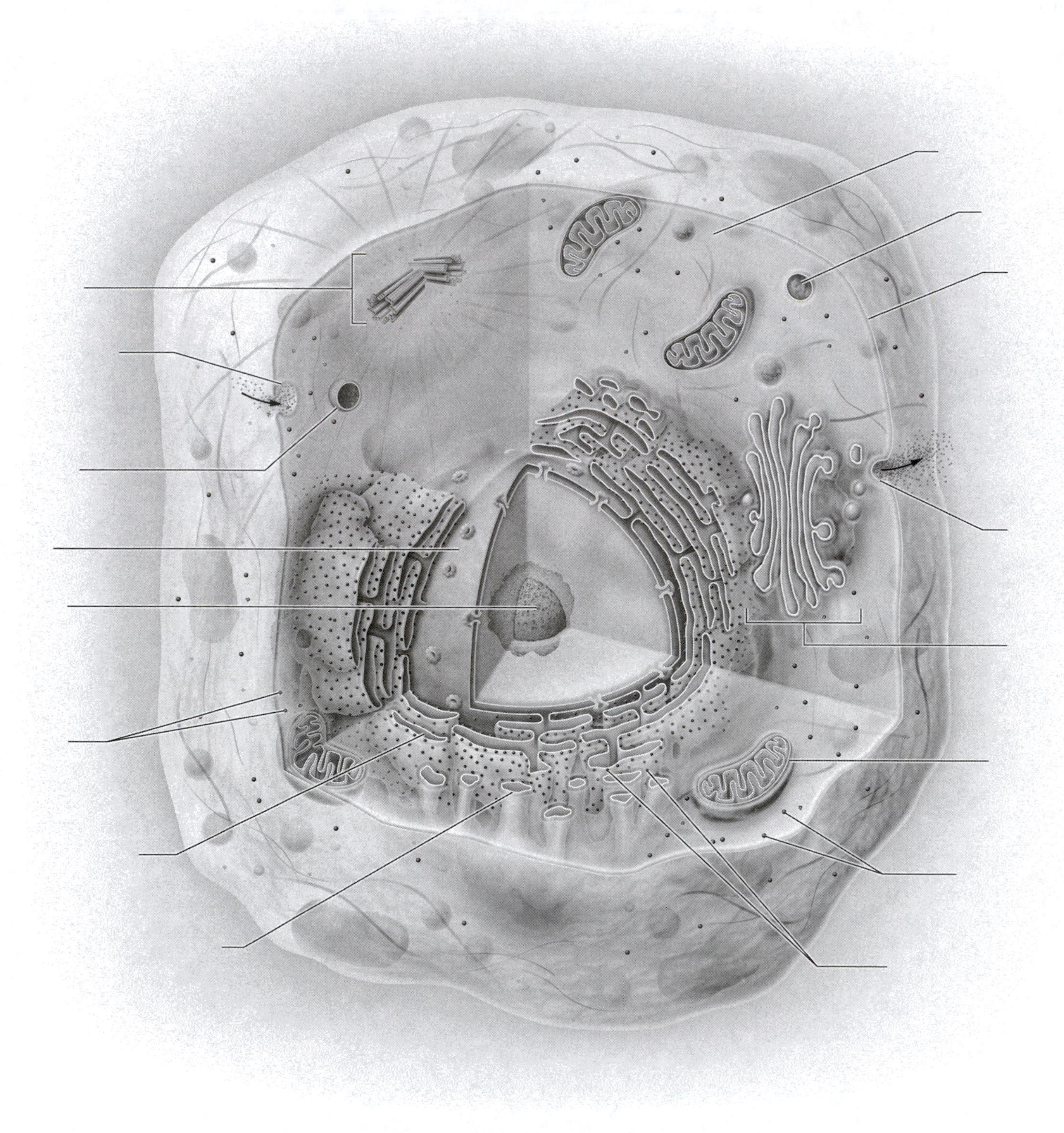

Figure 3.4 A generalized cell.

Describe It: The Endomembrane System

Write a paragraph describing the properties and functions of the endomembrane system as if you were explaining it to an audience of nonscientists.

Module 3.5: The Cytoskeleton

This module turns to the next component of the cell: the cytoskeleton. At the end of this module, you should be able to do the following:

1. Describe the structure and function of the three components of the cytoskeleton.
2. Describe the structure and function of centrioles, cilia, and flagella.
3. Explain the role of the cytoskeleton in cellular motion.

Build Your Own Glossary

Following is a table listing key terms from Module 3.5. Before you read the module, use the glossary at the back of your book or look through the module to define the following terms.

Key Terms of Module 3.5

Term	Definition
Actin filaments	
Intermediate filaments	
Microtubules	
Centriole	
Centrosome	
Microvilli	
Cilia	
Flagella	

Survey It: Form Questions

Before you read the module, survey it and form at least three questions for yourself. When you have finished reading the module, return to these questions and answer them.

Question 1: ______________________________

Answer: ______________________________

Question 2: ______________________________

Answer: ______________________________

Question 3: ______________________________

Answer: ______________________________

Build Your Own Summary Table: Cytoskeletal Elements and Structures

As you read Module 3.6, build your own summary table about the different types of cytoskeletal filaments and structures by filling in the information in the table below.

Summary of the Cytoskeletal Elements and Structures

Cytoskeletal Elements	**Structure**	**Main Functions**
Actin filaments		
Intermediate filaments		
Microtubules		
Cytoskeletal Structures	**Structure**	**Main Functions**
Microvilli		
Cilia		
Flagella		

Key Concept: Why would a disruption of the cytoskeleton possibly lead to death of the cell?

____

____

Key Concept: Why do cells that are involved in rapid absorption often have microvilli?

____

____

Module 3.6: The Nucleus

Module 3.6 in your text explores the structure and function of the nucleus and the DNA. By the end of the module, you should be able to do the following:

1. Describe the structure and function of the nucleus.
2. Explain the structure of chromatin and chromosomes.
3. Analyze the interrelationships among chromatin, chromosomes, and sister chromatids.
4. Describe the structure and function of the nucleolus.

Build Your Own Glossary

Below is a table listing key terms from Module 3.6. Before you read the module, use the glossary at the back of your book or look through the module to define the following terms.

Key Terms for Module 3.6

Term	Definition
Nuclear envelope	
Chromatin	
Chromosome	
Nucleolus	

Survey It: Form Questions

Before you read the module, survey it and form at least three questions for yourself. When you have finished reading the module, return to these questions and answer them.

Question 1: ______________________________

Answer: ______________________________

Question 2: ______________________________

Answer: ______________________________

Question 3: ______________________________

Answer: ______________________________

Draw It: The Nucleus

Draw and color-code a diagram of the nucleus. You can use Figure 3.25 in your text as a guide, but make the figure on your own and ensure that it makes sense to you.

Key Concept: Why is a nucleus necessary for the long-term survival of a cell?

____

____

Complete It: DNA and Chromatin

Fill in the blanks to complete the following paragraph that describes the properties of DNA and chromatin.

The human ____________ consists of about 25,000 genes. DNA in the cell is in the form of ____________, in which the DNA is wound around proteins called ____________. During ____________ ____________, the genetic material condenses to form chromosomes. Pairs of chromosomes are called ____________ ____________, which are joined at a region called the ____________.

Module 3.7: Protein Synthesis

Module 3.7 in your text introduces you to the process by which the code in DNA is translated into a protein via the process of protein synthesis. By the end of the module, you should be able to do the following:

1. Describe the genetic code, and explain how DNA codes for specific amino acid sequences.
2. Describe the processes of transcription and translation.
3. Explain the roles of rRNA, mRNA, and tRNA and of ribosomes in protein synthesis.

Build Your Own Glossary

Below is a table listing key terms from Module 3.7. Before you read the module, use the glossary at the back of your book or look through the module to define the following terms.

Key Terms for Module 3.7

Term	Definition
Gene	
Triplet	
Codon	
Transcription	
Messenger RNA (mRNA)	
Translation	
Transfer RNA (tRNA)	
Anticodon	

Survey It: Form Questions

Before you read the module, survey it and form at least three questions for yourself. When you have finished reading the module, return to these questions and answer them.

Question 1: ______________________________

Answer: ______________________________

Question 2: ______________________________

Answer: ______________________________

Question 3: ______________________________

Answer: ______________________________

Key Concept: What is the genetic code? How can changes in the genetic code alter the structure and function of a cell?

____

____

Draw It: Transcription

Draw yourself a diagram of the process of transcription and describe each step. You may use Figure 3.29 in your text as a reference, but make sure that it is your own diagram and words so that it makes the most sense to you.

Key Concept: What is the overarching goal of transcription? Why is this process necessary to make a protein?

____

____

Practice It: Complementary Base Pairing and Building a Polypeptide

Following is a series of DNA triplets. First, transcribe the correct complementary sequence of mRNA codons. Then, write out the tRNA anticodon that will bind to the mRNA codon during translation. Finally, use Figure 3.28 in your text to find the amino acid that corresponds with each mRNA codon.

GAC TTT TAG CGG AAC GGT TTA CAT GGC

mRNA sequence ____

____

tRNA anticodons ____

____

Amino acids ____

____

Key Concept: What is the overarching goal of translation? How does an mRNA codon relate to an amino acid on a tRNA molecule during translation?

____

____

Sequence the Events: Steps of Protein Synthesis

Fill in the steps of protein synthesis in Figure 3.5. You may use Figure 3.32 in your text for reference, but try to write the steps in your own words.

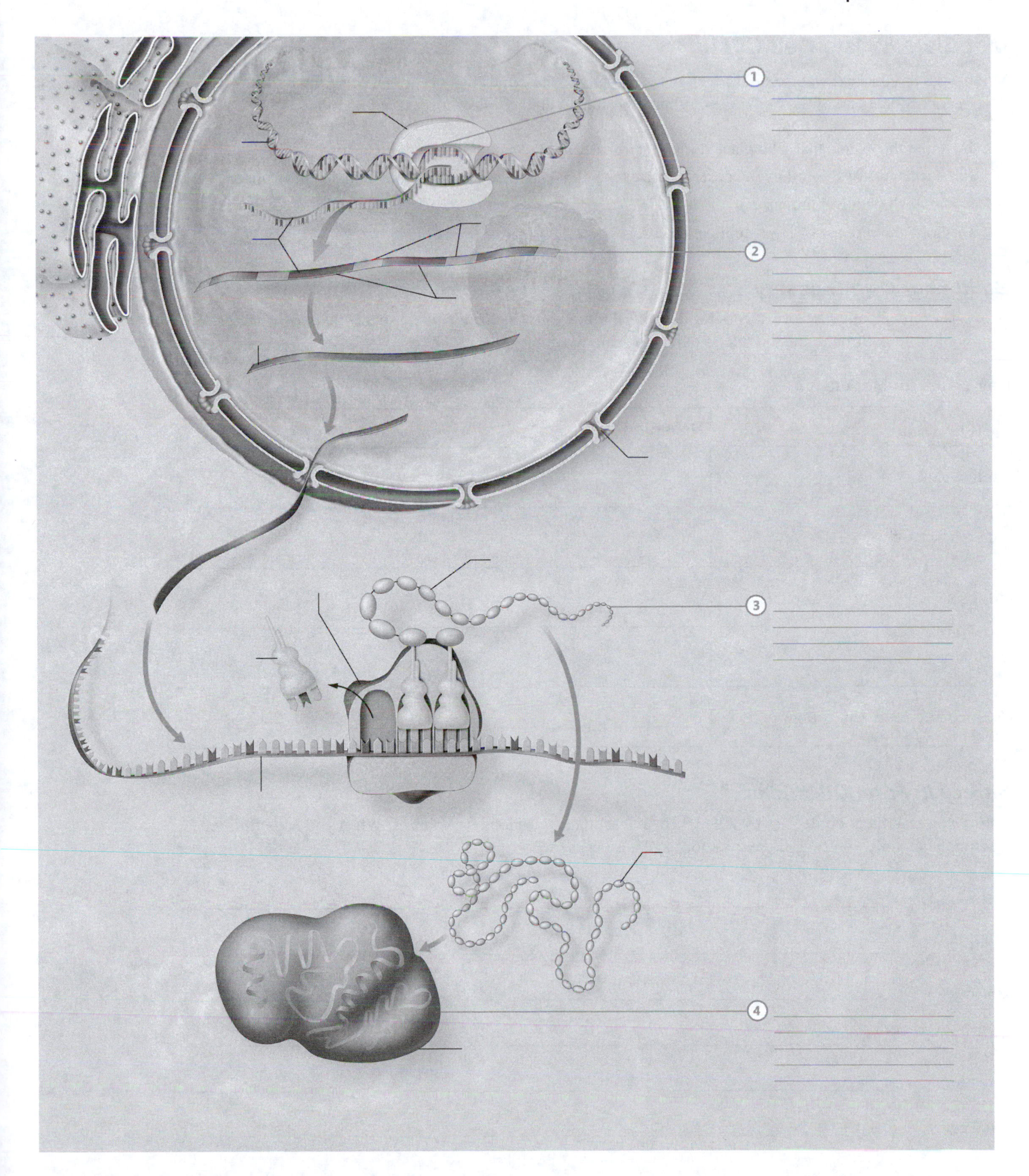

Figure 3.5 Summary of the events of protein synthesis.

Module 3.8: The Cell Cycle

Module 3.8 in your text walks you through the cell cycle, or the events that occur in a cell from its formation to its division. By the end of the module, you should be able to do the following:

1. Describe the events that take place during interphase and their functional significance.
2. For each stage of the cell cycle, describe the events that take place and their functional significance.
3. Distinguish between mitosis and cytokinesis.
4. Describe the process of DNA replication.

Build Your Own Glossary

Below is a table listing key terms from Module 3.8. Before you read the module, use the glossary at the back of your book or look through the module to define the following terms.

Key Terms for Module 3.8

Term	Definition
Cell cycle	
Interphase	
Mitosis	
Cytokinesis	
Semiconservative replication	

Survey It: Form Questions

Before you read the module, survey it and form at least three questions for yourself. When you have finished reading the module, return to these questions and answer them.

Question 1: ____________________

Answer: ____________________

Question 2: ____________________

Answer: ____________________

Question 3: ____________________

Answer: ____________________

Identify It: Phases of the Cell Cycle

Identify the phases of the cell cycle in Figure 3.6. Then, describe what happens during each phase.

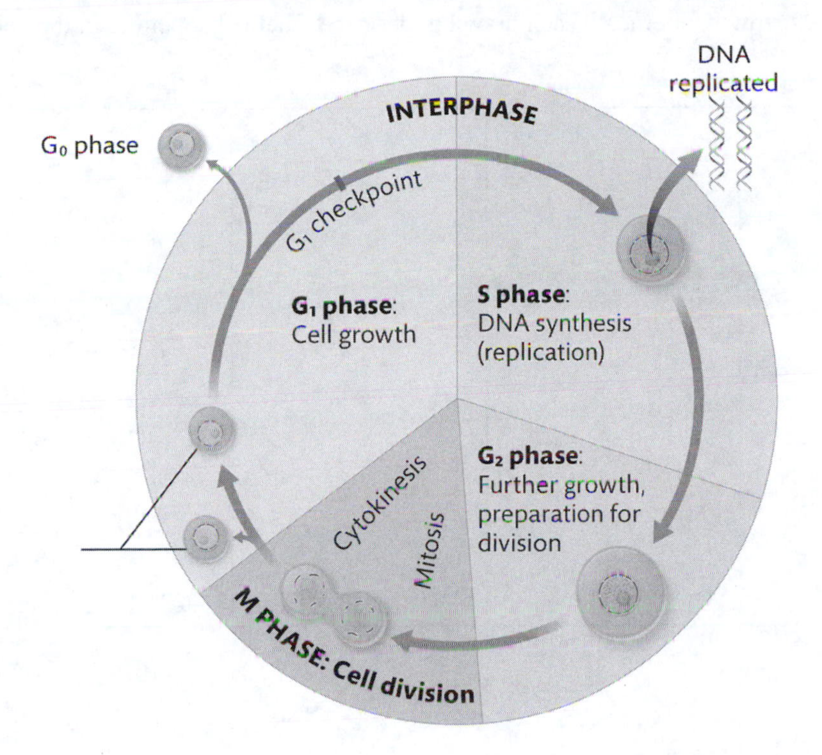

Figure 3.6 The cell cycle.

Key Concept: What parts of the cell cycle make up interphase? What happens during interphase?

____

____

Draw It: Stages of Mitosis

Draw a cell with three pairs of chromosomes undergoing division in each stage of mitosis. Label your drawings with the following terms, where applicable: nuclear membrane, chromosomes, sister chromatids, centromere, centrioles, spindle fibers, and cleavage furrow. Underneath each drawing, describe what is happening in the cell during the stage.

Cell Appearance	Stage ________	Stage ________	Stage ________	Stage ________
Description				

Key Concept: How does cytokinesis differ from mitosis? Why are both important?

____

____

What Do You Know Now?

Let's now revisit the questions you answered in the beginning of this chapter. How have your answers changed now that you've worked through the material?

- What are the basic components of a cell?

  ____

- What is DNA, and what is its function?

  ____

- How does a cell in the human body reproduce itself? Can all cells reproduce in this manner?

  ____

4 Histology

This chapter introduces you to the tissue level of organization. This is another important foundation chapter, as tissues are discussed in nearly every chapter in the rest of this book. So pay close attention to the structure and function of each tissue, and remember that the structure-function principle applies to tissues just as much as it does to organs.

What Do You Already Know?

Try to answer the following questions before proceeding to the next section. If you're unsure of the correct answers, give it your best attempt based on previous courses, previous chapters, or just your general knowledge.

- What is the definition of a tissue?

  ____

- What are the types of tissue in the body?

  ____

- How do tissues relate to organs and cells?

  ____

Module 4.1: Introduction to Tissues

Module 4.1 in your text introduces you to the histological level of organization, or *tissues*. By the end of the module, you should be able to do the following:

1. Define the term histology.
2. Explain where tissues fit in the levels of organization of the human body.
3. Compare and contrast the general features of the four major tissue types.
4. Describe the components of the extracellular matrix.
5. Describe the types of junctions that unite cells in a tissue.

Build Your Own Glossary

Following is a table listing key terms from Module 4.1. Before you read the module, use the glossary at the back of your book or look through the module to define the terms.

Key Terms for Module 4.1

Term	Definition
Epithelial tissue	
Connective tissue	
Muscle tissue	
Nervous tissue	
Extracellular matrix	
Ground substance	
Collagen fibers	
Elastic fibers	
Reticular fibers	
Tight junction	
Desmosome	
Gap junction	

Survey It: Form Questions

Before you read the module, survey it and form at least three questions for yourself. When you have finished reading the module, return to these questions and answer them.

Question 1: ______________________________

Answer: ______________________________

Question 2: ______________________________

Answer: ______________________________

Question 3: ______________________________

Answer: ______________________________

Key Concept: What is the importance of the extracellular matrix in tissue function?

____

____

Identify It: Components of the Extracellular Matrix

Identify and color-code the components of the extracellular matrix in Figure 4.1. Then, describe the main role each component plays in a tissue.

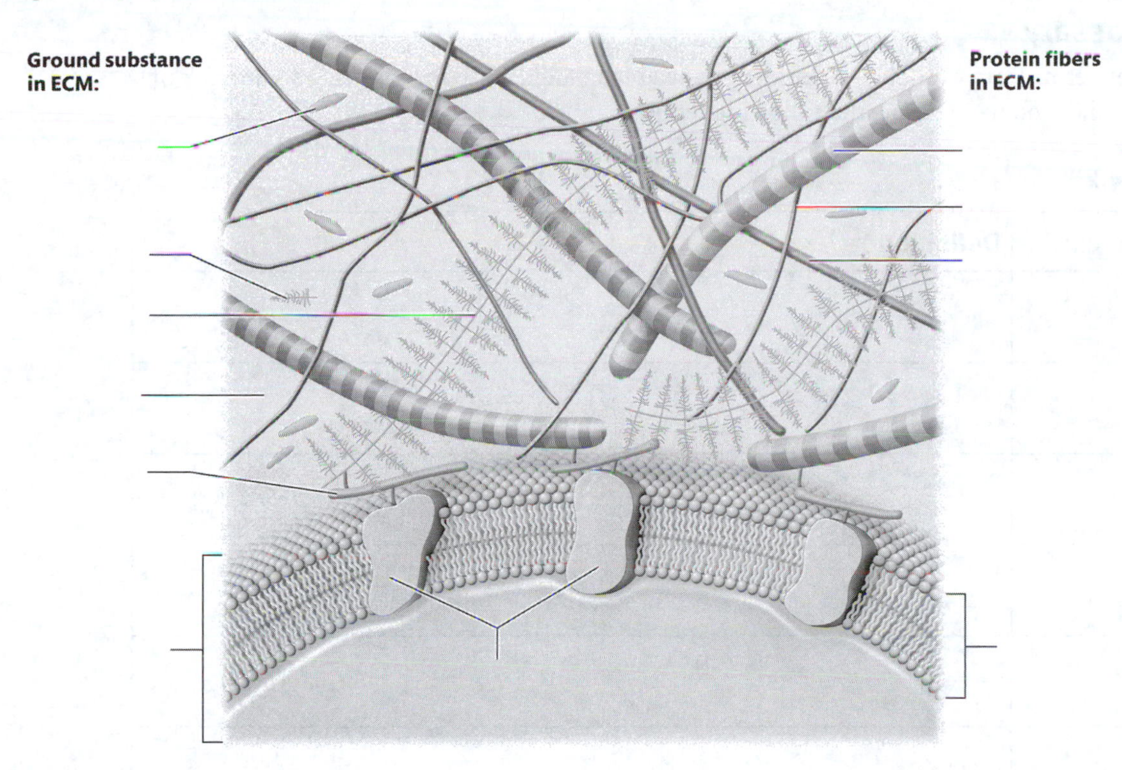

Figure 4.1 The extracellular matrix.

Draw It: Cell Junctions

In the boxes provided, draw the different types of cellular junctions. Label your drawings with the terms from Figure 4.2 in your text. Underneath your drawing, describe the properties of each junction.

Type of Junction			
Properties			

Module 4.2: Epithelial Tissue

The first tissue type we discuss in your text is epithelial tissues that cover and line body surfaces and form glands. When you finish this module, you should be able to do the following:

1. Classify and identify the different types of epithelial tissues.
2. Describe the location and function of each type of epithelial tissue and correlate that function with structure.
3. Describe and classify the structural and functional properties of exocrine and endocrine glands.

Build Your Own Glossary

Below is a table listing key terms from Module 4.2. Before you read the module, use the glossary at the back of your book or look through the module to define the following terms.

Key Terms for Module 4.2

Term	Definition
Basement membrane	
Simple epithelia	
Stratified epithelia	
Squamous cells	
Cuboidal cells	
Columnar cells	
Pseudostratified	
Gland	
Endocrine gland	
Exocrine gland	
Merocrine secretion	
Holocrine secretion	

Survey It: Form Questions

Before you read the module, survey it and form at least four questions for yourself. When you have finished reading the module, return to these questions and answer them.

Question 1: ____________________

Answer: ____________________

Question 2: ____________________

Answer: ____________________

Question 3: ____________________

Answer: ____________________

Question 4: ____________________

Answer: ____________________

Identify It: Types of Epithelial Tissue

When you finish reading about the types of epithelial tissue, identify and color-code each type in Figure 4.2. Label the following in each image: the type of epithelium, basement membrane, surrounding/underlying connective tissue, and cilia (where present).

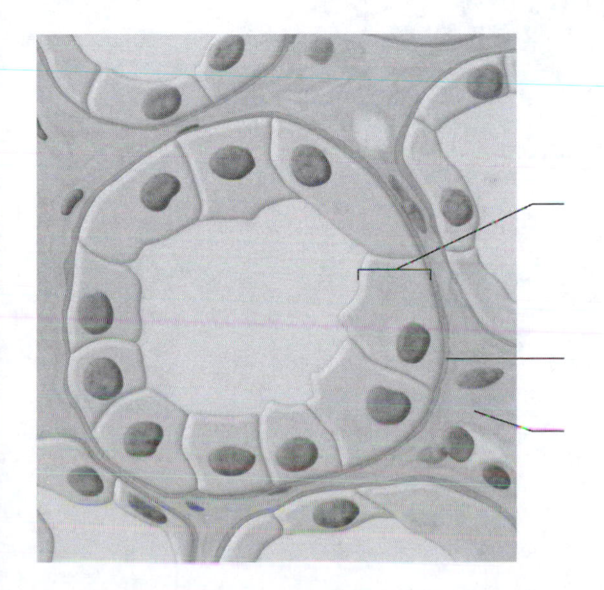

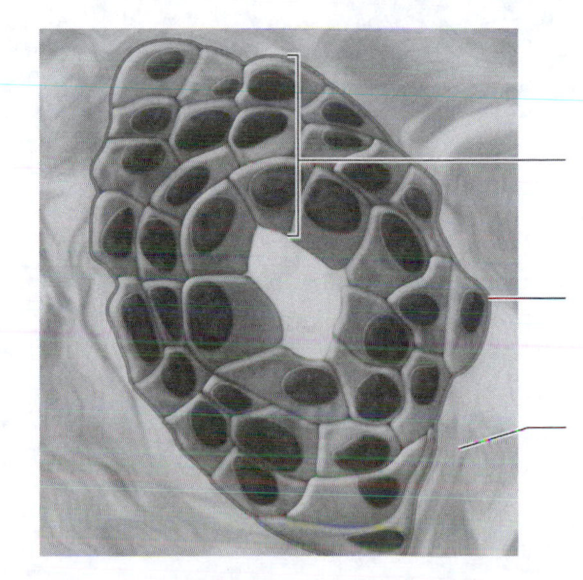

Figure 4.2 Types of epithelium.

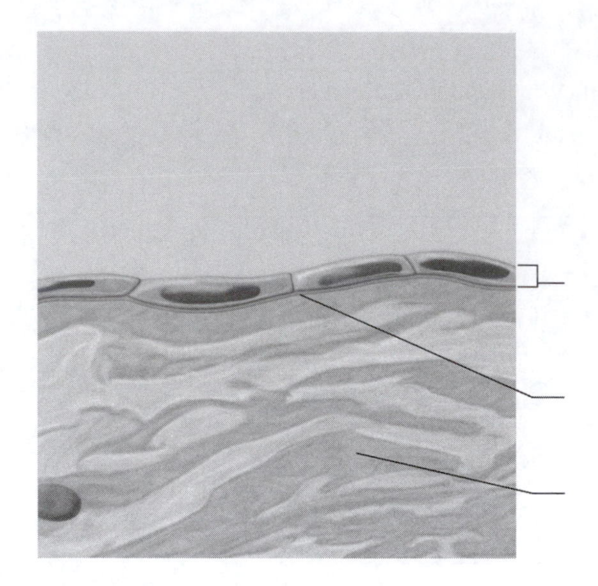

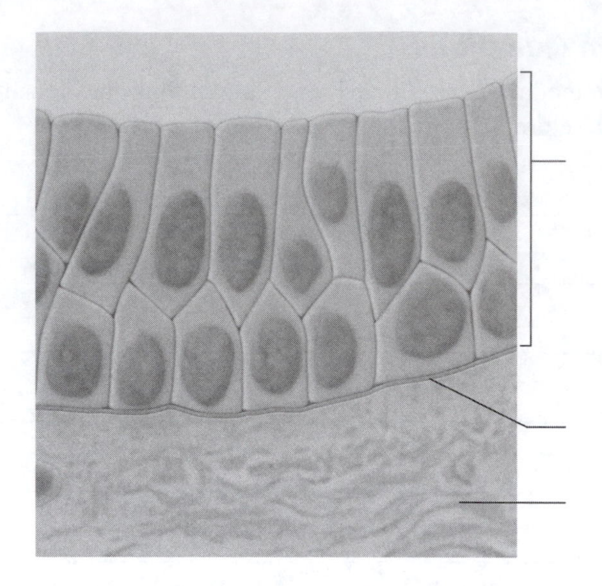

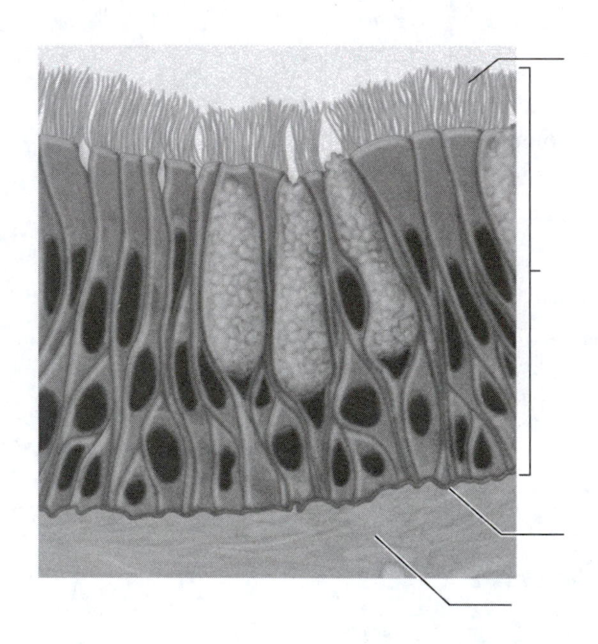

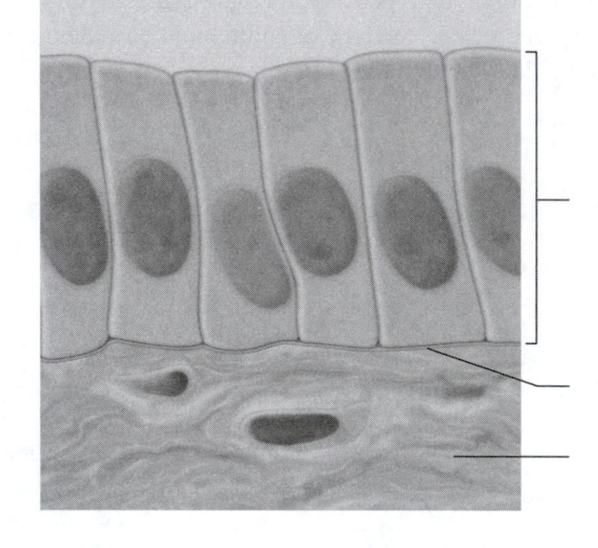

Figure 4.2 (*continued*)

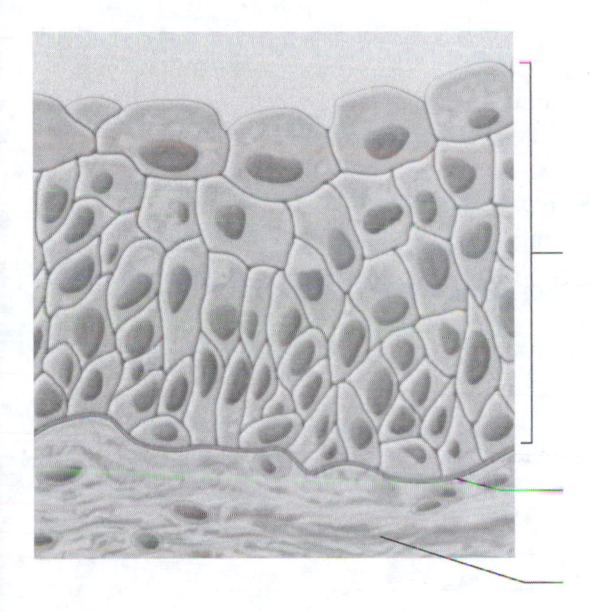

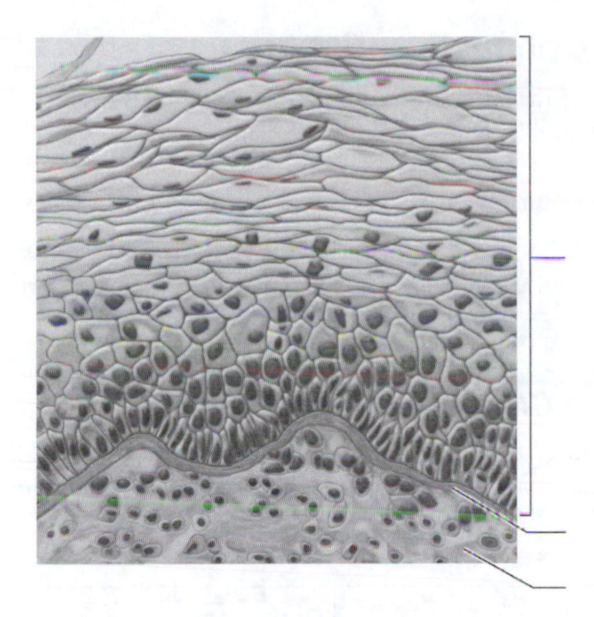

Figure 4.2 (*continued*)

Key Concept: How is the structure of simple epithelium related to its function? How is the structure of stratified epithelium related to its function?

____

____

Complete It: Glandular Epithelium

Fill in the blanks to complete the following paragraph that describes the types and functions of glandular epithelium.

An endocrine gland is one that releases its product, called a ____________, into the ____________. An exocrine gland releases its product through a ____________. Simple glands have ____________ ____________, whereas compound glands have ____________ ____________. Some glands release their product through ____________ ____________, in which the product is secreted by exocytosis. Sebaceous glands release their product through ____________ ____________, in which the cells rupture and dead cells are released with the product.

Module 4.3: Connective Tissues

This module explores the structure and function of the different types of connective tissues. When you complete this module, you should be able to do the following:

1. Compare and contrast the roles of individual cell and fiber types within connective tissues.
2. Identify the different types of connective tissue, and describe where in the body they are found.
3. Describe the functions of each type of connective tissue and correlate function with structure for each tissue type.

Build Your Own Glossary

Below is a table listing key terms from Module 4.3. Before you read the module, use the glossary at the back of your book or look through the module to define the following terms.

Key Terms of Module 4.3

Term	Definition
Connective tissue proper	
Fibroblast	
Adipocyte	
Mast cell	
Phagocyte	
Specialized connective tissue	
Cartilage	
Bone	
Blood	

Survey It: Form Questions

Before you read the module, survey it and form at least four questions for yourself. When you have finished reading the module, return to these questions and answer them.

Question 1: ______________________________

Answer: ______________________________

Question 2: ______________________________

Answer: ______________________________

Question 3: ______________________________

Answer: ______________________________

Question 4: ______________________________

Answer: ______________________________

Key Concept: What are some key differences between epithelial and connective tissues?

____

____

Complete It: Cells of Connective Tissues

Fill in the blanks to complete the following paragraphs that describe the types and functions of cells within connective tissues.

____________ are the most common cells of connective tissue, and they produce ____________ ____________ and ____________ ____________. Another common resident cell of connective tissue is an immune cell called a ____________ cell, which releases chemicals called ____________ ____________. Within fat tissue we find cells called ____________, which contain a large ____________ inclusion in their cytoplasm.

The primary cells of the specialized connective tissue cartilage are the immature ____________ and mature ____________. The specialized connective tissue bone contains three cell types: osteoblasts, which ____________ ____________; osteoclasts, which ____________ ____________; and mature ____________, which maintain the ECM. Blood is a specialized connective tissue that contains an ECM called ____________, cells called ____________ and ____________, and cellular fragments called ____________.

Key Concept: How do the extracellular matrices of different types of connective tissues relate to their functions? Give some specific examples.

____

____

Identify It: Types of Connective Tissue

When you finish reading the module, identify and color-code each type of connective tissue in Figure 4.3. Label the following in each image where applicable: the type of connective tissue, the types of protein fibers visible, ground substance, and the types of cells visible.

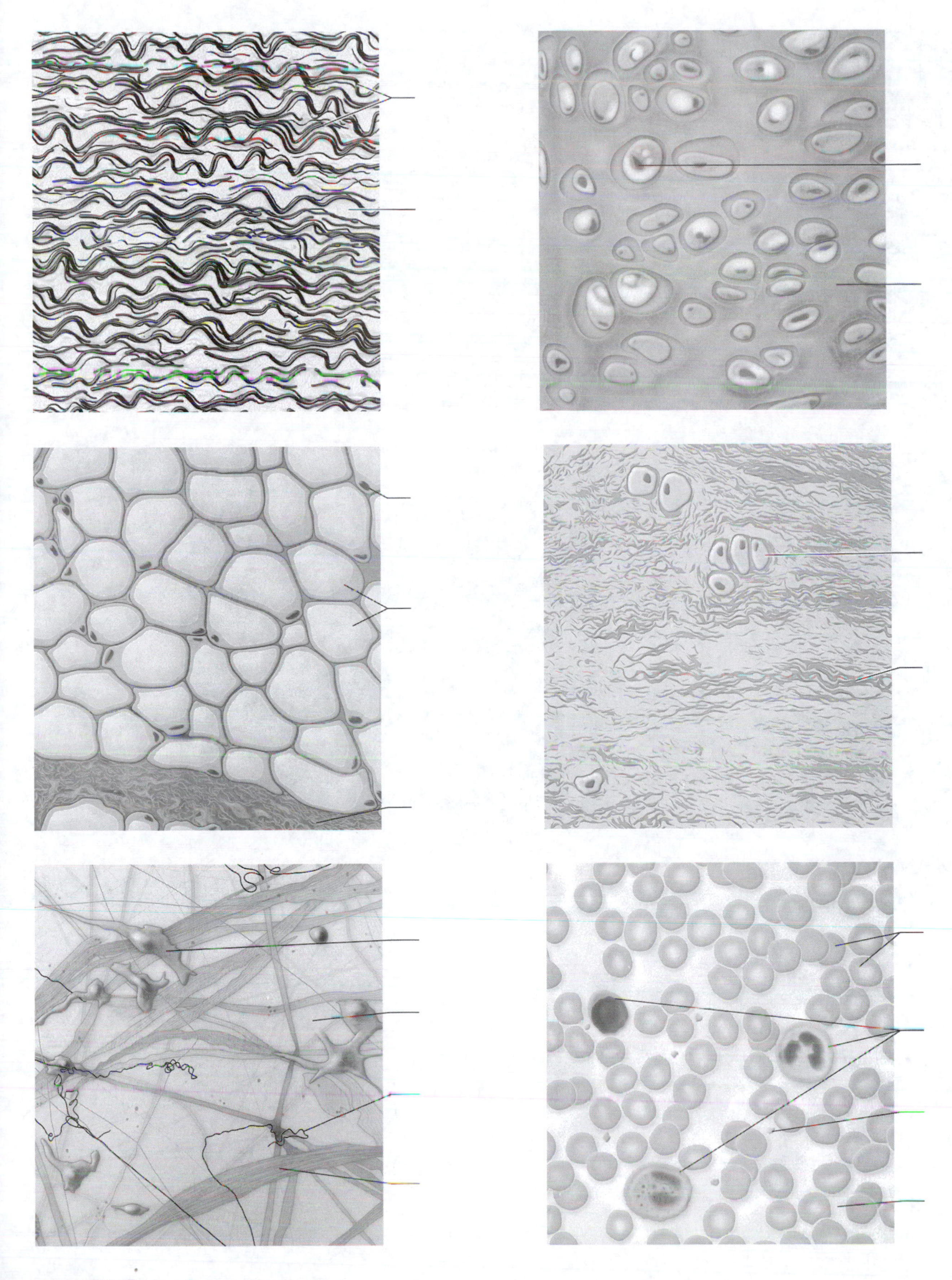

Figure 4.3 Types of connective tissue.

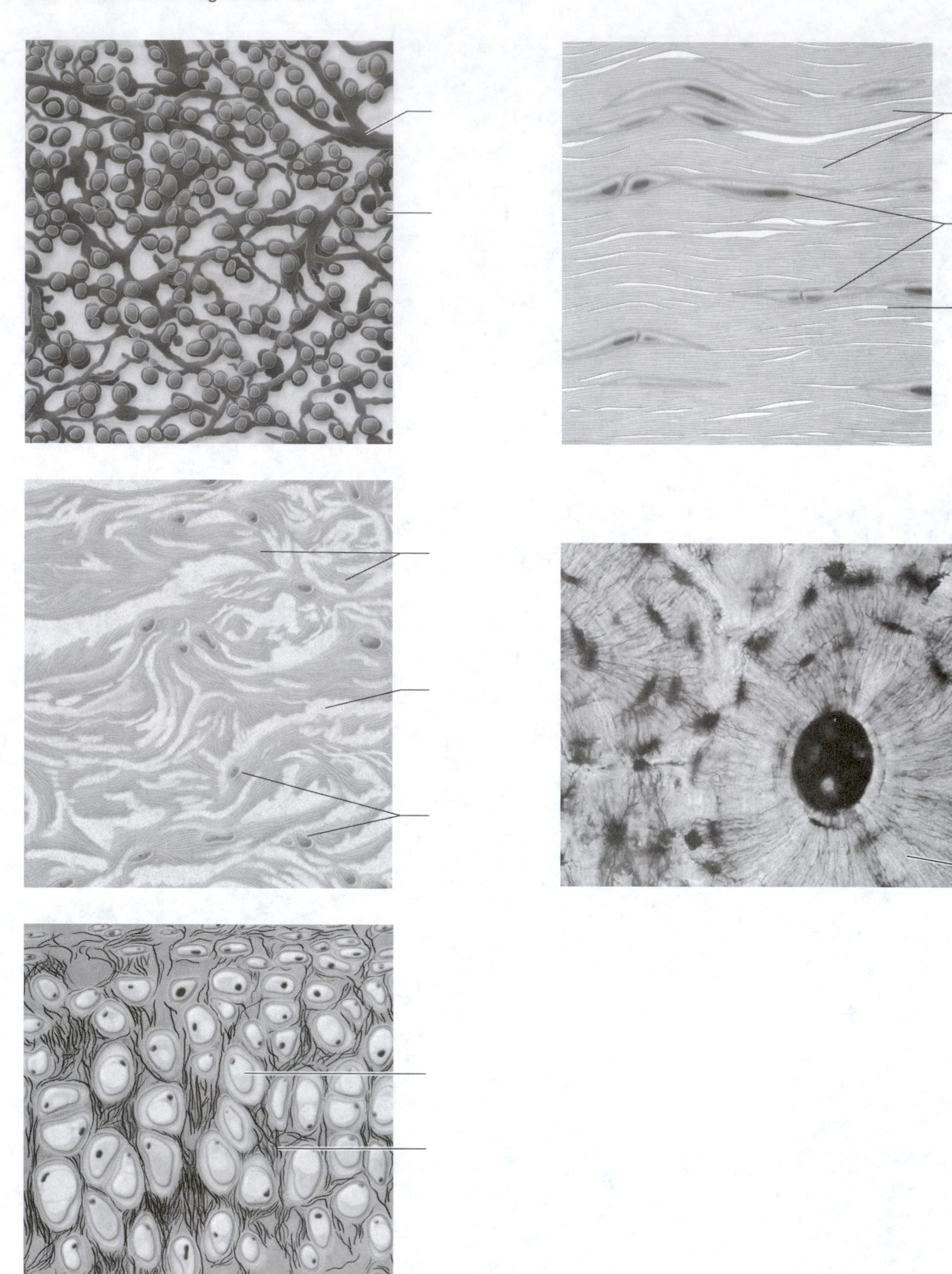

Figure 4.3 (*continued*)

Module 4.4: Muscle Tissues

Module 4.4 in your text explores muscle tissues, teaching you about their basic properties and roles. At the end of this module, you should be able to do the following:

1. Classify and identify the different types of muscle tissue based on distinguishing structural characteristics and location in the body.
2. Describe the functions of each type of muscle tissue, and correlate function with structure for each tissue type.

Build Your Own Glossary

Following is a table listing key terms from Module 4.4. Before you read the module, use the glossary at the back of your book or look through the module to define the following terms.

Key Terms of Module 4.4

Term	Definition
Striated muscle cell	
Smooth muscle cell	
Endomysium	

Survey It: Form Questions

Before you read the module, survey it and form at least two questions for yourself. When you have finished reading the module, return to these questions and answer them.

Question 1: ____________________

Answer: ____________________

Question 2: ____________________

Answer: ____________________

Key Concept: What is one of the key properties of muscle tissue that allow it to perform its functions?

____

____

Identify It: Muscle Tissue

When you finish reading the module, identify and color-code each type of muscle tissue in Figure 4.4. Label the following in each image where applicable: the type of muscle tissue, muscle cells, nuclei, endomysium, striations, and intercalated discs.

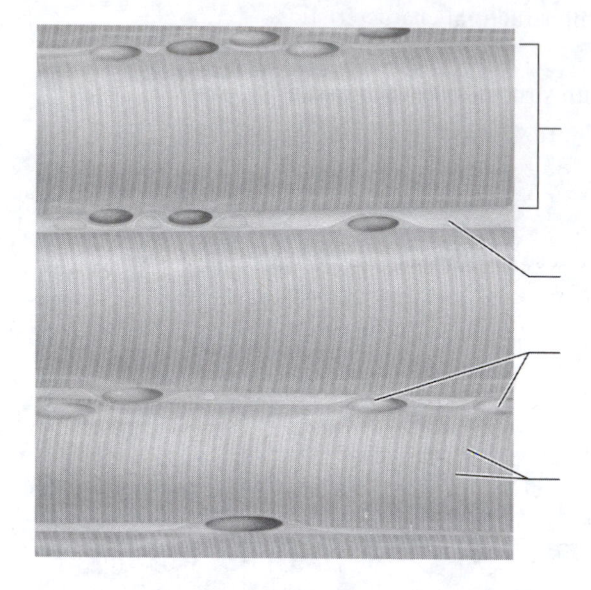

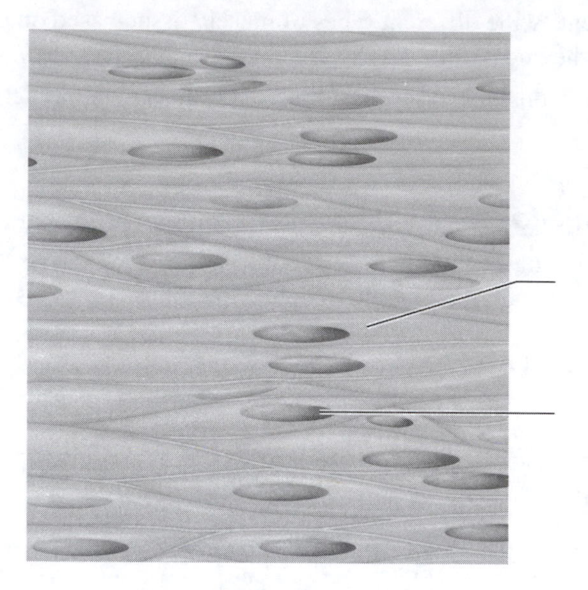

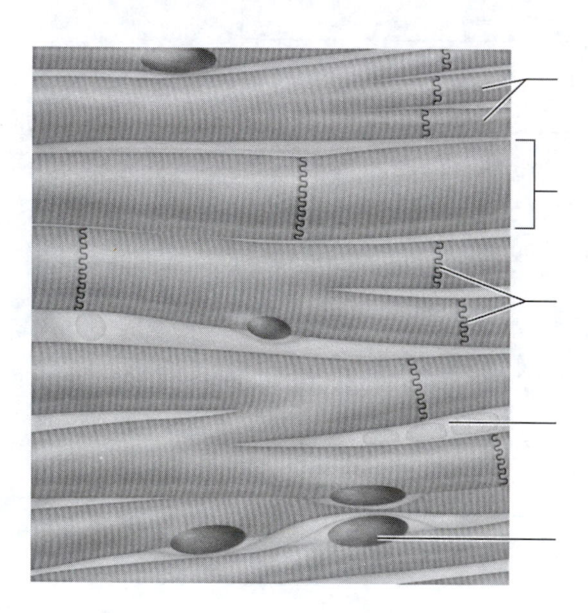

Figure 4.4 Types of muscle tissues.

Module 4.5: Nervous Tissue

This module examines the final type of tissue: nervous tissue. At the end of this module, you should be able to do the following:

1. Describe where in the body nervous tissue can be found and its general structural and functional characteristics.
2. Identify and describe the structure and function of neurons and neuroglial cells in nervous tissue.

Build Your Own Glossary

Below is a table listing key terms from Module 4.5. Before you read the module, use the glossary at the back of your book or look through the module to define the following terms.

Key Terms of Module 4.5

Term	Definition
Neuron	
Nerve impulse	
Cell body	
Axon	
Dendrite	
Neuroglial cell	

Survey It: Form Questions

Before you read the module, survey it and form at least two questions for yourself. When you have finished reading the module, return to these questions and answer them.

Question 1: ______________________________

Answer: ______________________________

Question 2: ______________________________

Answer: ______________________________

Key Concept: What is a key property that neurons share with muscle cells? How does this relate to their function?

____

____

Identify It: Nervous Tissue

Color-code and label Figure 4.5 with the following terms: neuron, cell body, dendrite, axon, neuroglial cell, and extracellular matrix.

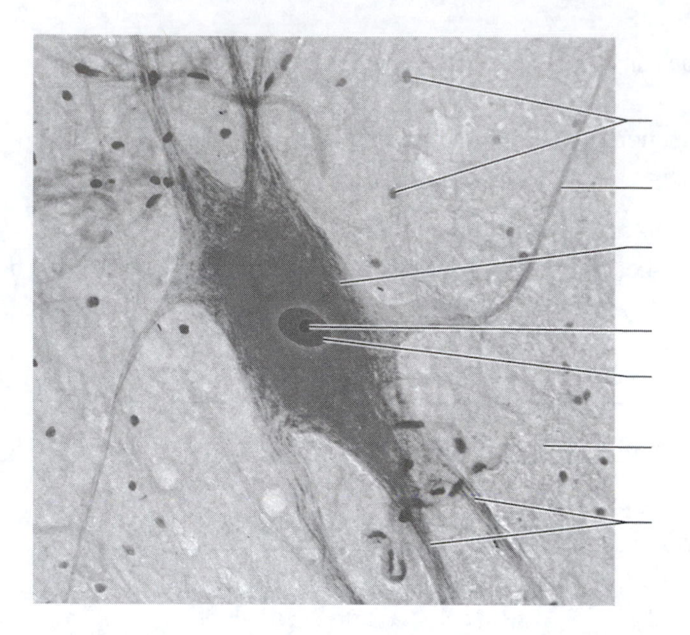

Figure 4.5 Nervous tissue.

Module 4.6: Putting It All Together: The Big Picture of Tissues in Organs

Module 4.6 in your text paints a big picture view of tissues by showing you how they come together to form an organ. By the end of the module, you should be able to do the following:

1. Describe how tissues work together to form organs.

Team Up

Work with a group of four students, and have each person choose one of the following organs: urinary bladder (Chapter 24), blood vessel (Chapter 18), heart (Chapter 17), or a bone (not bone *tissue*, but the actual entire bone, Chapter 6). Each person should look to the chapter in which this organ is discussed, then make a handout to teach the tissue layers of their organ and how to identify each type of tissue. As usual, each person should write a few quiz questions at the end of the handout. Study your partners' diagrams, and when you have finished, take the quiz at the end. When you and your group have finished taking all the quizzes, discuss the answers to determine places where you need additional study. After you've finished, compile the handouts and use them as a study guide for class and lab.

Key Concept: How do tissues contribute to the overall function of an organ?

____

____

Module 4.7: Membranes

Module 4.7 in your text introduces you to tissues that line other organs or body cavities: membranes. By the end of the module, you should be able to do the following:

1. Describe the general structure and function of membranes.
2. Explain the properties and locations of serous, synovial, mucous, and cutaneous membranes.

Build Your Own Glossary

Following is a table listing key terms from Module 4.7. Before you read the module, use the glossary at the back of your book or look through the module to define the following terms.

Key Terms for Module 4.7

Term	Definition
Membrane	
Serous membrane	
Synovial membrane	
Mucous membrane	
Cutaneous membrane	

Survey It: Form Questions

Before you read the module, survey it and form at least two questions for yourself. When you have finished reading the module, return to these questions and answer them.

Question 1: ______________________________

Answer: ______________________________

Question 2: ______________________________

Answer: ______________________________

Key Concept: What are the key differences between a serous membrane and a mucous membrane?

____

____

Describe It: Membranes and Their Functions

Imagine you are teaching a study group about the four different types of membranes. Write a paragraph or draw a diagram to explain to your group the basic structure and function of each membrane.

Module 4.8: Tissue Repair

Module 4.8 in your text examines what happens when a tissue is damaged: regeneration or fibrosis. By the end of the module, you should be able to do the following:

1. Describe how injuries affect epithelial, connective, muscular, and nervous tissues.
2. Describe the process of regeneration.
3. Explain the process of fibrosis.

Build Your Own Glossary

Below is a table listing key terms from Module 4.8. Before you read the module, use the glossary at the back of your book or look through the module to define the following terms.

Key Terms for Module 4.8

Term	Definition
Regeneration	
Fibrosis	

Survey It: Form Questions

Before you read the module, survey it and form at least two questions for yourself. When you have finished reading the module, return to these questions and answer them.

Question 1: ______________________________

Answer: ______________________________

Question 2: ______________________________

Answer: ______________________________

Key Concept: Why can fibrosis lead to a loss of function in an organ?

____

____

Build Your Own Summary Table: Tissues

Now that you have reached the end of the chapter, you can build your own summary table of all the tissue types by filling in the below template. You have been given enough space to draw in examples of the tissue types, and doing so will really help you understand and remember their key components.

Summary of Tissues

Type of Tissue	Structure	Function(s)	Location(s)	Heals by Fibrosis or Regeneration?
Epithelial tissue				
Simple squamous epithelium				
Simple cuboidal epithelium				
Simple columnar epithelium				
Pseudostratified columnar epithelium				
Stratified squamous epithelium				
Stratified cuboidal epithelium				

Type of Tissue	Structure	Function(s)	Location(s)	Heals by Fibrosis or Regeneration?
Stratified columnar epithelium				
Transitional epithelium				
Connective tissue				
Loose CT				
Dense irregular CT				
Dense regular collagenous CT				
Dense regular elastic CT				
Reticular CT				
Adipose CT				
Hyaline cartilage				
Fibrocartilage				
Elastic cartilage				
Bone				
Blood				

Type of Tissue	Structure	Function(s)	Location(s)	Heals by Fibrosis or Regeneration?
Muscle tissue				
Skeletal muscle tissue				
Cardiac muscle tissue				
Smooth muscle tissue				
Nervous Tissue				

What Do You Know Now?

Let's now revisit the questions you answered in the beginning of this chapter. How have your answers changed now that you've worked through the material?

- What is the definition of a tissue?

  ____

- What are the types of tissue in the body?

  ____

- How do tissues relate to organs and cells?

  ____

5 The Integumentary System

In previous chapters, we have examined the chemical, cellular, and tissue levels of organization. Now we move on to the organs and the systemic level of organization with an exploration of the integumentary system and its main organ: the skin.

What Do You Already Know?

Try to answer the following questions before proceeding to the next section. If you're unsure of the correct answers, give it your best attempt based on previous courses, previous chapters, or just your general knowledge.

- What is the basic structure of the skin?

  ____

- Which parts of the skin consist of living cells?

  ____

- What are the major functions of the skin?

  ____

Module 5.1: Overview of the Integumentary System

Module 5.1 in your text introduces you to the integumentary system, including the basic structure of the skin and its functions. By the end of the module, you should be able to do the following:

1. Describe the basic structure of the skin.
2. Describe the basic functions carried out by the components of the integumentary system.

Build Your Own Glossary

Following is a table listing key terms from Module 5.1. Before you read the module, use the glossary at the back of your book or look through the module to define the following terms.

Key Terms for Module 5.1

Term	Definition
Integument	
Epidermis	
Dermis	
Hypodermis	

Survey It: Form Questions

Before you read the module, survey it and form at least three questions for yourself. When you have finished reading the module, return to these questions and answer them.

Question 1: ______________________________

Answer: ______________________________

Question 2: ______________________________

Answer: ______________________________

Question 3: ______________________________

Answer: ______________________________

Identify It: Anatomy of the Skin

Identify and color-code the structures of the skin in Figure 5.1. Then, write the main function of each structure you have identified.

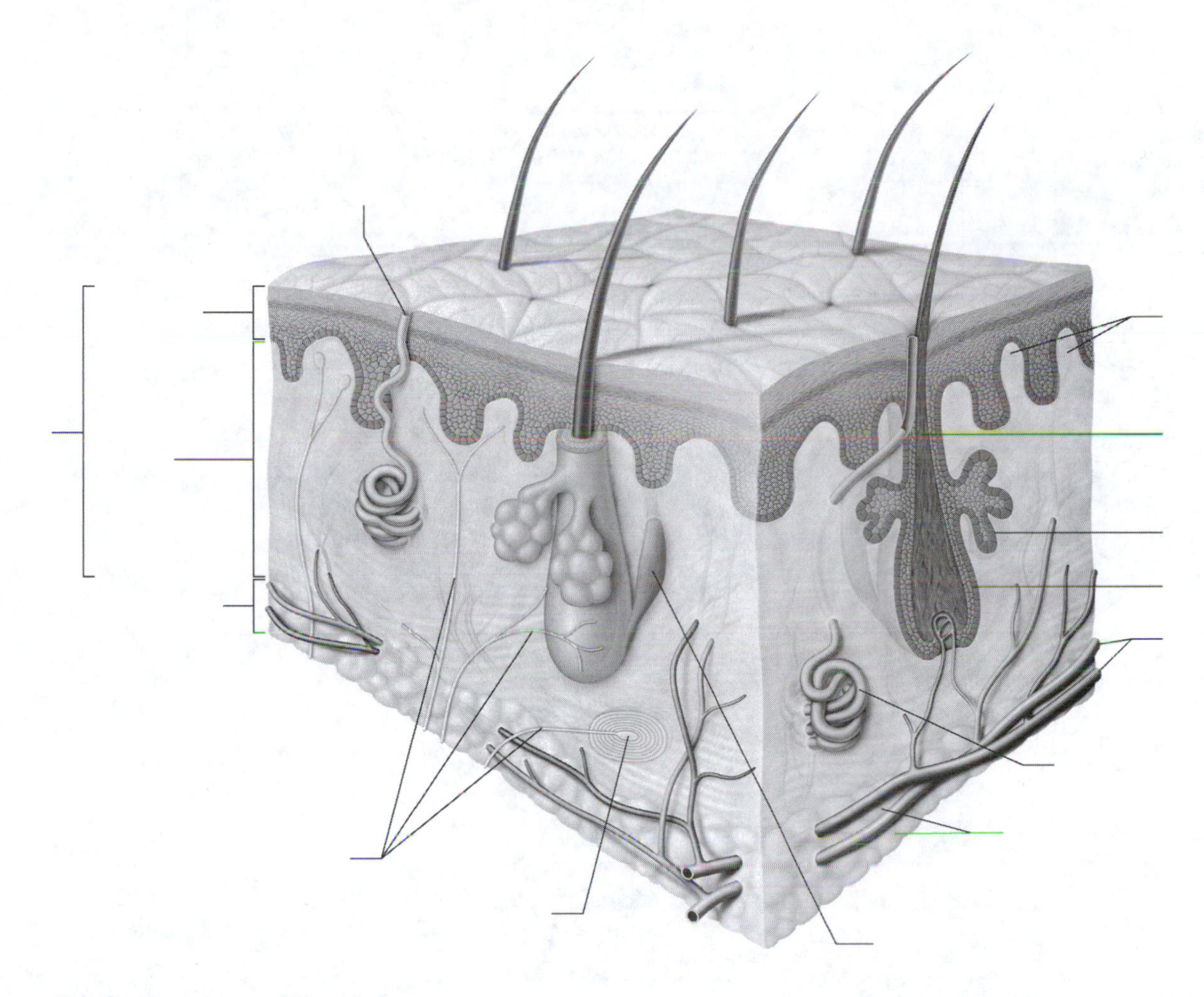

Figure 5.1 Basic anatomy of the skin.

Key Concept: What are the different ways that the integument offers protection? How does it perform these functions?

____

____

Sequence the Events: The Skin and Thermoregulation

Describe the steps in the boxes of each of the negative feedback loops in Figure 5.2, and label the important components of the loops. You may use text Figure 5.2 as reference, but your descriptions should be in your own words.

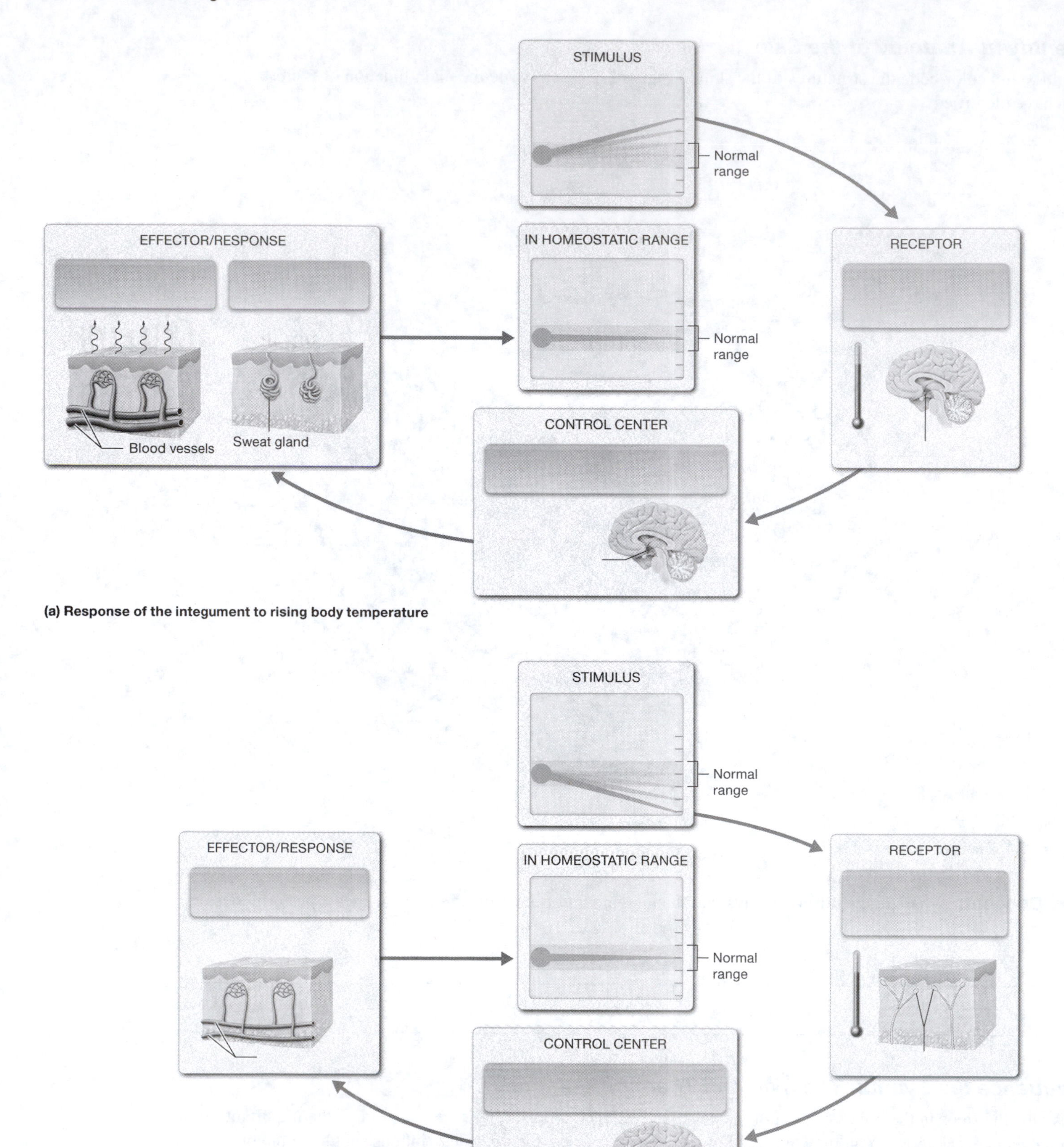

(a) Response of the integument to rising body temperature

(b) Response of the integument to falling body temperature

Figure 5.2 Homeostatic regulation of body temperature by the integumentary system.

Key Concept: How does altering blood flow to the skin help with thermoregulation?

____

____

Module 5.2: The Epidermis

Now we look at the epidermis, the skin's superficial tissue layer, in more detail. When you finish this module, you should be able to do the following:

1. Explain how the cells of the epidermis are arranged into layers.
2. Describe the cells of the epidermis and the life cycle of a keratinocyte.
3. Differentiate between thick skin and thin skin.

Build Your Own Glossary

Below is a table listing key terms from Module 5.2. Before you read the module, use the glossary at the back of your book or look through the module to define the following terms.

Key Terms for Module 5.2

Term	Definition
Keratinocyte	
Dendritic cell	
Merkel cell	
Melanocyte	
Thick skin	
Thin skin	

Survey It: Form Questions

Before you read the module, survey it and form at least three questions for yourself. When you have finished reading the module, return to these questions and answer them.

Question 1: ____

Answer: ____

Question 2: ____

Answer: ____

Question 3: ____

Answer: ____

Identify It: Layers of the Epidermis

Identify and color-code the structures of the epidermis in Figure 5.3. Then, describe the functions of the four cell types in the epidermis (keratinocytes, melanocytes, dendritic cells, and Merkel cells), and distinguish between the cells of the epidermis that are living and those that are dying or dead.

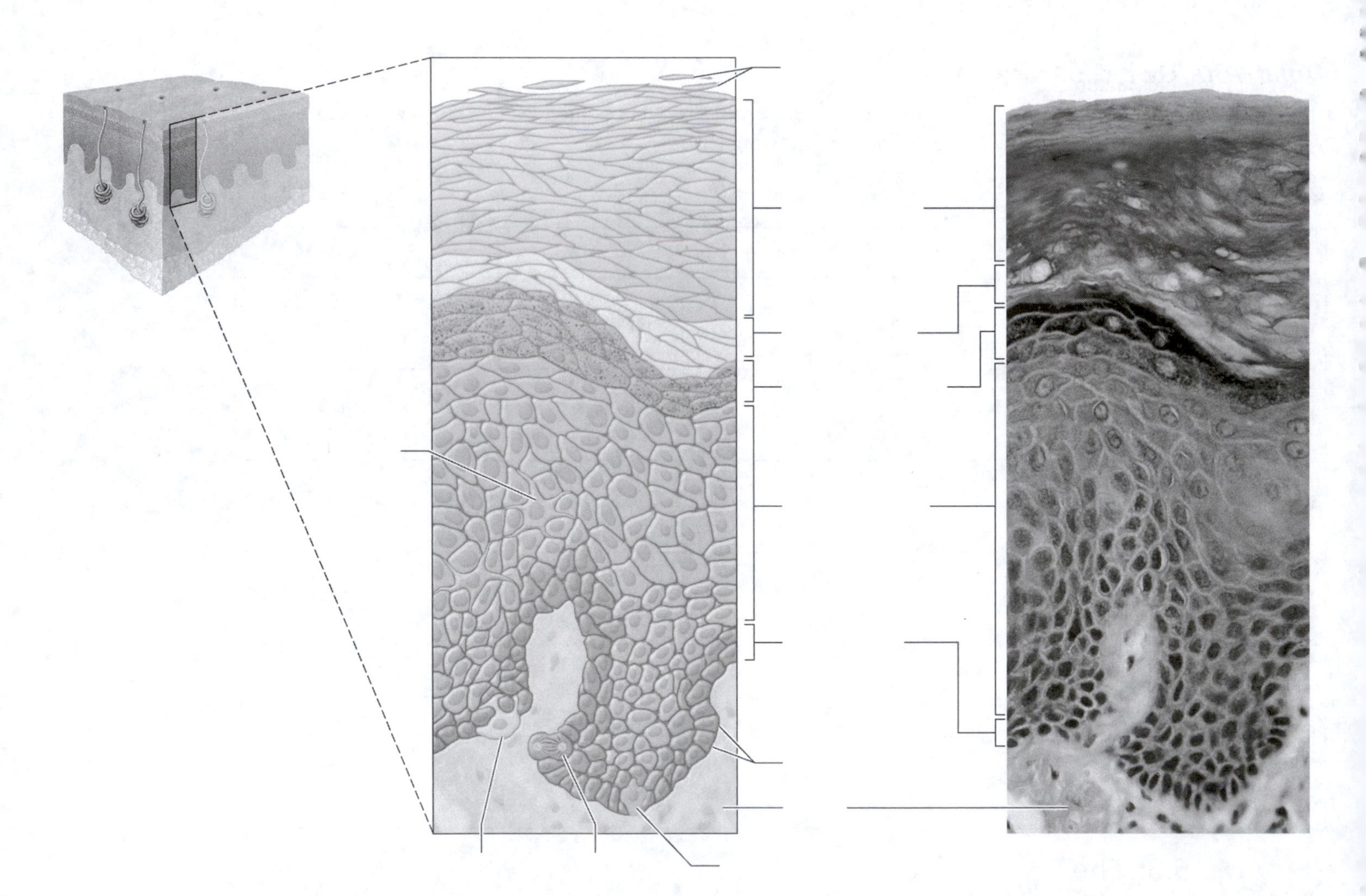

Figure 5.3 Structure of the epidermis.

Key Concept: Why are the cells of the superficial epidermis dead?

____

____

Key Concept: How does a keratinocyte that starts in the stratum basale eventually end up in the stratum corneum?

____

____

Identify It: Thin and Thick Skin

Identify each of the following properties as belonging to thin skin, thick skin, or both.

- Hair follicles are present. ______________________________
- Sweat glands are present. ______________________________
- Stratum corneum is very prominent with many layers. ______________________________
- Located on the palms of the hands and the soles of the feet. ______________________________
- Located in most places other than the palms and soles. ______________________________
- Composed of stratified squamous keratinized epithelium. ______________________________
- Contains a stratum lucidum. ______________________________
- Sebaceous glands are present. ______________________________

Key Concept: What do you think would happen if thin skin were present on the palms of the hands and the soles of the feet instead of thick skin?

____

____

Module 5.3: The Dermis

This module explores the structure and function of the deeper layer of skin, the dermis. When you complete this module, you should be able to do the following:

1. Describe the layers and basic structure and components of the dermis.
2. Explain the functions of the dermal papillae.
3. Explain how skin markings such as epidermal ridges are formed.

Build Your Own Glossary

Below is a table listing key terms from Module 5.3. Before you read the module, use the glossary at the back of your book or look through the module to define the following terms.

Key Terms of Module 5.3

Term	Definition
Papillary layer	
Dermal papillae	
Tactile corpuscles	
Reticular layer	
Lamellated corpuscles	
Epidermal ridges	
Flexure lines	

Survey It: Form Questions

Before you read the module, survey it and form at least three questions for yourself. When you have finished reading the module, return to these questions and answer them.

Question 1: ______________________________

Answer: ______________________________

Question 2: ______________________________

Answer: ______________________________

Question 3: ______________________________

Answer: ______________________________

Identify It: Structures of the Dermis

Identify and color-code the structures of the dermis in Figure 5.4. Then, describe the functions of the dermal papillae, tactile corpuscles, collagen fibers, and lamellated corpuscles located in the dermis.

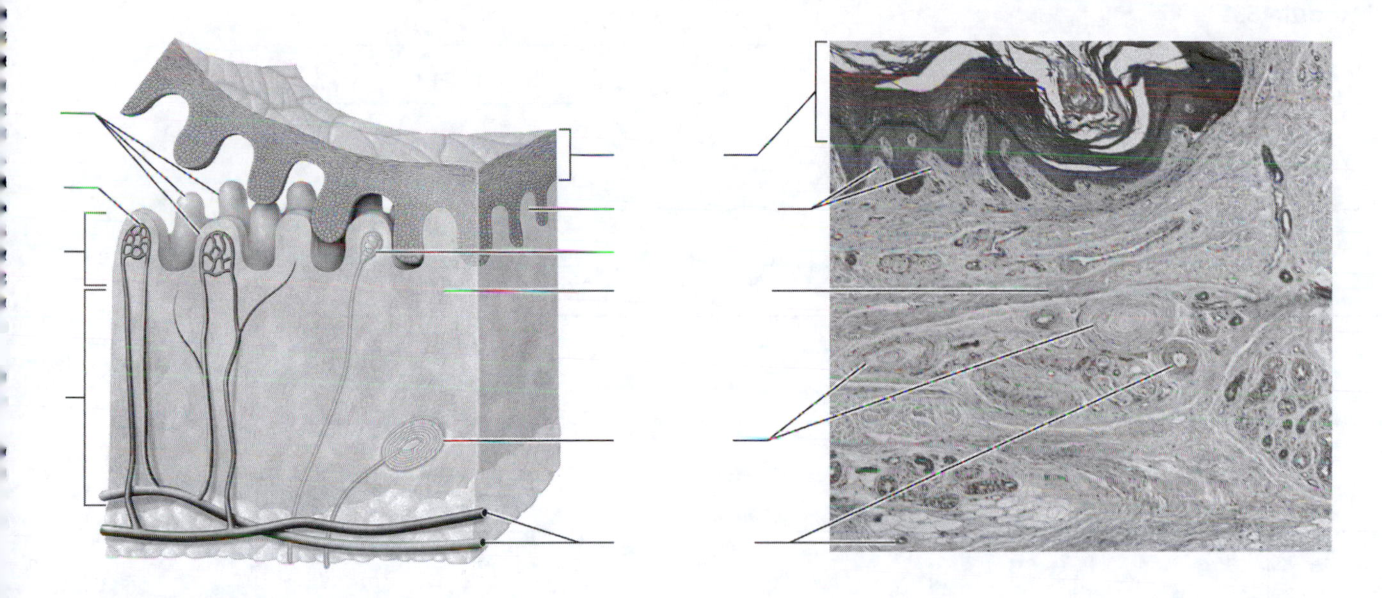

Figure 5.4 Structure of the dermis.

Key Concept: What do you think would happen to the epidermis if the dermal papillae were damaged in some way? Why?

Key Concept: How do epidermal ridges form, and why do we leave behind fingerprints when we touch certain surfaces? Why don't epidermal ridges form in thin skin?

Module 5.4: Skin Pigmentation

Module 5.4 in your text explores the chemical substances, or pigments, that give skin color. At the end of this module, you should be able to do the following:

1. Explain how melanin is produced and its role in the integument.
2. Describe the other pigments that contribute to skin color.
3. Explain how skin coloration may indicate pathology.

Build Your Own Glossary

Following is a table listing key terms from Module 5.4. Before you read the module, use the glossary at the back of your book or look through the module to define the following terms.

Key Terms of Module 5.4

Term	Definition
Melanin	
Carotene	
Hemoglobin	
Erythema	
Cyanosis	

Survey It: Form Questions

Before you read the module, survey it and form at least three questions for yourself. When you have finished reading the module, return to these questions and answer them.

Question 1: ______________________________

Answer: ______________________________

Question 2: ______________________________

Answer: ______________________________

Question 3: ______________________________

Answer: ______________________________

Describe It: Melanin Production and Function

Write a paragraph as if you were explaining to an audience of nonscientists how melanin is produced, how it gets into keratinocytes, and its main functions.

Key Concept: What are melanin's protective functions?

____

____

Key Concept: How can the skin pigment hemoglobin give clues to different pathologies?

____

____

Module 5.5: Accessory Structures of the Integument: Hair, Nails, and Glands

This module turns to the other structures of the integument: hair, nails, and glands. At the end of this module, you should be able to do the following:

1. Describe the structure and function of hair and nails.
2. Explain the process by which hair and nails grow.
3. Summarize the structural properties of sweat and sebaceous glands.
4. Explain the composition and function of sweat and sebum.

Build Your Own Glossary

Following is a table listing key terms from Module 5.5. Before you read the module, use the glossary at the back of your book or look through the module to define the terms.

Key Terms of Module 5.5

Term	Definition
Hair	
Hair shaft	
Hair root	
Hair follicle	
Nail	
Nail plate	
Nail bed	
Sweat gland	
Sebaceous gland	
Sebum	

Survey It: Form Questions

Before you read the module, survey it and form at least three questions for yourself. When you have finished reading the module, return to these questions and answer them.

Question 1: ____________________

Answer: ____________________

Question 2: ____________________

Answer: ____________________

Question 3: ____________________

Answer: ____________________

Identify It: Structures of a Hair

Identify and color-code the structures of a hair in Figure 5.5.

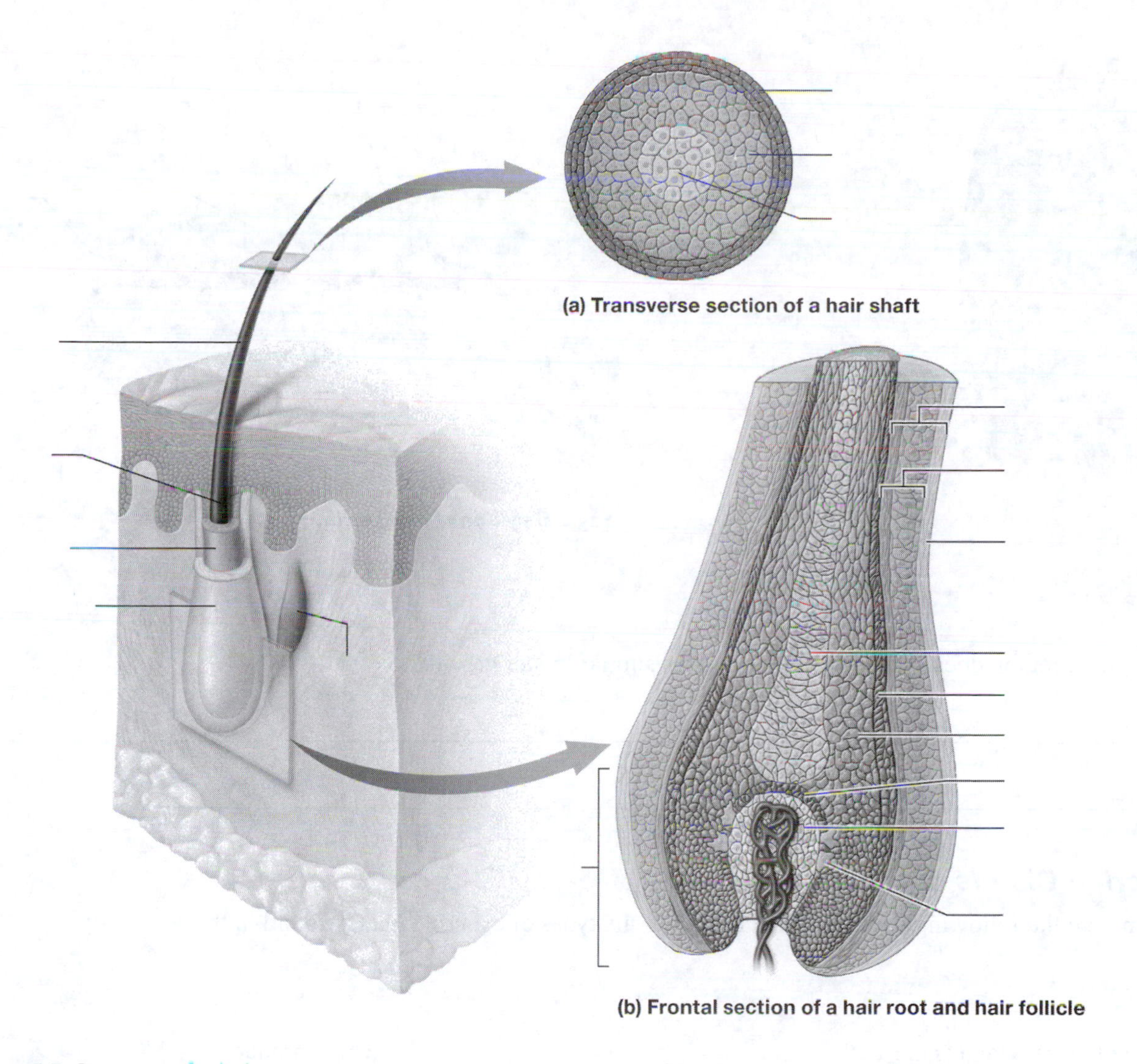

Figure 5.5 Structure of a hair.

Key Concept: What determines hair color, and how does it accomplish this?

____

____

Identify It: Structures of a Nail

Identify and color-code the structures of a nail in Figure 5.6.

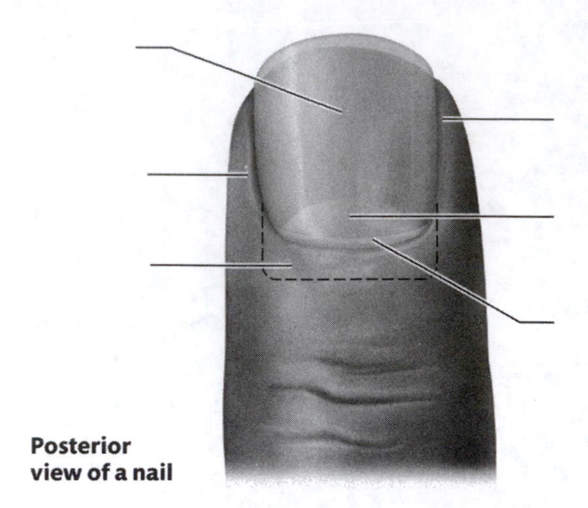

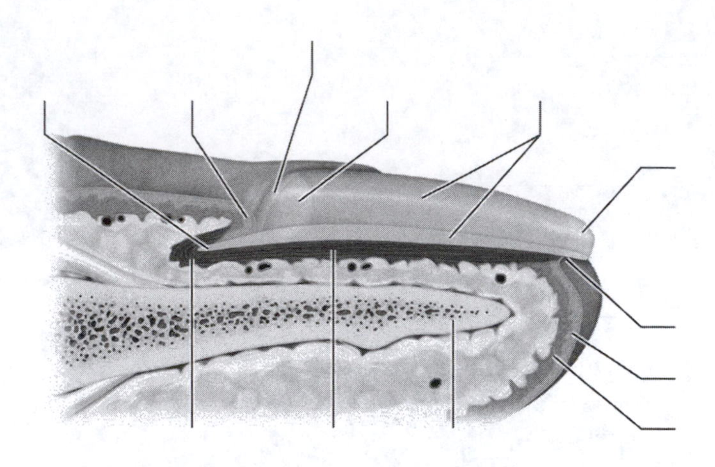

Figure 5.6 Structure of a nail.

Key Concept: From which region does a nail grow? How is this similar to hair growth?

____

____

Complete It: Exocrine Glands of the Skin

Fill in the blanks to complete the following paragraph that describes the types of exocrine glands found in the skin.

The most prevalent type of sweat gland is the ____________ ____________ ____________, which produces sweat that is released through ____________ ____________. ____________ glands are located in the axillae, areolae, and anal area, and they release a ____________ -rich sweat into a ____________ ____________. Sebaceous glands release ____________ into a ____________ ____________ through ____________ secretion. In contrast, sweat glands release sweat via ____________ secretion.

Key Concept: How do sweat and sebum differ in their protective functions of the skin?

____

____

Build Your Own Summary Table: Accessory Structures of the Skin

As you read Module 5.5, build your own summary table about the different types of accessory structures of the skin by filling in the information in the template provided below.

Summary of the Accessory Structure of the Skin

Accessory Structure	Structural Properties	Main Functions
Hair		
Nails		
Sweat glands		
Sebaceous glands		

Module 5.6: Pathology of the Skin

Module 5.6 in your text examines common pathological conditions that can impact the integument. By the end of the module, you should be able to do the following:

1. Explain how to classify burns and how to estimate their severity.
2. Describe the three main types of cancerous skin tumors.

Build Your Own Glossary

Below is a table listing key terms from Module 5.6. Before you read the module, use the glossary at the back of your book or look through the module to define the following terms.

Key Terms for Module 5.6

Term	Definition
Burn	
Rule of nines	
Cancer	

Survey It: Form Questions

Before you read the module, survey it and form at least two questions for yourself. When you have finished reading the module, return to these questions and answer them.

Question 1: ____________________

Answer: ____________________

Question 2: ____________________

Answer: ____________________

Build Your Own Summary Table: Skin Pathologies

Summarize each of the different types of skin pathologies discussed in Module 5.6 by filling in the template provided below.

Skin Pathologies

Skin Pathology	Description	Layer or Layers of Skin Involved	Treatment
First-degree burn			
Second-degree burn			
Third-degree burn			
Basal cell carcinoma			
Squamous cell carcinoma			
Malignant melanoma			

Key Concept: What disruptions to homeostasis can result from damage to the skin?

____

____

What Do You Know Now?

Let's now revisit the questions you answered in the beginning of this chapter. How have your answers changed now that you've worked through the material?

- What is the basic structure of the skin?

____

- Which parts of the skin consist of living cells?

____

- What are the major functions of the skin?

____

6 Bones and Bone Tissue

This chapter begins our exploration of the next organ system, the skeletal system. Here we introduce the basic structures of bones and the main tissue: bone or osseous tissue.

What Do You Already Know?

Try to answer the following questions before proceeding to the next section. If you're unsure of the correct answers, give it your best attempt based on previous courses, previous chapters, or just your general knowledge.

- What chemicals/substances make up bone tissue?

  ____

- What makes up a bone as an organ?

  ____

- Which substances are required in the diet in order to ensure healthy bones?

  ____

Module 6.1: Introduction to Bones as Organs

Module 6.1 in your text introduces you to the basic structure and function of the organs of the skeletal system: bones. By the end of the module, you should be able to do the following:

1. Describe the functions of the skeletal system.
2. Describe how bones are classified by shape.
3. Describe the gross structure of long, short, flat, irregular, and sesamoid bones.
4. Explain the differences between red and yellow bone marrow.

Build Your Own Glossary

Following is a table listing key terms from Module 6.1. Before you read the module, use the glossary at the back of your book or look through the module to define the following terms.

Key Terms for Module 6.1

Term	Definition
Periosteum	
Diaphysis	
Epiphysis	
Medullary cavity	
Compact bone	
Spongy bone	
Endosteum	
Red bone marrow	
Yellow bone marrow	

Survey It: Form Questions

Before you read the module, survey it and form at least three questions for yourself. When you have finished reading the module, return to these questions and answer them.

Question 1: ______________________________

Answer: ______________________________

Question 2: ______________________________

Answer: ______________________________

Question 3: ______________________________

Answer: ______________________________

Key Concept: How does the skeletal system help maintain homeostasis?

____

____

Draw It: Bone Shapes

In the space below, draw an example of a long, short, flat, irregular, and sesamoid bone. You may use Figure 6.2 in your text for reference, but try to use different examples. Under each of your drawings, explain the characteristics of the bone shape and list examples of each.

Identify It: Anatomy of a Long Bone

Identify and color-code each component of a long bone in Figure 6.1. Then, list the main function(s) of each component.

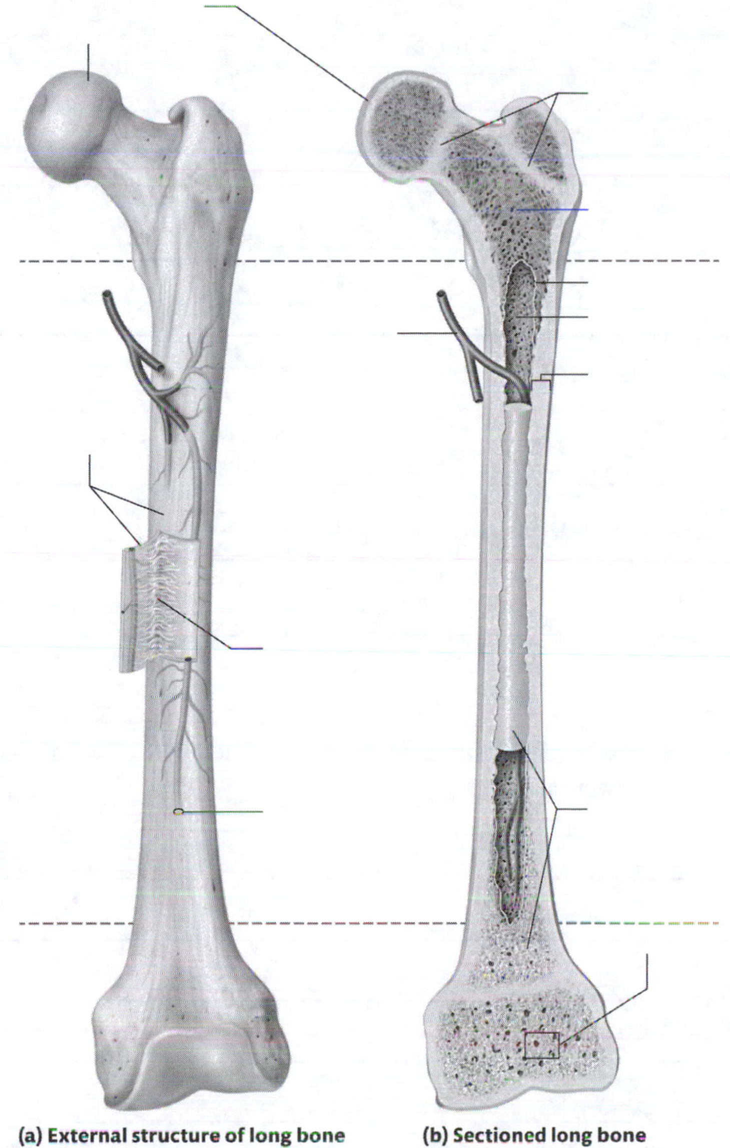

Figure 6.1 The femur, a long bone.

Module 6.2: Microscopic Structure of Bone Tissue

Now we will look at the structure of the main component of bone: bone or osseous tissue. When you finish this module, you should be able to do the following:

1. Describe the inorganic and organic components of the extracellular matrix of bone tissue.
2. Explain the functions of the three main cell types in bone tissue.
3. Describe the microscopic structure of compact bone and the components of the osteon.
4. Describe the microscopic structure of spongy bone.

Build Your Own Glossary

Below is a table listing key terms from Module 6.2. Before you read the module, use the glossary at the back of your book or look through the module to define the following terms.

Key Terms for Module 6.2

Term	Definition
Osseous (bone) tissue	
Inorganic matrix	
Organic matrix	
Osteoblast	
Osteocyte	
Osteoclast	
Osteon	
Trabeculae	

Survey It: Form Questions

Before you read the module, survey it and form at least three questions for yourself. When you have finished reading the module, return to these questions and answer them.

Question 1: ______________________________

Answer: ______________________________

Question 2: ______________________________

Answer: ______________________________

Question 3: ______________________________

Answer: ______________________________

Key Concept: Why are both organic and inorganic matrices required for bone tissue to function normally? What would happen if either one of the matrices were defective?

____

____

Build Your Own Summary Table: Components of Bone Tissue

As you read Module 6.2, build your own summary table of the components of bone tissue by filling in the template provided below.

Summary of the Components of Bone

Component	Structure	Main Functions
ECM and Cells		
Inorganic matrix		
Osteoid		
Osteoblast		
Osteocyte		
Osteoclast		
Osteon Components		
Concentric lamellae		
Central canal		
Lacunae		
Canaliculi		
Other Structures		
Perforating canal		

Component	Structure	Main Functions
Interstitial lamellae		
Trabeculae		

Identify It: Structure of Compact Bone and the Osteon

Identify and color-code each component of compact bone in Figure 6.2. Then, list the main function(s) of each component.

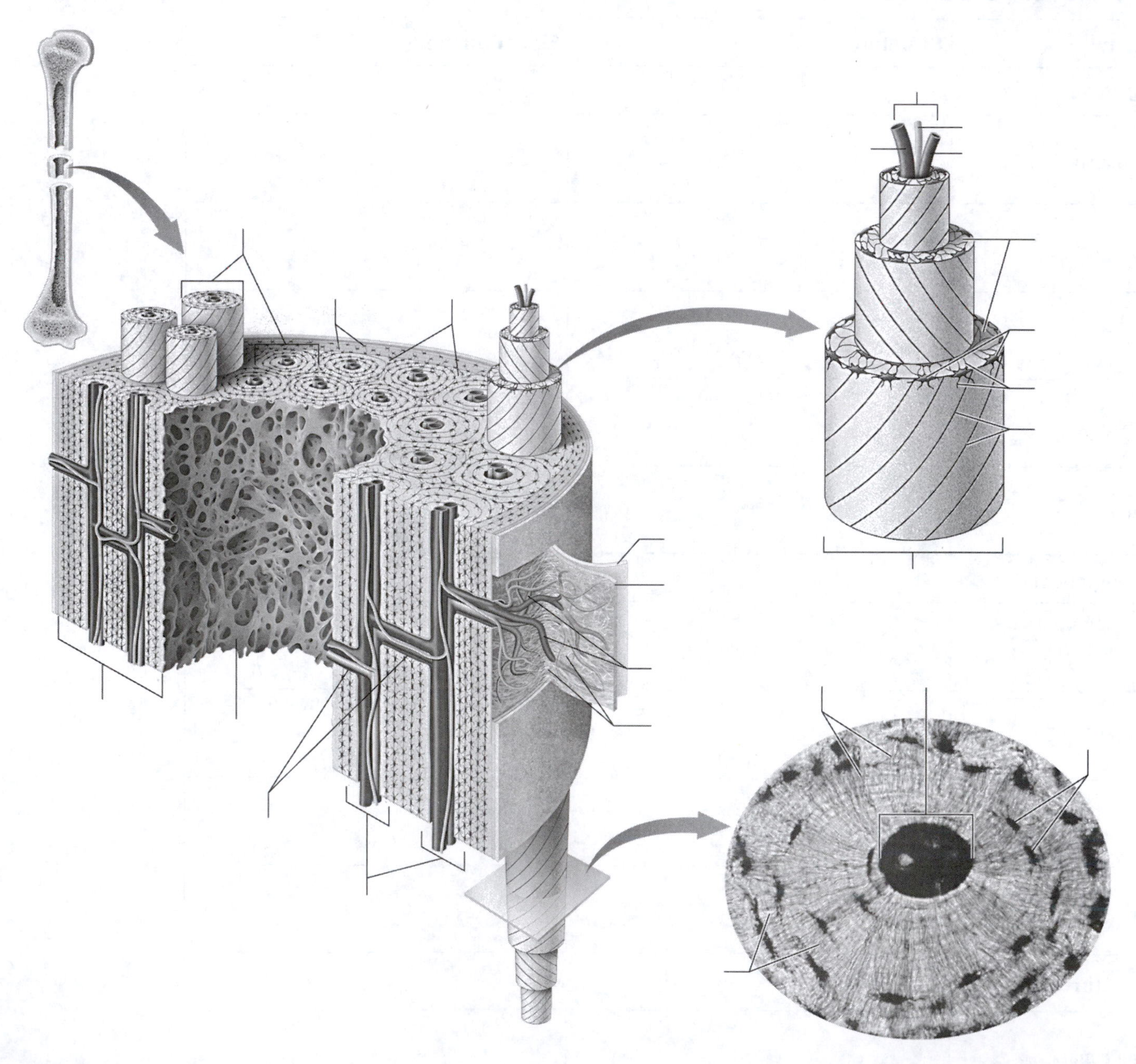

Figure 6.2 Structure of compact bone.

Module 6.3: Bone Formation: Ossification

The formation of bone, known as ossification, is discussed in this module. When you complete it, you should be able to do the following:

1. Explain the differences between primary and secondary bone.
2. Describe the process of intramembranous ossification.
3. Describe the process of endochondral ossification.

Build Your Own Glossary

Below is a table listing key terms from Module 6.3. Before you read the module, use the glossary at the back of your book or look through the module to define the following terms.

Key Terms of Module 6.3

Term	Definition
Osteogenesis	
Intramembranous ossification	
Primary ossification center	
Endochondral ossification	
Secondary ossification center	

Survey It: Form Questions

Before you read the module, survey it and form at least three questions for yourself. When you have finished reading the module, return to these questions and answer them.

Question 1: ____________________

Answer: ____________________

Question 2: ____________________

Answer: ____________________

Question 3: ____________________

Answer: ____________________

Key Concept: What are primary and secondary bones? How do they differ in structure and function?

___

___

Describe It: Stages of Intramembranous Ossification

Write out the steps of intramembranous ossification in the illustrations in Figure 6.3, and label the cells and structures involved in each step.

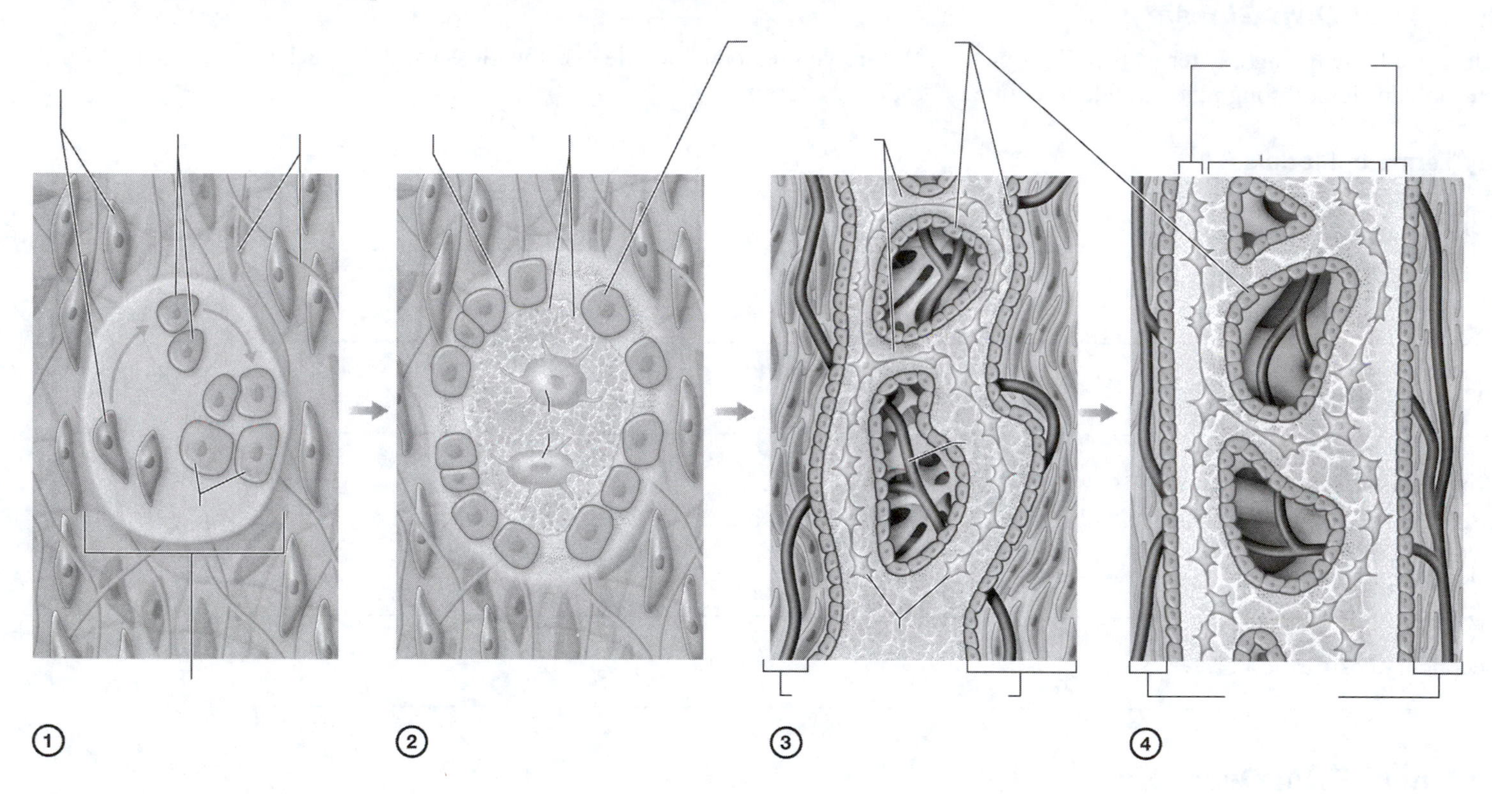

Figure 6.3 Intramembranous ossification.

Key Concept: What role do osteoblasts play in intramembranous ossification?

___

___

Describe It: Stages of Endochondral Ossification

Write out the steps of endochondral ossification in the illustrations in Figure 6.4, and label the cells and structures involved in each step.

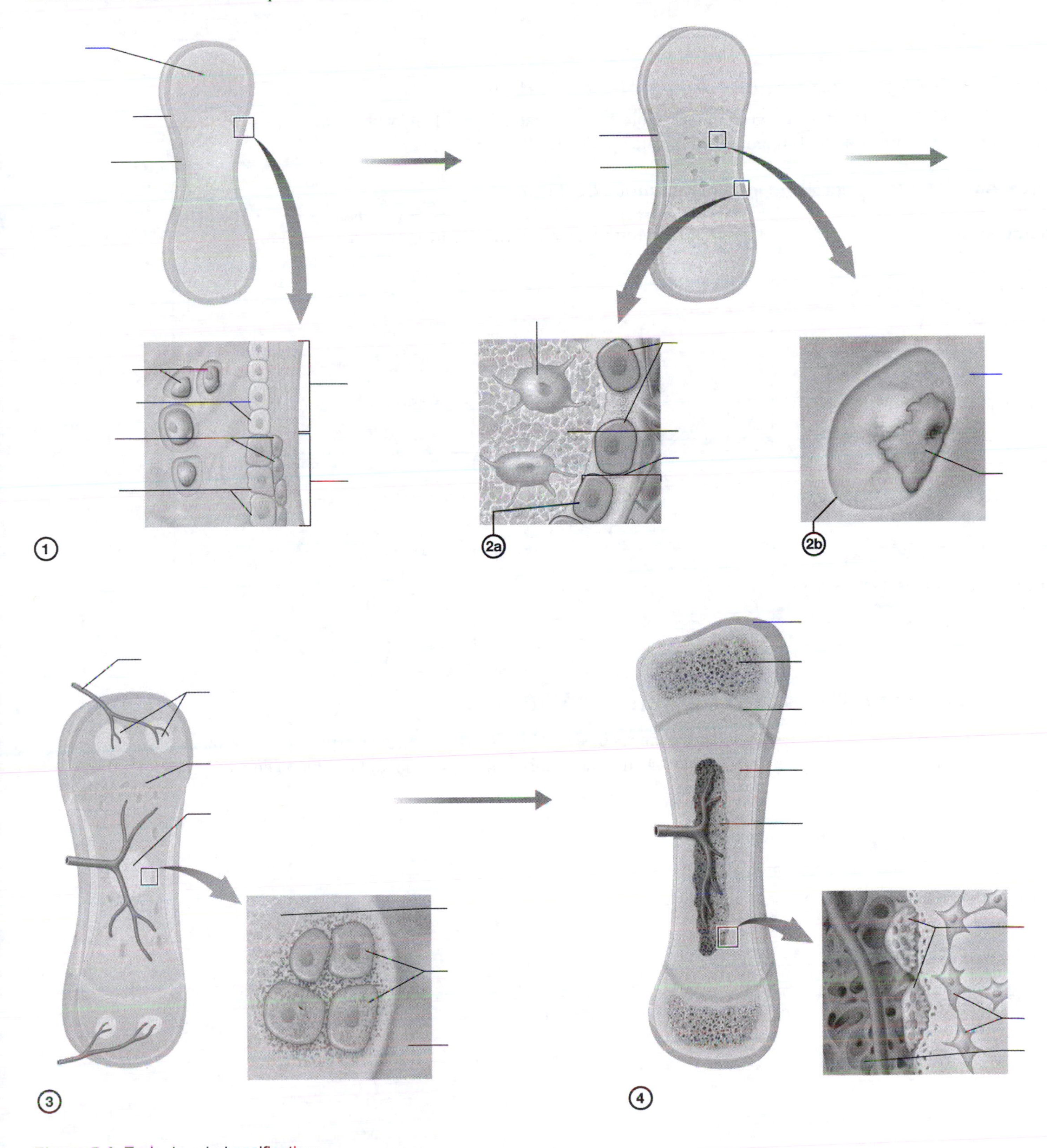

Figure 6.4 Endochondral ossification.

Key Concept: What role does the bone collar play in endochondral ossification?

____

____

Build Your Own Summary Table: Types of Ossification

As you read Module 6.3, build your own summary table of the characteristics of endochondral and intramembranous ossification by filling in the template provided below.

Comparison of Endochondral and Intramembranous Ossification

Characteristic	Intramembranous Ossification	Endochondral Ossification
Bones formed by this type of ossification		
Model used		
Where ossification begins		
Type of bone (compact or spongy) formed first		

Module 6.4: Bone Growth in Length and Width

Once a bone has ossified, it continues to undergo changes in size. Module 6.4 in your text explores bone growth, including how a bone grows in length and in width. At the end of this module, you should be able to do the following:

1. Describe how long bones grow in length.
2. Compare longitudinal and appositional bone growth.
3. Describe the hormones that play a role in bone growth.

Build Your Own Glossary

Below is a table listing key terms from Module 6.4. Before you read the module, use the glossary at the back of your book or look through the module to define the following terms.

Key Terms of Module 6.4

Term	Definition
Longitudinal growth	
Epiphyseal plate	
Appositional growth	

Survey It: Form Questions

Before you read the module, survey it and form at least two questions for yourself. When you have finished reading the module, return to these questions and answer them.

Question 1: ____________________

Answer: ____________________

Question 2: ____________________

Answer: ____________________

Key Concept: What type of cells are responsible for longitudinal bone growth? Where are these cells located?

____

____

Identify It: Zones of the Epiphyseal Plate

Identify and color-code each zone of the epiphyseal plate in Figure 6.5. Then, list the events that take place in each zone during longitudinal bone growth.

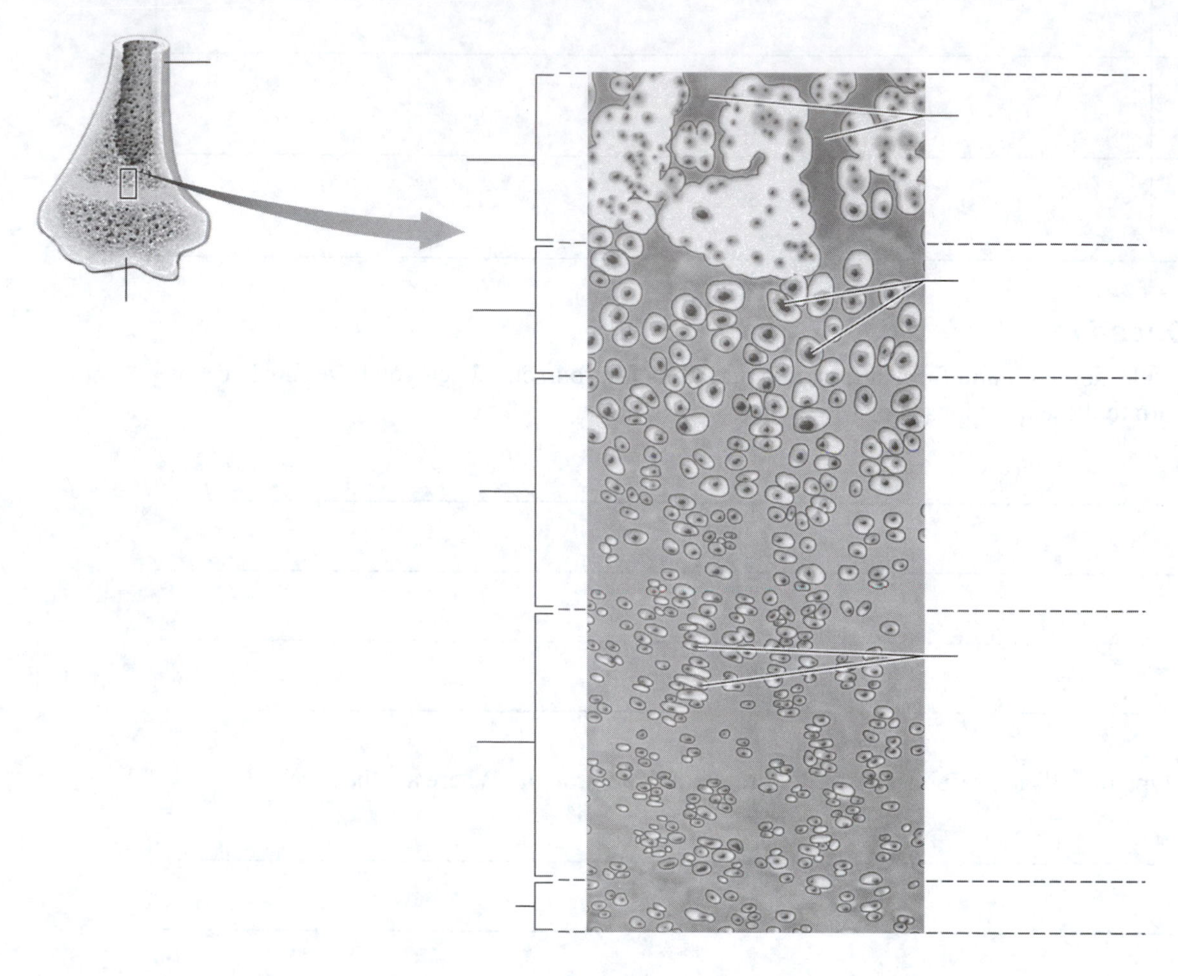

Figure 6.5 The epiphyseal plate.

Key Concept: What takes place during appositional bone growth?

____

____

Key Concept: What are three hormones that impact bone growth? How does each hormone impact this process?

____

____

Module 6.5: Bone Remodeling and Repair

Bone is a very dynamic tissue, constantly undergoing the process of bone remodeling, the topic we now examine. At the end of this module, you should be able to do the following:

1. Describe the processes of bone resorption and bone deposition.
2. Describe the physical, hormonal, and dietary factors that influence bone remodeling.
3. Explain the role of calcitonin, parathyroid hormone, and vitamin D in bone remodeling and calcium ion homeostasis.
4. Describe the general process of bone repair.

Build Your Own Glossary

Following is a table listing key terms from Module 6.5. Before you read the module, use the glossary at the back of your book or look through the module to define the following terms.

Key Terms of Module 6.5

Term	Definition
Bone remodeling	
Bone deposition	
Bone resorption	
Parathyroid hormone	
Calcitonin	
Fracture	
Bone callus	

Survey It: Form Questions

Before you read the module, survey it and form at least three questions for yourself. When you have finished reading the module, return to these questions and answer them.

Question 1: ____________________

Answer: ____________________

Question 2: ____

Answer: ____

Question 3: ____

Answer: ____

Key Concept: Does bone remodeling occur in healthy bone? Why or why not?

____

____

Complete It: Bone Remodeling

Fill in the blanks to complete the following paragraph that describes the properties of bone remodeling.

___________ ___________ is carried out by osteoblasts, while ___________ ___________ is carried out by osteoclasts. Osteoblasts are triggered by the force known as ___________, whereas osteoclasts are triggered by the force known as ___________. Other factors that influence bone remodeling include the intake of ___________, ___________, ___________, ___________, and ___________.

Key Concept: What happens when bone resorption is greater than bone deposition? What happens in the opposite case, when deposition is greater than absorption?

____

____

Team Up

Make a handout to teach the feedback loop involved in the regulation of calcium ion homeostasis. You can use Figure 6.15 on page 203 of your text as a guide, but the handout should be in your own words and with your own diagrams. At the end of the handout, write a few quiz questions. Once you have completed your handout, team up with one or more study partners and trade handouts. Study your partners' diagrams, and when you have finished, take the quiz at the end. When you and your group have finished taking all the quizzes, discuss the answers to determine places where you need additional study. After you've finished, combine the best elements of each handout to make one "master" diagram for the control of calcium ion homeostasis.

Describe It: Stages of Fracture Repair

Write out the steps of fracture repair in the illustrations in Figure 6.6, and label the cells and structures involved in each step.

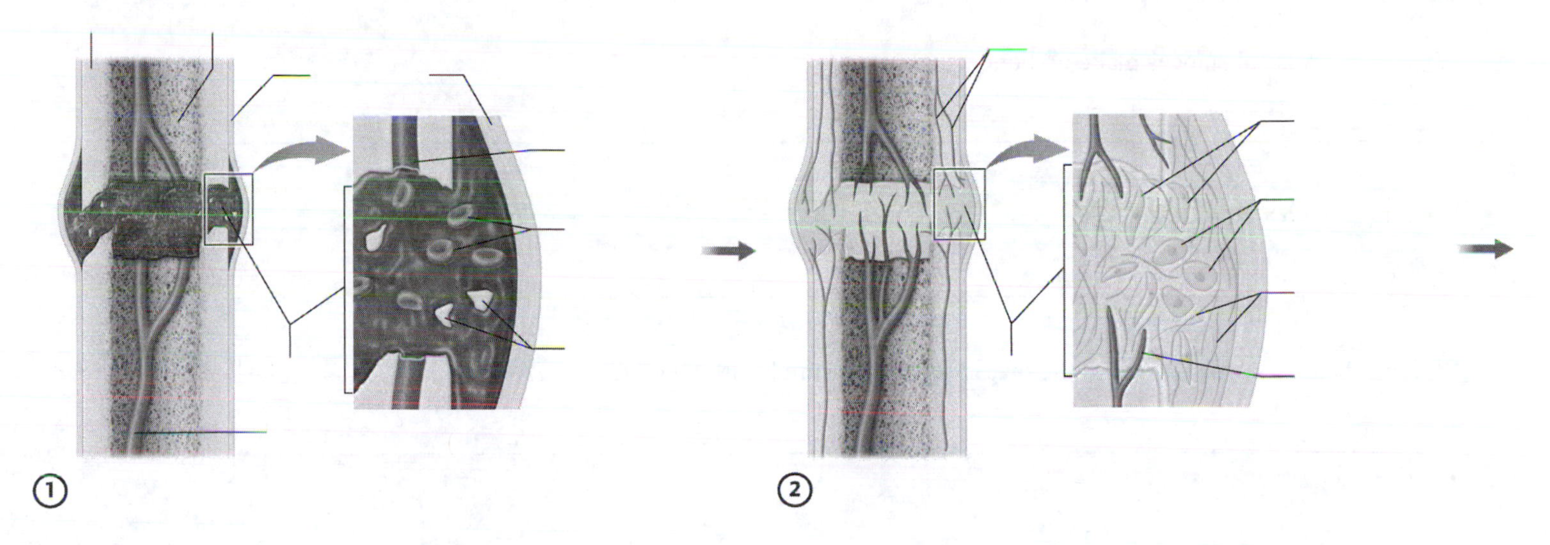

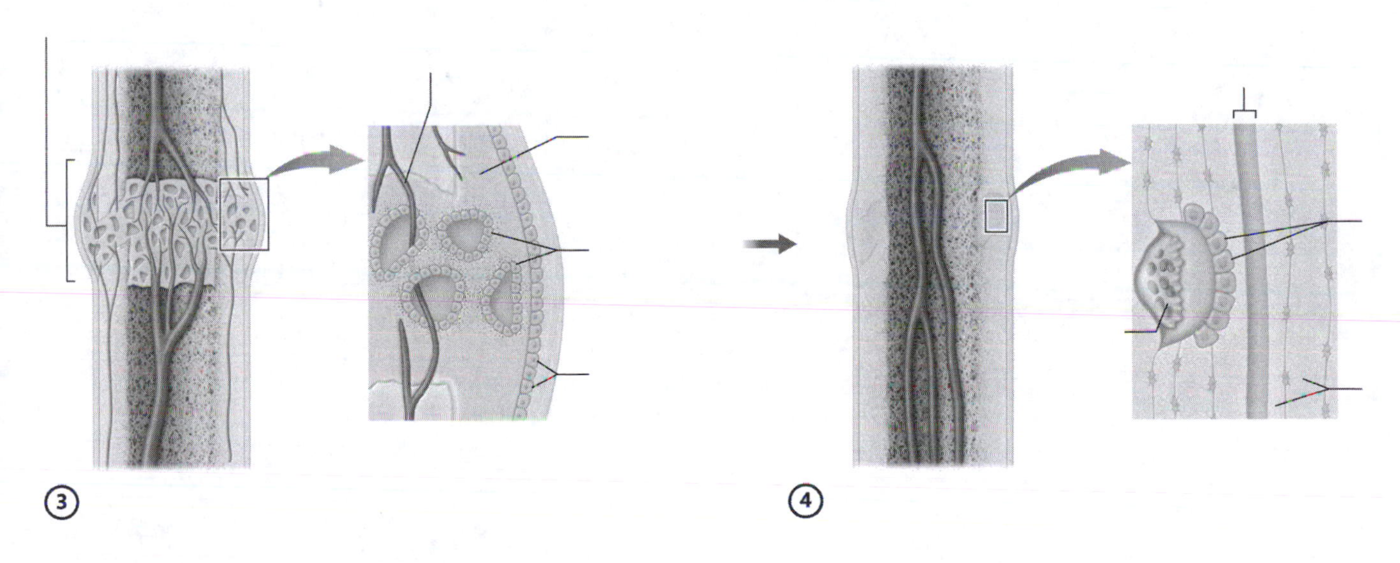

Figure 6.6 Fracture repair.

What Do You Know Now?

Let's now revisit the questions you answered in the beginning of this chapter. How have your answers changed now that you've worked through the material?

- What chemicals/substances make up bone tissue?

  ____

- What makes up a bone as an organ?

  ____

- Which substances are required in the diet in order to ensure healthy bones?

  ____

Module 7.2: The Skull

Now we look at the structure of the most complex collection of bones in the skeleton: the skull. When you finish this module, you should be able to do the following:

1. Describe the location, structural features, and functions of each of the cranial bones and the main sutures that unite them.
2. Describe the location, structural features, and functions of each of the facial bones and the hyoid bone.
3. Explain the structural features and functions of the orbit, nasal cavity, and paranasal sinuses.
4. Compare and contrast the skull of a fetus or infant with that of an adult.

Build Your Own Glossary

Below is a table listing key terms from Module 7.2. Before you read the module, use the glossary at the back of your book or look through the module to define the following terms.

Key Terms for Module 7.2

Term	Definition
Cranial bones	aKA
Facial bones	
Cranial vault	
Cranial base	
Orbit	
Nasal cavity	
Paranasal sinus	
Fontanel	
Hyoid bone	

Survey It: Form Questions

Before you read the module, survey it and form at least three questions for yourself. When you have finished reading the module, return to these questions and answer them.

Question 1: ______________________________

Answer: ______________________________

Question 2: ______________________________

Answer: ______________________________

Question 3: ______________________________

Answer: ______________________________

Build Your Own Summary Table: Bones of the Skull

As you read Module 7.2, build your own summary table of the bones of the skull by filling in the template provided below.

Summary of the Bones of the Skull

Bone	Description/Location	Main Features
Cranial Bones		
Frontal bone		
Parietal bones		
Occipital bones		
Temporal bone		
Sphenoid bone		
Ethmoid bone		
Facial Bones		
Nasal bones		
Lacrimal bones		
Zygomatic bones		

Palatine bones		
Mandible		
Maxillae		
Inferior nasal conchae		
Vomer		

Key Concept: Why do you think that the bones of the skull have so many canals and foramina?

____

____

Identify It: Anatomy of the Skull

Identify and color-code each component of the skull in Figure 7.1.

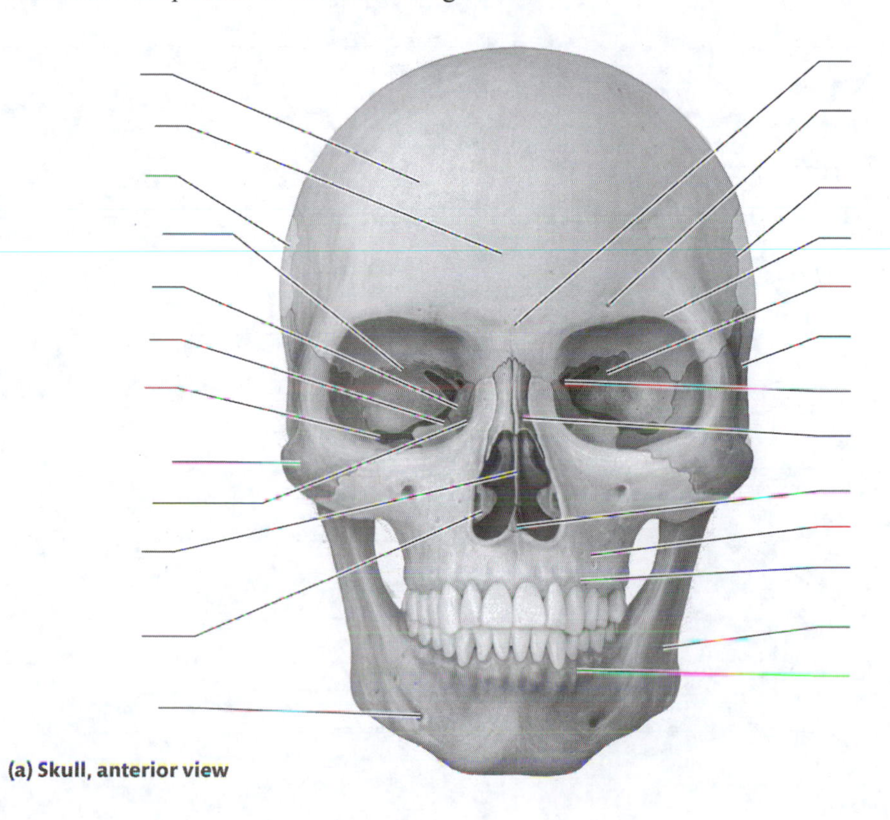

Figure 7.1 The skull.

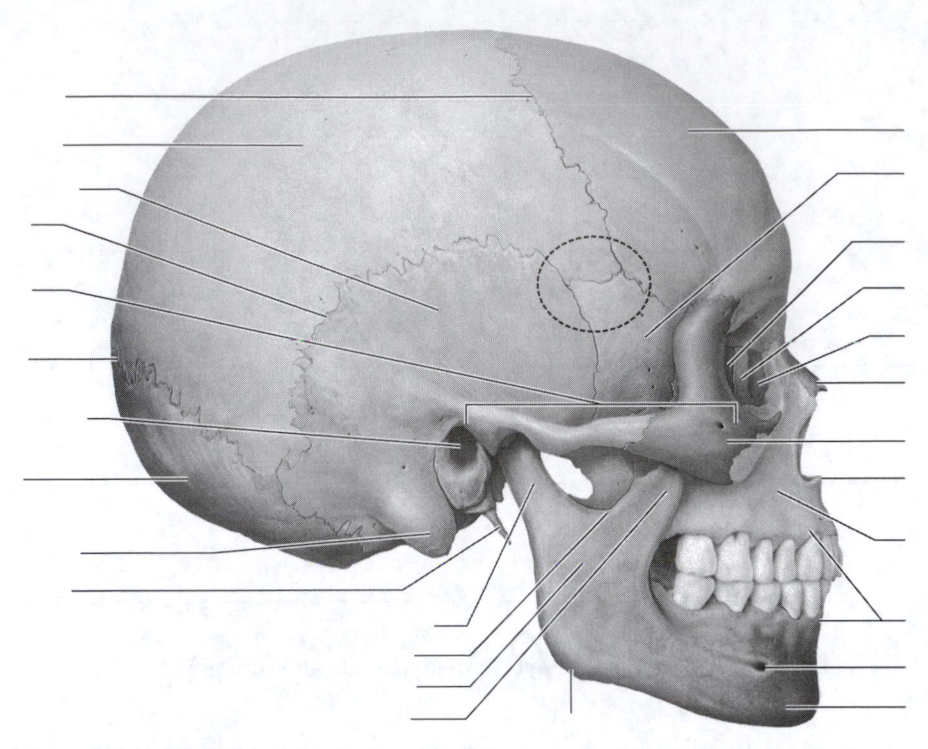

(b) External anatomy of skull, lateral view

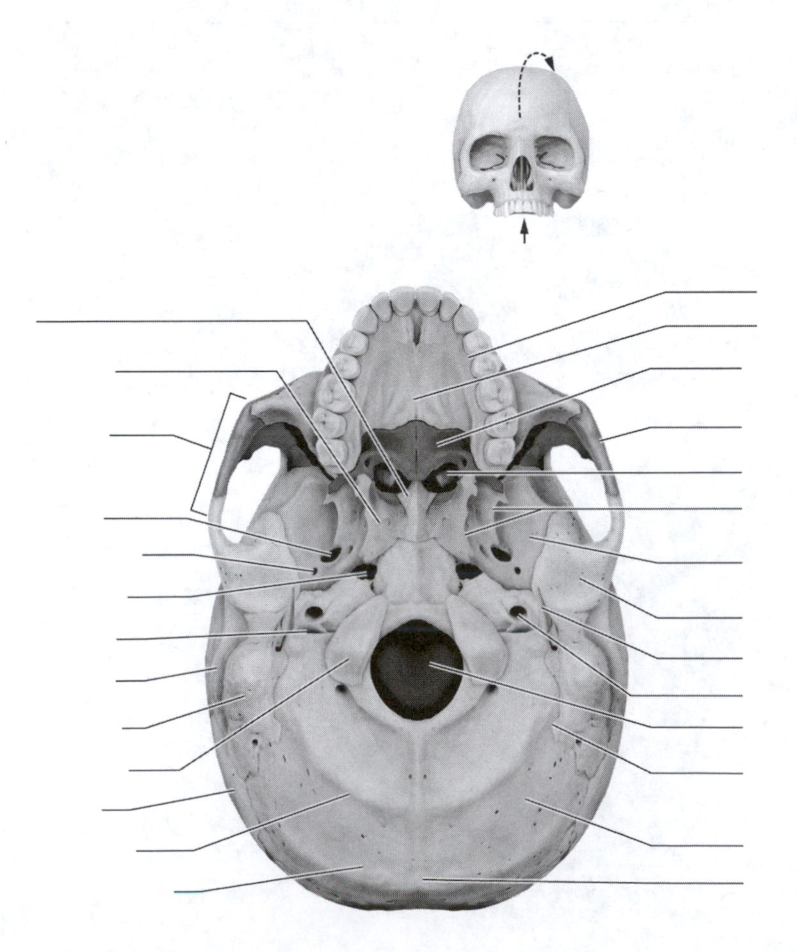

(c) Inferior view

Figure 7.1 (*continued*)

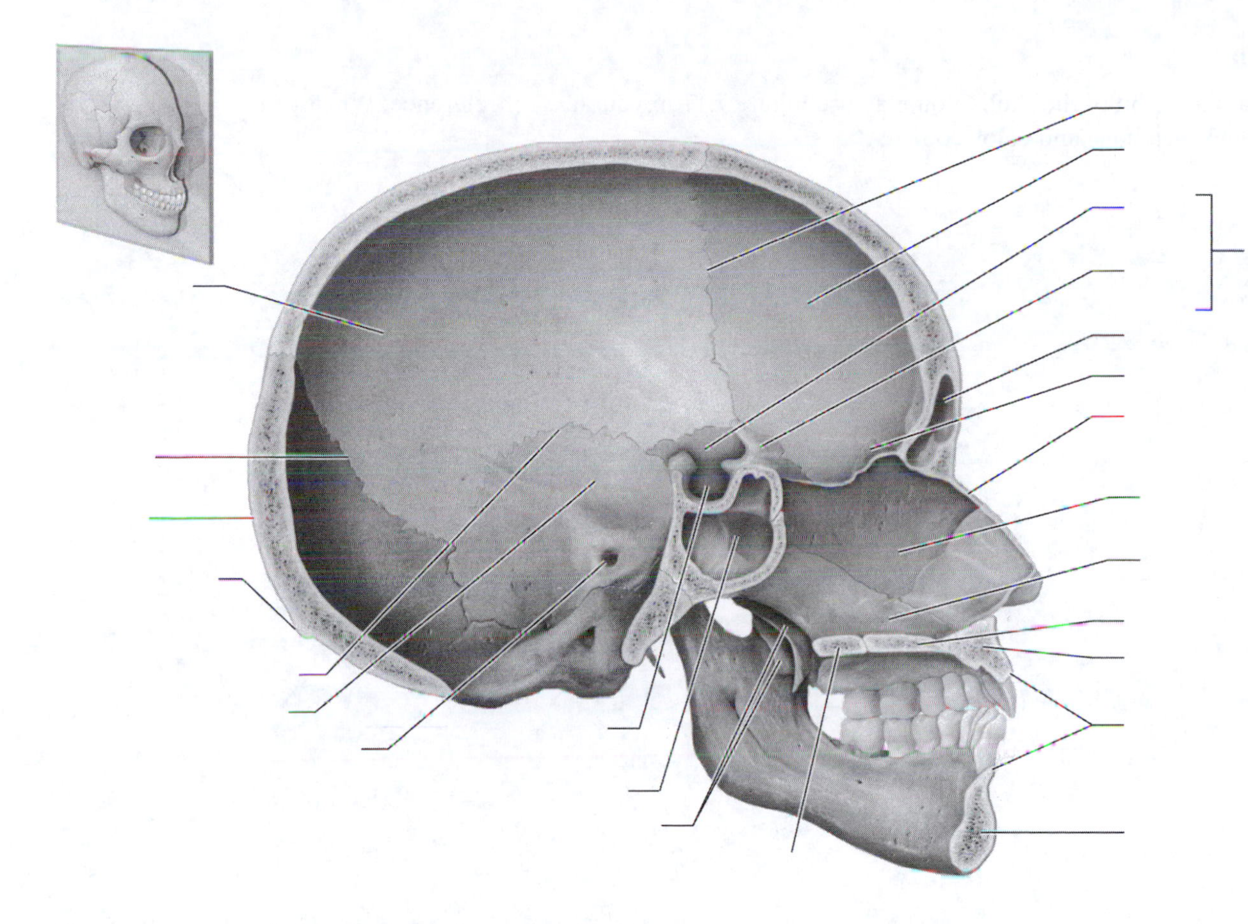

(d) Internal anatomy of the left half of skull, midsagittal section

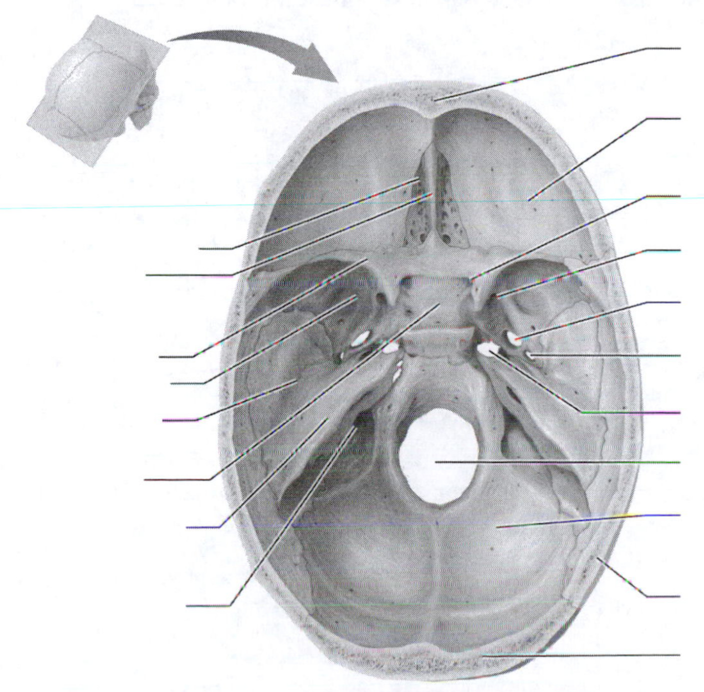

(e) Superior, interior view showing internal structures of skull

Figure 7.1 *(continued)*

Draw It: The Orbit

In the space below, draw the orbit of the skull. You may use Figure 7.11 in your text for reference. When you finish your drawing, label each bone and color-code it.

Identify It: The Nasal Cavity and Paranasal Sinuses

Identify and color-code each component of the nasal cavity and paranasal sinuses in Figure 7.2.

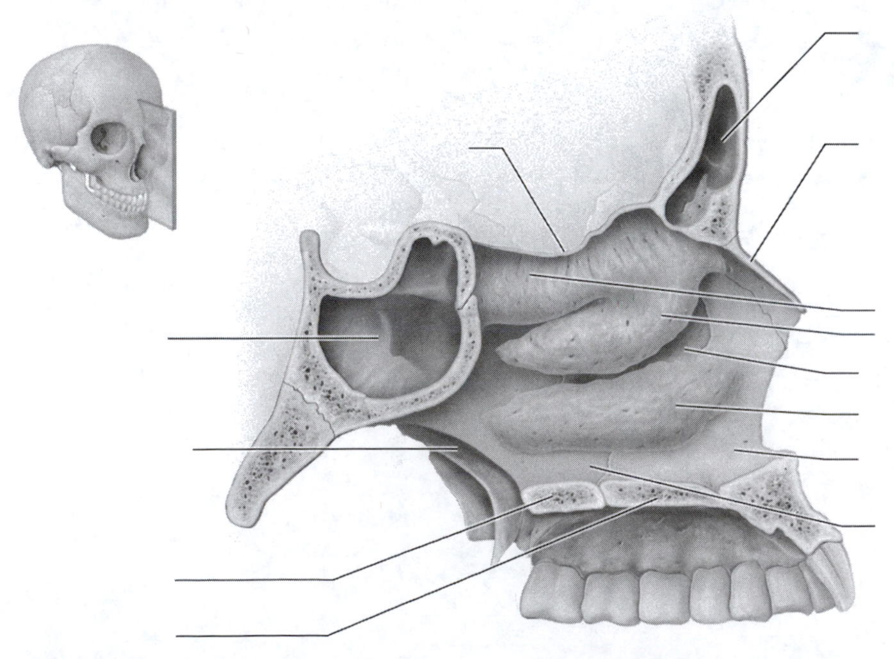

(a) Parasagittal section through nasal cavity (nasal septum removed)

Figure 7.2 The nasal cavity and paranasal sinuses.

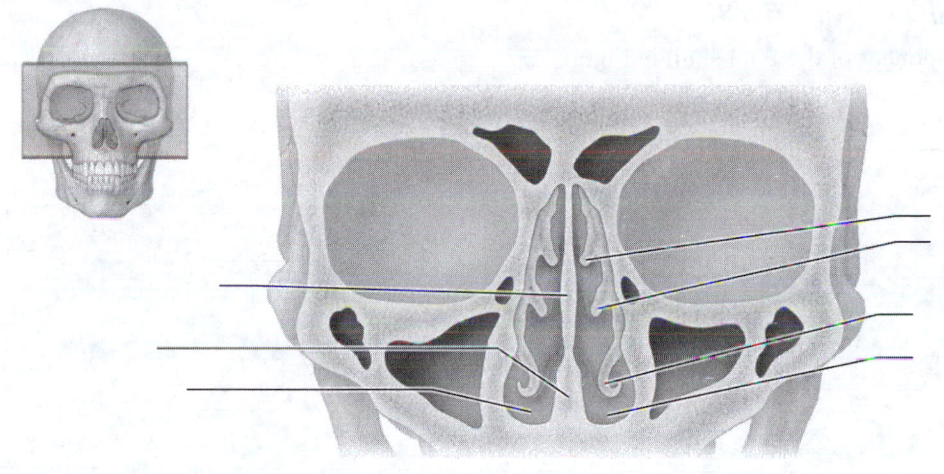

(b) Anterior view of nasal cavity

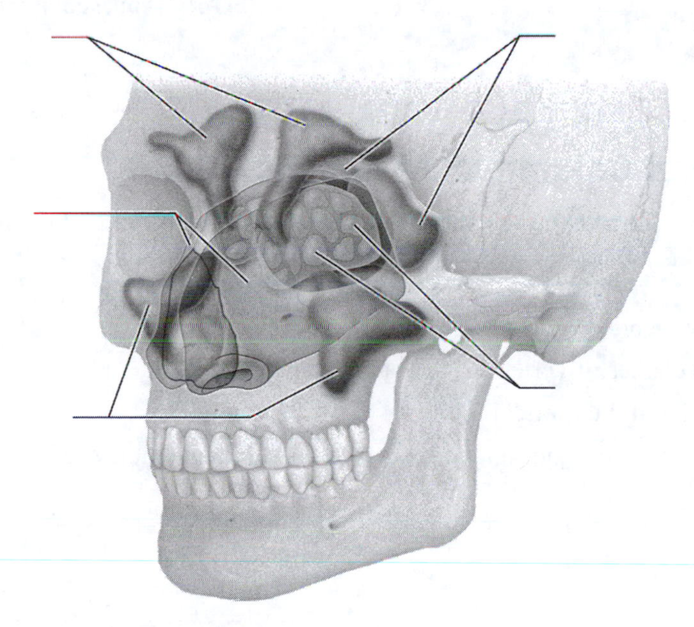

(c) Paranasal sinuses, anterolateral view

Figure 7.2 (*continued*)

Identify It: The Fetal Skull

Identify and color-code each component of the fetal skull in Figure 7.3.

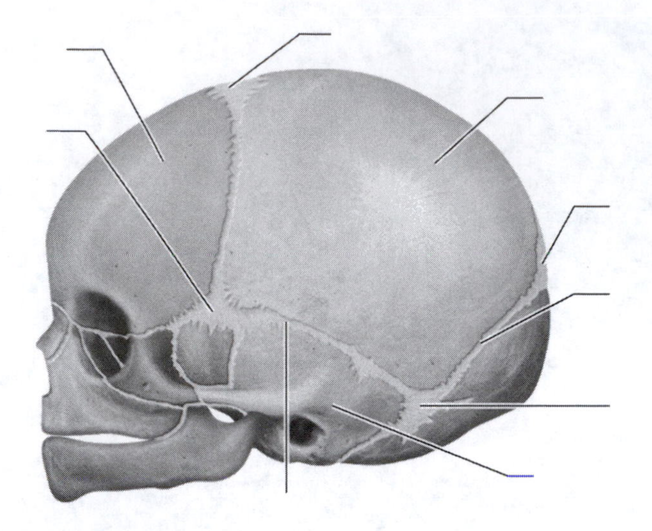

(a) Fetal skull, lateral view

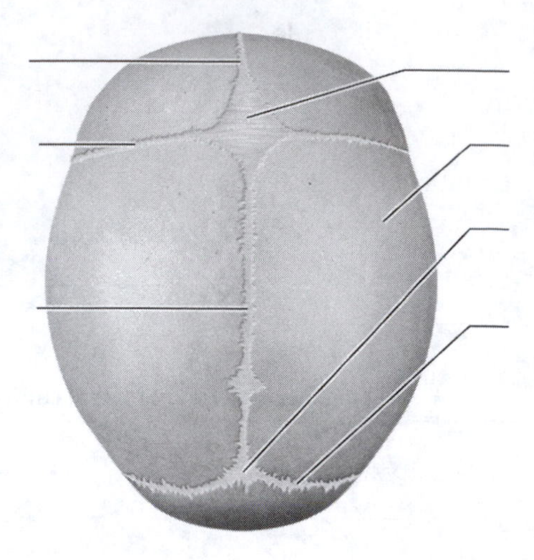

(b) Fetal skull, superior view

Figure 7.3 The fetal skull.

Module 7.3: The Vertebral Column and Thoracic Cage

This module discusses the structure of the remaining bones of the axial skeleton, the vertebral column, and thoracic cage. When you complete this module, you should be able to do the following:

1. Describe the curvatures of the vertebral column.
2. Compare and contrast the three classes of vertebrae, the sacrum, and the coccyx.
3. Explain the structure and function of intervertebral discs.
4. Describe the location, structural features, and functions of the bones of the thoracic cage.

Build Your Own Glossary

Following is a table listing key terms from Module 7.3. Before you read the module, use the glossary at the back of your book or look through the module to define the following terms.

Key Terms of Module 7.3

Term	Definition
Vertebral column	
Cervical curvature	
Lumbar curvature	
Thoracic curvature	
Sacral curvature	
Body (vertebral)	
Vertebral foramen	
Transverse process	
Spinous process	
Atlas	
Axis	
Intervertebral disc	
Thoracic cage	
Manubrium	
Body (sternal)	
Xiphoid process	
Intercostal space	

Survey It: Form Questions

Before you read the module, survey it and form at least three questions for yourself. When you have finished reading the module, return to these questions and answer them.

Question 1: ______________________________

Answer: ______________________________

Question 2: ______________________________

Answer: ______________________________

Question 3: ______________________________

Answer: ______________________________

Identify It: Basic Structure of a Vertebra

Identify and color-code each component of the vertebra in Figure 7.4.

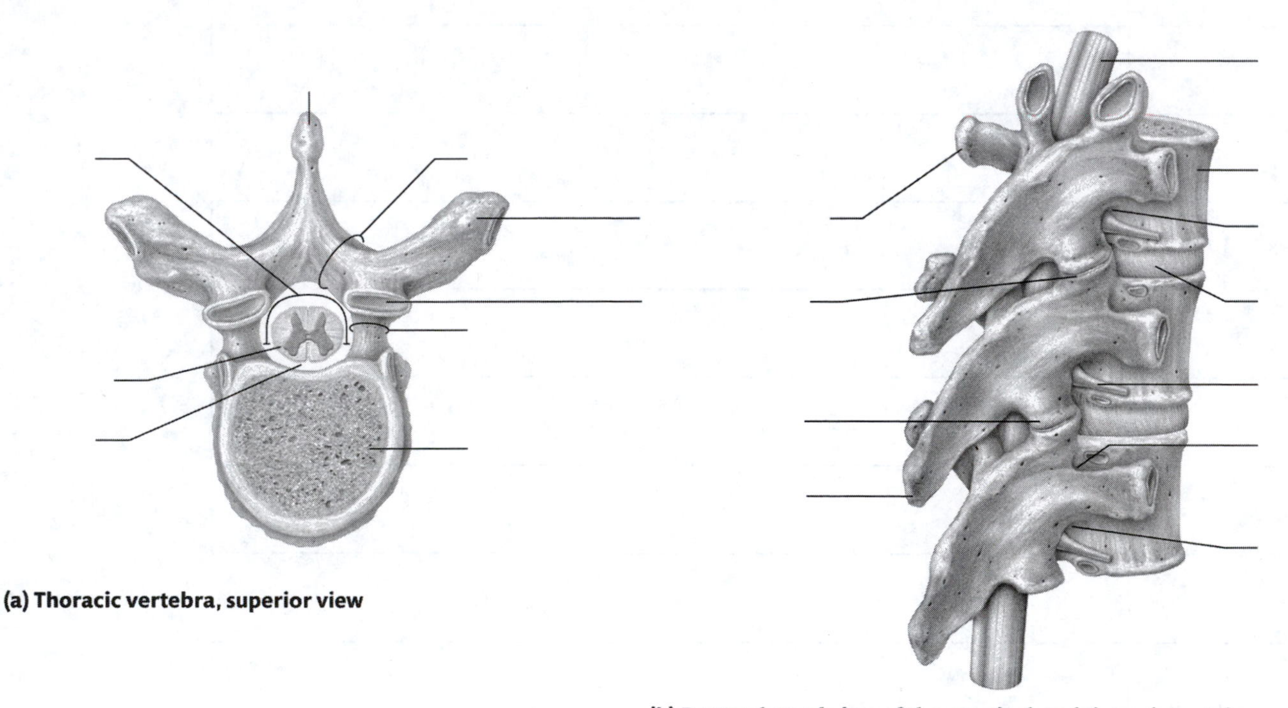

(a) Thoracic vertebra, superior view

(b) Posterolateral view of three articulated thoracic vertebrae

Figure 7.4 Basic structure of a vertebra.

Complete It: The Vertebral Column

Fill in the blanks to complete the following paragraphs that describe the properties of the vertebral column.

Cervical vertebrae can be identified by their ____________ ____________ that permit the passage of the vertebral arteries. The first cervical vertebra, the ____________, lacks a ____________. The second cervical vertebra, the ____________, is identifiable by a superior projection called the ____________. Thoracic vertebrae can be identified by their spinous processes and their ____________ ____________, which articulate with the ribs. ____________ ____________ can be identified by their thick bodies and spinous processes.

An intervertebral disc is composed of an outer ring of fibrous connective tissue called the ____________ ____________. The inner gelatinous mass of the disc is the ____________ ____________. A disc functions primarily in ____________ ____________.

Key Concept: How does the structure of each class of vertebra follow its function?

____

____

Identify It: Remainder of the Axial Skeleton

Identify and color-code each component of the sacrum, coccyx, and thoracic cage in Figure 7.5.

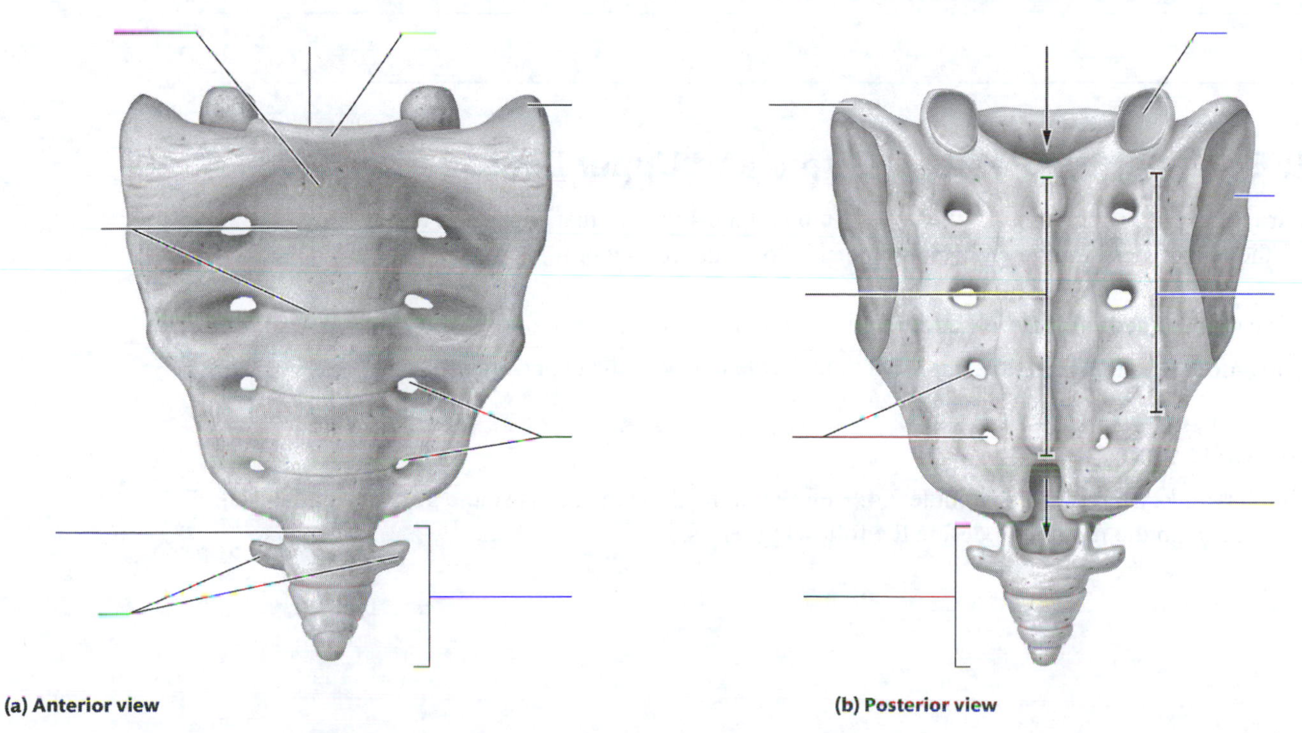

Figure 7.5 Remainder of the axial skeleton: the sacrum, coccyx, and thoracic cage.

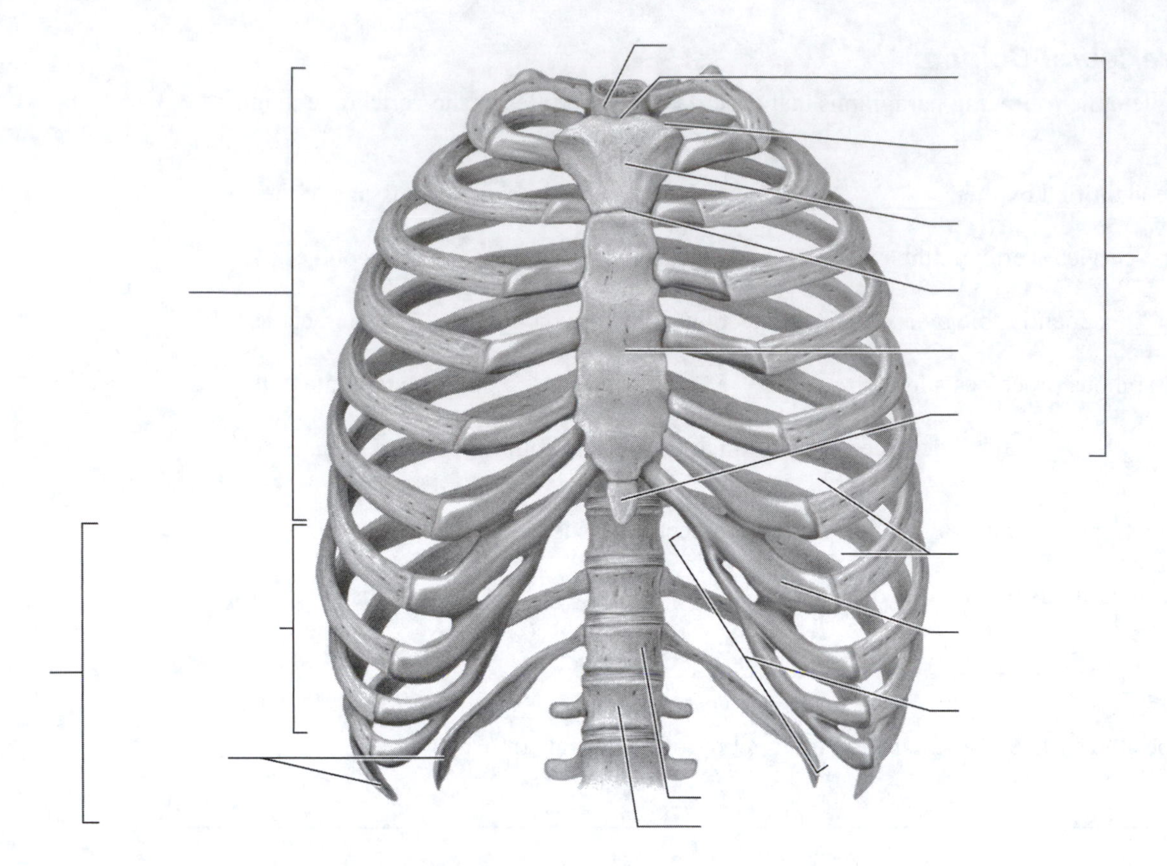

Figure 7.5 *(continued)*

Key Concept: What is the main function of the thoracic cage?

Module 7.4: Bones of the Pectoral Girdle and Upper Limb

Module 7.4 in your text explores the bones of the upper limb and the structure that anchors it to the body, the pectoral girdle. At the end of this module, you should be able to do the following:

1. Describe the location, structural features, and functions of the bones of the pectoral girdle.
2. Describe the location, structural features, and functions of the bones of the upper limb.

Build Your Own Glossary

Following is a table listing key terms from Module 7.4. Before you read the module, use the glossary at the back of your book or look through the module to define the following terms.

Key Terms of Module 7.4

Term	Definition
Clavicle	
Scapula	
Coracoid process	
Acromion	
Glenoid cavity	
Spine (of scapula)	
Humerus	
Trochlea	
Capitulum	
Olecranon fossa	
Radius	
Ulna	
Trochlear notch	
Olecranon	
Carpals	
Metacarpals	
Phalanges	

Survey It: Form Questions

Before you read the module, survey it and form at least two questions for yourself. When you have finished reading the module, return to these questions and answer them.

Question 1: ______________________________

Answer: ______________________________

Question 2: ______________________________

Answer: ______________________________

Key Concept: How do the structure and functions of the bones of the upper limb differ from those of the axial skeleton?

____

____

Identify It: Structure of the Scapula

Identify and color-code each component of the scapula in Figure 7.6.

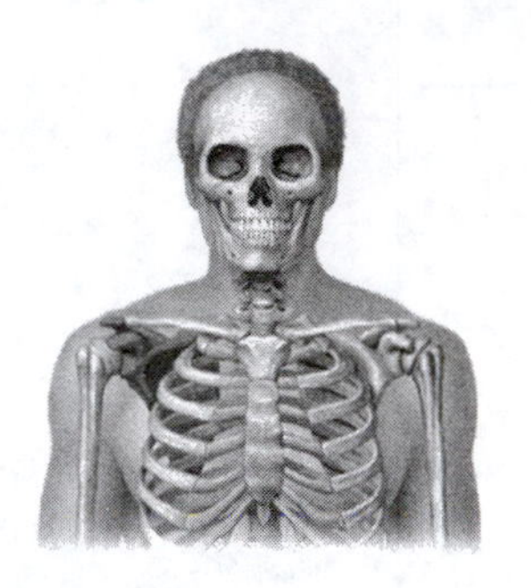

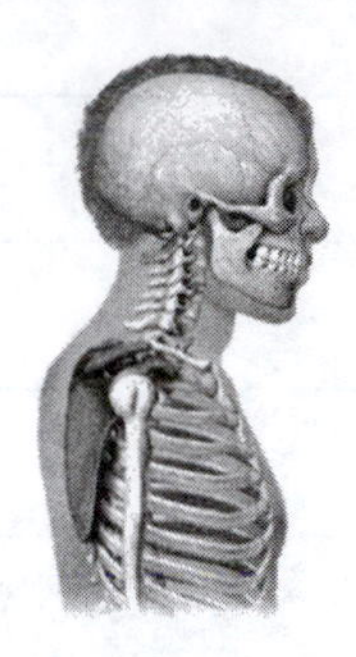

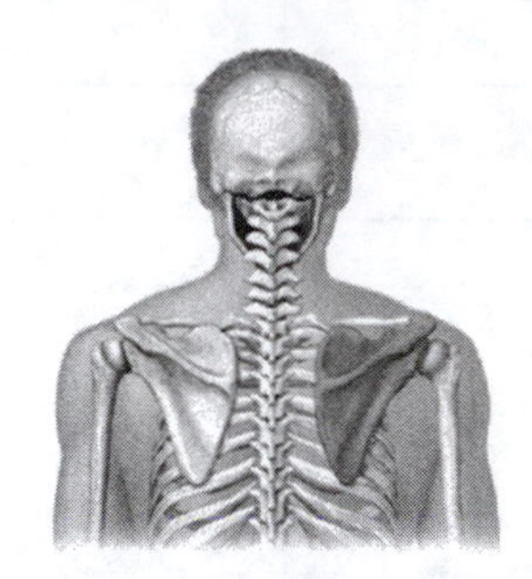

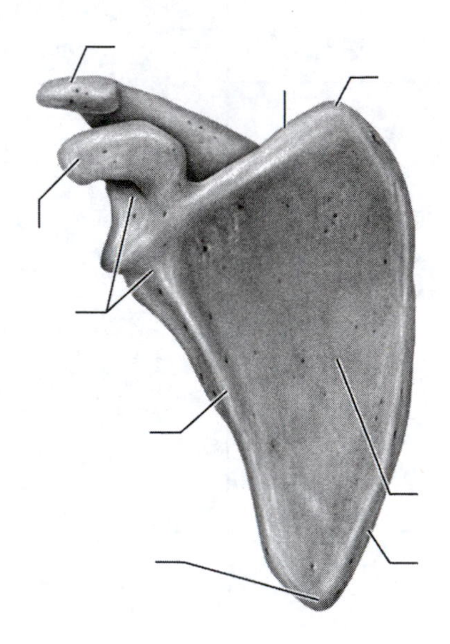

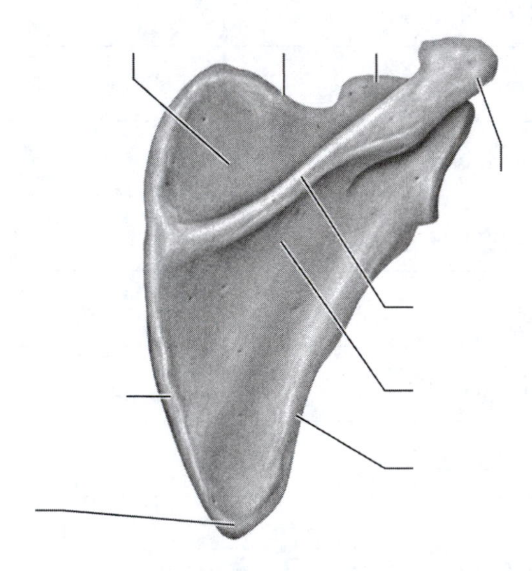

(a) Right scapula, anterior view **(b) Right scapula, lateral view** **(c) Right scapula, posterior view**

Figure 7.6 The scapula.

Identify It: Bones of the Upper Limb

Identify and color-code each component of the humerus, radius, and ulna in Figure 7.7.

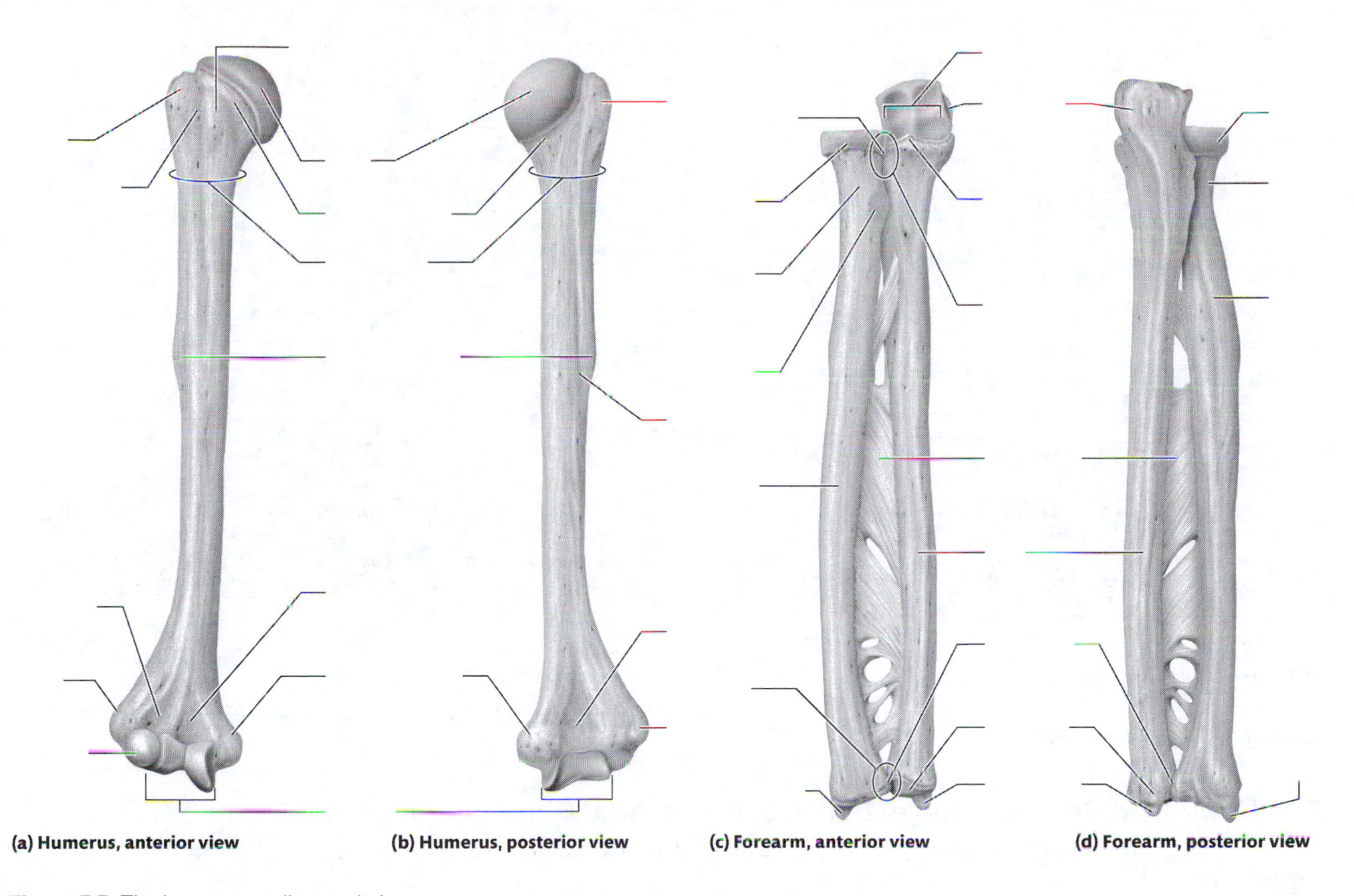

Figure 7.7 The humerus, radius, and ulna.

Identify It: Bones of the Wrist and Hand

Identify and color-code each component of the wrist and hand in Figure 7.8.

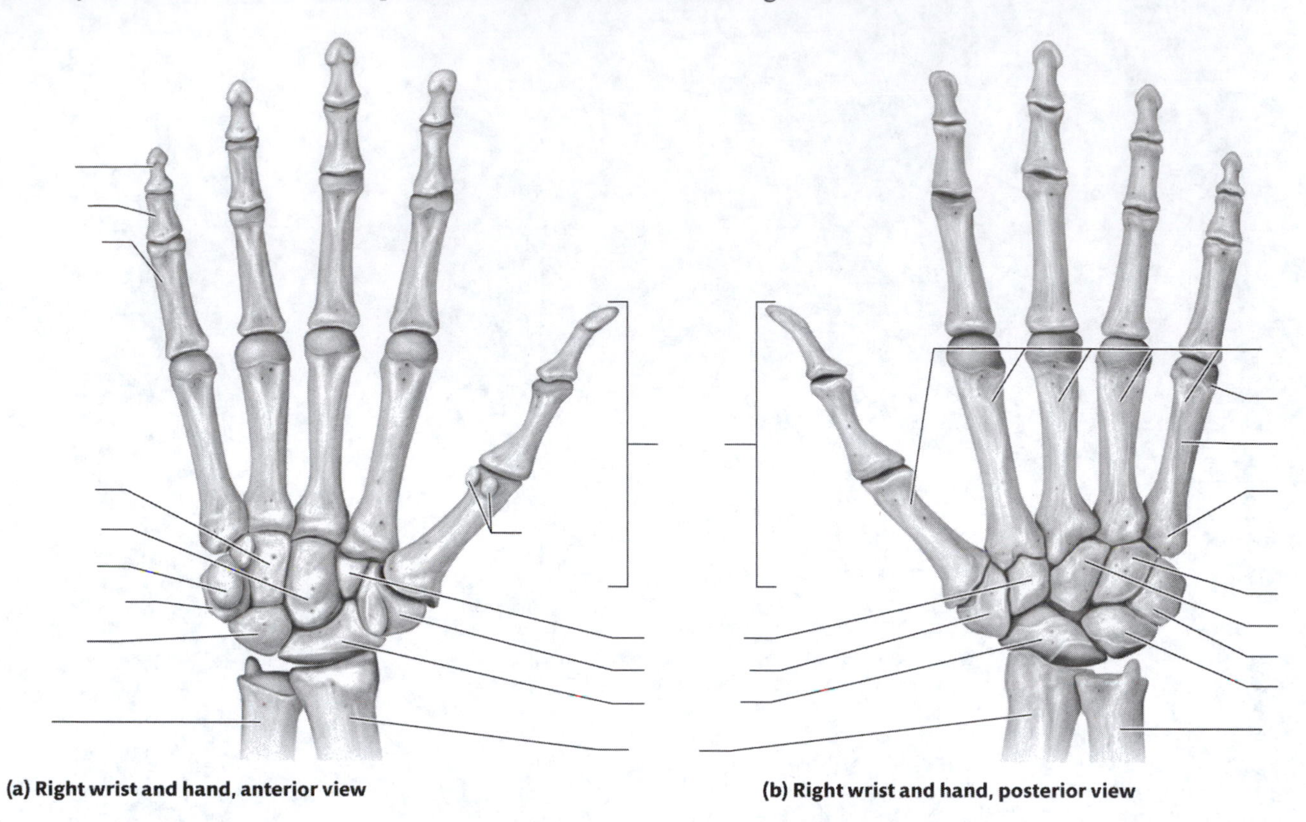

Figure 7.8 The wrist and hand.

Build Your Own Summary Table: Bones of the Pectoral Girdle and Upper Limb

As you read Module 7.4, build your own summary table of the characteristics of the pectoral girdle and upper limb by filling in the template provided below.

Summary of Bones of the Pectoral Girdle and Upper Limb

Bone	Location/Description	Important Features
Pectoral Girdle		
Scapula		
Clavicle		

Upper Limb		
Humerus		
Radius		
Ulna		
Carpals		
Metacarpals		
Phalanges		

Team Up

Write 10 quiz questions about the bones of the pectoral girdle and upper limb. Once you have completed your quiz, team up with one or more study partners and trade quizzes. When you and your group have finished taking all the quizzes, discuss the answers to determine places where you need additional study.

Module 7.5: Bones of the Pelvic Girdle and Lower Limb

The bones of the pelvic girdle and lower limb, the topic we now examine, round out our coverage of the skeleton. At the end of this module, you should be able to do the following:

1. Describe the location, structural features, and functions of the bones of the pelvic girdle.
2. Compare and contrast the adult male and female pelvic bones.
3. Describe the location, structural features, and functions of the bones of the thigh, the leg, the ankle, and the foot.

Build Your Own Glossary

Following is a table listing key terms from Module 7.5. Before you read the module, use the glossary at the back of your book or look through the module to define the following terms.

Key Terms of Module 7.5

Term	Definition
Acetabulum	
Ilium	
Iliac crest	
Ischium	
Ischial tuberosity	
Pubis	
Pubic symphysis	
Femur	
Greater trochanter	
Tibia	
Medial malleolus	
Fibula	
Lateral malleolus	
Tarsals	
Metatarsals	

Survey It: Form Questions

Before you read the module, survey it and form at least three questions for yourself. When you have finished reading the module, return to these questions and answer them.

Question 1: ______________________________

Answer: ______________________________

Question 2: ______________________________

Answer: ______________________________

Question 3: ______________________________

Answer: ______________________________

Key Concept: How do the bones of the lower limb differ structurally and functionally from those of the upper limb?

____

____

Identify It: The Pelvis and Pelvic Bones

Identify and color-code each component of the pelvis and pelvic bones in Figure 7.9.

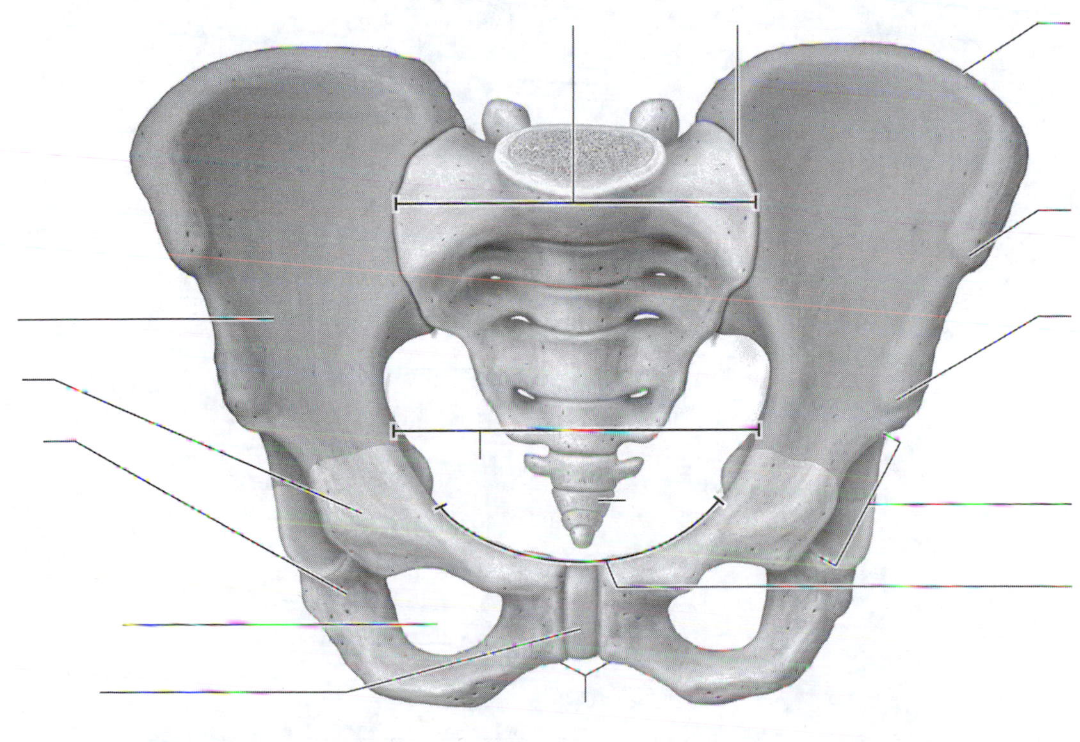

(a) Pelvis, anterior view

Figure 7.9 The pelvis and pelvic bones.

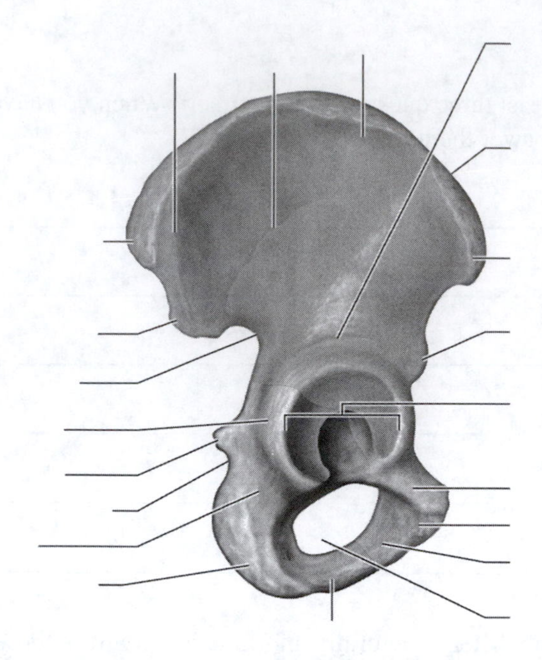

(b) Pelvic bone, anterior view

Figure 7.9 *(continued)*

Identify It: The Lower Limb

Identify and color-code each component of the femur, tibia, and fibula in Figure 7.10.

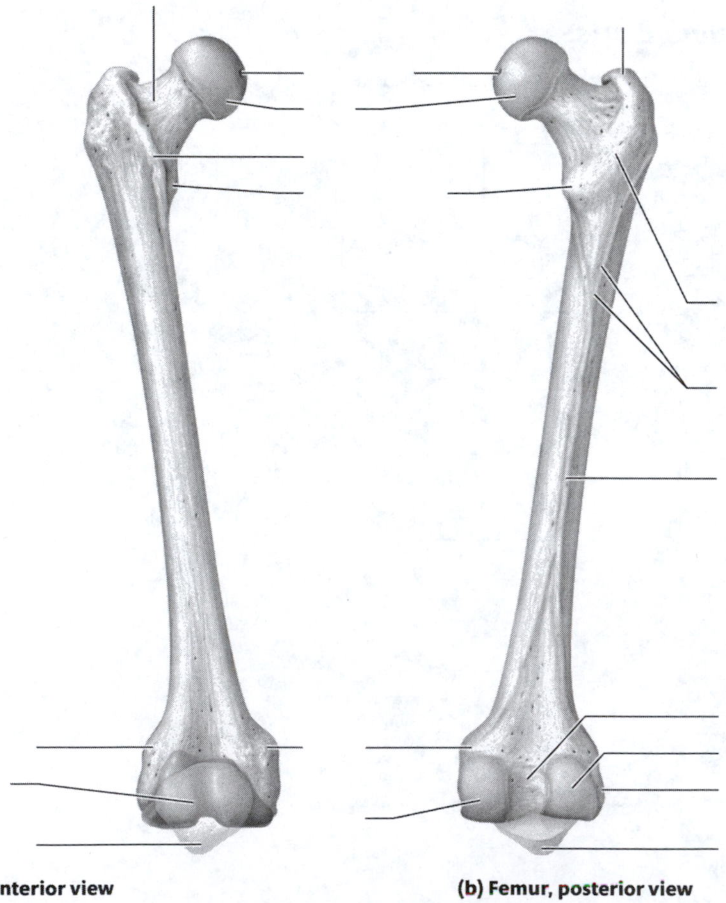

(a) Femur, anterior view

(b) Femur, posterior view

Figure 7.10 The femur, tibia, and fibula.

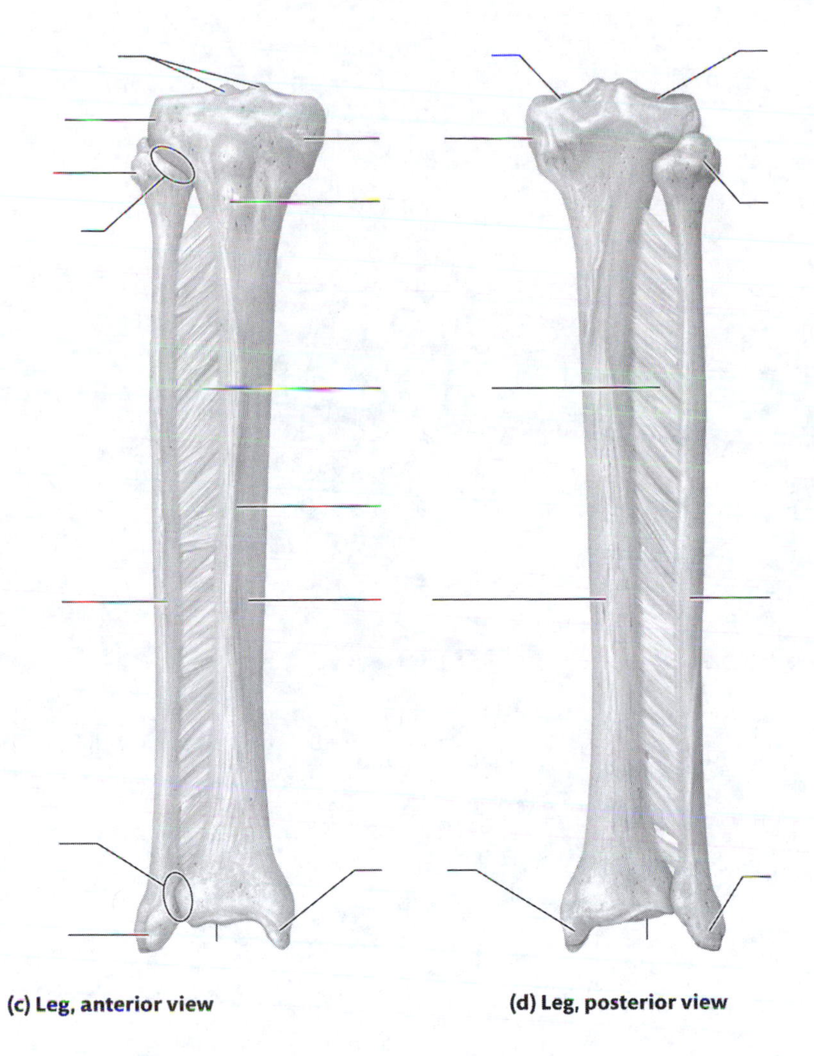

Figure 7.10 *(continued)*

Identify It: Bones of the Ankle and Foot

Identify and color-code each component of the ankle and foot in Figure 7.11.

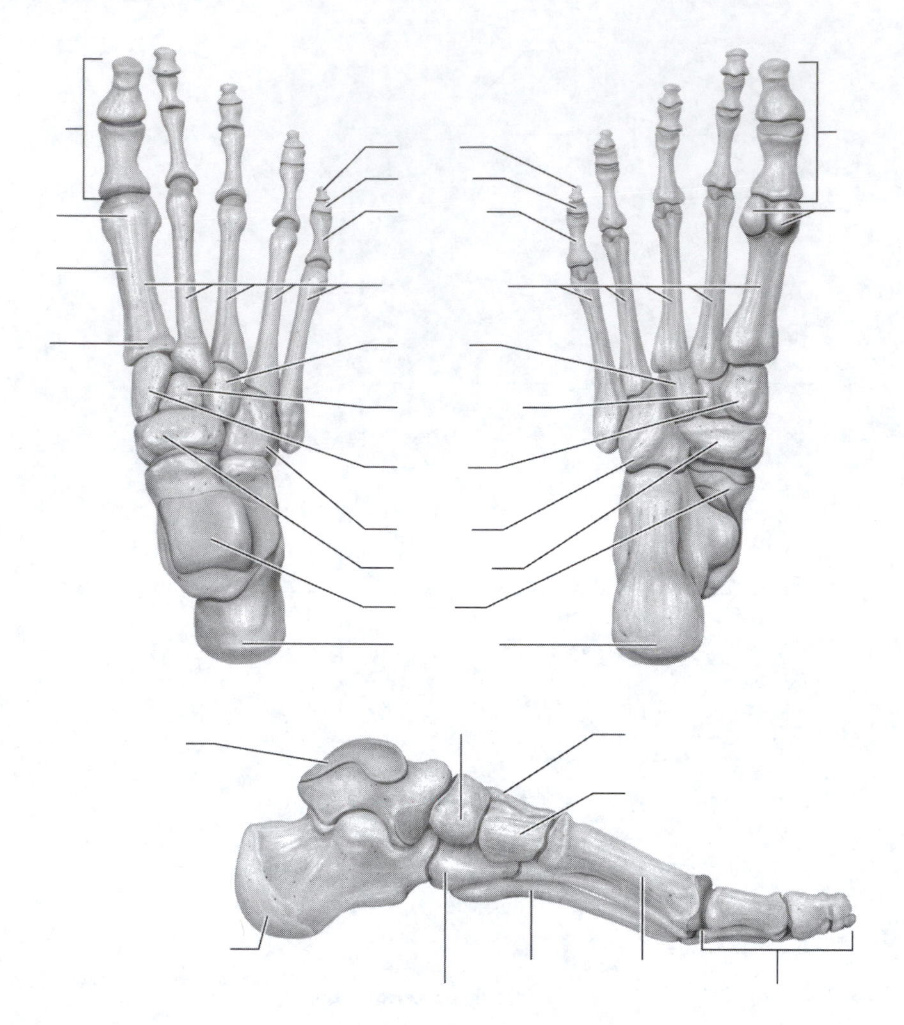

Figure 7.11 The ankle and foot.

Build Your Own Summary Table: Bones of the Pelvic Girdle and Lower Limb

As you read Module 7.5, build your own summary table of the characteristics of the pelvic girdle and lower limb by filling in the template provided below.

Summary of Bones of the Pelvic Girdle and Lower Limb

Bone	Location/Description	Important Features
Pelvic Girdle		
Ilium		
Ischium		

Pubis		
Lower Limb		
Femur		
Patella		
Tibia		
Fibula		
Tarsals		
Metatarsals		
Phalanges		

Team Up

Write 10 quiz questions about the bones of the pelvic girdle and lower limb. Once you have completed your quiz, team up with one or more study partners and trade quizzes. When you and your group have finished taking all the quizzes, discuss the answers to determine places where you need additional study.

What Do You Know Now?

Let's now revisit the questions you answered in the beginning of this chapter. How have your answers changed now that you've worked through the material?

- How many bones are in the body, on average?

  ____

- What are the two bones that make up the leg (remember that the leg is the distal portion of the lower limb, including the shin and the calf)?

  ____

- What are the functions of the skull?

  ____

8 Articulations

We now examine the last topic of the skeletal system: where bones come together to form joints, or *articulations*. This chapter looks at the three different structural and functional classes of joints and concludes with an exploration of the shoulder, elbow, hip, and knee joints.

What Do You Already Know?

Try to answer the following questions before proceeding to the next section. If you're unsure of the correct answers, give it your best attempt based on previous courses, previous chapters, or just your general knowledge.

- Where are joints found in the skull?

____

- What is more important for a joint: stability or motion?

____

- What structures cushion and protect a joint?

____

Module 8.1: Overview of Joints

Module 8.1 in your text introduces you to the different structural and functional classification of joints. By the end of the module, you should be able to do the following:

1. Describe the basic functions of joints.
2. Describe how joints are classified both structurally and functionally.

Build Your Own Glossary

Following is a table listing key terms from Module 8.1. Before you read the module, use the glossary at the back of your book or look through the module to define the following terms.

Key Terms for Module 8.1

Term	Definition
Synarthrosis	
Amphiarthrosis	
Diarthrosis	
Fibrous joint	
Cartilaginous joint	
Synovial joint	

Survey It: Form Questions

Before you read the module, survey it and form at least two questions for yourself. When you have finished reading the module, return to these questions and answer them.

Question 1: ______________________________

Answer: ______________________________

Question 2: ______________________________

Answer: ______________________________

Key Concept: Which structural and functional class of joints has the most stability?

____

____

Key Concept: Which structural and functional class of joints has the most mobility? How does this affect stability?

____

____

Module 8.2: Fibrous and Cartilaginous Joints

Now we look at the structure and function of the simplest articulations: fibrous joints. When you finish this module, you should be able to do the following:

1. Compare and contrast the three subclasses of fibrous joints.
2. Give examples of fibrous joints, and describe how they function.
3. Compare and contrast the two subclasses of cartilaginous joints.
4. Give examples of cartilaginous joints, and describe their function.

Build Your Own Glossary

Following is a table listing key terms from Module 8.2. Before you read the module, use the glossary at the back of your book or look through the module to define the following terms.

Key Terms for Module 8.2

Term	Definition
Suture	
Gomphosis	
Syndesmosis	
Synchondrosis	
Epiphyseal plate	
Symphysis	
Intervertebral joint	
Pubic symphysis	

Survey It: Form Questions

Before you read the module, survey it and form at least two questions for yourself. When you have finished reading the module, return to these questions and answer them.

Question 1: ______________________________

Answer: ______________________________

Question 2: ______________________________

Answer: ______________________________

Identify It: Structure of Fibrous Joints

Identify and color-code each component of the three types of fibrous joints in Figure 8.1.

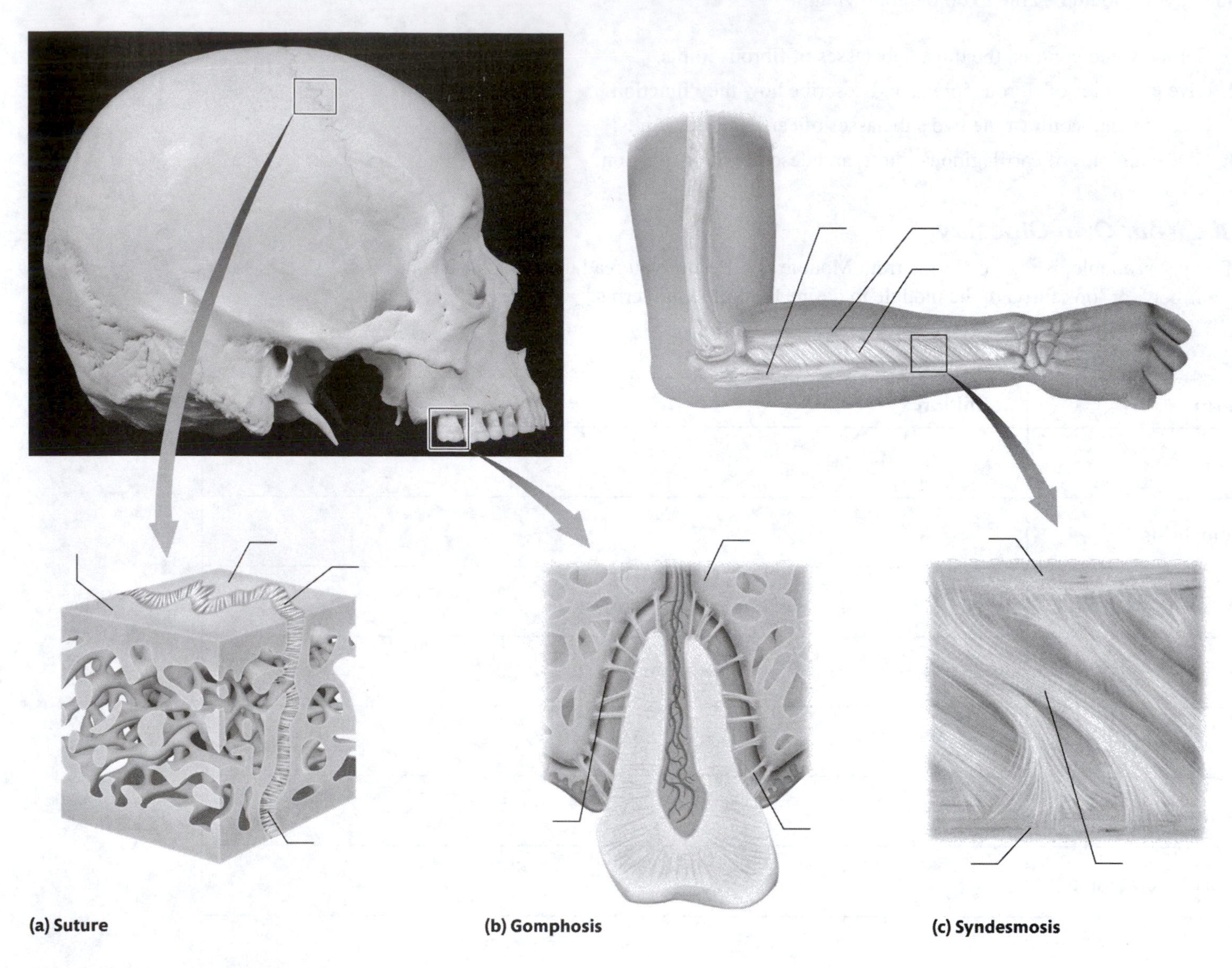

Figure 8.1 Fibrous joints.

Key Concept: Why is it important that the bones in fibrous joints are united by collagen fibers? What do you think would happen if they were joined by elastic fibers instead?

____

____

Identify It: Cartilaginous Joints

Identify each type of cartilaginous joint and the example illustrated in Figure 8.2.

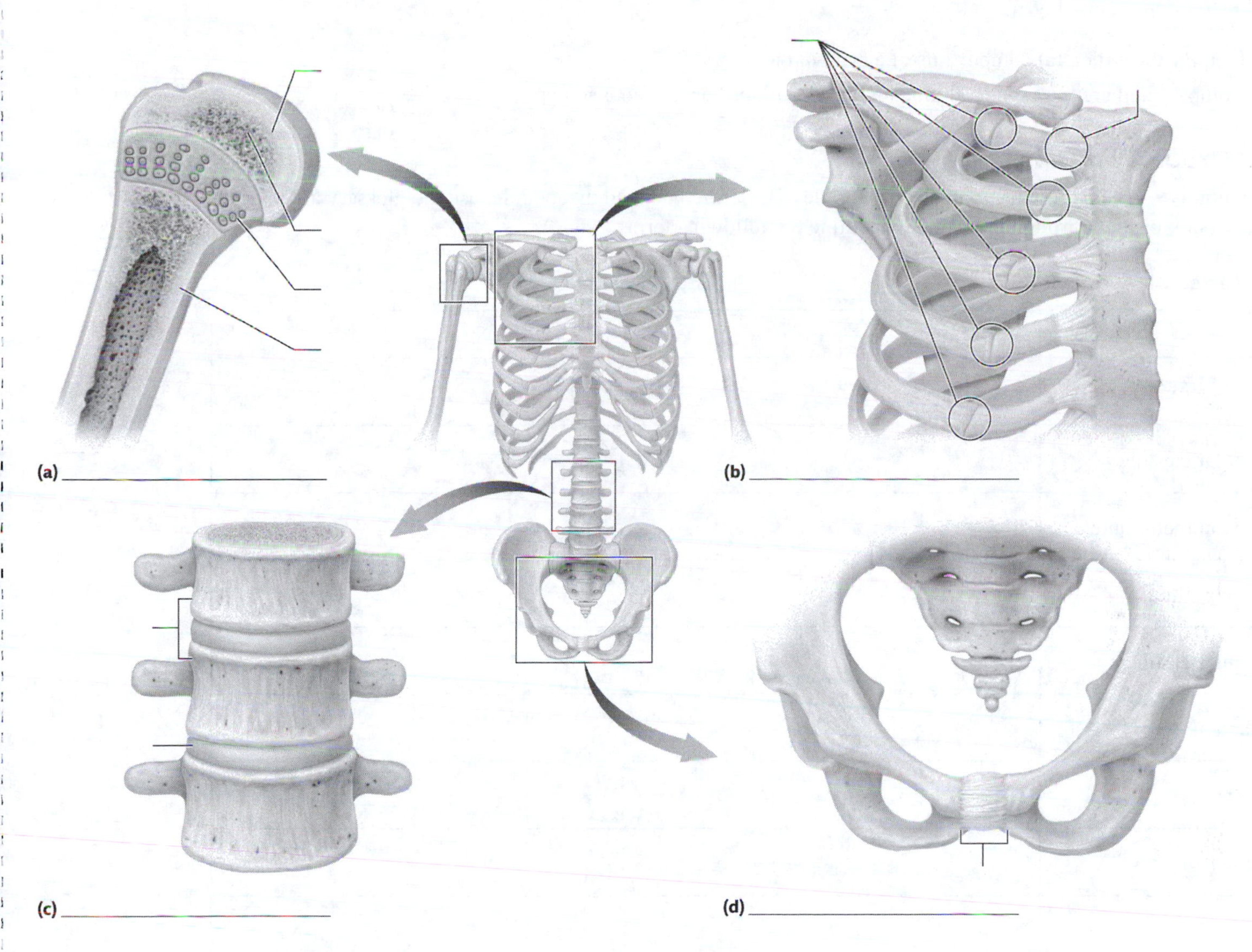

Figure 8.2 Cartilaginous joints.

Key Concept: Which type of synchondrosis allows the least motion? Why is this important for this joint's function?

______

______

Module 8.3: Structure of Synovial Joints

Module 8.3 in your text explores the structures common to synovial joints. At the end of this module, you should be able to do the following:

1. Identify the structural components of a synovial joint.
2. Compare and contrast synovial joints with fibrous and cartilaginous joints.

Build Your Own Glossary

Following is a table listing key terms from Module 8.3. Before you read the module, use the glossary at the back of your book or look through the module to define the following terms.

Key Terms of Module 8.3

Term	Definition
Joint (synovial) cavity	
Articular capsule	
Synovial membrane	
Synovial fluid	
Articular cartilage	
Ligament	
Tendon	
Bursa	
Tendon sheath	
Arthritis	

Survey It: Form Questions

Before you read the module, survey it and form at least three questions for yourself. When you have finished reading the module, return to these questions and answer them.

Question 1: ____________________

Answer: ____________________

Question 2: ____________________

Answer: ____________________

Question 3: ____

Answer: ____

Identify It: Structure of a Synovial Joint

Identify and color-code each component of the synovial joint illustrated in Figure 8.3. Then, write the function of each part under its label.

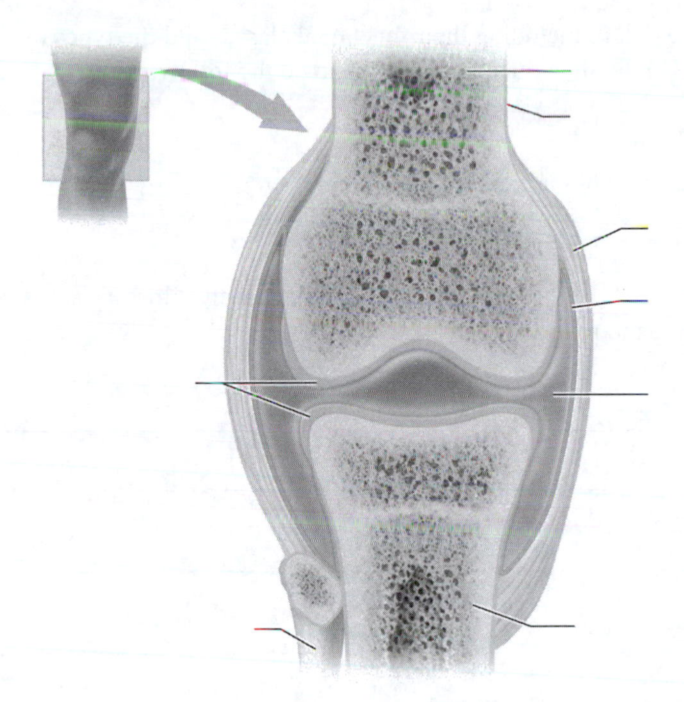

Figure 8.3 Structure of a typical synovial joint.

Key Concept: What is the function of synovial fluid? Why do cartilaginous and fibrous joints lack synovial fluid?

____

____

Complete It: Structures of a Synovial Joint

Fill in the blanks to complete the following paragraph that describes the properties of synovial joints.

Bones are held together in a synovial joint by fibrous cords called ____________ that are composed of ____________ ____________ ____________ connective tissue. Tendons are structures that connect a ____________ to a ____________. ____________ are fluid-filled sacs often found between tendons and joints, and they function to ____________ ____________. Some tendons in high-stress areas are surrounded by ____________ ____________, which are long ____________.

Key Concept: How does destruction of articular cartilage in osteoarthritis affect the function of a synovial joint?

____

____

Module 8.4: Function of Synovial Joints

This module examines synovial joints in more detail, including their functional classes and the types of movement each joint allows. At the end of this module, you should be able to do the following:

1. Define the functional classes of synovial joints.
2. Describe and demonstrate the movements of synovial joints.

Build Your Own Glossary

Below is a table listing key terms from Module 8.4. Before you read the module, use the glossary at the back of your book or look through the module to define the following terms.

Key Terms of Module 8.4

Term	Definition
Nonaxial joint	
Uniaxial joint	
Biaxial joint	
Multiaxial joint	
Gliding movement	
Angular movement	
Range of motion	

Survey It: Form Questions

Before you read the module, survey it and form at least three questions for yourself. When you have finished reading the module, return to these questions and answer them.

Question 1: ____

Answer: ____

Question 2: ____

Answer: ____

Question 3: ____________________

Answer: ____________________

Key Concept: Why, specifically, do multiaxial joints have a greater range of motion than do biaxial, uniaxial, and nonaxial joints?

____

____

Build Your Own Summary Table: Motions That Occur at Synovial Joints

As you read Module 8.4, build your own table of the motions that occur at or around synovial joints by filling in the template provided.

Summary of Motions That Occur at Synovial Joints

Motion	Definition	Example(s) of Motion at a Specific Joint
Gliding		
Rotation		
Angular Movements		
Flexion		
Extension		
Hyperextension		
Abduction		

Motion	Definition	Example(s) of Motion at a Specific Joint
Adduction		
Circumduction		
Special Movements		
Opposition		
Reposition		
Depression		
Elevation		
Protraction		
Retraction		
Inversion		
Eversion		

Dorsiflexion		
Plantarflexion		
Supination		
Pronation		

Try It

Perform a common movement, such as answering your phone, clicking your mouse, or standing up from your sitting position. As you perform the movement, write in the space below each synovial joint that is moving and the motion that is taking place.

Module 8.5: Types of Synovial Joints

The many synovial joints of the body are classified by the way the bones of the articulation fit together. We examine these classes in this module, as well as the structure of the elbow, knee, shoulder, and hip. At the end of this module, you should be able to do the following:

1. Describe the anatomical features of each structural type of synovial joint.
2. Describe where each structural type can be found.
3. Predict the kinds of movements that each structural type of synovial joint will allow.
4. Compare and contrast the structural features of the knee and elbow and of the shoulder and hip.

Build Your Own Glossary

Following is a table listing key terms from Module 8.5. Before you read the module, use the glossary at the back of your book or look through the module to define the following terms.

Key Terms of Module 8.5

Term	Definition
Plane joint	
Hinge joint	
Pivot joint	
Condylar joint	
Saddle joint	
Ball-and-socket joint	
Elbow	
Knee	
Menisci (medial and lateral)	
Collateral ligaments (tibial and fibular)	
Cruciate ligaments (anterior and posterior)	
Glenohumeral (shoulder) joint	
Rotator cuff	
Hip (coxal) joint	

Survey It: Form Questions

Before you read the module, survey it and form at least four questions for yourself. When you have finished reading the module, return to these questions and answer them.

Question 1: ______________________________

Answer: ______________________________

Question 2: ____________________

Answer: ____________________

Question 3: ____________________

Answer: ____________________

Question 4: ____________________

Answer: ____________________

Draw It: Classes of Synovial Joints

In the space below, draw the six different classes of synovial joint. You may use Figure 8.11 in your text for reference, but try not to copy it—your original drawings will make more sense to you. When you finish your drawing, label each bone and color-code it.

a. Plane joint

b. Hinge joint

c. Pivot joint

d. Condylar joint

e. Saddle joint

f. Ball-and-socket joint

Key Concept: Which class of synovial joint is the most stable? Why? Which is least stable, and why?

___

___

Identify It: The Elbow

Identify and color-code each component of the elbow joint in Figure 8.4.

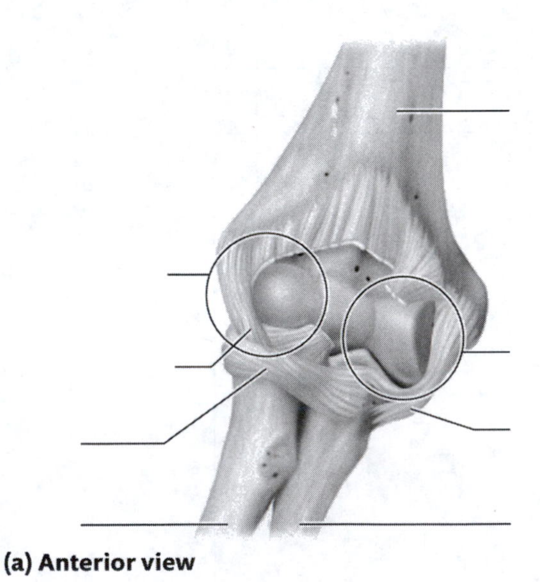

(a) Anterior view

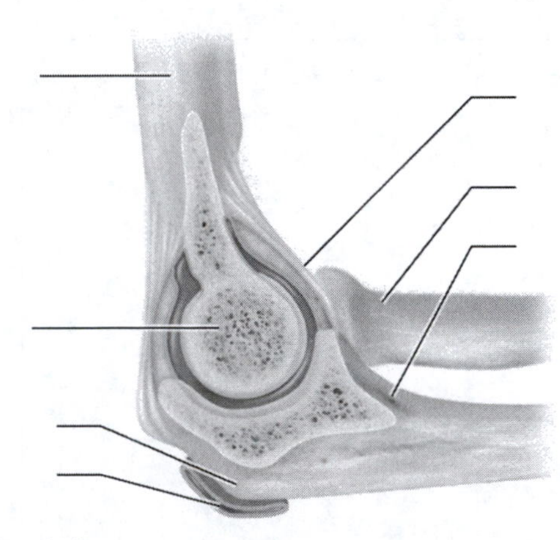

(b) Sagittal view

Figure 8.4 The structure of the elbow.

Identify It: The Knee

Identify and color-code each component of the knee joint in Figure 8.5.

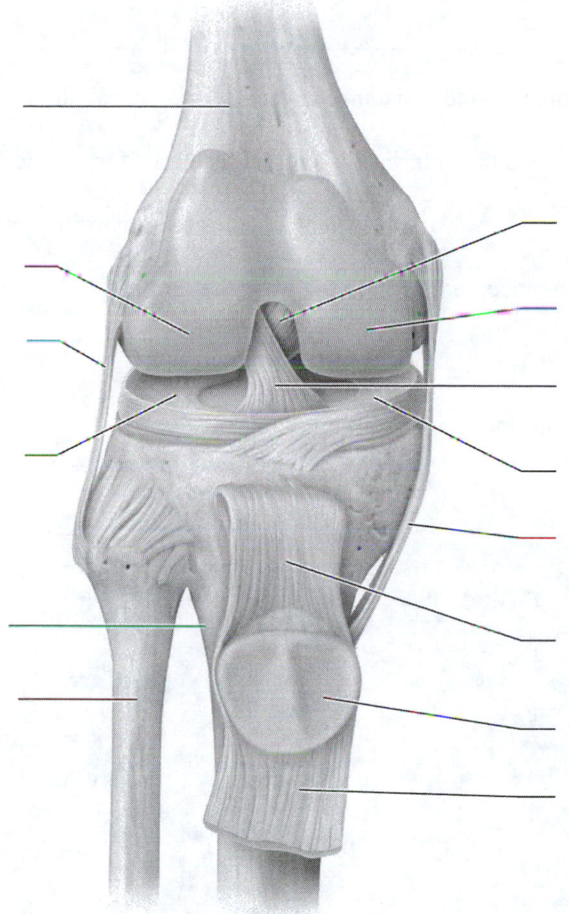

(a) Anterior view, right knee

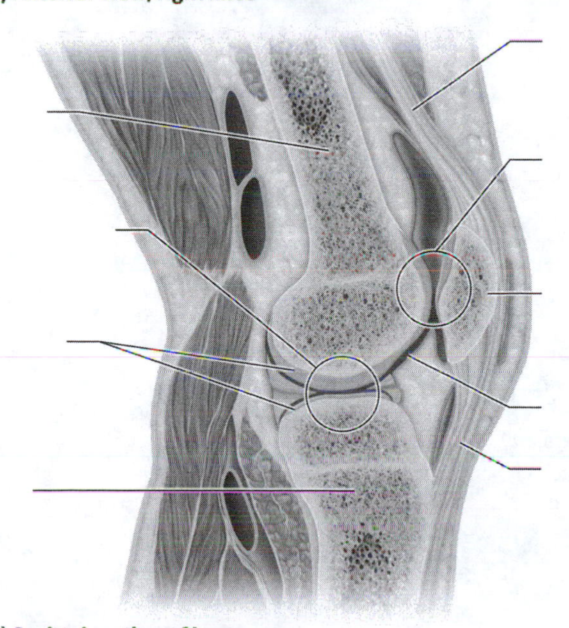

(b) Sagittal section of knee

Figure 8.5 The structure of the knee joint.

Complete It: The Knee Joint

Fill in the blanks to complete the following paragraph that describes the properties of the knee joint.

The medial and lateral menisci rest on the ____________ ____________ and provide ____________ ____________. The two ____________ ligaments provide resistance to medial and lateral stresses. The ____________ ____________ ligament attaches the anterior tibia to the posterior femur and prevents ____________. In contrast, the ____________ ____________ ligament attaches the posterior tibia to the anterior femur, and it prevents ____________ displacement of the tibia from the femur.

Identify It: The Shoulder

Identify and color-code each component of the shoulder joint in Figure 8.6.

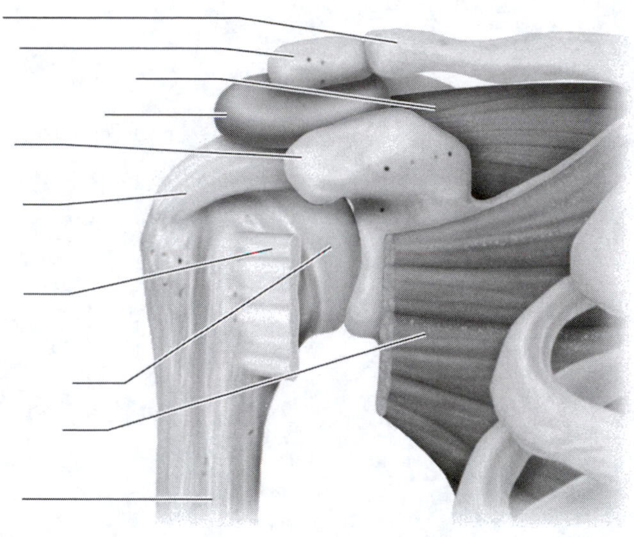

(a) Anterior view

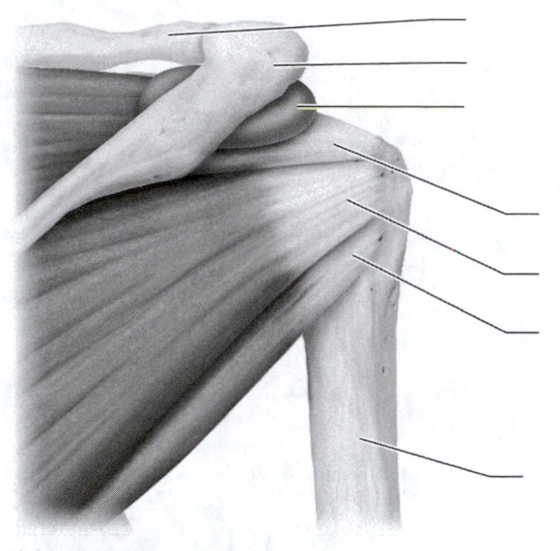

(b) Posterior view

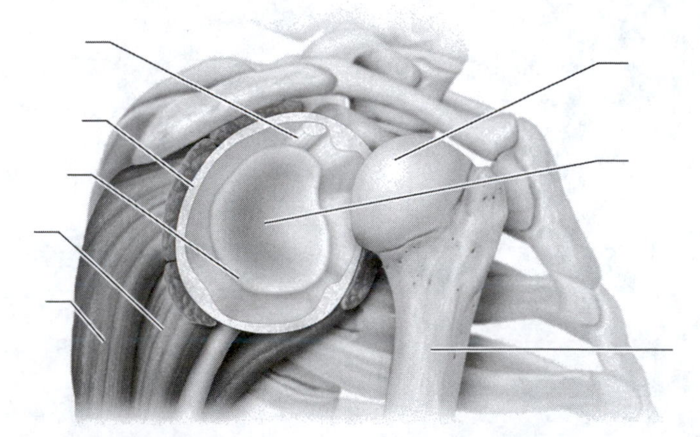

(c) Lateral view with head of humerus removed from glenoid cavity

Figure 8.6 The structure of the shoulder joint.

Identify It: The Hip

Identify and color-code each component of the hip joint in Figure 8.7.

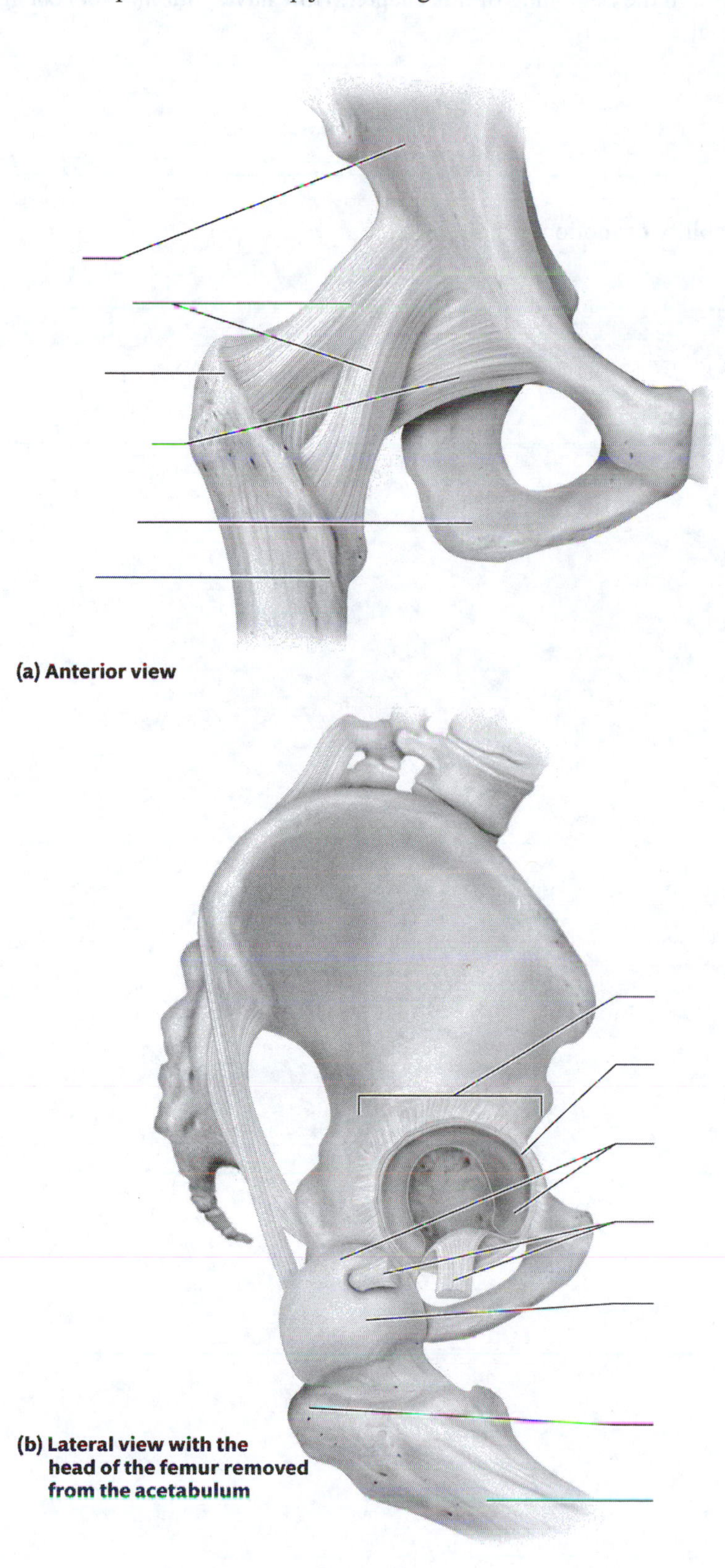

Figure 8.7 The structure of the hip joint.

What Do You Know Now?

Let's now revisit the questions you answered in the beginning of this chapter. How have your answers changed now that you've worked through the material?

- Where are joints found in the skull?

____

- What is more important for a joint: stability or motion?

____

- What structures cushion and protect a joint?

____

The Muscular System

We can now build on your knowledge gained from the previous chapters as we explore the organs of the muscular system: skeletal muscles. As implied by the name, most skeletal muscles attach to different parts of the skeleton and cause the movement of bones at joints. The structure, location, and actions of skeletal muscles on bones and other tissues are explored in this chapter.

What Do You Already Know?

Try to answer the following questions before proceeding to the next section. If you're unsure of the correct answers, give it your best attempt based on previous courses, previous chapters, or just your general knowledge.

- What are the functions of skeletal muscles in the body?

  ____

- Of which tissue type(s) is a skeletal muscle composed?

  ____

- What structures attach a skeletal muscle to bone or another tissue?

  ____

Module 9.1: Overview of Skeletal Muscles

Module 9.1 in your text introduces you to the basic structure and function of skeletal muscles. By the end of the module, you should be able to do the following:

1. Explain how the name of a muscle can help identify its action, appearance, or location.
2. Summarize the major functions of skeletal muscles.
3. Define the terms agonist, antagonist, synergist, and fixator.
4. Differentiate among the three classes of levers in terms of their structure and function, and give examples of each class in the human body.

Build Your Own Glossary

Following is a table listing key terms from Module 9.1. Before you read the module, use the glossary at the back of your book or look through the module to define the following terms.

Key Terms for Module 9.1

Term	Definition
Parallel muscle	
Convergent muscle	
Pennate muscle	
Circular muscle	
Spiral muscle	
Agonist	
Antagonist	
Origin	
Insertion	
First-class lever	
Second-class lever	
Third-class lever	

Survey It: Form Questions

Before you read the module, survey it and form at least two questions for yourself. When you have finished reading the module, return to these questions and answer them.

Question 1: ____________________

Answer: ____________________

Question 2: ____________________

Answer: ____________________

Key Concept: How does the arrangement of fascicles influence muscle shape?

____

____

Key Concept: What are some of the functions of skeletal muscles?

____

____

Complete It: Skeletal Muscle Functions

Fill in the blanks to complete the following paragraph that describes the functional properties and groups of skeletal muscles.

The muscle known as the ____________, also known as the ____________ ____________, provides most of the force involved in performing a movement. Muscles known as ____________ assist with a movement and provide guidance, making the movement more efficient. A muscle known as an ____________, which usually resides on the ____________ ____________ of the main muscle, opposes or slows the action of the main muscle. Finally, a ____________ is a muscle that holds its body part in place.

Key Concept: What is the difference between a muscle's origin and its insertion?

____

____

Draw It: Levers

In the space below, draw an example of a first-, second-, and third-class lever. Label the load, fulcrum, force, and arc of motion in each drawing and explain, in your own words, why each is at a mechanical advantage or disadvantage. You may use Figure 9.5 in your text for reference, but try not to copy it—your original drawings will make more sense to you.

a. First-class lever

b. Second-class lever

c. Third-class lever

Module 9.2: Muscles of the Head, Neck, and Vertebral Column

Now we look at the structure and function of the muscles of the head, neck, and vertebral column. When you finish this module, you should be able to do the following:

1. Name, describe, and identify the muscles of the head, neck, and vertebral column.
2. Identify the origin, insertion, and actions of these muscles, and demonstrate their actions.

Survey It: Form Questions

Before you read the module, survey it and form at least three questions for yourself. When you have finished reading the module, return to these questions and answer them.

Question 1: ______________________________

Answer: ______________________________

Question 2: ______________________________

Answer: ______________________________

Question 3: ______________________________

Answer: ______________________________

Identify It: Muscles of the Head

Identify and color-code each muscle illustrated in Figure 9.1. Then, write the muscle's main action(s) under its name.

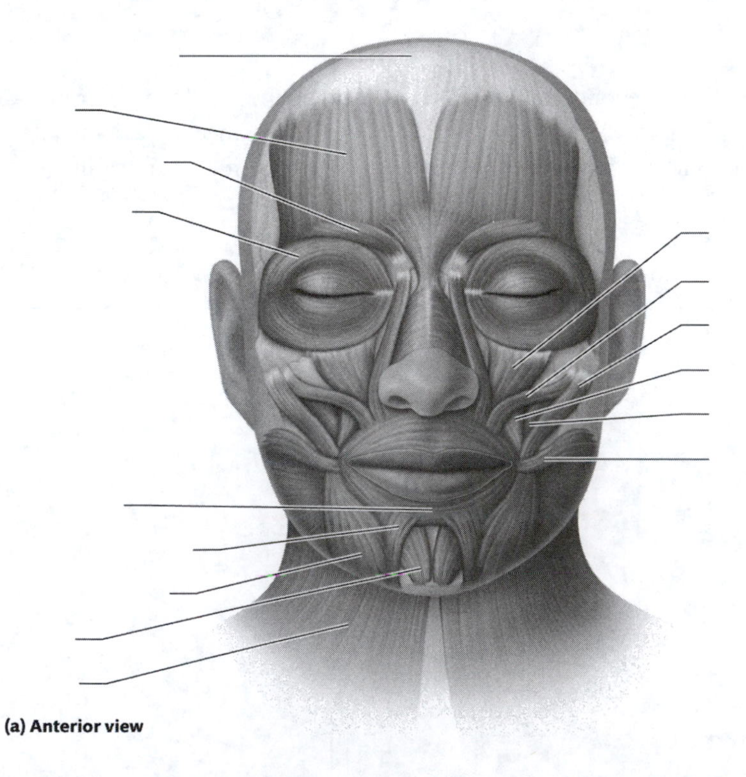

Figure 9.1 Muscles of the head.

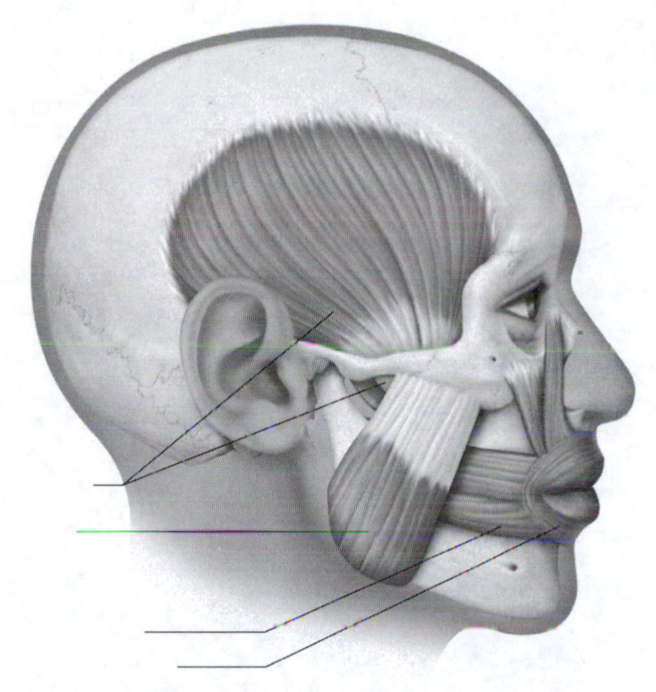

(b) Muscles of the face and mastication, lateral view

Figure 9.1 (*continued*)

Key Concept: How are the muscles of facial expression different from most other muscles?

____

____

Identify It: Muscles of the Neck

Identify and color-code each muscle illustrated in Figure 9.2. Then, write the muscle's main action(s) under its name.

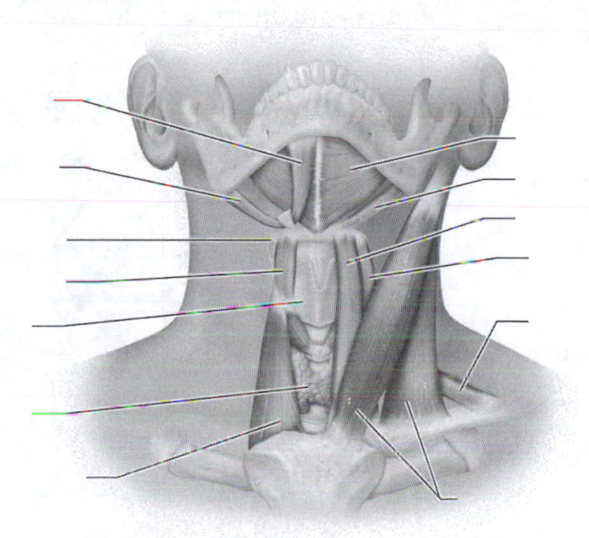

(a) Muscles of swallowing, anterior view

Figure 9.2 Muscles of the neck.

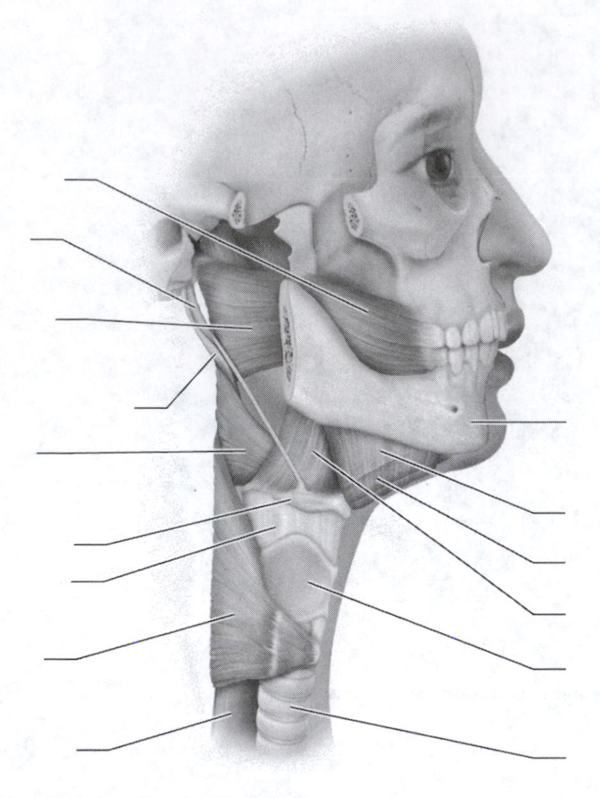

(b) Muscles of swallowing, lateral view

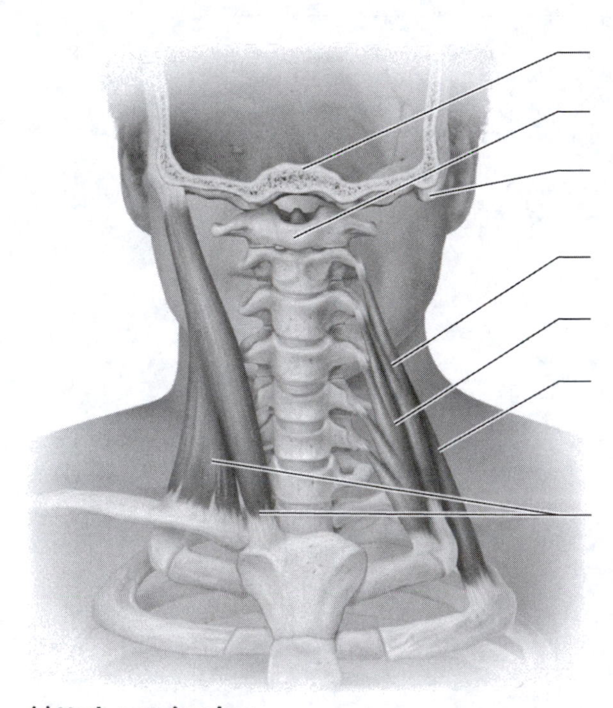

(c) Neck, anterior view

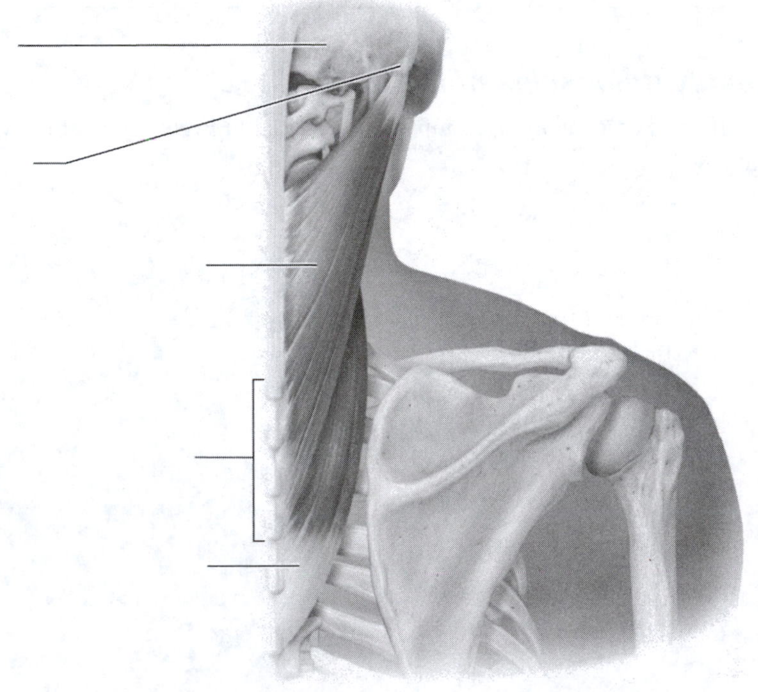

(d) Neck, posterior view

Figure 9.2 (*continued*)

Identify It: Deep Muscles of the Back

Identify and color-code each muscle illustrated in Figure 9.3. Then, write the muscle's main action(s) under its name.

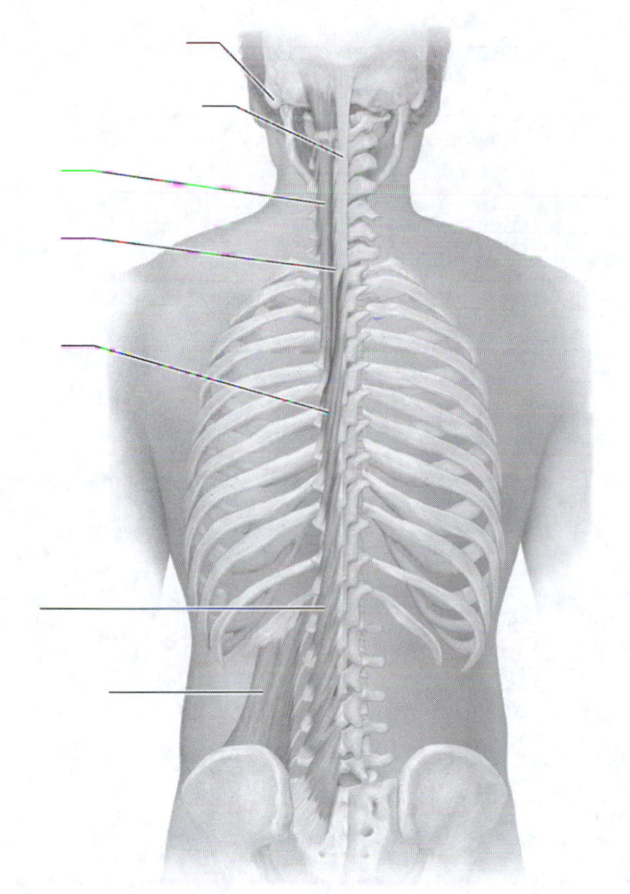

(a) The erector spinae group, posterior deep view

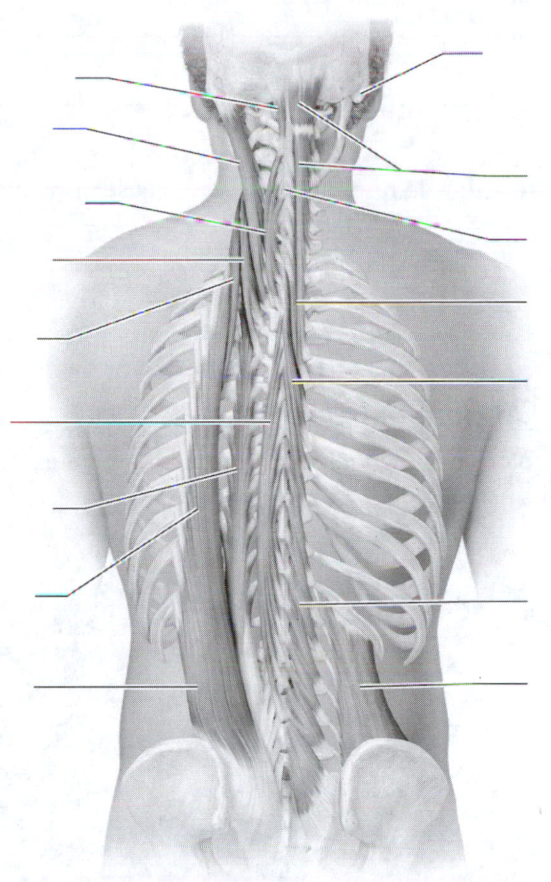

(b) The transversospinalis group and quadratus lumborum (note that the quadratus lumborum is not a deep back muscle)

Figure 9.3 Deep muscles of the back.

Module 9.3: Muscles of the Trunk and Pelvic Floor

The structure and function of the muscles of the trunk and pelvic floor are discussed in this module. When you complete the module, you should be able to do the following:

1. Identify and describe the muscles of the trunk and pelvic floor.
2. Identify the origin, insertion, and actions of these muscles, and demonstrate their actions.

Survey It: Form Questions

Before you read the module, survey it and form at least three questions for yourself. When you have finished reading the module, return to these questions and answer them.

Question 1: ____________________

Answer: ____________________

Question 2: ______________________________

Answer: ______________________________

Question 3: ______________________________

Answer: ______________________________

Key Concept: How do respiratory muscles cause inspiration?

____

____

Identify It: Abdominal Muscles

Identify and color-code each muscle illustrated in Figure 9.4. Then, write the muscle's main action(s) in the blank under its name.

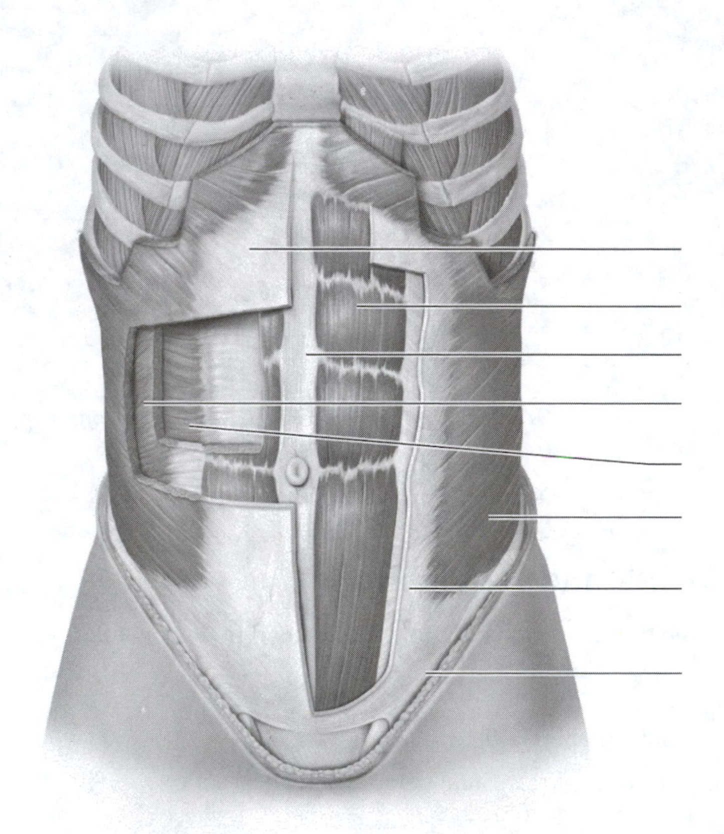

Figure 9.4 Abdominal muscles.

Identify It: Muscles of the Pelvic Floor

Identify and color-code each muscle illustrated in Figure 9.5. Then, write the muscle's main action(s) under its name.

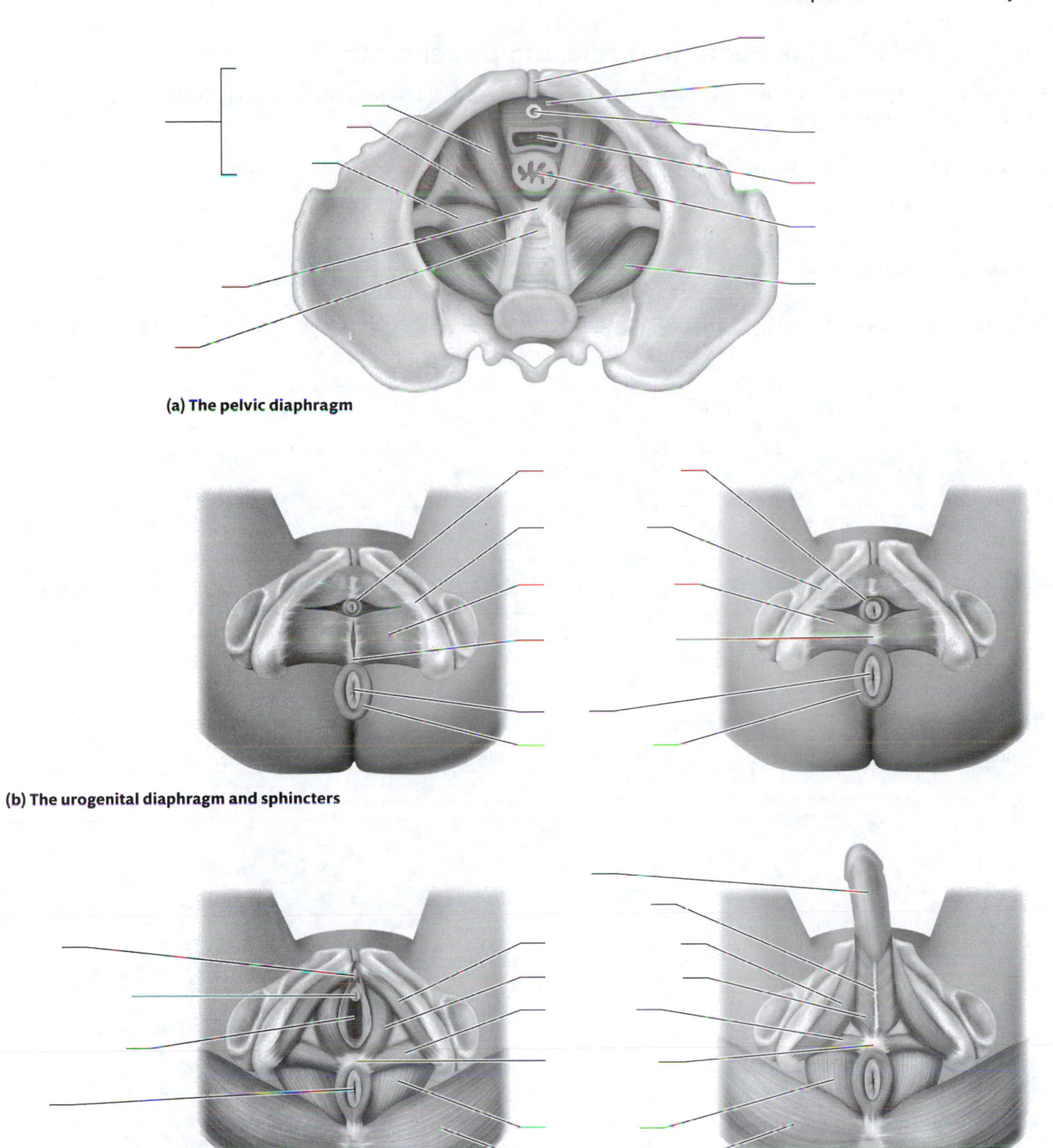

Figure 9.5 Muscles of the pelvic floor.

Module 9.4: Muscles of the Pectoral Girdle and Upper Limb

Module 9.4 in your text explores the structure and functions of the muscles of the pectoral girdle and upper limb. At the end of this module, you should be able to do the following:

1. Identify and describe the muscles that move the pectoral girdle and upper limb.
2. Identify the origin, insertion, and actions of these muscles, and demonstrate their actions.

Survey It: Form Questions

Before you read the module, survey it and form at least three questions for yourself. When you have finished reading the module, return to these questions and answer them.

Question 1: ______________________________

Answer: ______________________________

Question 2: ______________________________

Answer: ______________________________

Question 3: ______________________________

Answer: ______________________________

Identify It: Muscles of the Chest and Back

Identify and color-code each muscle illustrated in Figure 9.6. Then, write the muscle's main action(s) under its name.

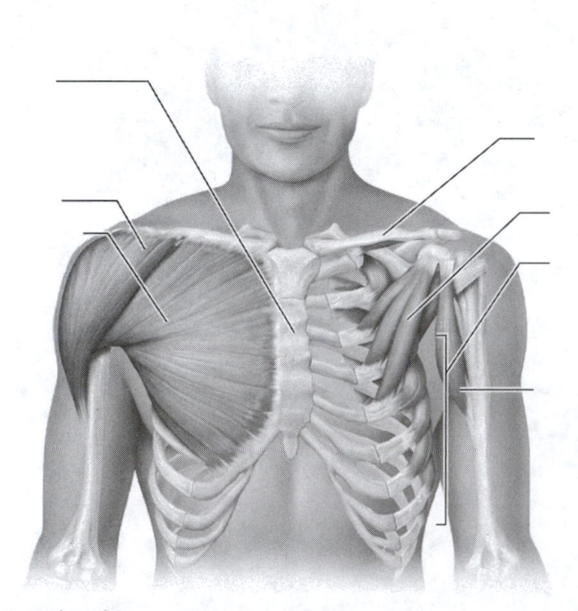

(a) Anterior view

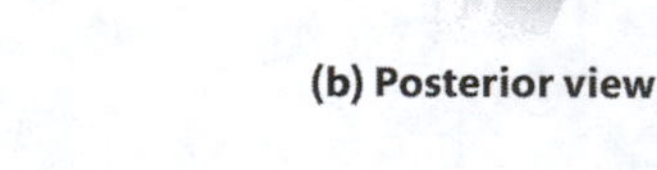

(b) Posterior view

Figure 9.6 Muscles of the chest and back.

Identify It: Muscles of the Arm

Identify and color-code each muscle illustrated in Figure 9.7. Then, write the muscle's main action(s) under its name.

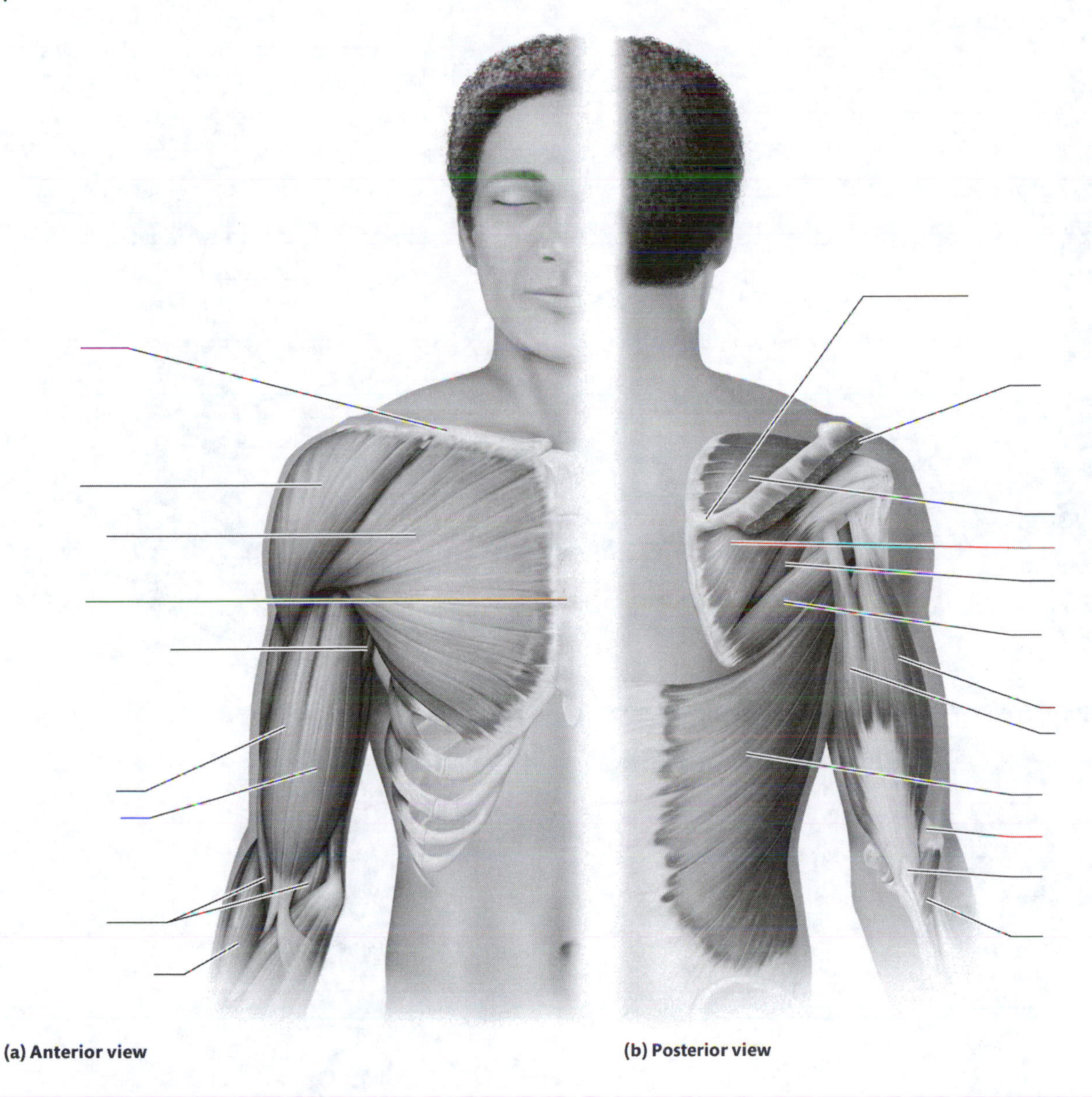

Figure 9.7 Muscles of the arm.

Key Concept: What is the function of the rotator cuff? What are the rotator cuff muscles?

____

____

Identify It: Muscles of the Forearm

Identify and color-code each muscle illustrated in Figure 9.8. Then, write the muscle's main action(s) under its name.

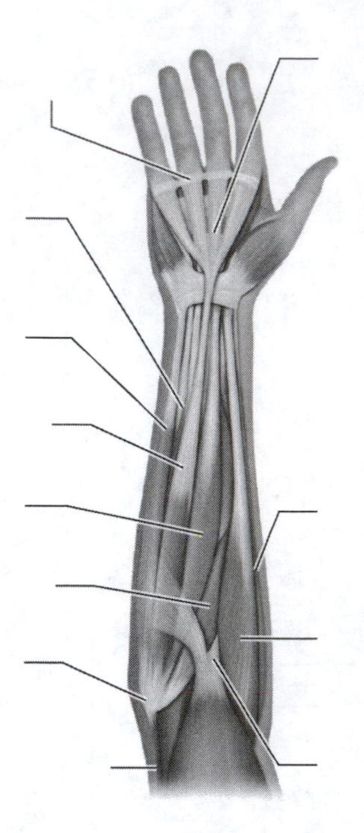

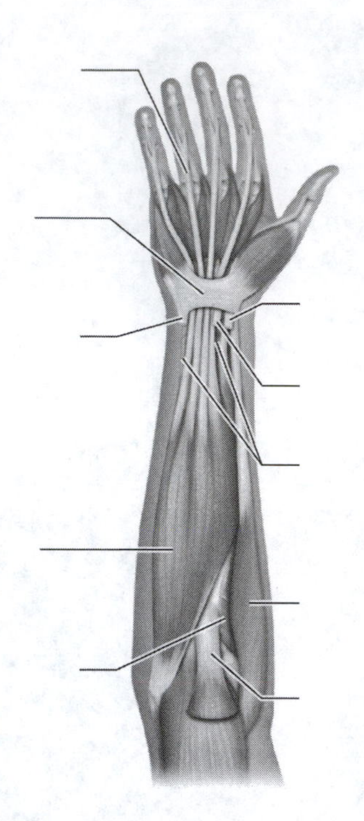

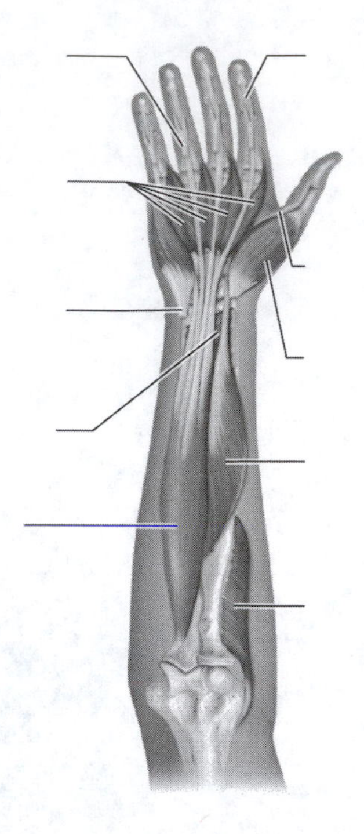

(a) Anterior view with muscles removed as necessary to reveal deeper muscles

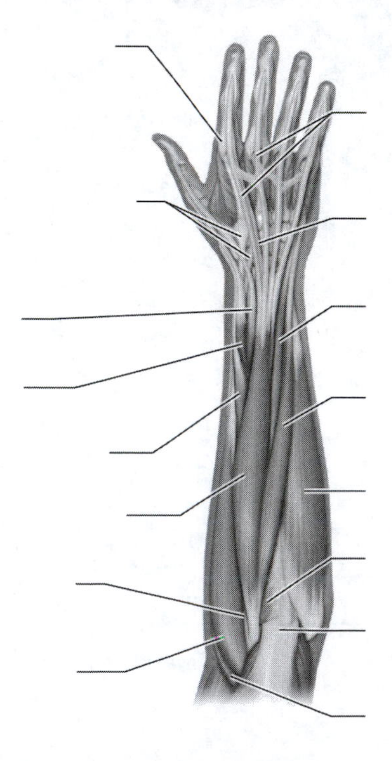

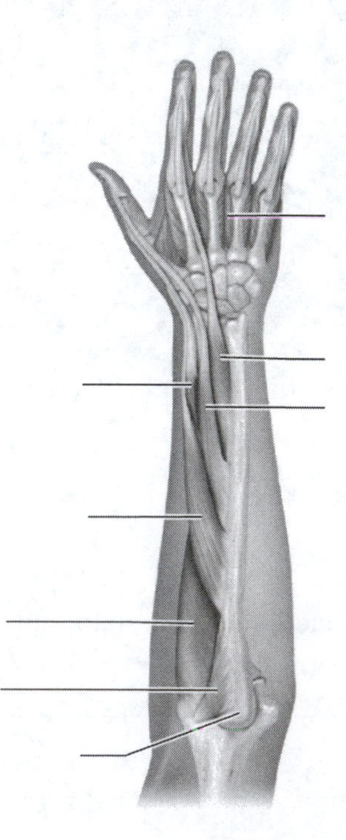

(b) Posterior view with muscles removed as necessary to reveal deeper muscles

Figure 9.8 Muscles of the forearm.

Module 9.5: Muscles of the Hip and Lower Limb

The muscles of the hip and lower limb, the topic we now examine, round out our coverage of the skeletal muscles. At the end of this module, you should be able to do the following:

1. Identify and describe the muscles that move the hip and lower limb.
2. Identify the origin, insertion, and actions of these muscles, and demonstrate their actions.

Survey It: Form Questions

Before you read the module, survey it and form at least three questions for yourself. When you have finished reading the module, return to these questions and answer them.

Question 1: ______________________________

Answer: ______________________________

Question 2: ______________________________

Answer: ______________________________

Question 3: ______________________________

Answer: ______________________________

Identify It: Muscles of the Hip and Thigh

Identify and color-code each muscle illustrated in Figure 9.9. Then, write the muscle's main action(s) under its name.

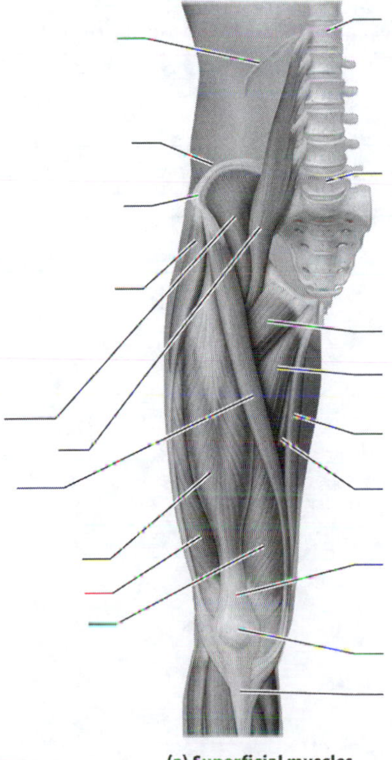

(a) Superficial muscles

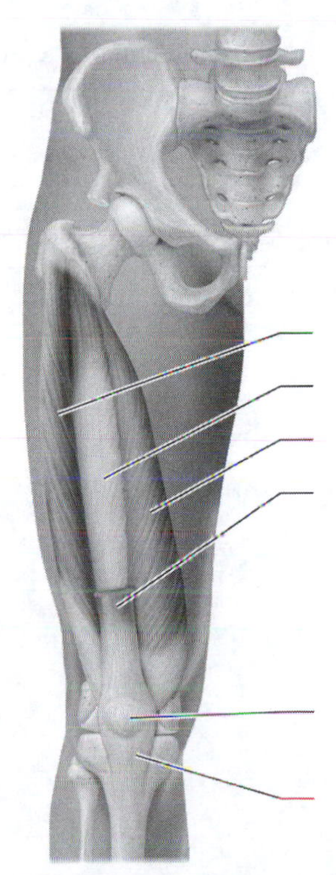

(b) Deeper muscles under the rectus femoris

Figure 9.9 Muscles of the hip and thigh.

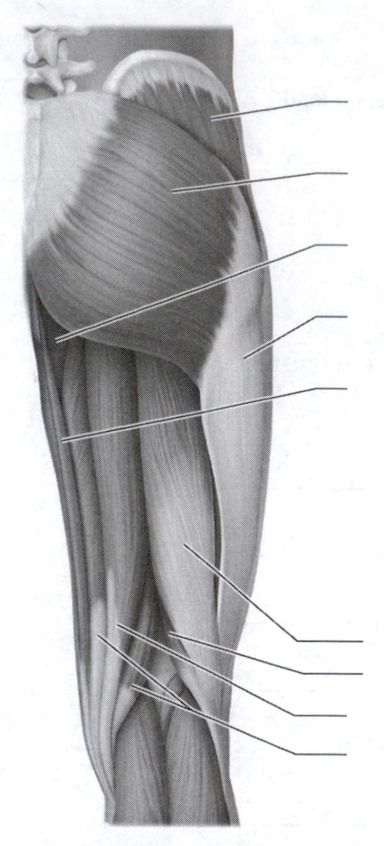
(c) Superficial muscles, posterior view

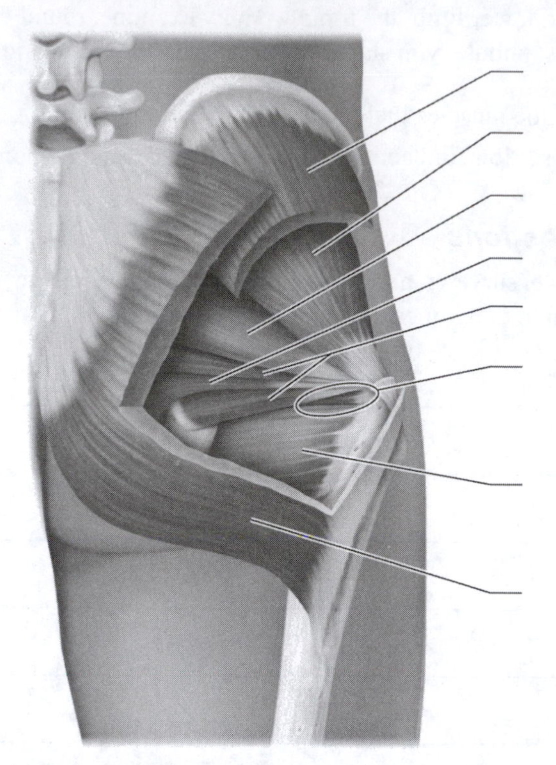
(d) Muscles under the gluteus maximus and gluteus medius

Figure 9.9 (*continued*)

Key Concept: Why is the rectus femoris the only muscle of the quadriceps femoris group to have an action on the hip joint?

____

____

Identify It: Muscles of the Leg and Foot

Identify and color-code each muscle illustrated in Figure 9.10. Then, write the muscle's main action(s) under its name.

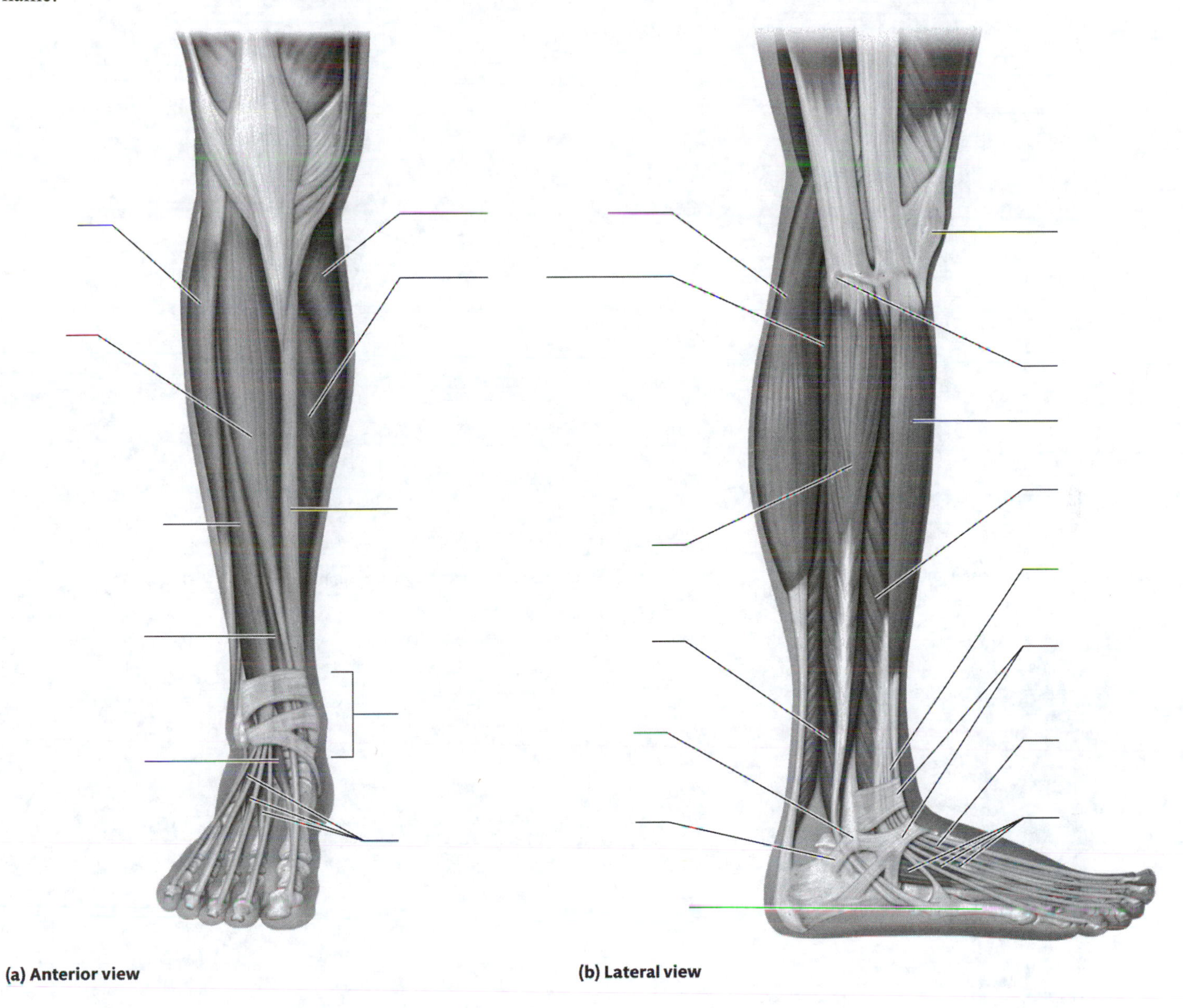

Figure 9.10 Muscles of the leg and foot.

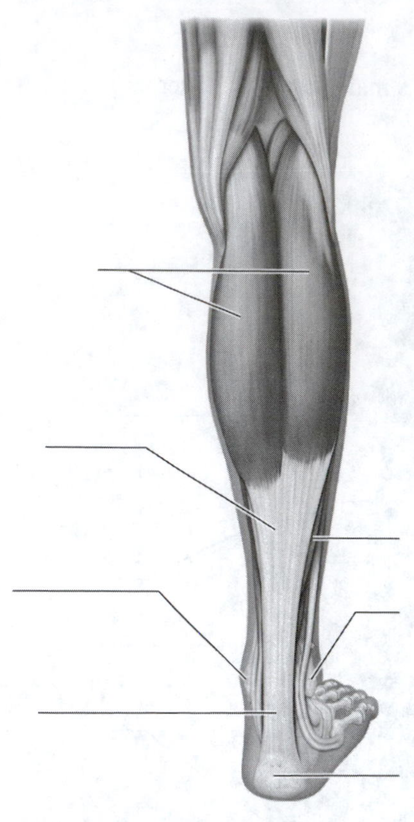

(c) Superficial muscles, posterior view

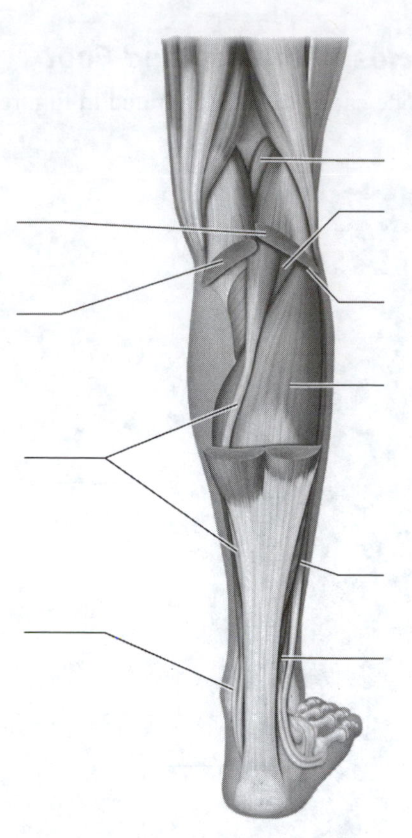

(d) Muscles under the gastrocnemius

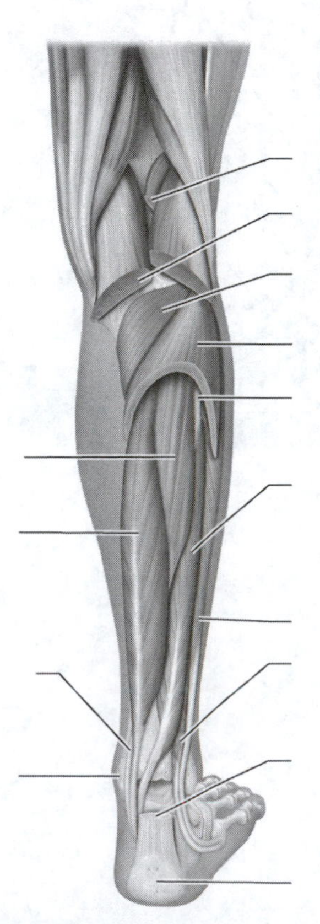

(e) Deeper muscles under the gastrocnemius and soleus

Figure 9.10 *(continued)*

Module 9.6: Putting It All Together: The Big Picture of Muscle Movement

Now we put all the muscles we have learned about together in order to form a big picture of all the muscles involved in performing common movements. At the end of this module, you should be able to do the following:

1. Identify the locations of the major skeletal muscles, and demonstrate their actions.

Identify It: Muscles of the Whole Body

Identify and color-code each muscle illustrated in Figure 9.11. Note that these diagrams show only the superficial muscles—most deep muscles are not visible.

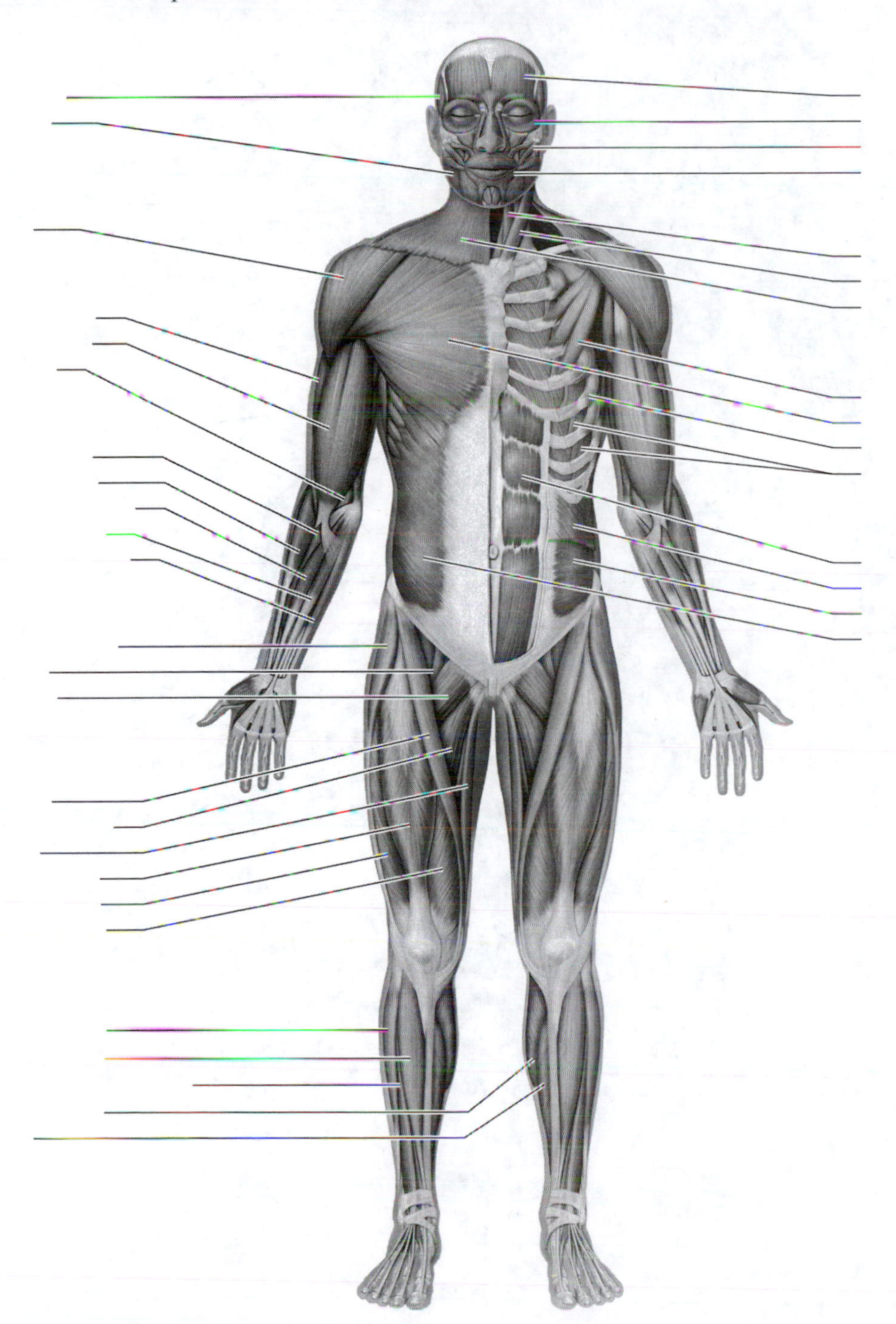

Figure 9.11 Muscles of the whole body.

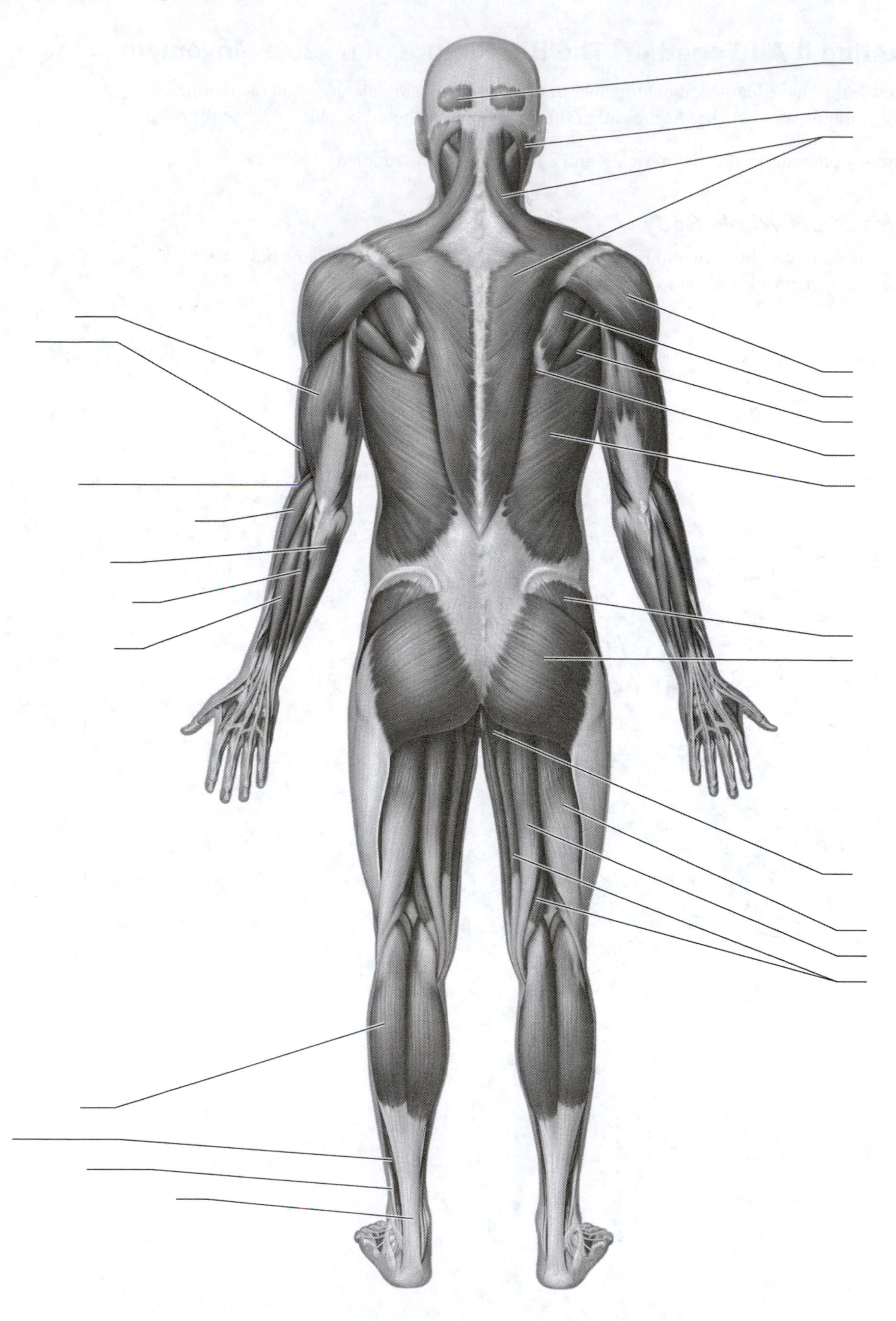

Figure 9.11 (*continued*)

Practice It: Muscle Movements

List the muscles involved in performing each of the following movements, and describe the actions each muscle is undergoing during the movement.

1. Lifting a coffee cup to your mouth.

2. Doing a sit-up.

3. Squatting down to the ground and returning to a standing position.

Build Your Own Summary Table: Motions That Occur at Synovial Joints

Build your own table of the main muscles that move the major synovial joints by filling in the table below. This table is far from complete, as there are so many muscles in the body. For this reason, some blank cells have been added to the end if you want to list other joints and motions.

Summary of Main Muscles That Act on Joints

Joint/Motion	Muscles Responsible for the Action
Head/Neck	
Flexion	
Extension	
Lateral flexion	
Rotation	
Shoulder	
Flexion	

Joint/Motion	Muscles Responsible for the Action
Extension	
Abduction	
Adduction	
Medial rotation	
Lateral rotation	
Elbow	
Flexion	
Extension	
Wrist/Hand	
Flexion	
Extension	
Abduction	
Adduction	

Hip	
Flexion	
Extension	
Abduction	
Adduction	
Medial rotation	
Lateral rotation	
Knee	
Flexion	
Extension	
Ankle/Foot	
Dorsiflexion	
Plantarflexion	
Inversion	

Joint/Motion	Muscles Responsible for the Action
Eversion	
Other Joints/Tissues	

What Do You Know Now?

Let's now revisit the questions you answered in the beginning of this chapter. How have your answers changed now that you've worked through the material?

- What are the functions of skeletal muscles in the body?

  ____

- Of which tissue type(s) is a skeletal muscle composed?

  ____

- What structures attach a skeletal muscle to bone or another tissue?

  ____

10 Muscle Tissue and Physiology

We continue our exploration of muscles at the tissue and cellular levels of organization. Although the bulk of this chapter is devoted to skeletal muscle tissue, we also discuss smooth muscle tissue and provide an overview of cardiac muscle tissue in the last module.

What Do You Already Know?

Try to answer the following questions before proceeding to the next section. If you're unsure of the correct answers, give it your best attempt based on previous courses, previous chapters, or just your general knowledge.

- How do the three types of muscle tissue differ?

  ____

- What is the primary function of all types of muscle tissue?

  ____

- What happens to a muscle during a muscle contraction?

  ____

- What is the role of ATP in a muscle contraction?

  ____

Module 10.1: Overview of Muscle Tissue

Module 10.1 in your text looks at the basic properties that are common to all types of muscle tissue. By the end of this module, you should be able to do the following:

1. Describe the major functions of muscle tissue.
2. Name and describe the structural elements and properties common to all types of muscle cells.
3. Compare and contrast the characteristics of skeletal, cardiac, and smooth muscle tissue.

Build Your Own Glossary

Following is a table with a list of key terms from Module 10.1. Before you read the module, use the glossary at the back of your book or look through the module to define the following terms.

Key Terms for Module 10.1

Term	Definition
Endomysium	
Muscle fiber	
Muscle tension	
Sarcoplasm	
Sarcolemma	
Myofibril	
Sarcoplasmic reticulum	
Contractility	
Conductivity	
Elasticity	

Survey It: Form Questions

Before you read the module, survey it and form at least three questions for yourself. When you have finished reading the module, return to these questions and answer them.

Question 1: ______________________________

Answer: ______________________________

Question 2: ______________________________

Answer: ______________________________

Question 3: ______________________________

Answer: ______________________________

Identify It: The Types of Muscle Tissue

Identify the type of muscle tissue shown in each of the following micrographs in Figure 10.1. Explain for each micrograph what features of the tissue led you to the correct identification.

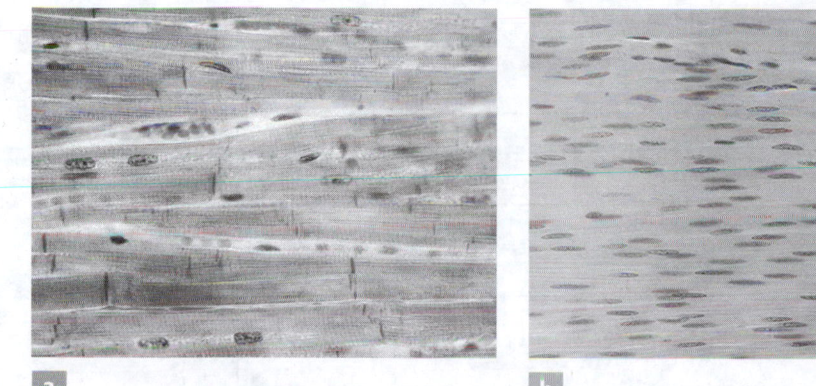

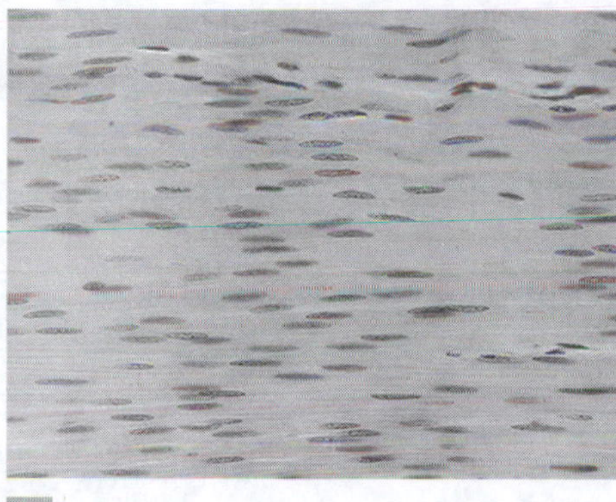

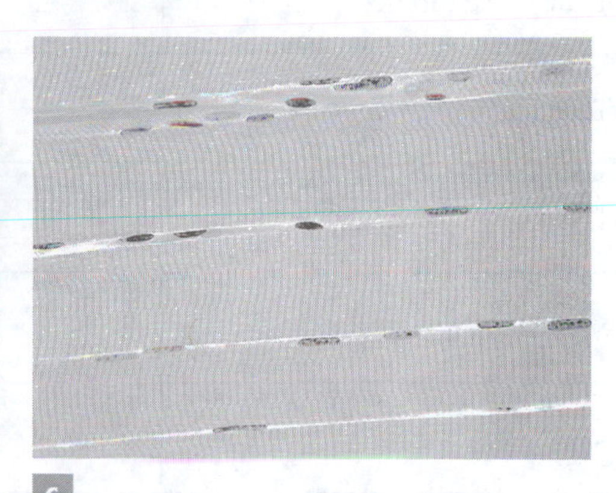

Figure 10.1 Three types of muscle tissue.

Key Concept: What is the function of all types of muscle tissue? What structural properties do they all share?

____

____

Module 10.2: Structure and Function of Skeletal Muscle Fibers

This module in your text examines the structure of skeletal muscle fibers at the cellular and chemical levels of organization. When you finish this module, you should be able to do the following:

1. Describe the structural properties and components of a skeletal muscle fiber.
2. Explain the organization of a myofibril.
3. Describe the structure and components of thick, thin, and elastic filaments.
4. Name and describe the function of each of the contractile, regulatory, and structural protein components of a sarcomere.
5. Explain the sliding-filament mechanism of muscle contraction.

Build Your Own Glossary

Following is a table listing key terms from Module 10.2. Before you read the module, use the glossary at the back of your book or look through the module to define the following terms.

Key Terms for Module 10.2

Term	Definition
Transverse tubules	
Terminal cisternae	
Triad	
Myofilament	
Sarcomere	
Sliding-filament mechanism	

Survey It: Form Questions

Before you read the module, survey it and form at least three questions for yourself. When you have finished reading the module, return to these questions and answer them.

Question 1: ______________________________

Answer: ______________________________

Question 2: ______________________________

Answer: ______________________________

Question 3: ______________________________

Answer: ______________________________

Identify It: The Structure of the Skeletal Muscle Fiber

Color and label Figure 10.2a with the structures of the skeletal muscle fiber. Label part b of the figure with the names of the myofilaments and their individual proteins.

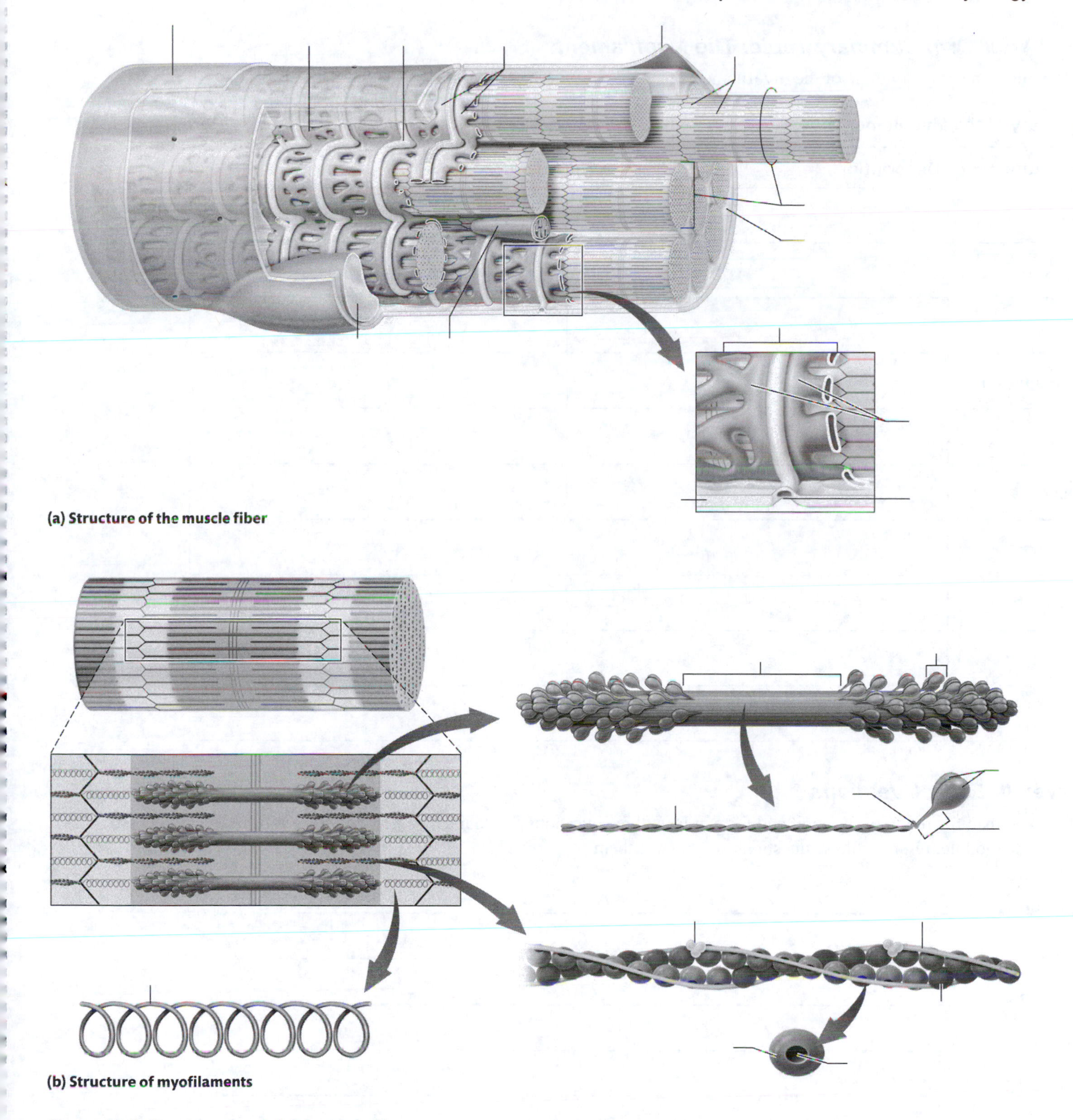

Figure 10.2 Structure of a skeletal muscle fiber.

Build Your Own Summary Table: The Myofilaments

Build your own summary table of the myofilaments using the table template provided.

Summary of the Myofilaments

Structure	Description	Function
Actin		
Myosin		
Troponin		
Tropomyosin		
Titin		
I band		
Z disc		
A band		
H zone		
M line		

Survey It: Form Questions

Before you read the module, survey it and form at least three questions for yourself. When you have finished reading the module, return to these questions and answer them.

Question 1: ______________________________

Answer: ______________________________

Question 2: ______________________________

Answer: ______________________________

Question 3: ______________________________

Answer: ______________________________

Key Concept: What produces the characteristic striations of skeletal muscle fibers?

____

____

Draw It: The Sarcomere

Draw a sarcomere using the following outline in Figure 10.3 (which includes the Z-discs) and label each filament, band, and component.

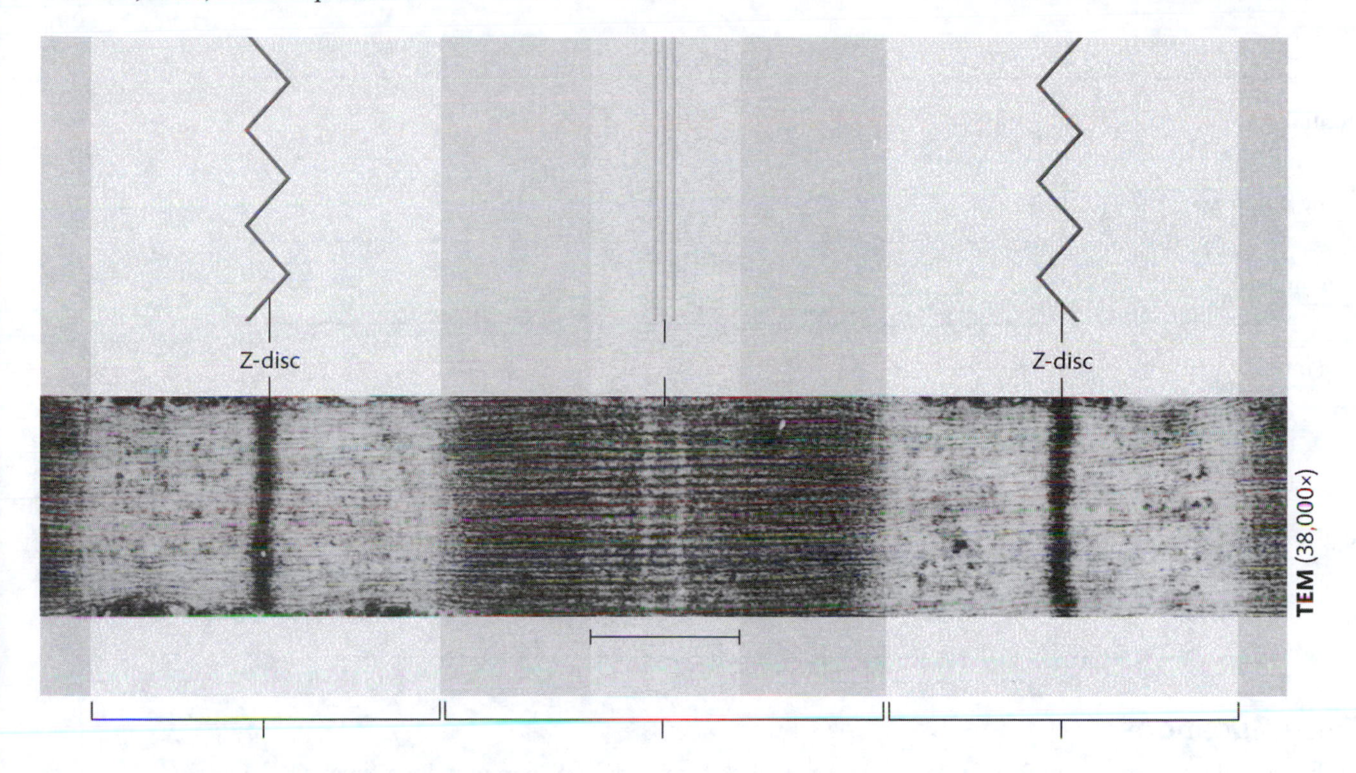

Figure 10.3 Structure and bands of the sarcomere.

Key Concept: What is a sarcomere? Why is it considered the functional unit of contraction?

____

____

Module 10.3: Skeletal Muscle Fibers as Electrically Excitable Cells

This module in your text examines the introductory principles of electrophysiology—the study of electrical changes across plasma membranes and the physiological processes that accompany these changes. When you complete this module, you should be able to do the following:

1. Contrast the relative concentrations of sodium and potassium ions inside and outside a cell.
2. Differentiate between a concentration gradient and an electrochemical gradient.
3. Describe how the resting membrane potential is generated.
4. Describe the sequence of events of a skeletal muscle fiber action potential.

Build Your Own Glossary

Following is a table listing key terms from Module 10.3. Before you read the module, use the glossary at the back of your book or look through the module to define the following terms.

Key Terms of Module 10.3

Term	Definition
Voltage	
Membrane potential	
Resting membrane potential	
Ligand-gated channel	
Voltage-gated channel	
Action potential	
Depolarization	
Repolarization	

Survey It: Form Questions

Before you read the module, survey it and form at least three questions for yourself. When you have finished reading the module, return to these questions and answer them.

Question 1: ____________________

Answer: ____________________

Question 2: ____________________

Answer: ____________________

Question 3: ____________________

Answer: ____________________

Key Concept: What is a membrane potential? What is the value of a skeletal muscle fiber's resting membrane potential?

____

____

Draw It: The Action Potential

Draw and label the stages of an action potential on the graph in Figure 10.4. Then, draw, label, and color the ion channels and the ions that move through them in each stage of the action potential.

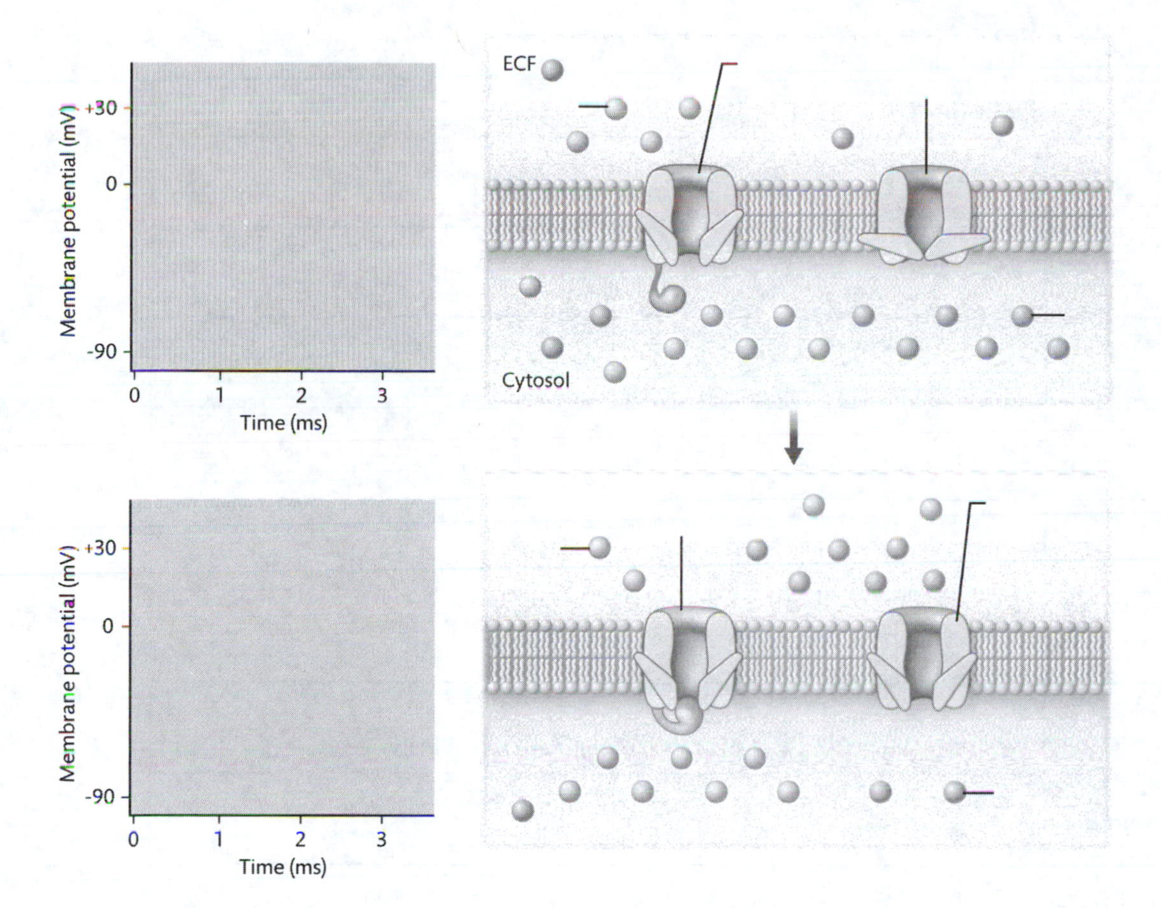

Figure 10.4 Stages of an action potential.

Key Concept: What is an action potential?

____

____

Module 10.4: The Process of Skeletal Muscle Contraction and Relaxation

Module 10.4 in your text teaches you the precise events of the sliding-filament mechanism of contraction and examines the relationship between the muscular system and the nervous system. At the end of this module, you should be able to do the following:

1. Describe the anatomy of the neuromuscular junction.
2. Describe the events at the neuromuscular junction that elicit an action potential in the muscle fiber.
3. Explain excitation-contraction coupling.
4. Describe the sequence of events involved in the contraction cycle of a skeletal muscle fiber.
5. Explain the process of skeletal muscle fiber relaxation.

Build Your Own Glossary

Following is a table listing key terms from Module 10.4. Before you read the module, use the glossary at the back of your book or look through the module to define the following terms.

Key Terms of Module 10.4

Term	Definition
Neuromuscular junction	
Axon terminal	
Synaptic vesicle	
Acetylcholine	
Synaptic cleft	
Motor end plate	
End-plate potential	
Excitation-contraction coupling	
Crossbridge cycle	
Power stroke	
Acetylcholinesterase	

Survey It: Form Questions

Before you read the module, survey it and form at least two questions for yourself. When you have finished reading the module, return to these questions and answer them.

Question 1: ______________________________

Answer: ______________________________

Question 2: ______________________________

Answer: ______________________________

Identify It: The Neuromuscular Junction

Color and label the neuromuscular junction illustrated in Figure 10.5.

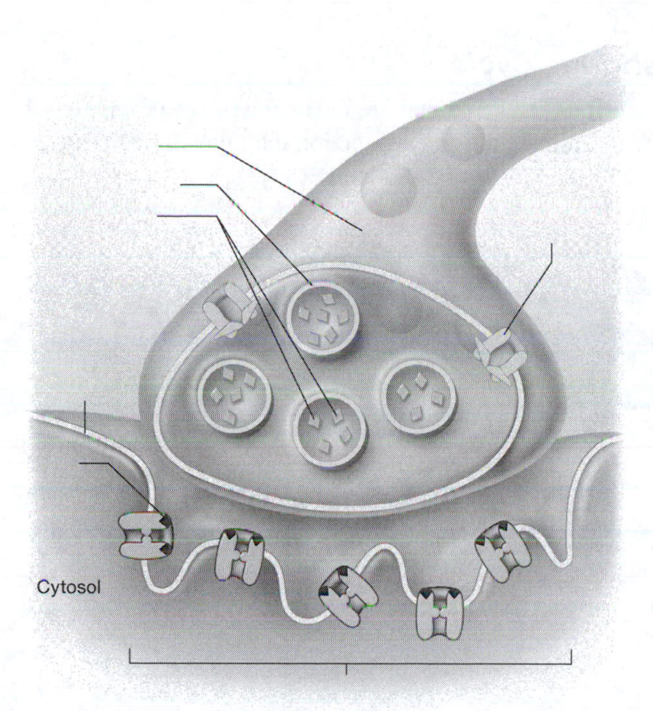

Figure 10.5 Structures of the neuromuscular junction.

Key Concept: What is the structural and functional relationship between neurons and skeletal muscle fibers?

____

____

Trace It: Excitation and Excitation-Contraction Coupling

Trace the events of skeletal muscle fiber excitation and excitation-contraction coupling. Number the first event, the arrival of an action potential at the axon terminal of the motor neuron, as step 1, and continue numbering through the end of the process.

___ The action potential is propagated down the T-tubule.

___ Entry of Na^+ depolarizes the sarcolemma locally, producing an end-plate potential.

___ An action potential arrives at the axon terminal of the motor neuron.

___ T-tubule depolarization triggers the opening of calcium ion channels in the SR, and calcium ions flood the cytosol.

___ Synaptic vesicles release ACh into the synaptic cleft.

___ Ion channels open and Na^+ enters the muscle fiber.

___ ACh binds to ligand-gated ion channels in the motor end plate.

___ The end-plate potential stimulates an action potential.

Key Concept: Why does an increase in calcium ion concentration in the cytosol lead to a muscle contraction?

____

____

Identify It: The Stages of the Crossbridge Cycle

Figure 10.6 diagrams the crossbridge cycle. Describe the events that are occurring next to each step, label and color-code key components, and draw an arrow to indicate in which direction the myosin head moves.

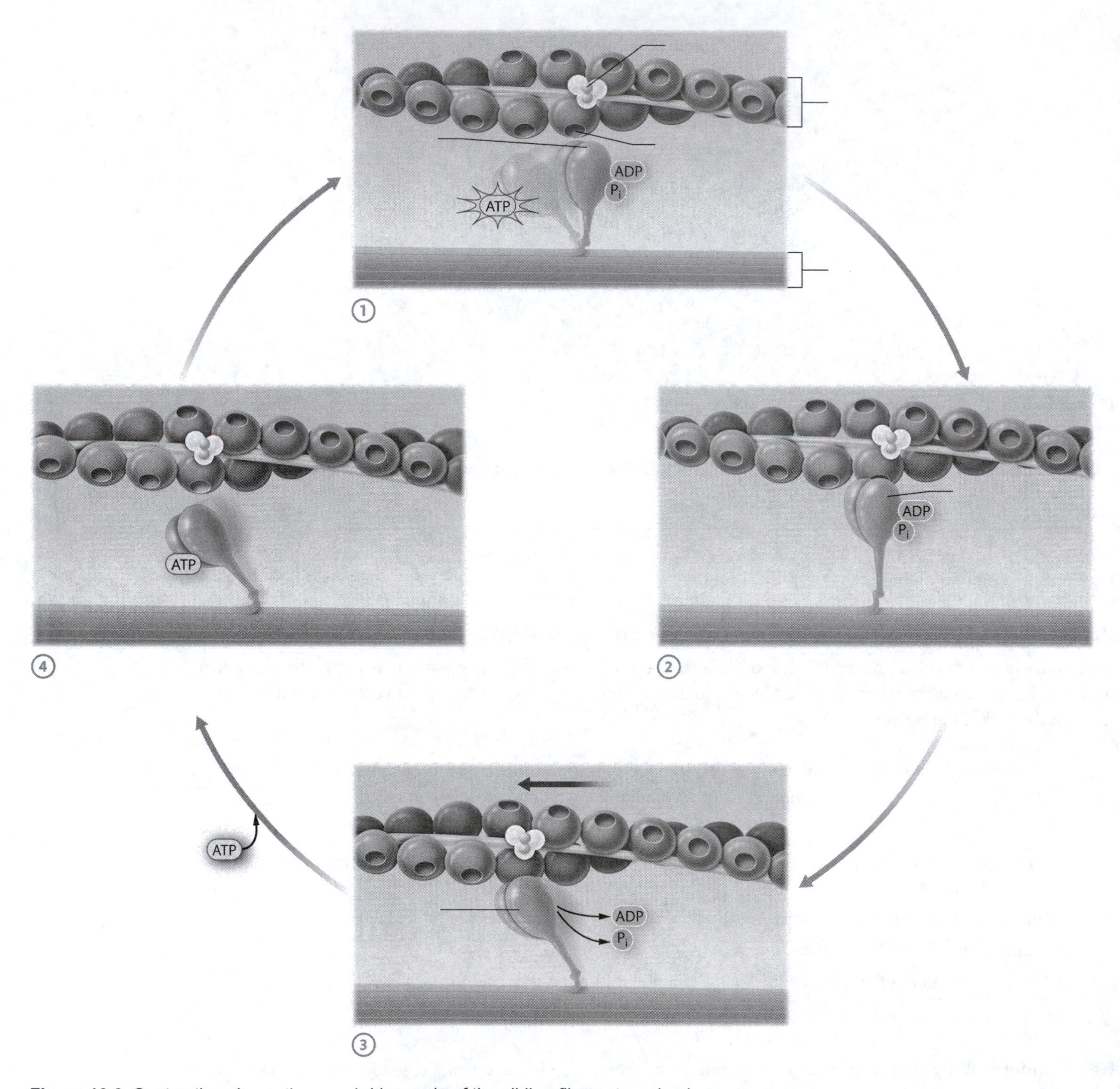

Figure 10.6 Contraction phase: the crossbridge cycle of the sliding-filament mechanism.

Team Up

Develop a handout for your fellow students that teaches the process of skeletal muscle contraction, beginning with excitation and concluding with muscle contraction by the sliding-filament mechanism. Be sure to include diagrams and text to detail the steps.

Key Concept: How does a muscle fiber generate tension via a series of crossbridge cycles?

____

____

Key Concept: What happens when the motor neuron stops releasing ACh? Why?

____

____

Module 10.5: Energy Sources for Skeletal Muscle

Module 10.5 covers the sources of ATP for skeletal muscle contraction, relaxation, and its other many processes. At the end of this module, you should be able to do the following:

1. Describe the immediate energy sources available to muscle fibers.
2. Describe the glycolytic and oxidative mechanisms that muscle fibers use to obtain ATP for muscle contraction.
3. Explain the duration of activity that each ATP source can fuel.

Build Your Own Glossary

Below is a table listing key terms from Module 10.5. Before you read the module, use the glossary at the back of your book or look through the module to define the following terms.

Key Terms of Module 10.5

Term	Definition
Creatine phosphate	
Creatine kinase	
Glycolytic catabolism	
Glycogen	
Lactic acid	
Oxidative catabolism	
Myoglobin	

Survey It: Form Questions

Before you read the module, survey it and form at least three questions for yourself. When you have finished reading the module, return to these questions and answer them.

Question 1: ____________________

Answer: ____________________

Question 2: ____________________

Answer: ____________________

Question 3: ____________________

Answer: ____________________

Key Concept: Why do skeletal muscles need ATP?

____

____

Build Your Own Summary Table: Muscle Fiber Energy Sources

Build your own summary table about the different sources of energy for muscle fibers by filling in the information below.

Summary of Energy Sources for Skeletal Muscle Tissue

Energy Source	Duration of Energy Production	Oxygen Requirement	ATP Production
Immediate energy sources			
Glycolytic energy sources			
Oxidative energy sources			

Key Concept: What energy source(s) must one use to sustain muscle activity for several minutes? Why?

____

____

Module 10.6: Muscle Tension at the Fiber Level

As you learned earlier, the basic function of muscle tissue is to contract to produce tension. This is explored in this module at the level of the individual muscle fiber. At the end of this module, you should be able to do the following:

1. Describe the stages of a twitch contraction, and explain how a twitch is affected by the frequency of stimulation.
2. Relate tension production to the length of a sarcomere.
3. Compare and contrast the anatomical and metabolic characteristics of type I and type II muscle fibers.

Build Your Own Glossary

Following is a table listing key terms from Module 10.6. Before you read the module, use the glossary at the back of your book or look through the module to define the following terms.

Key Terms for Module 10.6

Term	Definition
Muscle twitch	
Length-tension relationship	
Wave summation	
Unfused tetanus	
Fused tetanus	
Type I fibers	
Type II fibers	

Survey It: Form Questions

Before you read the module, survey it and form at least three questions for yourself. When you have finished reading the module, return to these questions and answer them.

Question 1: ______________________________

Answer: ______________________________

Question 2: ______________________________

Answer: ______________________________

Question 3: ______________________________

Answer: ______________________________

Draw It: A Muscle Twitch

Draw and label the stages of a twitch contraction on the graph in Figure 10.7.

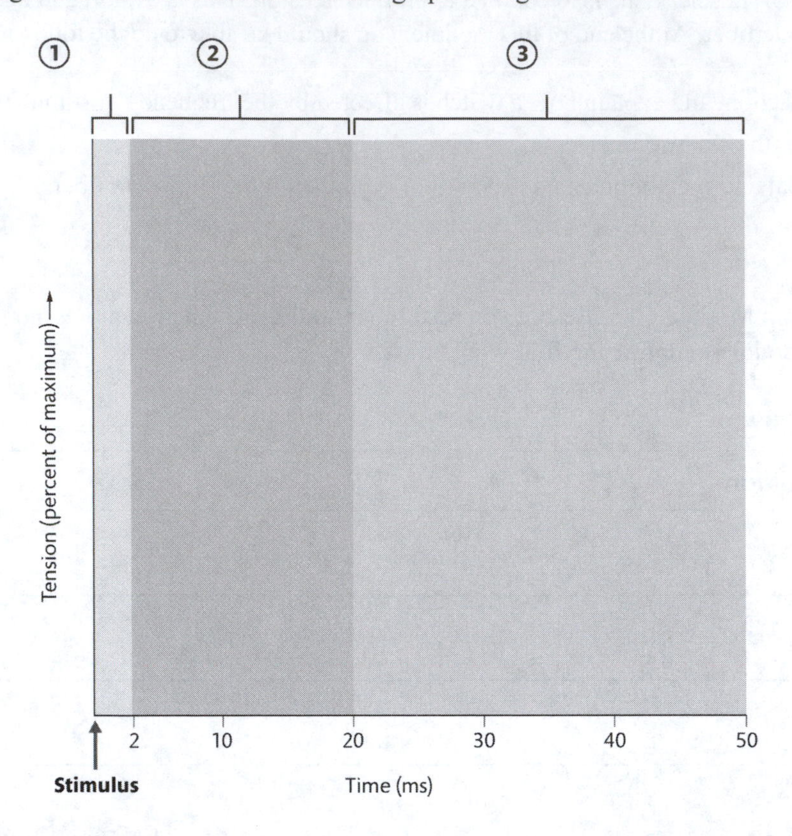

Figure 10.7 Myogram of a twitch contraction.

Identify It: Wave Summation

Identify each of the graphs in Figure 10.8 as being either fused tetanus or unfused tetanus. Which contraction (a or b) will produce the most tension? Why?

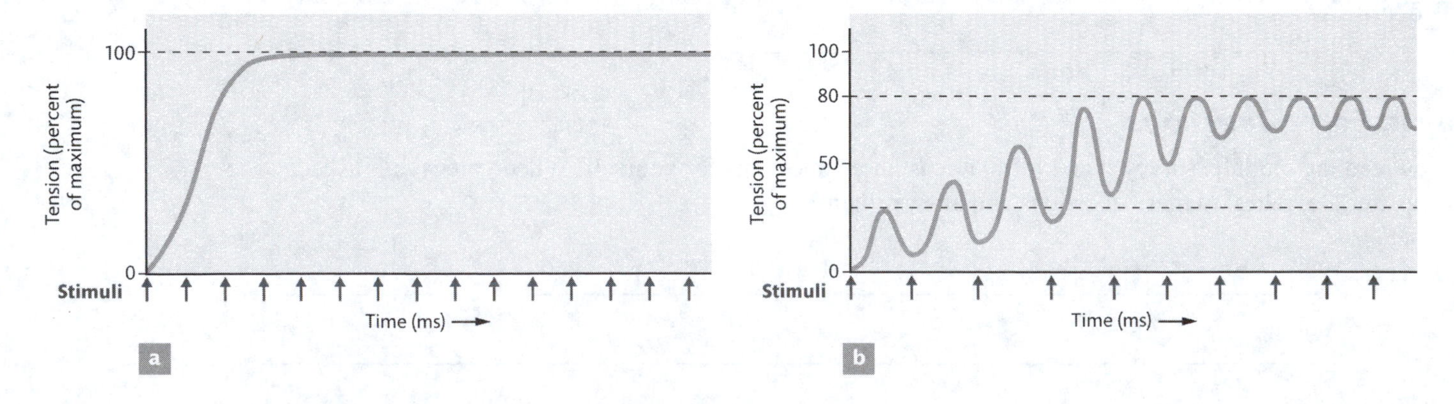

Figure 10.8 Wave summation: unfused and fused tetanus.

Key Concept: How does the frequency of stimulation of a muscle fiber impact tension production? Why?

____

____

Try It: The Length-Tension Relationship

It is easy to see for yourself the impact that muscle starting length has on the amount of tension that can be produced by a muscle contraction. First, grip a pen or other small object in your hand with your wrist extended and squeeze the object as hard as you are able. Then, flex your wrist and grip the same object, again as hard as you are able. Finally, hyperextend your wrist and repeat the procedure. In which position were you able to grip the object the most strongly? Why?

____

____

Key Concept: How does the starting length of a muscle fiber impact tension production, and why?

____

____

Module 10.7: Muscle Tension at the Organ Level

Now we look at muscle tension production on the organ level. At the end of this module, you should be able to do the following:

1. Describe the structure and function of a motor unit.
2. Explain how muscle tone is produced.
3. Compare and contrast the three types of contractions.

Build Your Own Glossary

Below is a table listing key terms from Module 10.7. Before you read the module, use the glossary at the back of your book or look through the module to define the following terms.

Key Terms for Module 10.7

Term	Definition
Motor unit	
Recruitment	
Isometric contraction	
Isotonic eccentric contraction	
Isotonic concentric contraction	
Muscle tone	

Survey It: Form Questions

Before you read the module, survey it and form at least three questions for yourself. When you have finished reading the module, return to these questions and answer them.

Question 1: ______________________________

Answer: ______________________________

Question 2: ______________________________

Answer: ______________________________

Question 3: ______________________________

Answer: ______________________________

Identify It: The Motor Unit

Color and label the components of the motor unit illustrated in Figure 10.9.

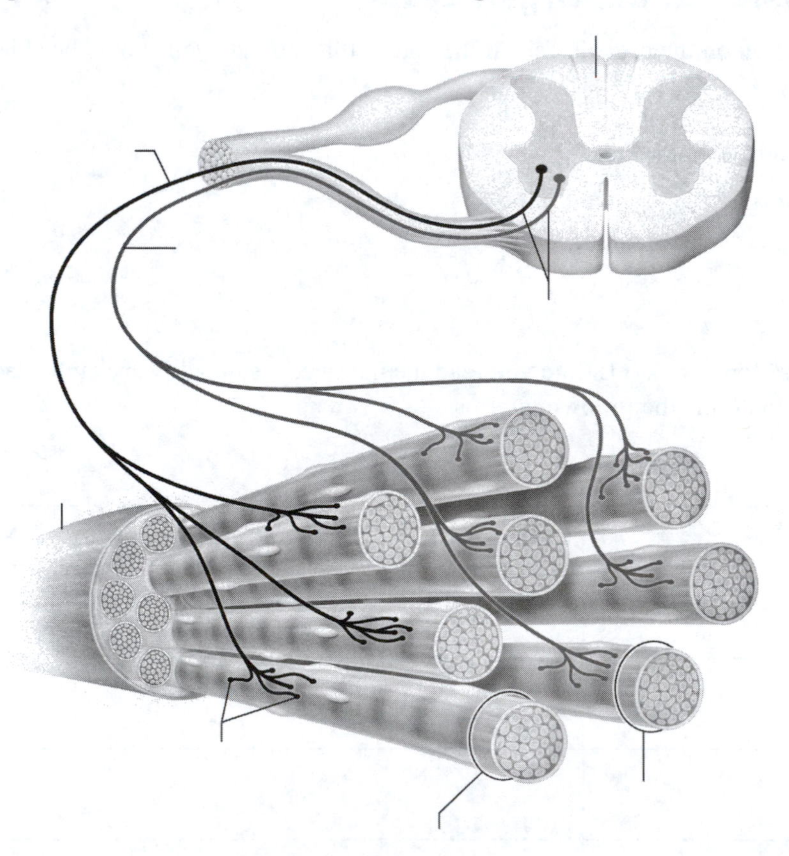

Figure 10.9 The motor unit. Two motor units are shown here.

Key Concept: What factors influence tension production at the organ level?

____

____

Identify It: Types of Muscle Contraction

Identify each of the following muscle contractions in Figure 10.10 as either isotonic eccentric, isotonic concentric, or isometric. Explain how you arrived at your conclusion for each contraction.

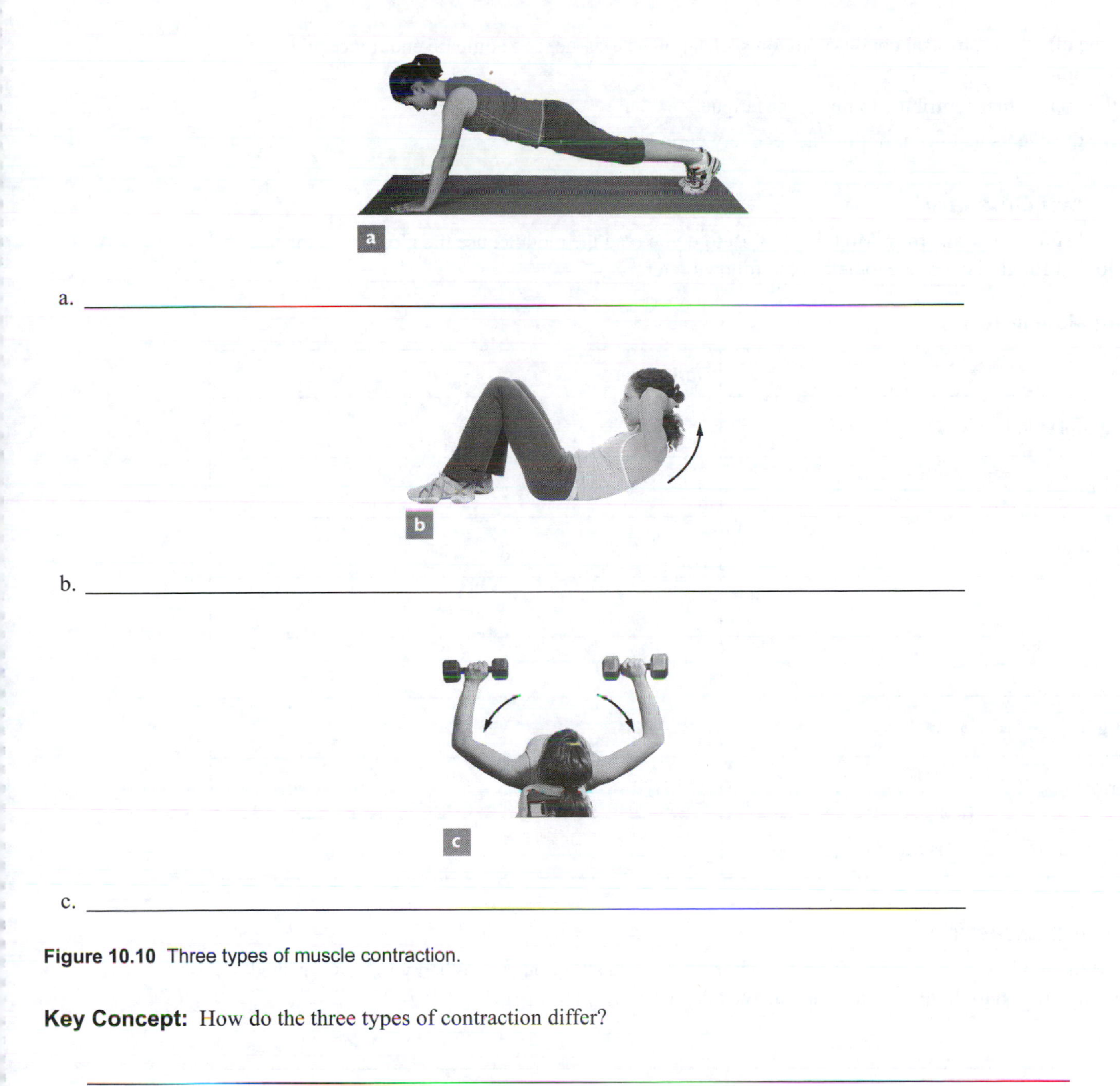

a. ______________________________

b. ______________________________

c. ______________________________

Figure 10.10 Three types of muscle contraction.

Key Concept: How do the three types of contraction differ?

____

____

Team Up

Pair up with a partner, and each of you write a 10-question quiz for the modules on muscle tension (Modules 10.6 and 10.7) using multiple choice, short-answer, and true/false questions. Then, trade quizzes, and take the quiz your partner wrote. Once you have both finished your quizzes, exchange them again, and grade the quizzes to determine areas where you need further study.

Module 10.8: Skeletal Muscle Performance

This module examines how the body responds to exercise in the short and long term. By the end of the module, you should be able to do the following:

1. Describe the effects of physical conditioning on skeletal muscle tissue, and compare endurance and resistance training.
2. Explain the factors that contribute to muscular fatigue.
3. Summarize the events that occur during the recovery period.

Build Your Own Glossary

Below is a table listing key terms from Module 10.8. Before you read the module, use the glossary at the back of your book or look through the module to define the following terms.

Key Terms for Module 10.8

Term	Definition
Principle of myoplasticity	
Endurance training	
Resistance training	
Hypertrophy	
Atrophy	
Muscular fatigue	
Recovery period	
Excess postexercise oxygen consumption	

Survey It: Form Questions

Before you read the module, survey it and form at least three questions for yourself. When you have finished reading the module, return to these questions and answer them.

Question 1: ______________________________

Answer: ______________________________

Question 2: ______________________________

Answer: ______________________________

Question 3: ______________________________

Answer: ______________________________

Build Your Own Summary Table: Muscle Fiber Changes with Training and Disuse

Build your own summary table about the changes to a muscle fiber with training and disuse, using the template provided.

Summary of Changes Due to Training and Disuse

Variable	Endurance Training	Resistance Training	Disuse
Diameter of fiber			
Number of myofibrils			
Number of blood vessels			
Number of oxidative enzymes			
Number of mitochondria			

Key Concept: How does a muscle fiber change with endurance training, resistance training, and disuse?

___

___

Team Up

Work with a partner, and have one of you develop a diagram or concept map explaining the causes of muscular fatigue. The other partner should develop a diagram or concept map explaining the reasons for excess postexercise oxygen consumption. When you have completed your diagrams, use them to teach each other about these two concepts.

Key Concept: What are the causes of muscular fatigue? Which homeostatic imbalances does excess postexercise oxygen consumption help to correct?

___

___

Module 10.9: Smooth and Cardiac Muscle

This module discusses the structural and functional properties of smooth muscle and provides an overview of cardiac muscle tissue. At the end of this module, you should be able to do the following:

1. Describe the structure, location in the body, and functions of smooth and cardiac muscle tissue.
2. Describe the contraction process of smooth muscle fibers, and contrast it with skeletal muscle fiber contraction.

Build Your Own Glossary

Following is a table listing key terms from Module 10.9. Before you read the module, use the glossary at the back of your book or look through the module to define the following terms.

Key Terms for Module 10.9

Term	Definition
Dense bodies	
Pacemaker cells	
Calmodulin	
Myosin light-chain kinase	
Single-unit smooth muscle	
Multi-unit smooth muscle	
Intercalated disc	

Survey It: Form Questions

Before you read the module, survey it and form at least three questions for yourself. When you have finished reading the module, return to these questions and answer them.

Question 1: ______________________________

Answer: ______________________________

Question 2: ______________________________

Answer: ______________________________

Question 3: ______________________________

Answer: ______________________________

Describe It: Smooth Muscle Contraction

Using the boxes below, write a brief description of the four events of smooth muscle contraction.

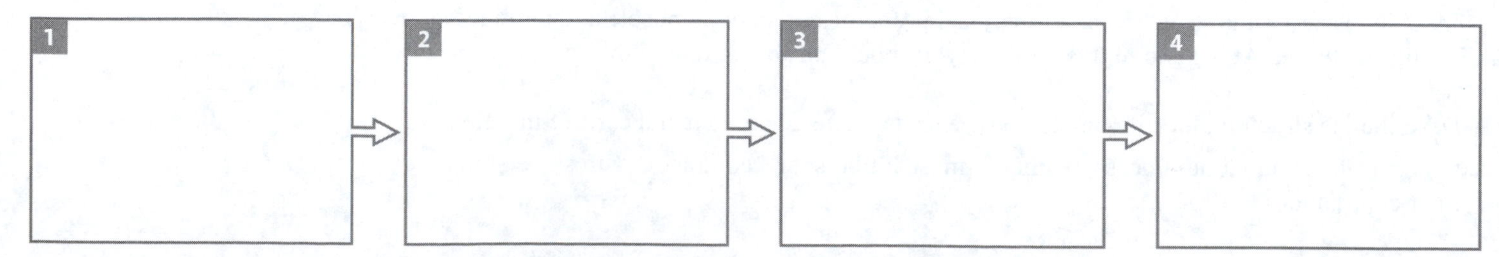

Key Concept: How does smooth muscle contraction differ from skeletal muscle contraction? Why?

___

___

Build Your Own Summary Table: The Three Types of Muscle Tissue

Now that you have covered all three types of muscle tissue, let's compare and contrast them. The following template provides you with an outline for a summary table with the key properties of each type of muscle tissue. Complete the table as you finish Module 10.9 in your text.

Summary Table of the Three Types of Muscle Tissue

	Skeletal Muscle	Smooth Muscle	Cardiac Muscle
Structural Features			
Appearance			
Number of nuclei			
Arrangement of myofilaments			
Proteins in myofilaments			
Other structural features			
Functional Features			
Control			
Trigger for contraction			

Key Concept: What are important structural and functional differences between cardiac muscle tissue and skeletal muscle tissue?

___

___

What Do You Know Now?

Let's now revisit the questions you answered in the beginning of this chapter. How have your answers changed now that you've worked through the material?

- How do the three types of muscle tissue differ?

  ____

- What is the primary function of all types of muscle tissue?

  ____

- What happens to a muscle during a muscle contraction?

  ____

- What is the role of ATP in a muscle contraction?

  ____

11 Introduction to the Nervous System and Nervous Tissue

The nervous system is one of the body's main homeostatic systems, meaning that its organs work to maintain homeostasis by impacting the functions of other tissues and cells. We begin our discussion of the nervous system in this chapter with a look at how it is organized and the structure and function of nervous tissue.

What Do You Already Know?

Try to answer the following questions before proceeding to the next section. If you're unsure of the correct answers, give it your best attempt based on previous courses, previous chapters, or just your general knowledge.

- What are the organs of the nervous system?

  ____

- What do the organs of the nervous system do?

  ____

- What are the types of cells within the nervous system?

  ____

Module 11.1: Overview of the Nervous System

Module 11.1 in your text gives you an overview of the structures and functions of the nervous system and its divisions. By the end of the module, you should be able to do the following:

1. Describe the major functions of the nervous system.
2. Describe the structures and basic functions of each organ of the central and peripheral nervous systems.
3. Explain the major differences between the two functional divisions of the peripheral nervous system.

Build Your Own Glossary

Following is a table listing key terms from Module 11.1. Before you read the module, use the glossary at the back of your book or look through the module to define the following terms.

Key Terms for Module 11.1

Term	Definition
Central nervous system	
Peripheral nervous system	
Brain	
Spinal cord	
Nerve	
Sensory (afferent) division	
Motor (efferent) division	

Survey It: Form Questions

Before you read the module, survey it and form at least two questions for yourself. When you have finished reading the module, return to these questions and answer them.

Question 1: ______________________________

Answer: ______________________________

Question 2: ______________________________

Answer: ______________________________

Identify It: Structures and Divisions of the Nervous System

Identify and color-code the structures and divisions of the nervous system shown in Figure 11.1 as you read Module 11.1.

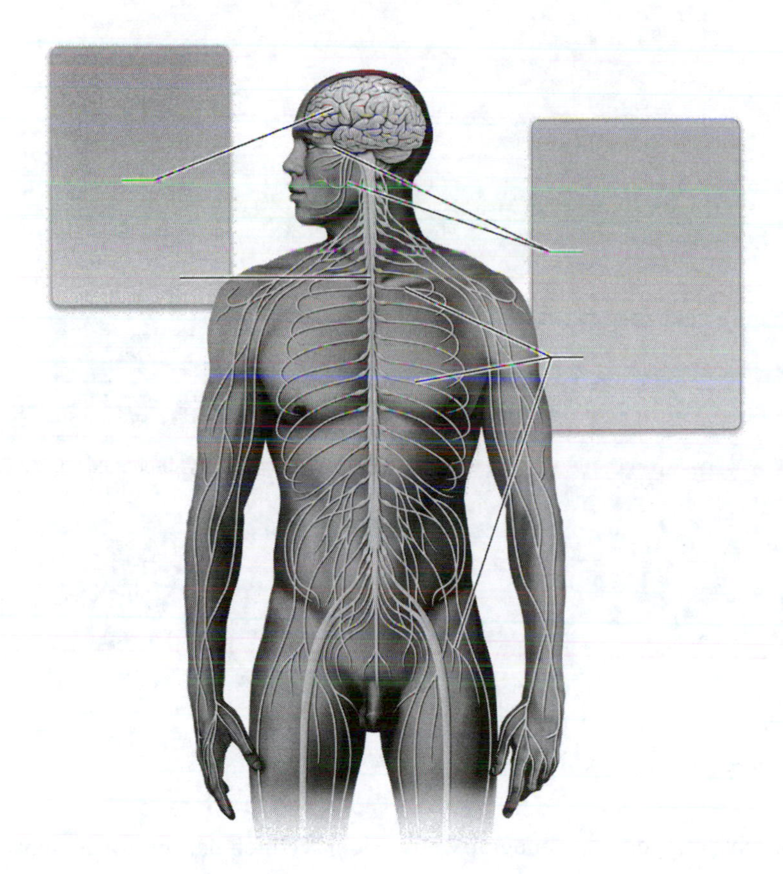

Figure 11.1 Structures and divisions of the nervous system.

Key Concept: What are the main functions of the nervous system?

____

____

Describe It: The Divisions of the Nervous System

Fill in Figure 11.2 with the main functions of each of the divisions of the nervous system. Then, draw arrows to indicate the flow of information between divisions.

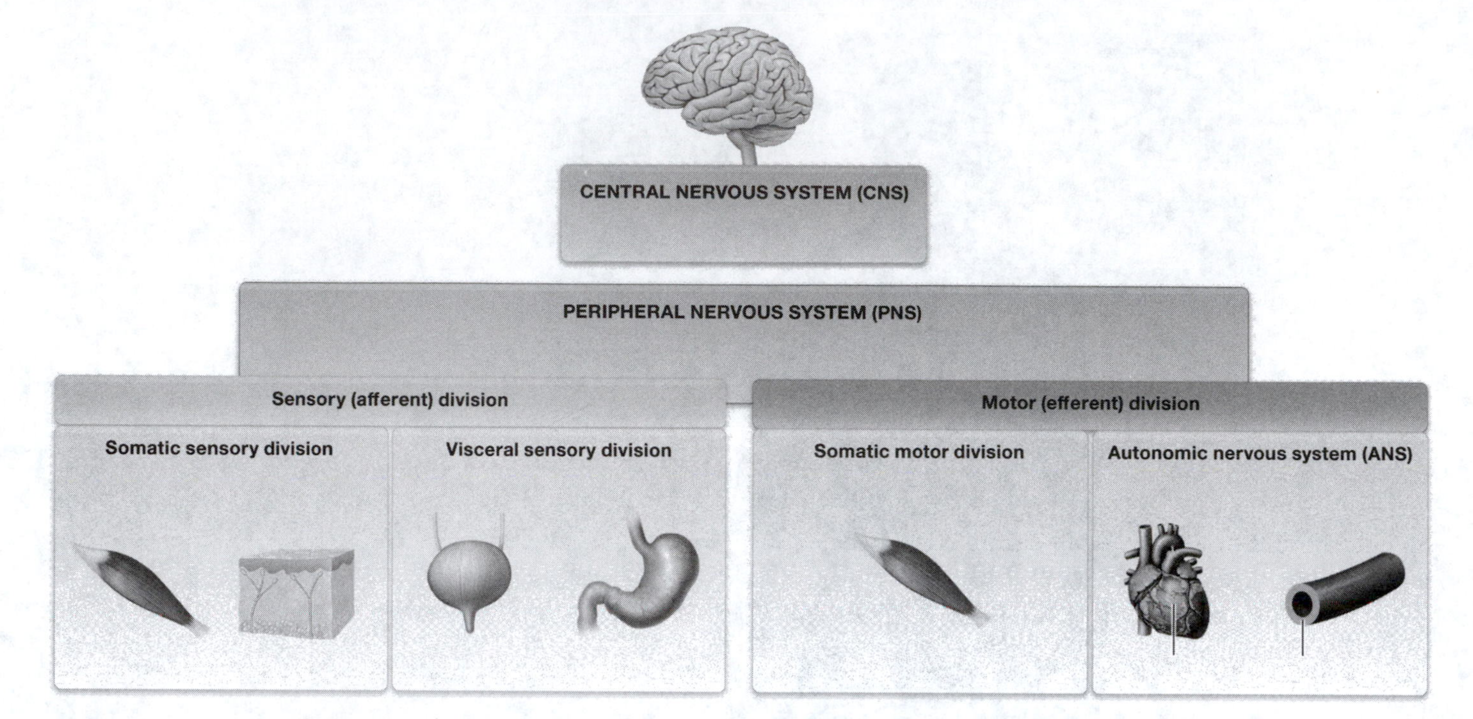

Figure 11.2 The divisions of the nervous system.

Key Concept: Which part of the nervous system performs integrative functions? Which part performs sensory and motor functions?

____

____

Key Concept: How do the motor and sensory divisions of the nervous system differ?

____

____

Module 11.2: Nervous Tissue

This module in your text examines the structure of nervous tissue, including the two main cell types: neurons and neuroglial cells. When you finish this module, you should be able to do the following:

1. Describe the structure and function of each component of the neuron.
2. Describe the structure and function of each type of neuron.
3. Describe how the structure of each type of neuron supports its function.
4. Describe the structure and function of the four types of CNS neuroglial cells and the two types of PNS neuroglial cells.
5. Explain how the structure of each neuroglial cell supports its function.

Build Your Own Glossary

Below is a table listing key terms from Module 11.2. Before you read the module, use the glossary at the back of your book or look through the module to define the following terms.

Key Terms for Module 11.2

Term	Definition
Neuron	
Cell body	
Dendrite	
Axon	
Axon terminal	
Neuroglial cells	
Astrocyte	
Oligodendrocyte	
Microglial cell	
Ependymal cell	
Schwann cell	
Satellite cell	
Myelin sheath	

Survey It: Form Questions

Before you read the module, survey it and form at least three questions for yourself. When you have finished reading the module, return to these questions and answer them.

Question 1: ______________________________

Answer: ______________________________

Question 2: ______________________________

Answer: ______________________________

Question 3: ______________________________

Answer: ______________________________

Identify It: The Structure of a Neuron

Color and label Figure 11.3 with the parts of the neuron as you read Module 11.2.

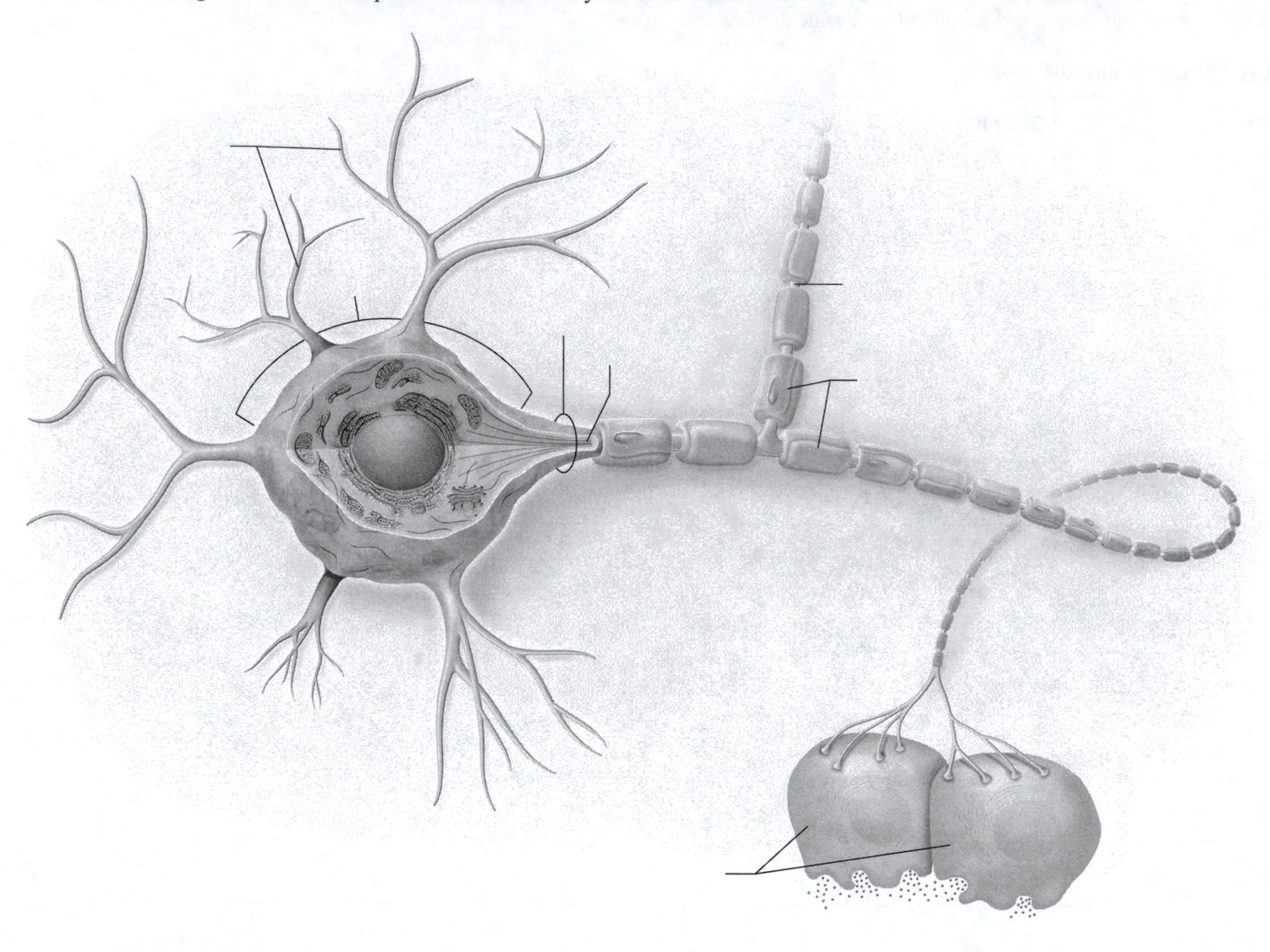

Figure 11.3 Neuron structure.

Key Concept: What are the three structural classes of neurons, and how do they differ? How do the structural classes relate to the three functional classes?

___

___

___

Build Your Own Summary Table: Neuroglial Cells

As you read Module 11.2, build your own summary table about the location, structural features, and functions of the neuroglial cells by filling in the information in the template provided below.

Summary of the Structure and Functions of Neuroglial Cells

Cell	Location	Structural Features	Functions
Astrocyte			
Oligodendrocyte			
Microglial cell			
Ependymal cell			
Schwann cell			
Satellite cell			

Key Concept: What is myelin? How does myelination differ in the CNS and PNS?

____

____

Key Concept: What conditions must be present in order for an axon to regenerate?

____

____

Module 11.3: Electrophysiology of Neurons

This module applies the principles of electrophysiology that you learned in previous chapters to the functioning of neurons. When you complete this module, you should be able to do the following:

1. Describe the voltage-gated ion channels that are essential for the development of the action potential.
2. Interpret a graph of an action potential, and describe the depolarization, repolarization, and hyperpolarization phases of an action potential.
3. Explain the physiological basis of the absolute and relative refractory periods.
4. Compare and contrast continuous and saltatory conduction.
5. Explain how axon diameter and myelination affect conduction speed.

Build Your Own Glossary

Following is a table listing key terms from Module 11.3. Before you read the module, use the glossary at the back of your book or look through the module to define the following terms.

Key Terms of Module 11.3

Term	Definition
Membrane potential	
Resting membrane potential	
Electrochemical gradient	
Depolarization	
Repolarization	
Hyperpolarization	
Local potential	
Action potential	
Refractory period	
Propagation	
Saltatory conduction	
Continuous conduction	

Survey It: Form Questions

Before you read the module, survey it and form at least three questions for yourself. When you have finished reading the module, return to these questions and answer them.

Question 1: ____________________

Answer: ____________________

Question 2: ____________________

Answer: ____________________

Question 3: ____________________

Answer: ____________________

Key Concept: What are the three main types of ion channels and how do they differ?

____

____

Draw It: Ion Gradients and Ion Movements

Following is an illustration of a plasma membrane with sodium and potassium ion channels (Figure 11.4). Draw the sodium and potassium ions in the correct distribution for a resting cell. Then, draw an arrow to show in which directions the ions will diffuse when the channels have opened.

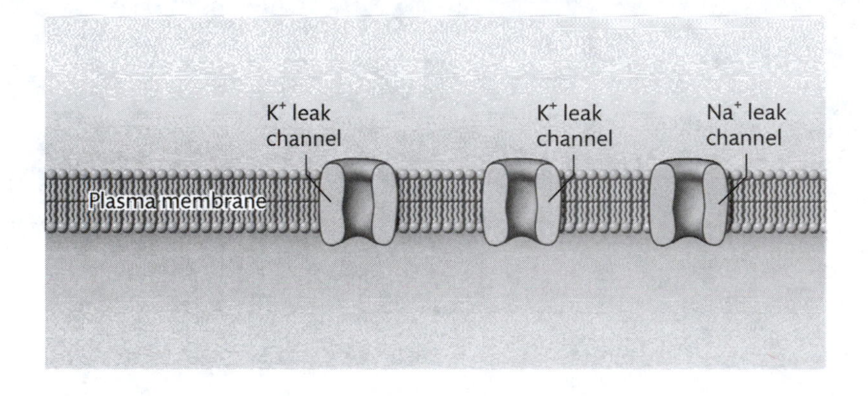

Figure 11.4 The plasma membrane of an excitable cell.

Key Concept: How does the movement of positive ions lead to a negative resting membrane potential?

____

____

Complete It: Changing the Membrane Potential

Fill in the blanks to complete the following paragraph that walks through what happens when we open and close ion channels to change the membrane potential of a neuron.

A small, local change in the membrane potential of the neuron is called a ____________ ____________. It may have one of two effects: It may make the membrane more positive, a change called ____________, or it may make the membrane potential more negative, a change called ____________. If a positive change in membrane potential reaches a value called ____________, ____________ channels open, initiating an event known as an ____________ ____________. This event has three phases in a neuron: (1) ____________, mediated by the influx of ____________ ions; (2) ____________, mediated by the outflow of ____________ ions; and (3) ____________, mediated by the continued outflow of ____________ ions.

Draw It: The Action Potential

Draw the line of a typical neuronal action potential in the graph provided in Figure 11.5. Then, identify each stage of the action potential and describe in your own words what happens during the stages.

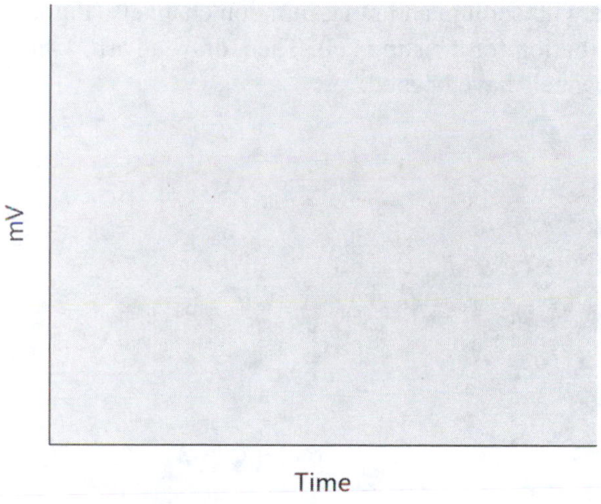

mV

Time

Figure 11.5 A graph for an action potential tracing.

Key Concept: How is a neuron prevented from firing action potentials continuously?

____

____

Describe It: Action Potential Propagation

Describe the events of action potential propagation in Figure 11.6. In addition, label and color-code key components of this process.

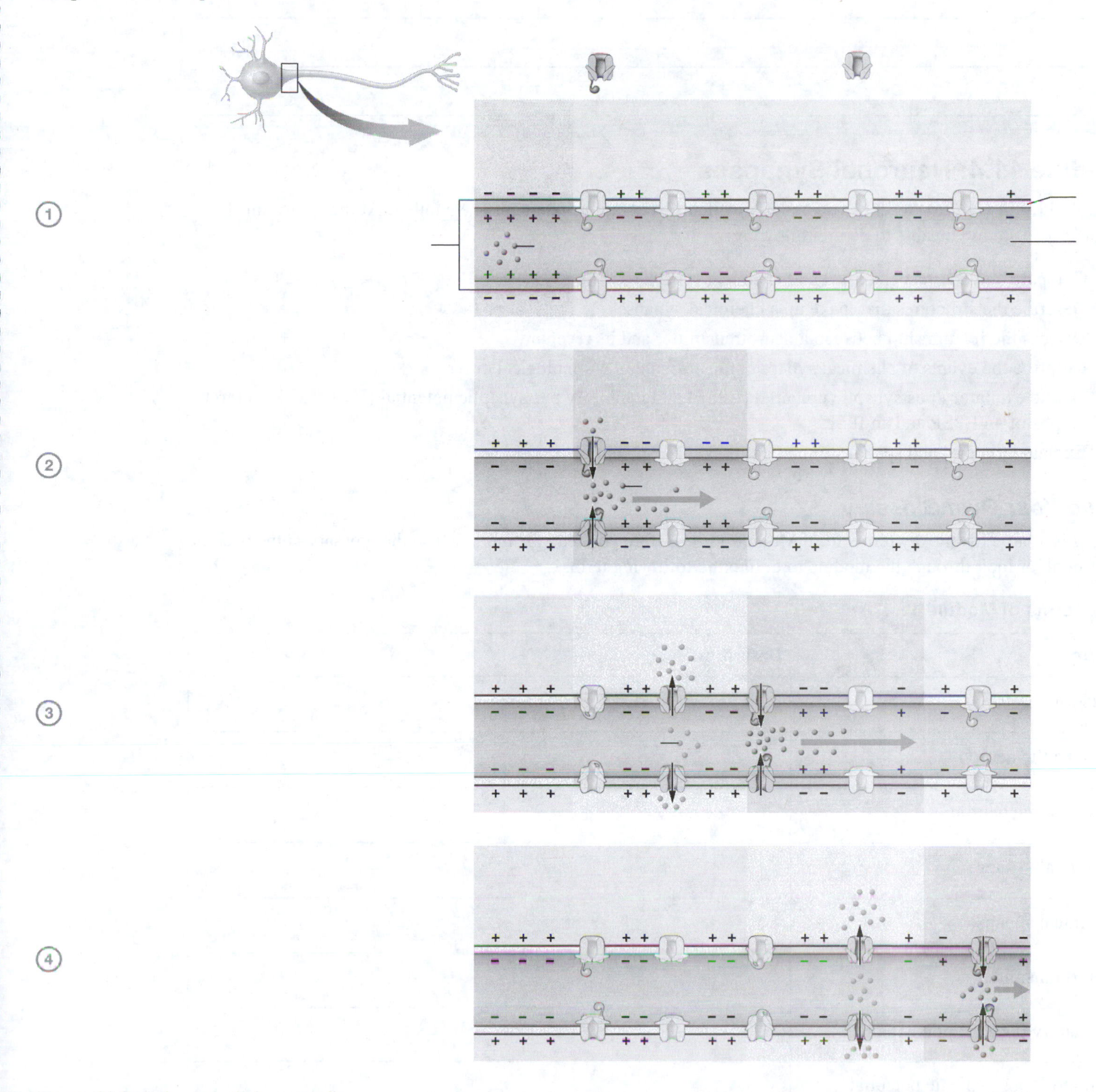

Figure 11.6 Action potential propagation.

Key Concept: Which type of conduction is used by myelinated neurons and why? What does myelin do to conduction velocity, and how does it do this?

____

____

____

Module 11.4: Neuronal Synapses

Module 11.4 in your text teaches you about how neurons communicate with one another. At the end of this module, you should be able to do the following:

1. Compare and contrast electrical and chemical synapses.
2. Describe the structures that make up a chemical synapse.
3. Discuss the relationship between a neurotransmitter and its receptor.
4. Describe the events of chemical synaptic transmission in chronological order.
5. Define excitatory postsynaptic potential (EPSP) and inhibitory postsynaptic potential (IPSP), and interpret graphs of an EPSP and an IPSP.
6. Explain temporal and spatial summation of synaptic potentials.

Build Your Own Glossary

Below is a table listing key terms from Module 11.4. Before you read the module, use the glossary at the back of your book or look through the module to define the following terms.

Key Terms of Module 11.4

Term	Definition
Synapse	
Presynaptic neuron	
Postsynaptic neuron	
Electrical synapse	
Chemical synapse	
Neurotransmitter	
Excitatory postsynaptic potential	
Inhibitory postsynaptic potential	
Temporal summation	
Spatial summation	

Survey It: Form Questions

Before you read the module, survey it and form at least two questions for yourself. When you have finished reading the module, return to these questions and answer them.

Question 1: ______________________________

Answer: ______________________________

Question 2: ______________________________

Answer: ______________________________

Identify It: Electrical and Chemical Synapses

Color and label the electrical and chemical synapses illustrated in Figure 11.7.

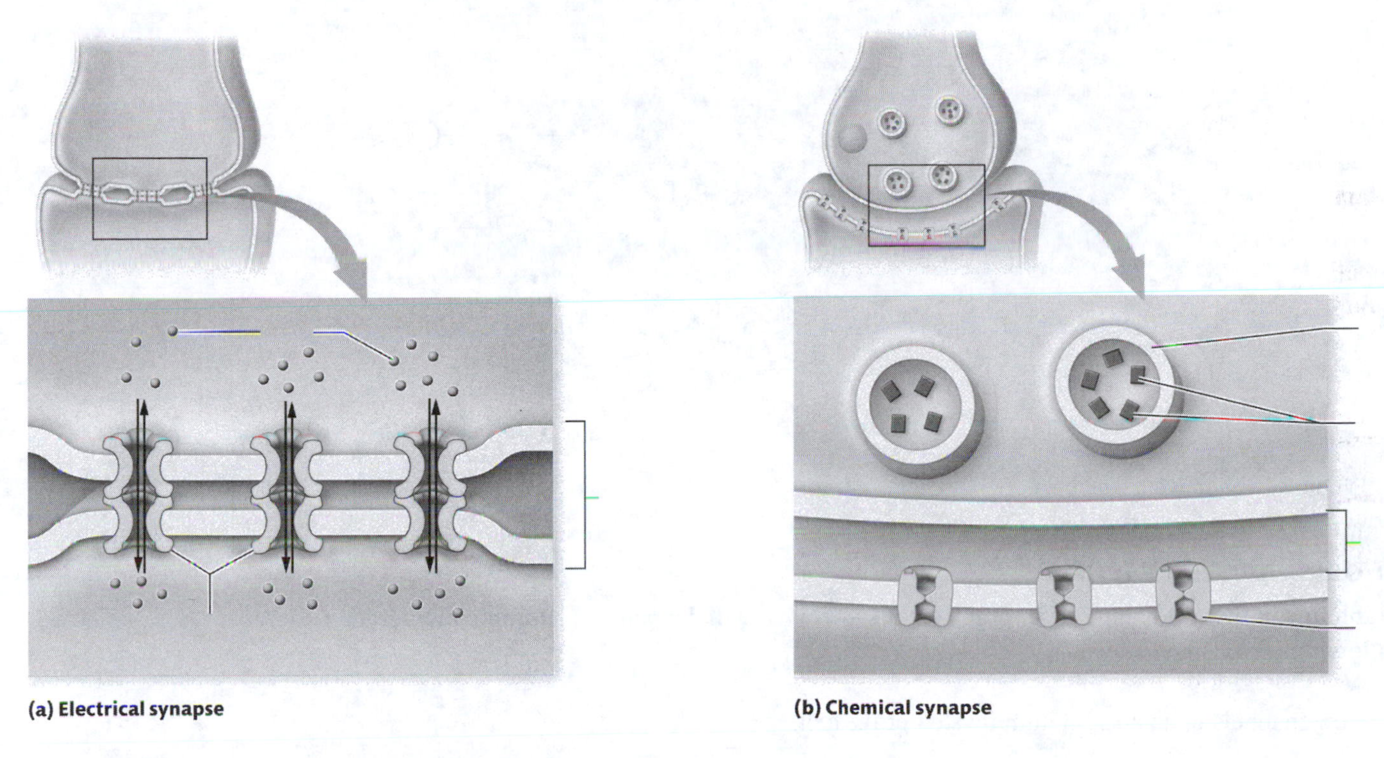

Figure 11.7 The structures of electrical and chemical synapses.

Key Concept: What are the key differences between chemical and electrical synapses? Why is transmission at electrical synapses bidirectional but unidirectional at chemical synapses?

____

____

Describe It: The Events at a Chemical Synapse

Describe the sequence of events of chemical synaptic transmission in Figure 11.8. Also, identify and color-code key components of the process.

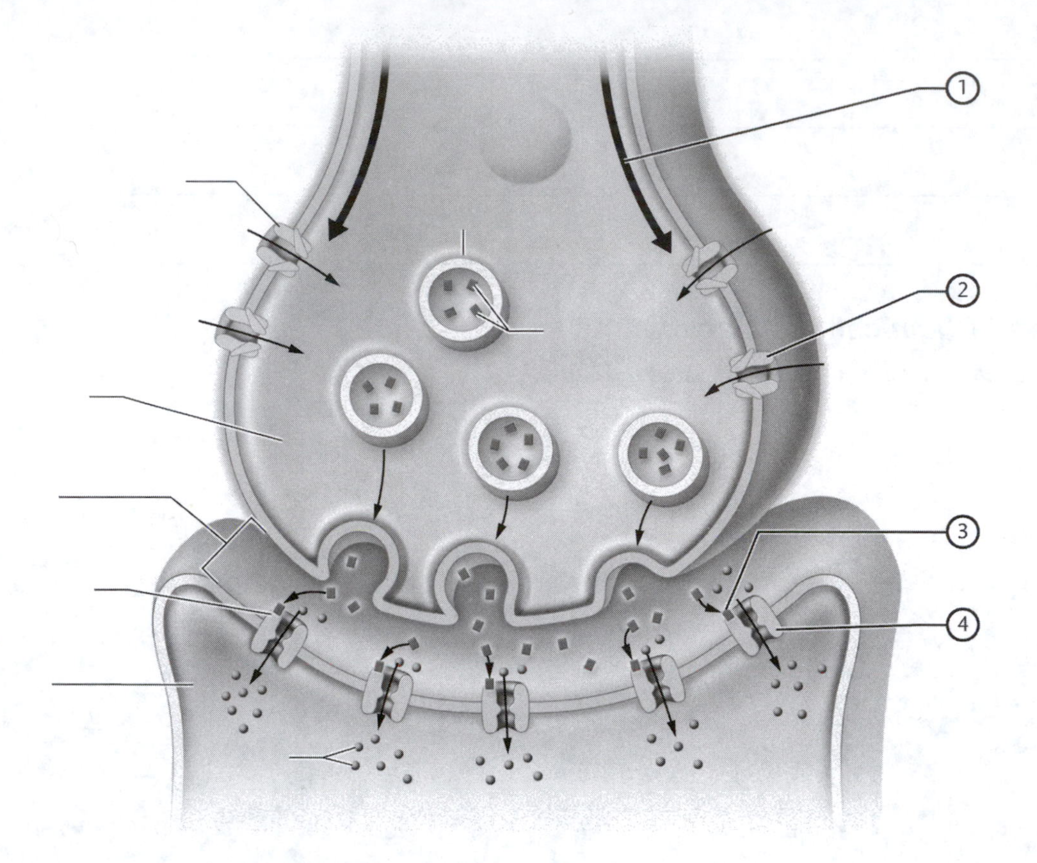

Figure 11.8 Events at a chemical synapse.

Draw It: Postsynaptic Potentials

Draw the graph line in the following empty graphs that show what will happen to a neuron's membrane potential when the following ion channels open.

a. Calcium ion channels open and calcium ions enter the cell.

b. Chloride ion channels open and chloride ions enter the cell.

c. Potassium ion channels open and potassium ions exit the cell.

Key Concept: Which type of postsynaptic potential makes an action potential more likely? Why? Which type makes an action potential less likely, and why?

Team Up

Make a handout to teach the big picture of chemical synaptic transmission. You can use Figure 11.25 in your text as a guide, but the handout should be in your own words and with your own diagram. At the end of the handout, write a few quiz questions. Once you have completed your handout, team up with one or more study partners, and trade handouts. Study your partners' diagrams, and when you have finished, take the quiz at the end. When you and your group have finished taking all the quizzes, discuss the answers to determine places where you need additional study. After you've finished, combine the best elements of each handout to make one "master" diagram for the big picture of chemical synaptic transmission.

Identify It: Temporal and Spatial Summation

Identify the two following graphs as showing either temporal or spatial summation (Figure 11.9). Explain how you arrived at your conclusions.

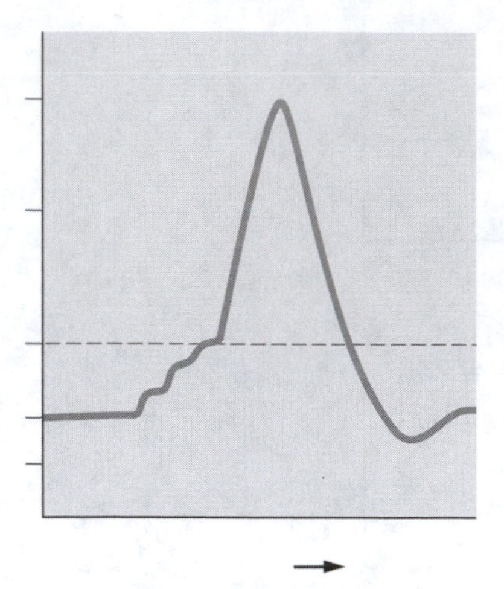

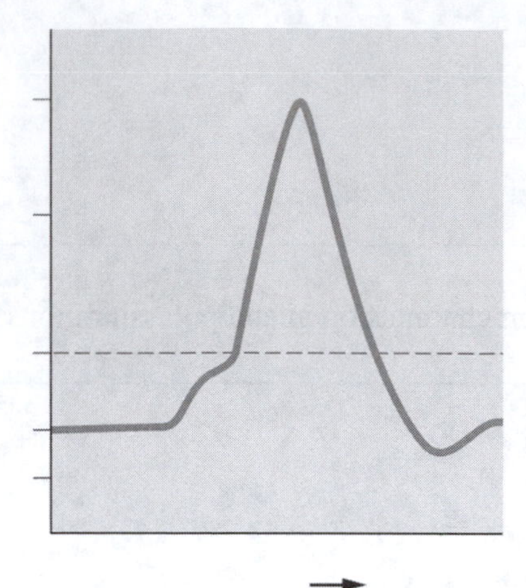

Figure 11.9 Types of summation.

Key Concept: How does summation connect local potentials and action potentials?

____

____

Module 11.5: Neurotransmitters

Module 11.5 covers the individual neurotransmitters that are involved in synaptic transmission. At the end of this module, you should be able to do the following:

1. Explain how a single neurotransmitter may be excitatory at one synapse and inhibitory at another.
2. Describe the structural and functional properties of the major classes of neurotransmitters.
3. Describe the most common excitatory and inhibitory neurotransmitters in the CNS.

Build Your Own Glossary

Following is a table listing key terms from Module 11.5. Before you read the module, use the glossary at the back of your book or look through the module to define the following terms.

Key Terms of Module 11.5

Term	Definition
Second messenger	
Excitatory neurotransmitter	
Inhibitory neurotransmitter	
Acetylcholine	
Biogenic amine	
Catecholamine	
Amino acid neurotransmitter	
Neuropeptide	

Survey It: Form Questions

Before you read the module, survey it and form at least three questions for yourself. When you have finished reading the module, return to these questions and answer them.

Question 1: ______________________________

Answer: ______________________________

Question 2: ______________________________

Answer: ______________________________

Question 3: ______________________________

Answer: ______________________________

Key Concept: What happens when a neurotransmitter binds a metabotropic receptor? How does this differ from the effects of binding an ionotropic receptor?

____

____

Key Concept: How can a neurotransmitter cause different effects on different neurons?

____

____

Build Your Own Summary Table: Neurotransmitters

Build your own summary table about the neurotransmitters by filling in the information in the template provided below. Don't simply copy Table 11.3 in the text; instead, fill in the information about each neurotransmitter as you read the module. When you have finished the module, compare your table to Table 11.3 and make any needed corrections.

Summary of Neurotransmitters

Neurotransmitter	Class	Location	Major Effects (Excitatory and/or inhibitory)	Receptor Type
Acetylcholine				
Epinephrine				
Norepinephrine				
Dopamine				
Serotonin				
Histamine				
Glutamate				
Glycine				
GABA				
Substance P				
Opioids				
Neuropeptide Y				

Module 11.6: Functional Groups of Neurons

This module examines how groups of neurons work together as groups called *neuronal pools*. At the end of this module, you should be able to do the following:

1. Define a neuronal pool, and explain its purpose.
2. Compare and contrast the two main types of neural circuits in the central nervous system.

Build Your Own Glossary

Following is a table listing key terms from Module 11.6. Before you read the module, use the glossary at the back of your book or look through the module to define the following terms.

Key Terms for Module 11.6

Term	Definition
Neuronal pool	
Neural circuit	
Diverging circuit	
Converging circuit	

Survey It: Form Questions

Before you read the module, survey it and form at least two questions for yourself. When you have finished reading the module, return to these questions and answer them.

Question 1: ______________________________

Answer: ______________________________

Question 2: ______________________________

Answer: ______________________________

Draw It: Neural Circuits

In the space provided, draw an example of a diverging circuit and a converging circuit. Then, label your drawings with the input and output neurons.

Key Concept: How is the structure of a neural circuit related to its function?

What Do You Know Now?

Let's now revisit the questions you answered in the beginning of this chapter. How have your answers changed now that you've worked through the material?

- What are the organs of the nervous system?

- What do the organs of the nervous system do?

- What are the types of cells within the nervous system?

12 The Central Nervous System

Our next chapter in the nervous system series covers the division known as the central nervous system (CNS). Here we explore the structure and functions of the two main organs of the CNS: the brain and the spinal cord.

What Do You Already Know?

Try to answer the following questions before proceeding to the next section. If you're unsure of the correct answers, give it your best attempt based on previous courses, previous chapters, or just your general knowledge.

- What percentage of our brains do we use?

  ____

- What protects the brain and spinal cord?

  ____

- Can people be "right-brained" or "left-brained?"

  ____

Module 12.1: Overview of the Central Nervous System

Module 12.1 in your text introduces you to the structural and functional features of the brain and the spinal cord and looks at the process of nervous system development. By the end of the module, you should be able to do the following:

1. Describe the structure and function of each major area of the brain.
2. Describe the five developmental regions of the brain, and identify the major areas of the adult brain that arise from each region.

Build Your Own Glossary

Following is a table listing key terms from Module 12.1. Before you read the module, use the glossary at the back of your book or look through the module to define the following terms.

Key Terms for Module 12.1

Term	Definition
Brain	
Cerebrum	
Diencephalon	
Cerebellum	
Brainstem	
Spinal cord	
White matter	
Gray matter	
Tract	
Nuclei	

Survey It: Form Questions

Before you read the module, survey it and form at least two questions for yourself. When you have finished reading the module, return to these questions and answer them.

Question 1: ______________________________

Answer: ______________________________

Question 2: ______________________________

Answer: ______________________________

Identify It: Divisions of the Brain

Identify and color-code each division of the brain in Figure 12.1. Then, in the boxes on the right, list the main functions of each division.

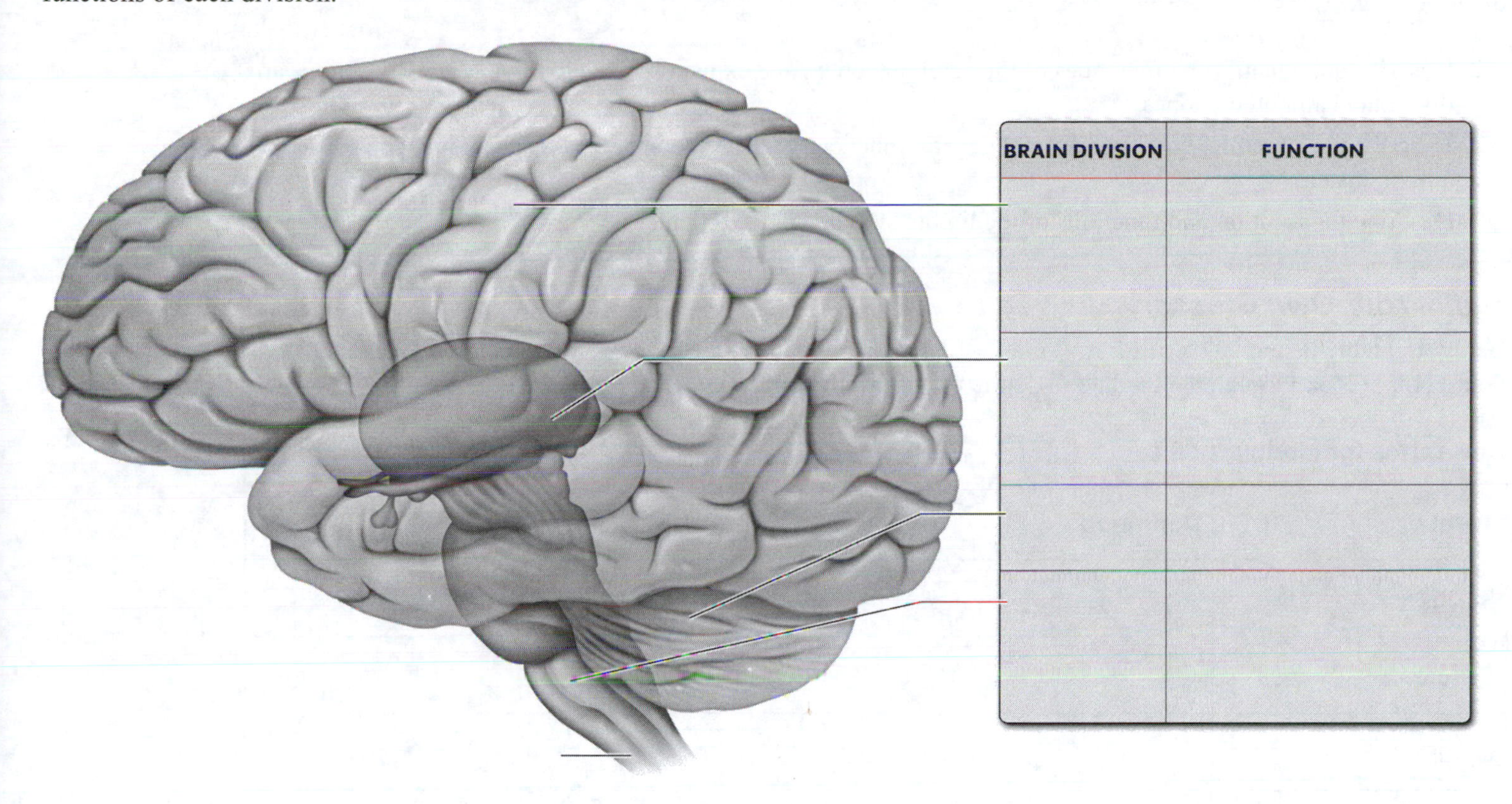

BRAIN DIVISION	FUNCTION

Figure 12.1 Divisions of the brain.

Key Concept: How do gray matter and white matter differ? Why are they different colors?

____

____

Key Concept: What are the early nervous system structures called, and how do they change as development takes place?

____

____

Module 12.2: The Brain

Now we take a closer look at the structure of the brain. When you finish this module, you should be able to do the following:

1. Describe and identify the five lobes of the cerebral cortex, and explain how motor and sensory functions are distributed among the lobes.
2. Describe the structure, components, and general functions of the regions of the diencephalon, cerebellum, and brainstem.
3. Describe the location and functions of the limbic system and the reticular formation.

Build Your Own Glossary

Below is a table listing key terms from Module 12.2. Before you read the module, use the glossary at the back of your book or look through the module to define the following terms.

Key Terms for Module 12.2

Term	Definition
Sulcus	
Fissure	
Gyrus	
Cerebral cortex	
Association area	
Basal nuclei	
Corpus callosum	
Limbic system	
Thalamus	
Hypothalamus	
Midbrain	
Pons	
Medulla oblongata	
Reticular formation	

Survey It: Form Questions

Before you read the module, survey it and form at least two questions for yourself. When you have finished reading the module, return to these questions and answer them.

Question 1: ______________________________

Answer: ______________________________

Question 2: ______________________________

Answer: ______________________________

Key Concept: What is the advantage of having the surface of the brain folded into gyri and sulci?

____

____

Identify It: Lobes and Area of the Cerebrum

Identify and color-code each lobe and area of the cerebrum in Figure 12.2. Also, add your own leaders and label the following: gyrus, sulcus, central sulcus, lateral fissure, precentral gyrus, and postcentral gyrus. Then, identify the main function(s) of each.

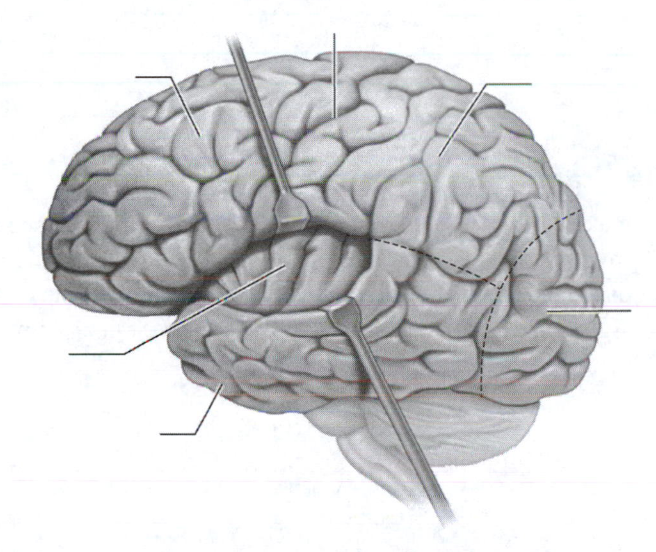

(a) Lateral view (frontal, parietal, and temporal lobes pulled back)

Figure 12.2 The cerebrum.

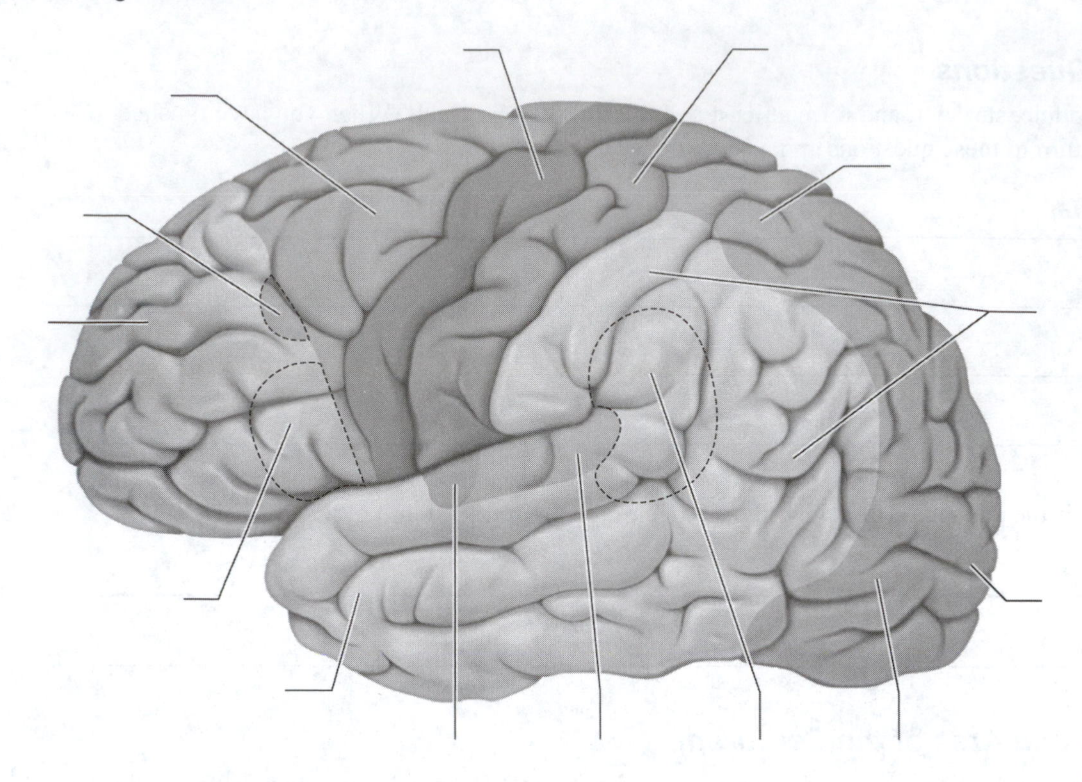

(b) Areas of the cerebrum

Figure 12.2 *(continued)*

Build Your Own Summary Table: The Brain

There are a *lot* of structures in the brain, and that makes it easy to get overwhelmed. To help yourself see the big picture and focus on the main points, fill in this summary table as you read. In the first column, label the brain structure and, in the second, list its functions. A completed table (text Figure 12.17) can be seen at the end of the module. When you finish, compare your table to Figure 12.17 to ensure that you didn't leave out any important details.

Summary of Brain Structures and Functions

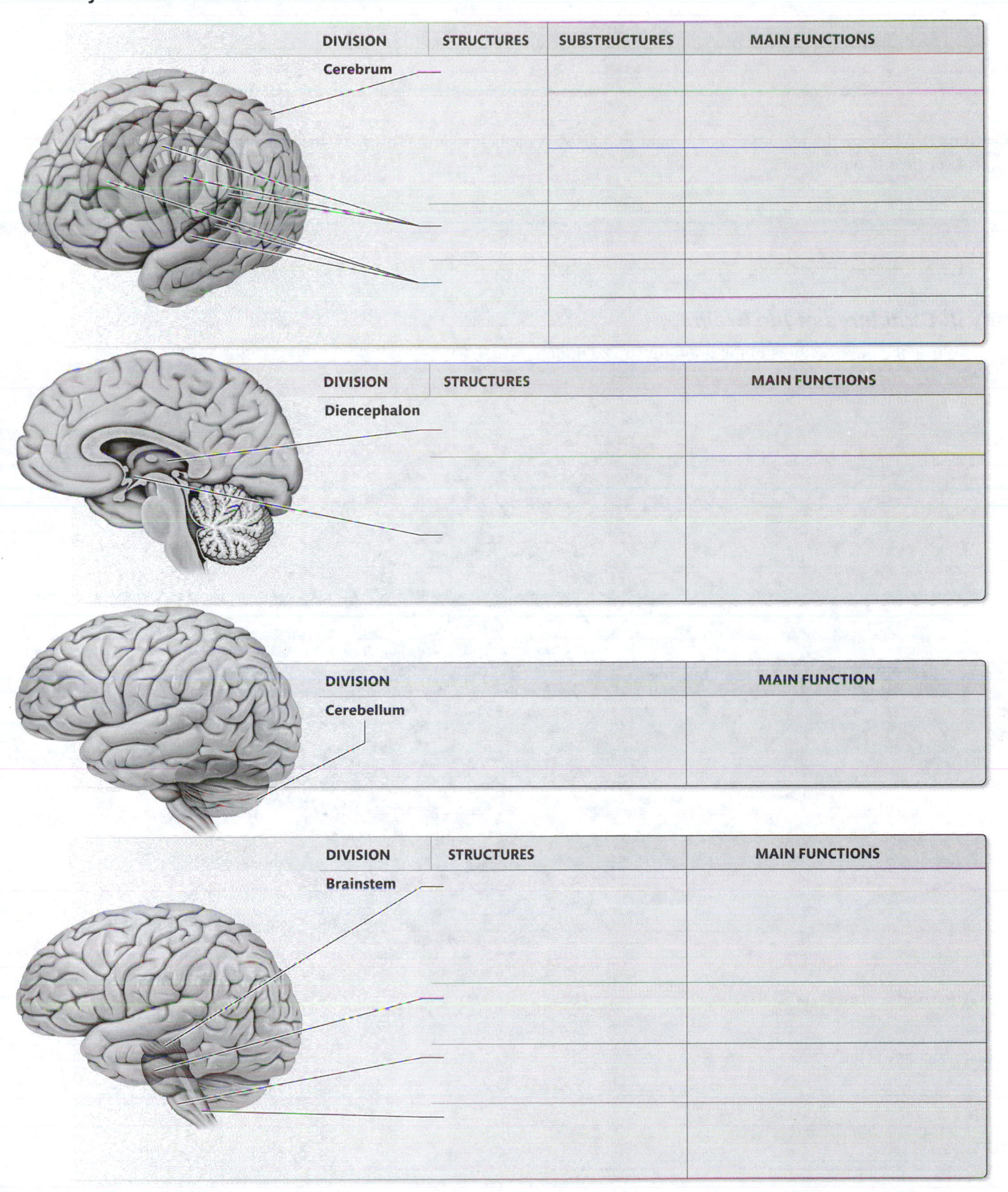

DIVISION	STRUCTURES	SUBSTRUCTURES	MAIN FUNCTIONS
Cerebrum			

DIVISION	STRUCTURES	MAIN FUNCTIONS
Diencephalon		

DIVISION	MAIN FUNCTION
Cerebellum	

DIVISION	STRUCTURES	MAIN FUNCTIONS
Brainstem		

Key Concept: Why is the thalamus often called the "gateway" to the cerebrum?

____

____

Key Concept: What are the main types of structures housed in the brainstem? Why does this make the brainstem so critical to our survival?

____

____

Identify It: Structures of the Brain

Label and color-code the structures of the brain illustrated in Figure 12.3.

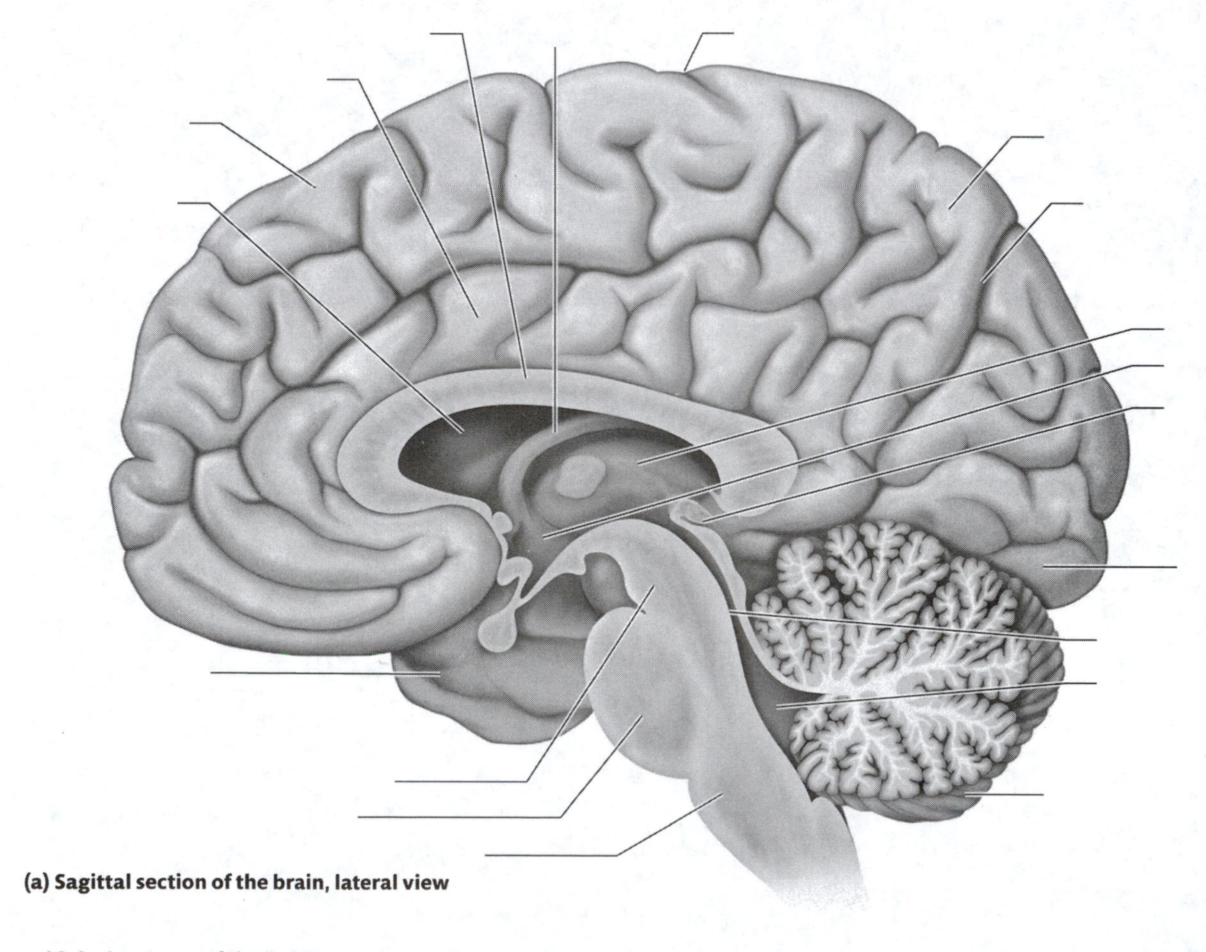

Figure 12.3 Anatomy of the brain.

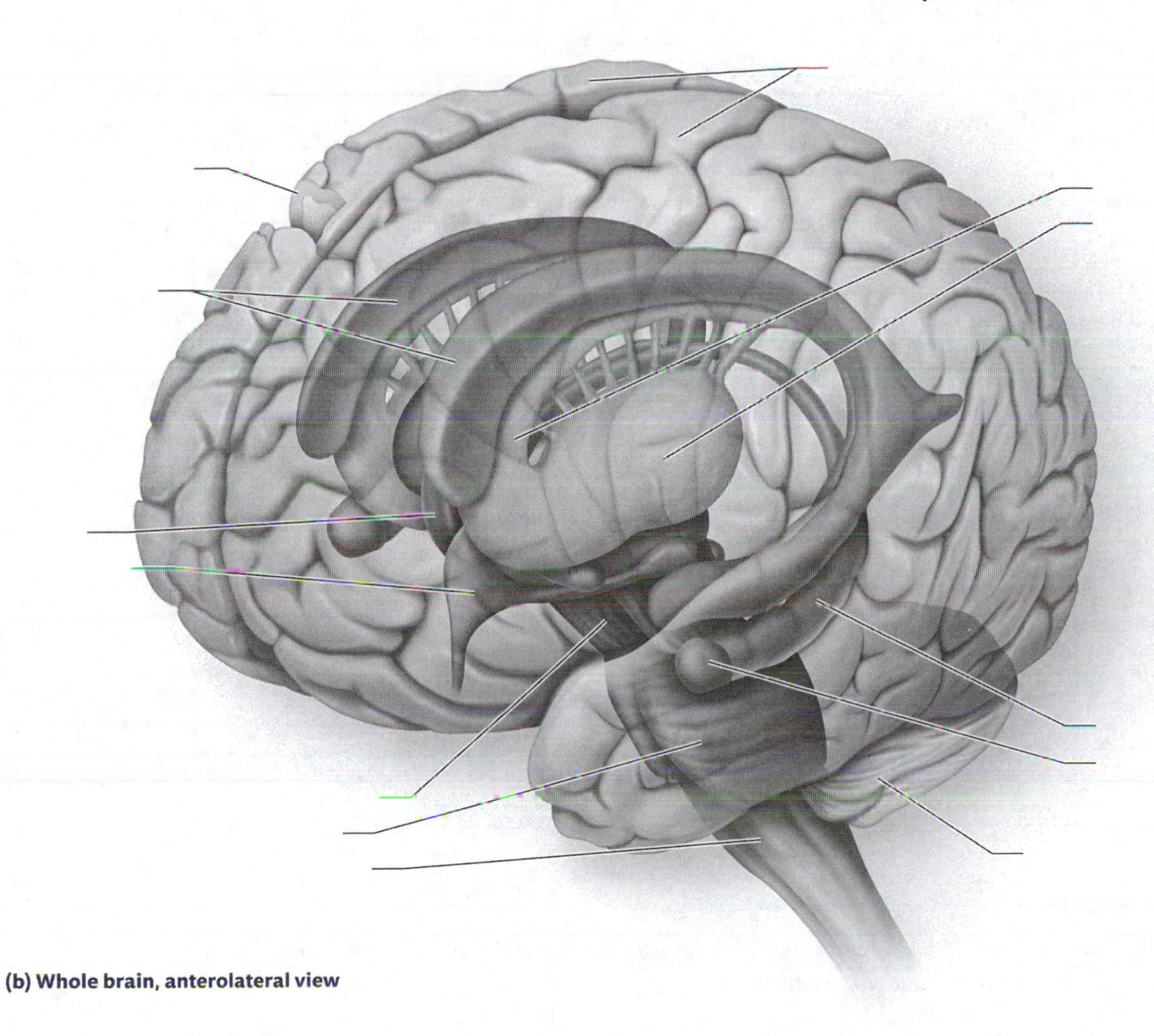

(b) Whole brain, anterolateral view

Figure 12.3 (*continued*)

Module 12.3: Homeostasis Part I: Role of the Brain in Maintenance of Homeostasis

The nervous system is one of the major regulatory systems in the body and is tasked with maintaining homeostasis of other organ systems. We take a look at the role of the CNS in these regulatory processes in this module. The discussion of homeostasis continues in the chapter on the autonomic nervous system. At the end of this module, you should be able to do the following:

1. Describe the differences between the endocrine system and the nervous system in terms of how they maintain homeostasis.
2. Provide specific examples demonstrating how the nervous system responds to maintain homeostasis in the body.

Build Your Own Glossary

Following is a table listing key terms from Module 12.3. Before you read the module, use the glossary at the back of your book or look through the module to define the following terms.

Key Terms of Module 12.3

Term	Definition
Autonomic nervous system	
Vasopressor center	
Vasodepressor center	
Sleep	
Circadian rhythm	
Beta waves	
Theta waves	
Delta waves	
REM sleep	

Survey It: Form Questions

Before you read the module, survey it and form at least three questions for yourself. When you have finished reading the module, return to these questions and answer them.

Question 1: ______________________________

Answer: ______________________________

Question 2: ______________________________

Answer: ______________________________

Question 3: ______________________________

Answer: ______________________________

Key Concept: Which two components of the CNS are largely responsible for homeostatic regulation?

____

____

Describe It: CNS Control of Homeostasis

Write a paragraph describing the mechanisms of control of homeostasis by the CNS as if you were teaching the topic to a group of your fellow students.

Key Concept: Which parts of the brain are involved in sleep and circadian rhythms? How is regulation of sleep and circadian rhythms connected to the level of light?

___

___

Identify It: Stages of Wakefulness and Sleep

Identify and color-code each component of the graph of the stages of wakefulness and sleep in Figure 12.4.

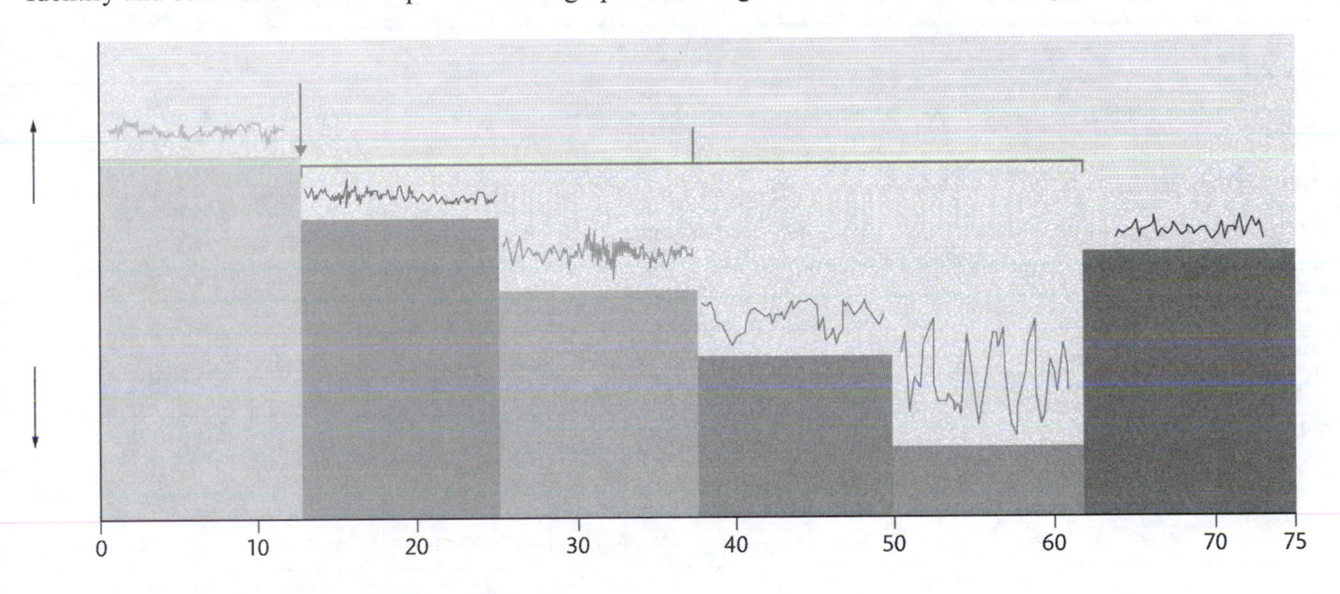

Figure 12.4 The stages of wakefulness and sleep.

Module 12.4: Higher Mental Functions

We now explore the higher brain functions, including cognition, language, learning, memory, and emotions. At the end of this module, you should be able to do the following:

1. Describe the areas of the cortex responsible for cognition and language.
2. Discuss the concept of cerebral hemispheric specialization.
3. Describe the parts of the brain involved in storage of long-term memory, and discuss possible mechanisms of memory consolidation.

Build Your Own Glossary

Below is a table listing key terms from Module 12.4. Before you read the module, use the glossary at the back of your book or look through the module to define the following terms.

Key Terms of Module 12.4

Term	Definition
Cognition	
Cerebral lateralization	
Language	
Declarative memory	
Nondeclarative memory	
Consolidation	
Long-term potentiation	

Survey It: Form Questions

Before you read the module, survey it and form at least two questions for yourself. When you have finished reading the module, return to these questions and answer them.

Question 1: ______________________________

Answer: ______________________________

Question 2: ______________________________

Answer: ______________________________

Key Concept: What are cognitive functions, and which parts of the brain are responsible for executing these functions?

____

____

Key Concept: What is cerebral lateralization? Can people be "right-brained" or "left-brained?"

____

____

Complete It: Language, Learning, Memory, and Emotion

Fill in the blanks to complete the following paragraphs that describe the role of the brain in language, learning, memory, and emotion.

Language functions tend to be lateralized to the ____________ ____________. The two language areas are ____________ ____________ and ____________ ____________. The first area is responsible for the ____________ ____________ ____________, whereas the second is responsible for the ____________ ____________ ____________.

Memories that are fact-based are called ____________ ____________. These memories are stored in the brain via the ____________ and a process called ____________ ____________ ____________. Memories that are procedure-based are called ____________ ____________. The storage of these memories seems to involve the ____________ ____________, the ____________, and the ____________ ____________.

The visceral response to emotion seems to involve the ____________, whereas the somatic response is mediated by this part of the brain and the ____________ ____________. The "feeling" component of emotion involves the part of the brain known as the ____________.

Module 12.5: Protection of the Brain

The delicate tissues of the brain are protected in several ways, which we examine in this module. When you complete it, you should be able to do the following:

1. Describe the functions of cerebrospinal fluid as well as the details of its production, its circulation within the CNS, and its ultimate reabsorption into the bloodstream.
2. Describe the structural basis for and the importance of the blood brain barrier.
3. Identify and describe the cranial meninges, and explain their functional relationship to the brain.

Build Your Own Glossary

Following is a table listing key terms from Module 12.5. Before you read the module, use the glossary at the back of your book or look through the module to define the following terms.

Key Terms of Module 12.5

Term	Definition
Cranial meninges	
Dura mater	
Arachnoid mater	
Pia mater	
Dural sinuses	
Lateral ventricles	
Third ventricle	
Fourth ventricle	
Cerebrospinal fluid	
Choroid plexuses	
Blood brain barrier	

Survey It: Form Questions

Before you read the module, survey it and form at least four questions for yourself. When you have finished reading the module, return to these questions and answer them.

Question 1: ______________________________

Answer: ______________________________

Question 2: ______________________________

Answer: ______________________________

Question 3: ______________________________

Answer: ______________________________

Question 4: ______________________________

Answer: ______________________________

Identify It: Cranial Meninges

Identify and color-code the structures associated with the cranial meninges illustrated in Figure 12.5.

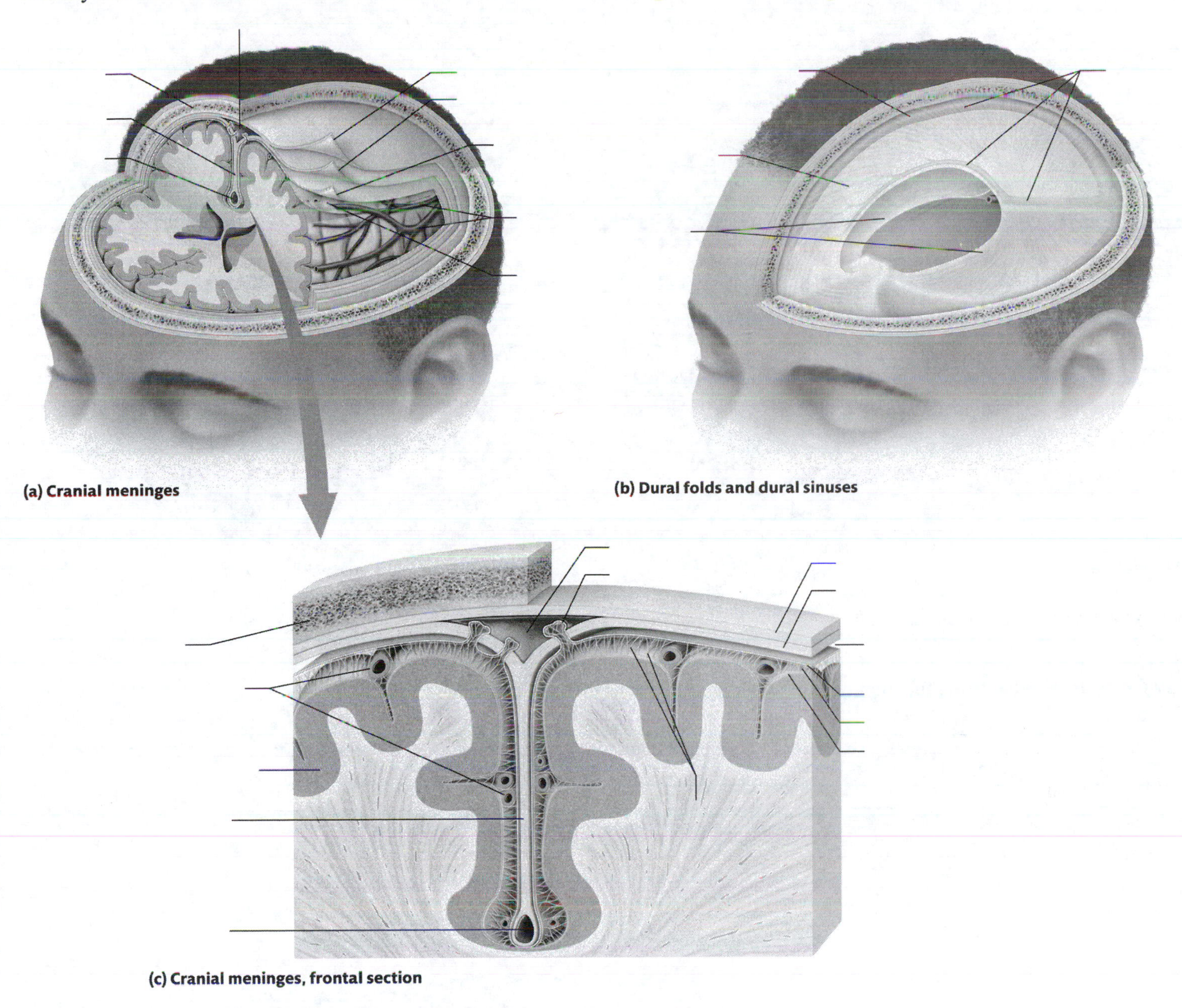

Figure 12.5 The cranial meninges and dural sinuses.

Key Concept: Why is there no true epidural space around the brain? What is located in the subdural and subarachnoid spaces?

____

____

Identify It: Ventricles of the Brain

Identify and color-code the ventricles of the brain illustrated in Figure 12.6.

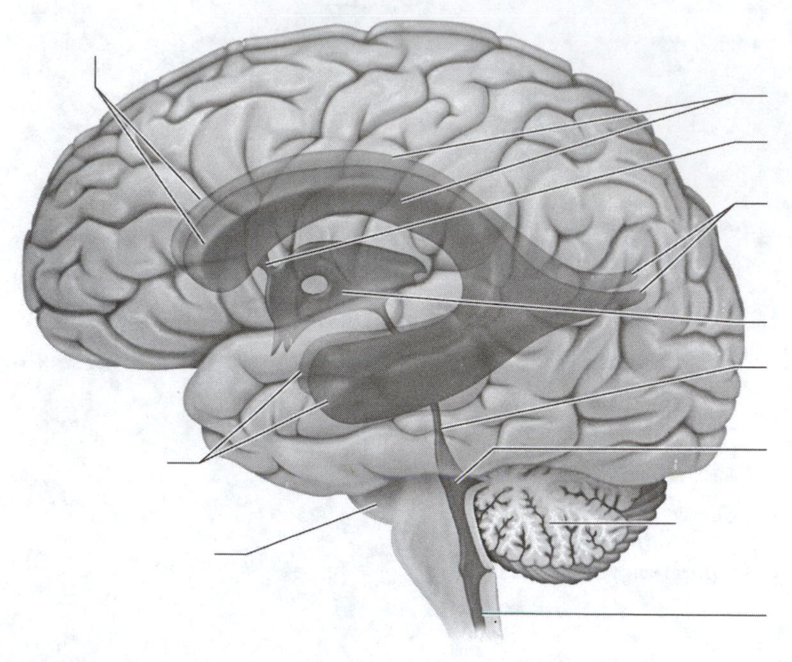

(a) Lateral view

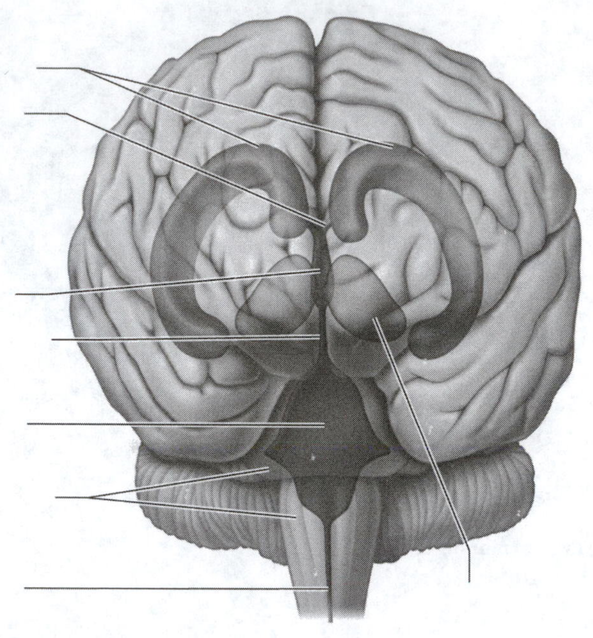

(b) Anterior view

Figure 12.6 Ventricles of the brain.

Key Concept: What is the function of CSF and how is it produced?

____

____

Draw It: CSF Circulation

Draw arrows to trace the pathway of CSF circulation through the brain illustrated in Figure 12.7. Then, label the structures and describe the steps of CSF formation, circulation, and reabsorption.

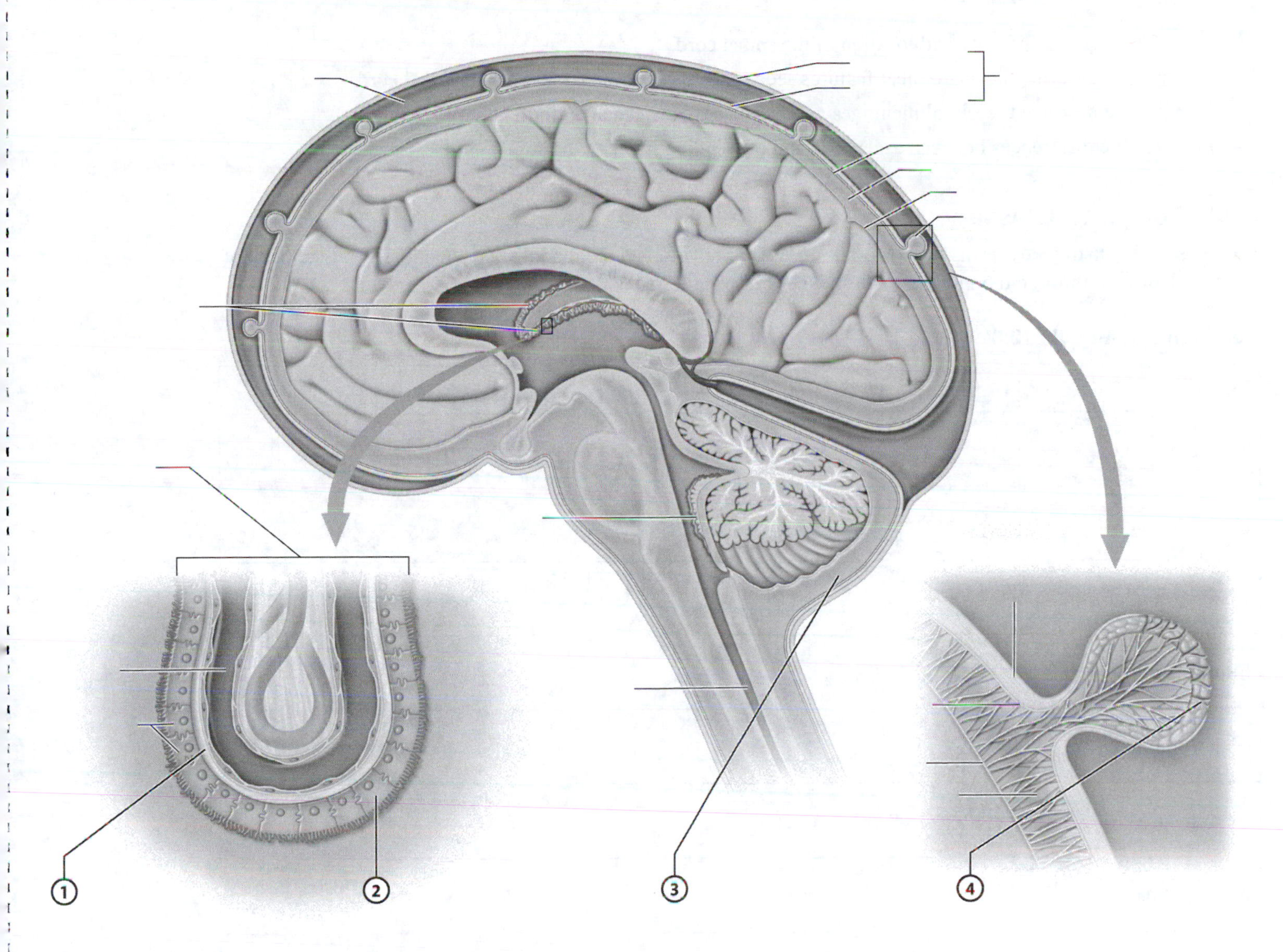

Figure 12.7 Formation and flow of cerebrospinal fluid.

Module 12.6: The Spinal Cord

Module 12.6 in your text explores the structure of the spinal cord. At the end of this module, you should be able to do the following:

1. Describe the gross anatomy and location of the spinal cord.
2. Identify and describe the anatomical features seen in a cross-sectional view of the spinal cord.
3. Identify and describe the spinal meninges and the spaces between and around them.
4. Describe the differences between ascending and descending tracts in the spinal cord.

Build Your Own Glossary

Below is a table listing key terms from Module 12.6. Before you read the module, use the glossary at the back of your book or look through the module to define the following terms.

Key Terms of Module 12.6

Term	Definition
Spinal meninges	
Epidural space	
Conus medullaris	
Filum terminale	
Cauda equina	
Central canal	
Anterior horn	
Lateral horn	
Posterior horn	
Funiculus	

Survey It: Form Questions

Before you read the module, survey it and form at least three questions for yourself. When you have finished reading the module, return to these questions and answer them.

Question 1: ______________________________

Answer: ______________________________

Question 2: ______________________________

Answer: ______________________________

Question 3: ______________________________

Answer: ______________________________

Identify It: Structure of the Spinal Meninges

Identify and color-code each component of the spinal meninges illustrated in Figure 12.8. Then, write the function of each part under its label.

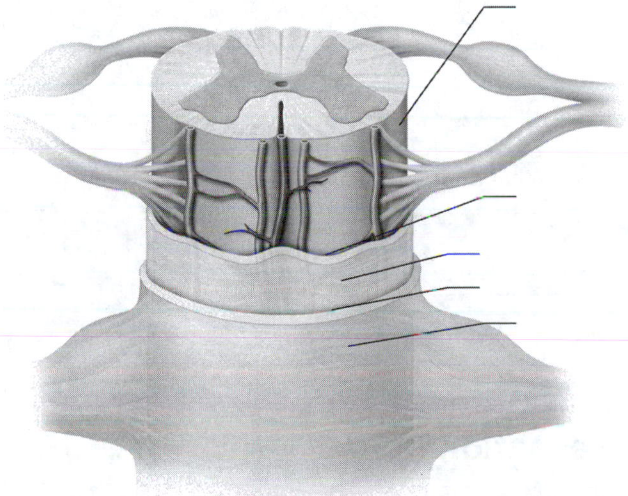

(a) Spinal meninges and spinal cord, anterior view

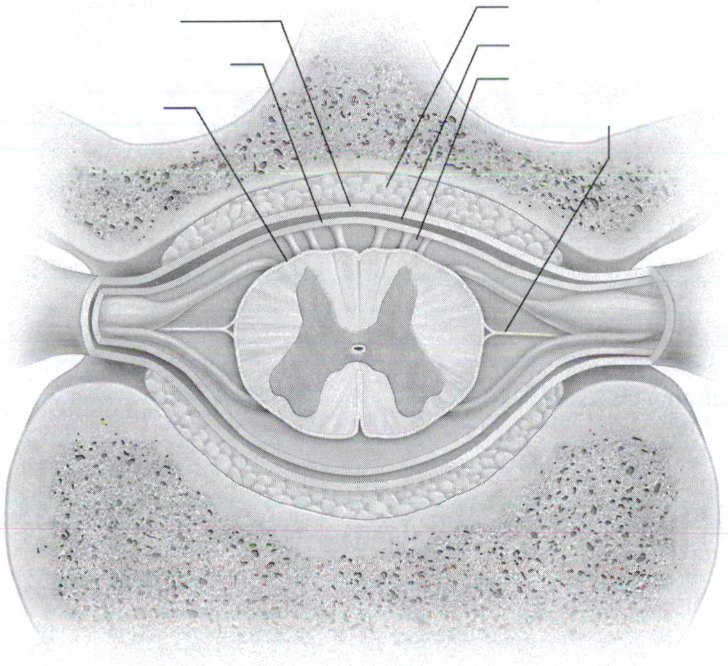

(b) Spinal meninges and spinal cord, transverse section

Figure 12.8 Structure of the spinal meninges.

Key Concept: What are the key differences between the cranial and spinal meninges?

____

____

Draw It: Internal Anatomy of the Spinal Cord

In the space below, draw an outline of a spinal cord transverse section, as shown in text Figures 12.28 and 12.29. Then, draw in the gray matter, central canal, nerve roots, and ascending and descending tracts. After you have drawn each part, color-code and label them.

Key Concept: What types of stimuli do the spinal cord's ascending tracts carry? What do the spinal cord's descending tracts carry?

____

____

Module 12.7: Sensation Part I: Role of the CNS in Sensation

Sensation is the process of detecting stimuli in the internal and external environments. In this module, we look at the role that the central nervous system plays in sensation; our discussion of sensation continues in the peripheral nervous system (PNS) chapter. At the end of this module, you should be able to do the following:

1. Describe the roles of the central and peripheral nervous systems in processing sensory stimuli.
2. Describe the locations and functions of first-, second-, and third-order neurons in a sensory pathway.
3. Explain the ways in which special sensory stimuli are processed by the CNS.

Build Your Own Glossary

Following is a table listing key terms from Module 12.7. Before you read the module, use the glossary at the back of your book or look through the module to define the following terms.

Key Terms of Module 12.7

Term	Definition
Sensory stimuli	
Perception	
Tactile senses	
Nondiscriminative touch	
Posterior columns/ medial lemniscus tracts	
Anterolateral system	
Spinothalamic tracts	
Somatotopy	
Nociception	

Survey It: Form Questions

Before you read the module, survey it and form at least three questions for yourself. When you have finished reading the module, return to these questions and answer them.

Question 1: ______________________________

Answer: ______________________________

Question 2: ______________________________

Answer: ______________________________

Question 3: ______________________________

Answer: ______________________________

Key Concept: What are discriminative touch and nondiscriminative touch? Which tract(s) carry these stimuli?

____

____

Describe It: The Posterior Columns

Label and color-code the main structures involved in the posterior columns/medial lemniscal system in Figure 12.9. Then, write out each step of information transfer from the initial stimulus to its delivery to the cerebral cortex.

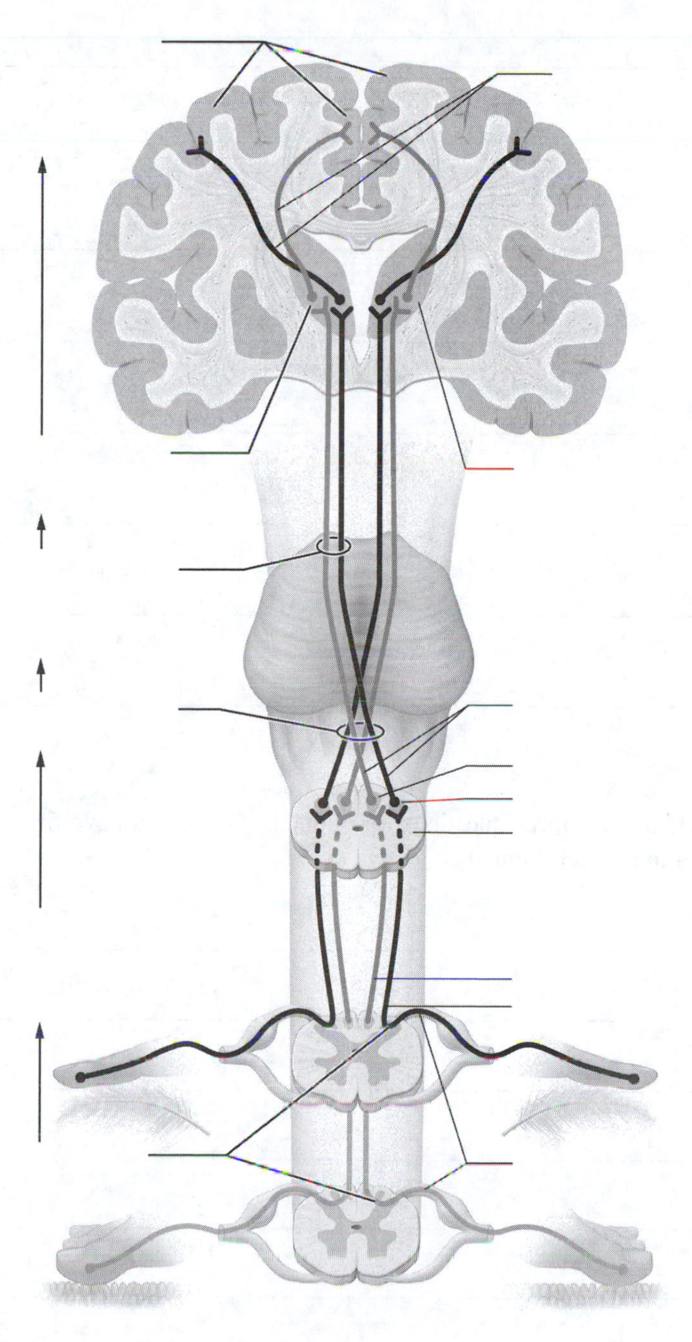

Figure 12.9 The posterior columns/medial lemniscal system.

Describe It: The Spinothalamic Tract (Anterolateral System)

Label and color-code the main structures involved in the spinothalamic tract (anterolateral system) in Figure 12.10. Then, write out each step of information transfer from the initial stimulus to its delivery to the cerebral cortex.

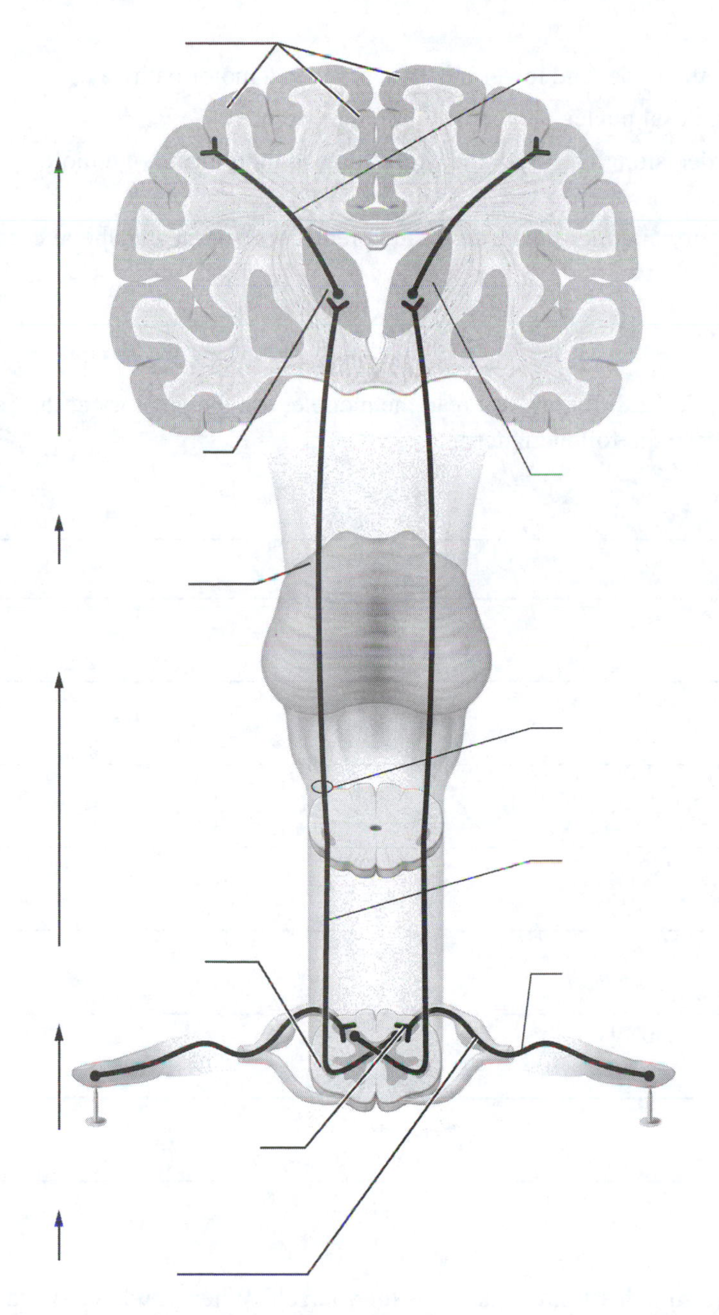

Figure 12.10 The spinothalamic tract (part of the anterolateral system).

Key Concept: What are the key structural and functional differences between the posterior columns and anterolateral system?

Module 12.8: Movement Part I: Role of the CNS in Voluntary Movement

Like sensation, movement involves both the PNS and CNS. We examine the role of the CNS in movement in this module and will continue this discussion in the PNS chapter. At the end of this module, you should be able to do the following:

1. Describe the locations and functions of the upper and lower motor neurons in a motor pathway.
2. Explain the roles of the cerebral cortex, basal nuclei, and cerebellum in movement.
3. Describe the overall pathway from the decision to move to the execution and monitoring of a motor program.
4. Explain how decussation occurs in sensory and motor pathways, and predict how brain and spinal cord injuries affect these pathways.

Build Your Own Glossary

Below is a table listing key terms from Module 12.8. Before you read the module, use the glossary at the back of your book or look through the module to define the following terms.

Key Terms of Module 12.8

Term	Definition
Corticospinal tracts	
Motor program	
Upper motor neurons	
Globus pallidus	
Caudate nucleus	
Putamen	
Motor learning	

Survey It: Form Questions

Before you read the module, survey it and form at least three questions for yourself. When you have finished reading the module, return to these questions and answer them.

Question 1: ______________________________

Answer: ______________________________

Question 2: ______________________________

Answer: ______________________________

Question 3: ______________________________

Answer: ______________________________

Key Concept: How do upper motor neurons and lower motor neurons differ?

____

____

Describe It: The Corticospinal Tract

Label and color-code the main structures involved in the corticospinal tract in Figure 12.11. Then, write out each step of information transfer from the initial stimulus to its delivery to the lower motor neuron.

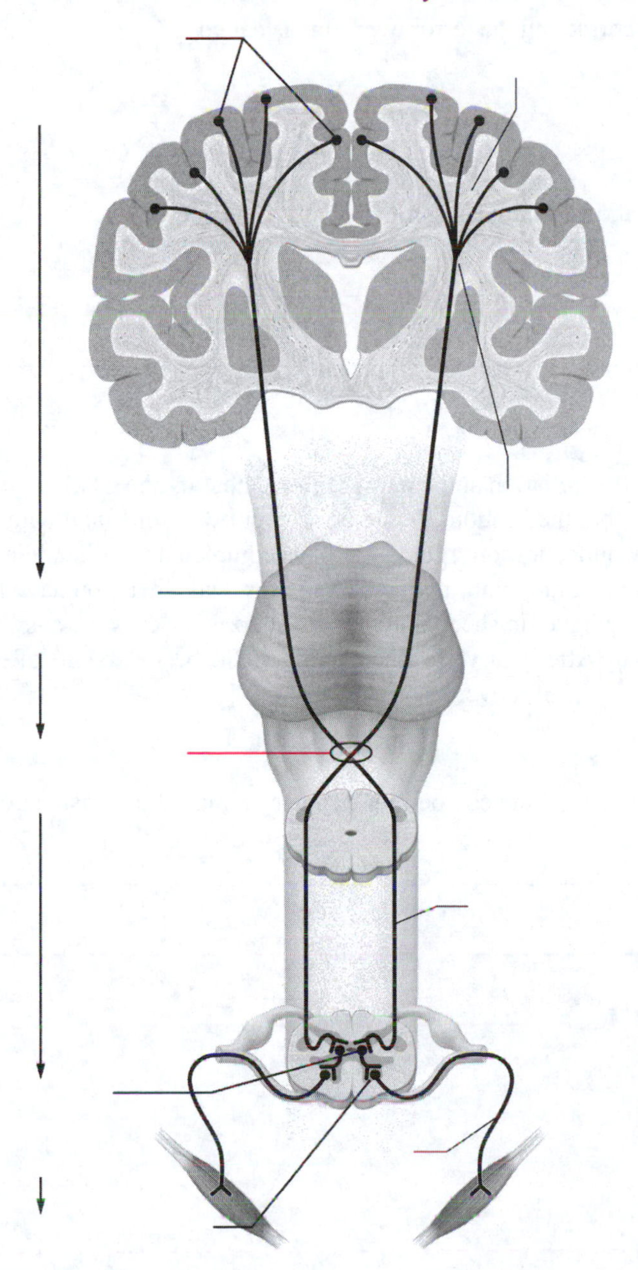

Figure 12.11 The corticospinal tract.

Complete It: The Brain and Movement

Fill in the blanks to complete the following paragraph that describes the role of the brain in movement.

Like the primary somatosensory cortex, the primary motor cortex is organized ____________ and can be represented by the ____________ ____________. Smooth, fluid movement requires the activity of the ____________ ____________, including the putamen, caudate nucleus, and putamen, and the ____________ ____________ of the midbrain. The basal nuclei modify the activity of ____________ ____________. The cerebellum is also required for smooth movement, as it determines ____________ ____________. The process of correcting this error over time is called ____________ ____________.

Key Concept: What are the two key roles of the basal nuclei in movement?

____

____

Team Up

Make a handout to teach the overall pathway of the initiation and control of movement by the CNS. You can use Figure 12.37 in your text on page 469 as a guide, but the handout should be in your own words and with your own diagrams. At the end of the handout, write a few quiz questions. Once you have completed your handout, team up with one or more study partners and trade handouts. Study your partners' diagrams, and when you have finished, take the quiz at the end. When you and your group have finished taking all the quizzes, discuss the answers to determine places where you need additional study. After you've finished, combine the best elements of each handout to make one "master" diagram for the control of movement by the CNS.

Key Concept: How does the control of movement by the cerebellum differ from that of the basal nuclei?

____

____

What Do You Know Now?

Let's now revisit the questions you answered in the beginning of this chapter. How have your answers changed now that you've worked through the material?

- What percentage of our brains do we use?

  ____

- What protects the brain and spinal cord?

  ____

- Can people be "right-brained" or "left-brained?"

  ____

13 The Peripheral Nervous System

This chapter examines the second major division of the nervous system: the peripheral nervous system, or PNS. Here we explore the structure and function of the spinal and cranial nerves, how the CNS and PNS work together to detect and perceive sensory stimuli, and how they work together to control movement. Finally, we examine how sensory and motor functions are connected via reflex arcs.

What Do You Already Know?

Try to answer the following questions before proceeding to the next section. If you're unsure of the correct answers, give it your best attempt based on previous courses, previous chapters, or just your general knowledge.

- What is the longest spinal nerve in the body?

  ____

- Why do you blink when something touches your eye?

  ____

- What is the purpose of the knee-jerk reflex?

  ____

Module 13.1: Overview of the Peripheral Nervous System

Module 13.1 in your text introduces you to the divisions of the PNS and the structure of its main organs, the peripheral nerves. By the end of the module, you should be able to do the following:

1. Explain the differences between the sensory and motor divisions of the peripheral nervous system.
2. Differentiate between the somatic motor and visceral motor (autonomic) divisions of the nervous system.
3. Describe the structure of a peripheral nerve, and explain the differences between spinal nerves and cranial nerves.

Build Your Own Glossary

Following is a table listing key terms from Module 13.1. Before you read the module, use the glossary at the back of your book or look through the module to define the following terms.

Key Terms for Module 13.1

Term	Definition
Somatic sensory division	
Visceral sensory division	
Somatic motor division	
Visceral motor division	
Peripheral nerve	
Mixed nerve	
Cranial nerve	
Spinal nerve	
Epineurium	
Fascicle	
Perineurium	
Endoneurium	

Survey It: Form Questions

Before you read the module, survey it and form at least two questions for yourself. When you have finished reading the module, return to these questions and answer them.

Question 1: ______________________________

Answer: ______________________________

Question 2: ______________________________

Answer: ______________________________

Key Concept: What is the fundamental difference between the terms "somatic" and "visceral"?

____

____

Identify It: Structure of a Nerve

Identify and color-code each component of the peripheral nerve in Figure 13.1.

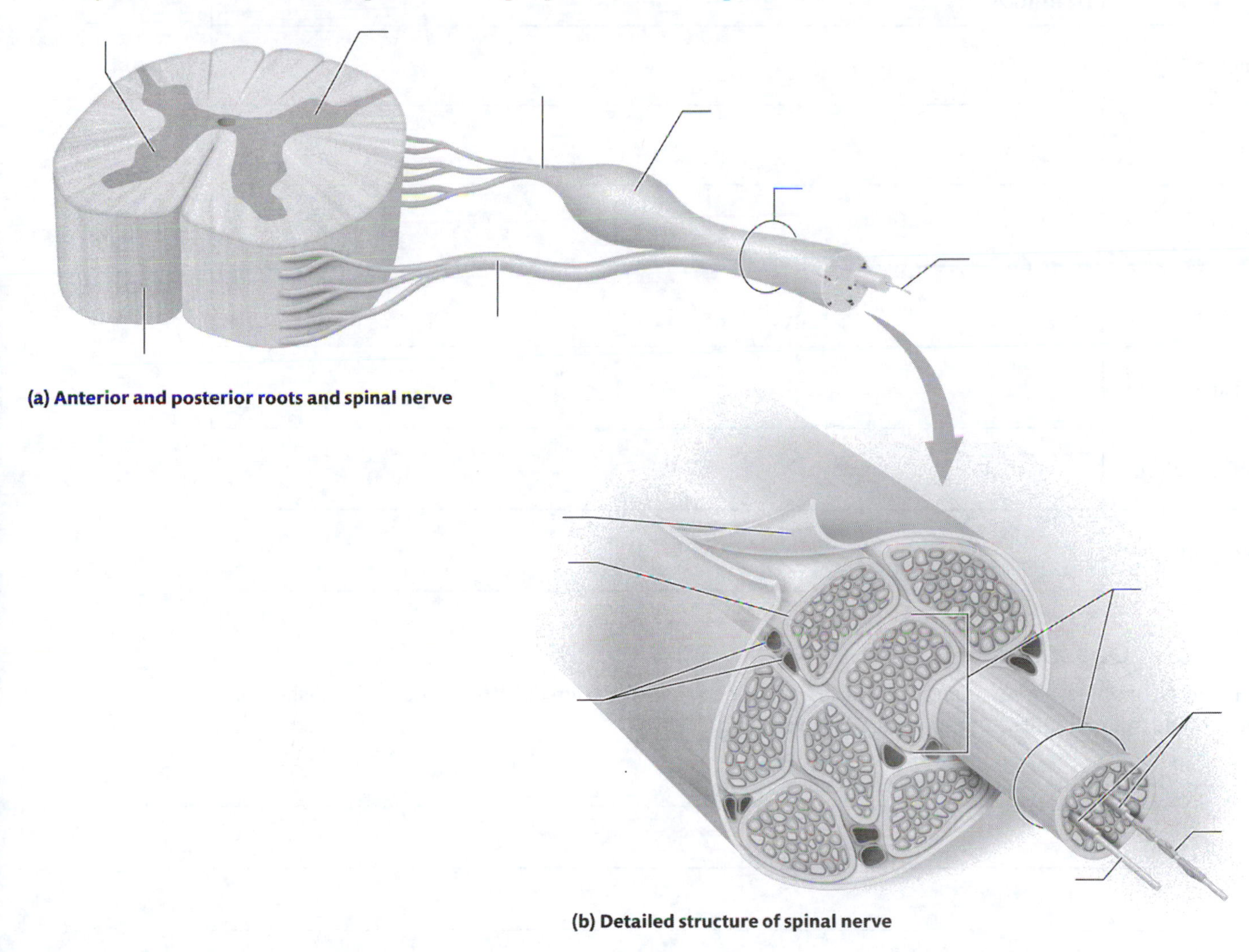

Figure 13.1 The structure of a peripheral nerve.

Module 13.2: The Cranial Nerves

Now we look at the anatomy and physiology of the 12 pairs of cranial nerves. When you finish this module, you should be able to do the following:

1. Identify the cranial nerves by name and number.
2. Describe the specific functions of each pair of cranial nerves, and classify each pair as sensory, motor, or mixed nerves.
3. Describe the locations of selected cranial nerve nuclei and the ganglia associated with the cranial nerves.

Build Your Own Glossary

Following is a table listing key terms from Module 13.2. Before you read the module, use the glossary at the back of your book or look through the module to define the following terms.

Key Terms for Module 13.2

Term	Definition
Sensory nerve	
Motor nerve	
Mixed nerve	
Olfactory bulb	
Optic chiasma	
Trigeminal ganglion	
Geniculate ganglion	

Survey It: Form Questions

Before you read the module, survey it and form at least four questions for yourself. When you have finished reading the module, return to these questions and answer them.

Question 1: ____________________

Answer: ____________________

Question 2: ____________________

Answer: ____________________

Question 3: ____________________

Answer: ____________________

Question 4: ____________________

Answer: ____________________

Identify It: The Cranial Nerves

Identify and color-code each cranial nerve in Figure 13.2. Then, list the main structure(s) innervated by each nerve. Then, add arrows along each nerve showing the direction of information flow consistent with designation of the nerve as sensory, motor, or mixed.

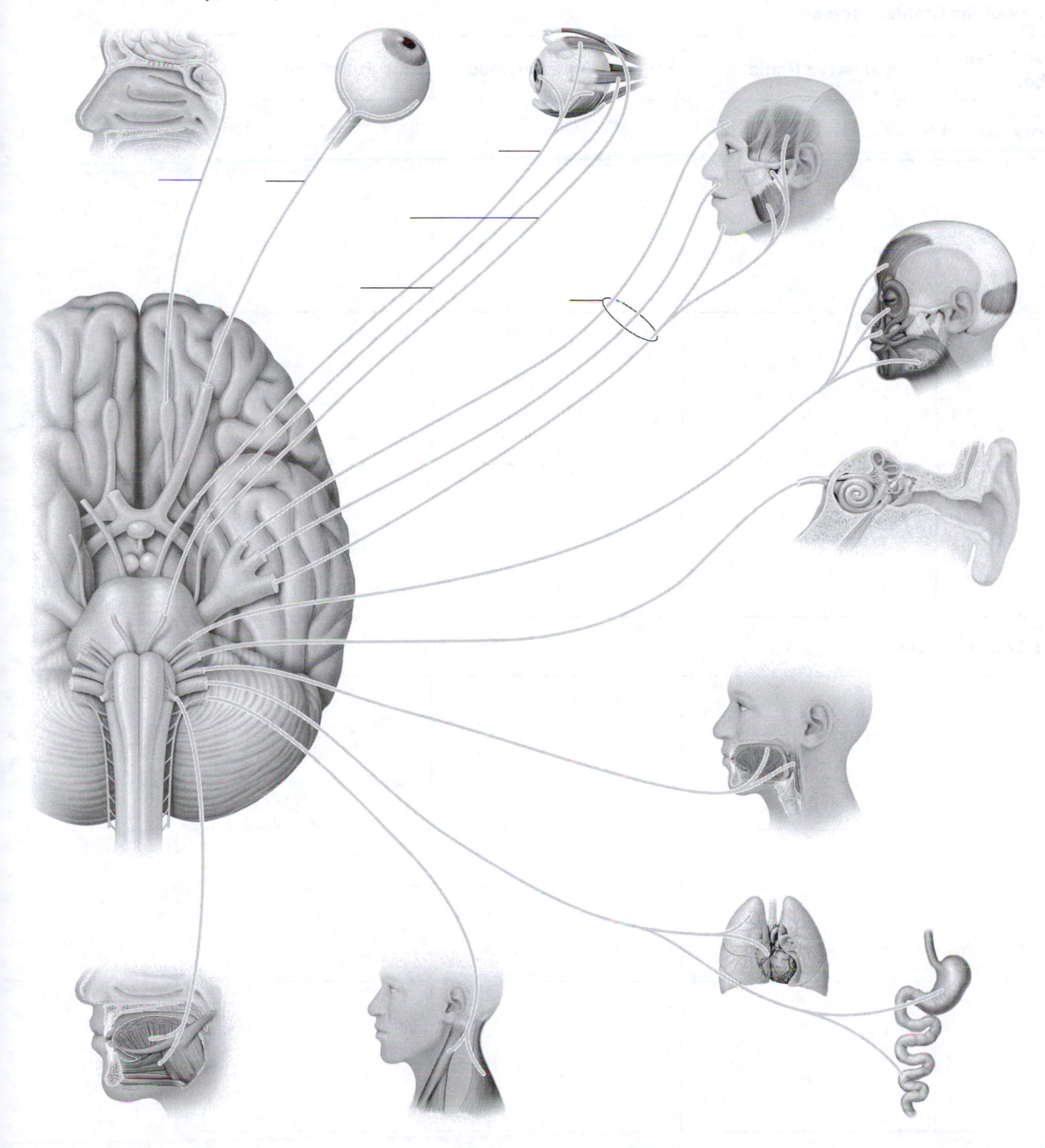

Figure 13.2 Overview of cranial nerves.

Build Your Own Summary Table: The Cranial Nerves

As you read Module 13.2, build your own summary table about the cranial nerves by filling in the information in the template provided below.

Summary of the Cranial Nerves

Cranial Nerve Number	Cranial Nerve Name	Structure(s) Innervated	Main Functions
Sensory Cranial Nerves			
I			
II			
VIII			
Motor Cranial Nerves			
III			
IV			
VI			

Cranial Nerve Number	Cranial Nerve Name	Structure(s) Innervated	Main Functions
XI			
XII			
Mixed Cranial Nerves			
V			
VII			
IX			
X			

Team Up

Make a handout to teach the 12 pairs of cranial nerves. You can use the Study Boost and art in your text as a guide, but the handout should be in your own words and with your own diagrams. At the end of the handout, write a few quiz questions. Once you have completed your handout, team up with one or more study partners and trade handouts. Study your partners' diagrams and, when you have finished, take the quiz at the end. When you and your group have finished taking all the quizzes, discuss the answers to determine places where you need additional study. After you've finished, combine the best elements of each handout to make one "master" diagram for the cranial nerves.

Module 13.3: The Spinal Nerves

This module explores the structure and function of the spinal nerves. When you complete this module, you should be able to do the following:

1. Discuss the relationships between structures of the spinal nerves: root, nerve, ramus, plexus, tract, and ganglion.
2. Identify and describe the four spinal nerve plexuses, and give examples of nerves that emerge from each.
3. Describe the functions of the major spinal nerves.

Build Your Own Glossary

Following is a table listing key terms from Module 13.3. Before you read the module, use the glossary at the back of your book or look through the module to define the following terms.

Key Terms of Module 13.3

Term	Definition
Posterior ramus	
Anterior ramus	
Cervical plexuses	
Brachial plexuses	
Intercostal nerves	
Lumbar plexuses	
Sacral plexuses	

Survey It: Form Questions

Before you read the module, survey it and form at least three questions for yourself. When you have finished reading the module, return to these questions and answer them.

Question 1: ______________________________

Answer: ______________________________

Question 2: ______________________________

Answer: ______________________________

Question 3: ______________________________

Answer: ______________________________

Key Concept: How long is an actual spinal nerve? Explain.

____

____

Key Concept: What forms nerve plexuses? What is the only group of nerves that does not stem from a nerve plexus?

____

____

Identify It: Spinal Nerves

Identify and color-code the major nerves and nerve plexuses in Figure 13.3. Next to the name of each plexus, write the main nerve or nerves that emerge from it.

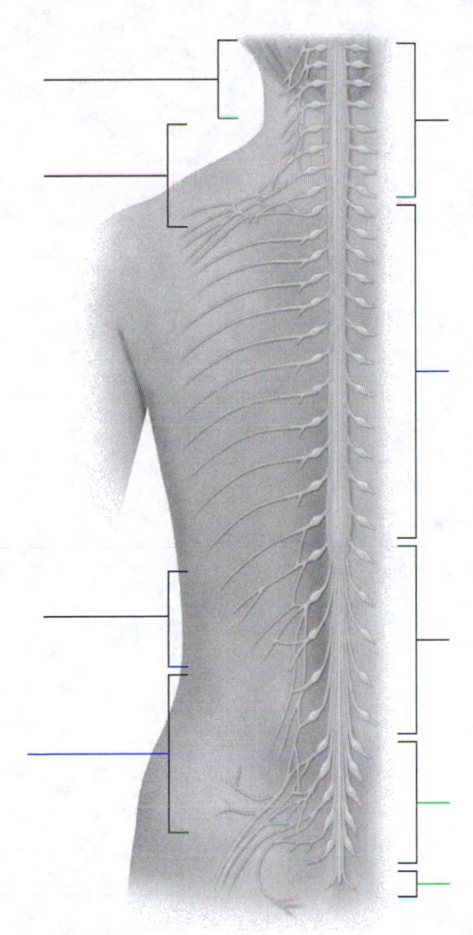

Figure 13.3 Overview of spinal nerves.

Key Concept: From which nerve roots does the phrenic nerve form? Why is this nerve so important to survival?

____

____

Identify It: The Brachial Plexus and Its Nerves

Identify and color-code the trunks, cords, and main nerves of the brachial plexus in Figure 13.4. Then, write the main functions of each of the nerves of the brachial plexus.

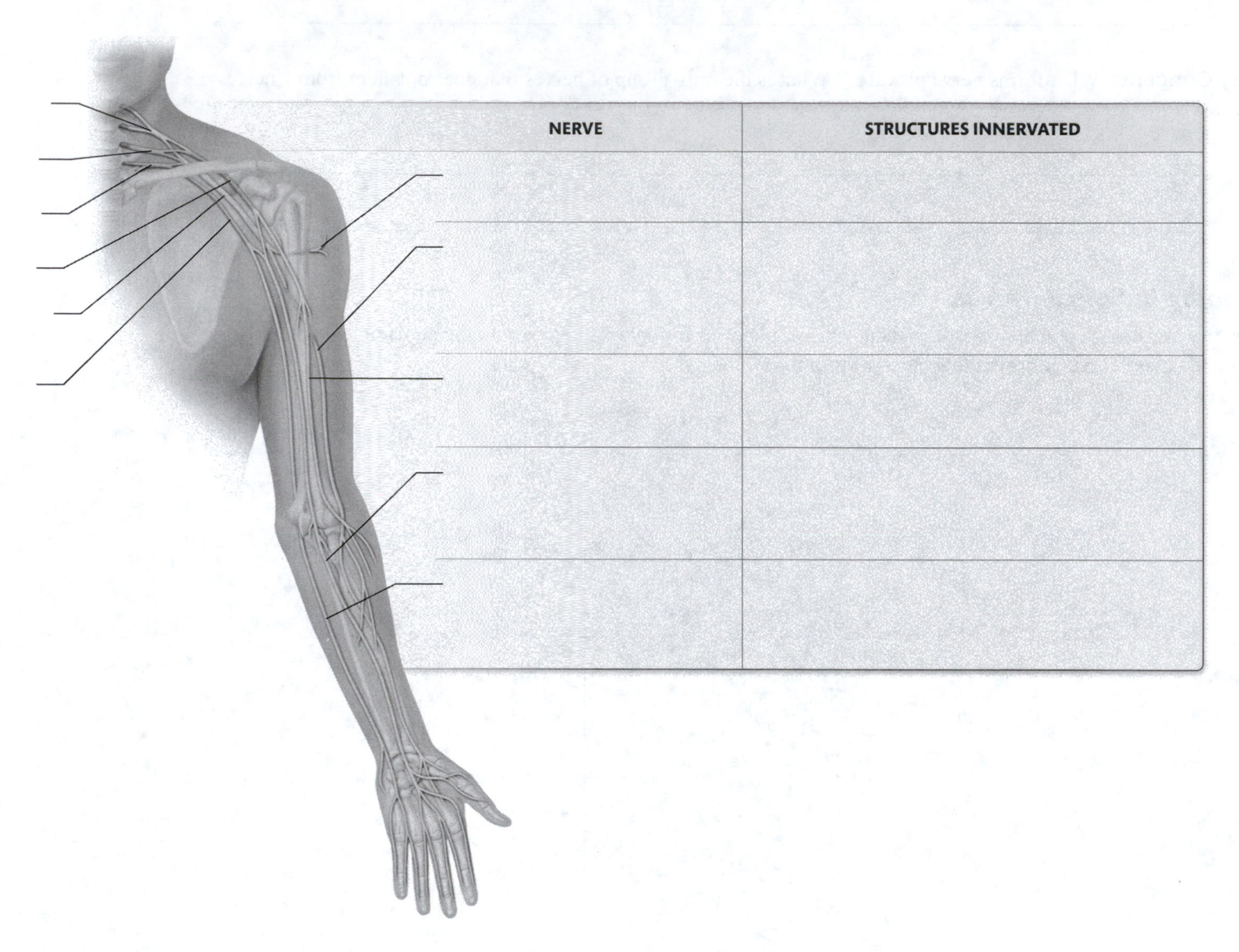

NERVE	STRUCTURES INNERVATED

Figure 13.4 The brachial plexus and its nerves.

Identify It: The Lumbar Plexus and Its Nerves

Identify and color-code the nerve roots and main nerves of the lumbar plexus in Figure 13.5. Then, write the main functions of each of the nerves of the lumbar plexus.

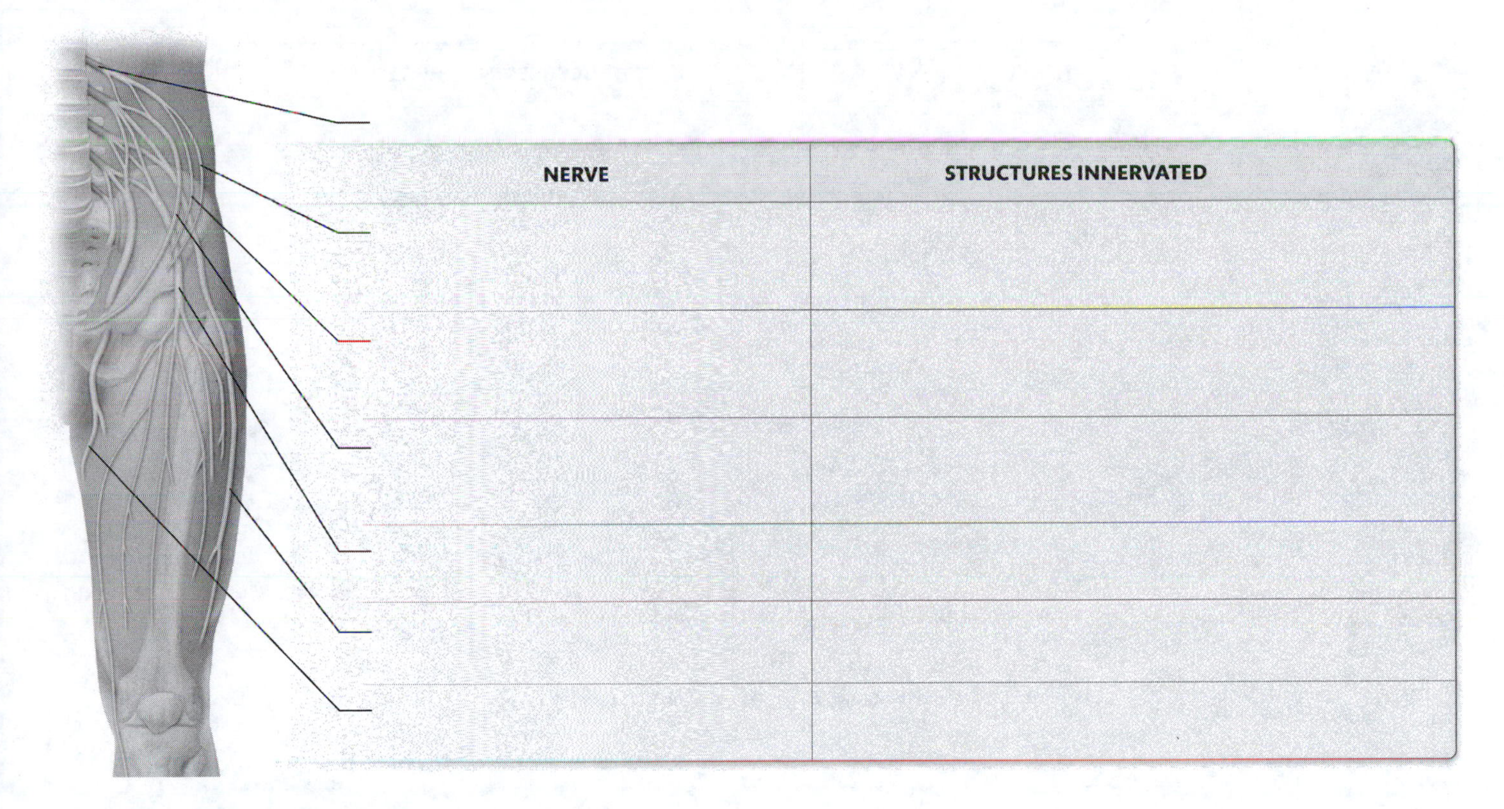

NERVE	STRUCTURES INNERVATED

Figure 13.5 The lumbar plexus and its nerves.

Identify It: The Sacral Plexus and Its Nerves

Identify and color-code the main nerves of the sacral plexus in Figure 13.6. Then, write the main functions of each of the nerves of the sacral plexus.

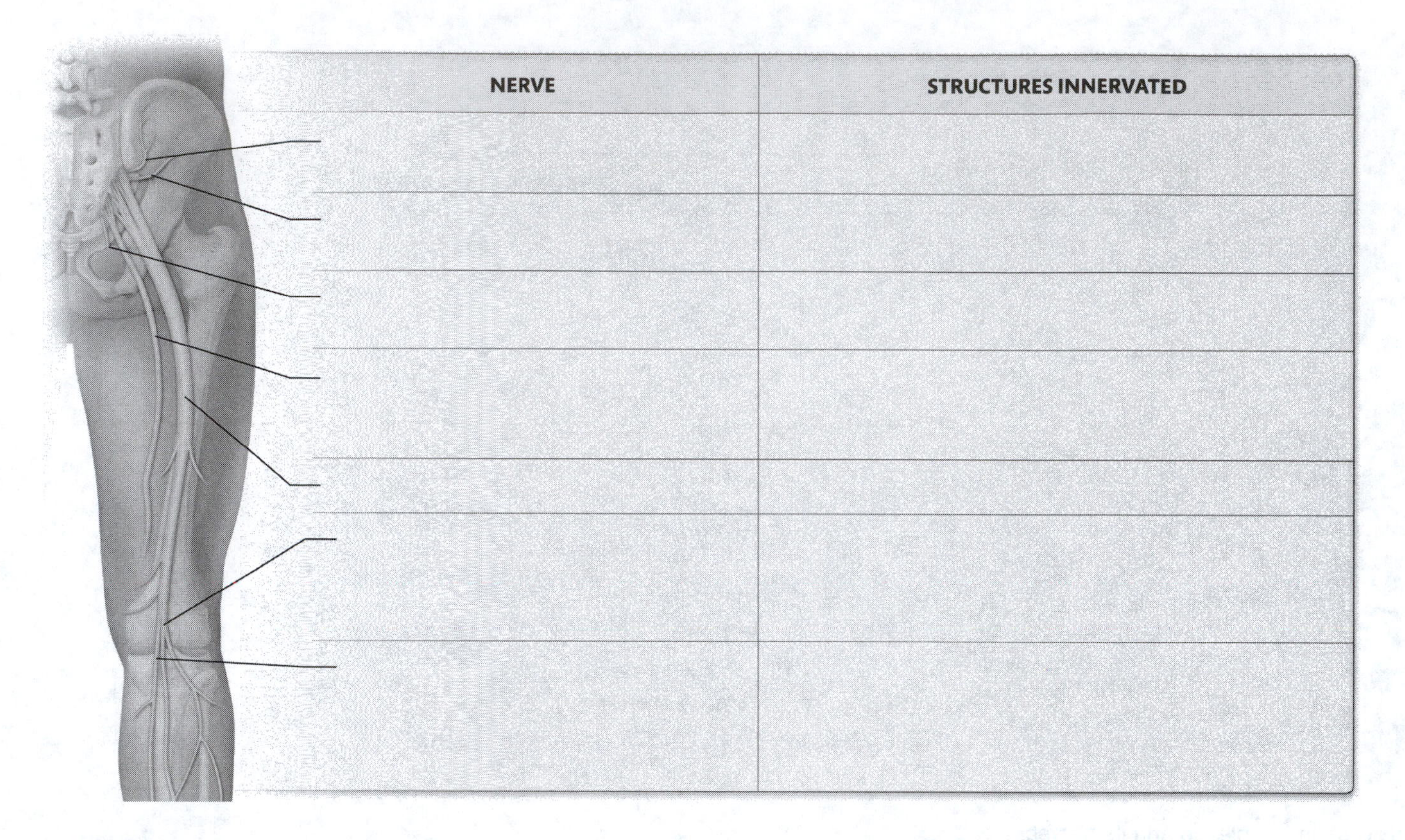

Figure 13.6 The sacral plexus and its nerves.

Module 13.4: Sensation Part 2: Role of the PNS in Sensation

In the CNS chapter, we explored the role of the CNS in sensation and perception. Now we complete the picture by looking at the role of the PNS. At the end of this module, you should be able to do the following:

1. Compare and contrast the structure and functions of exteroceptors, interoceptors, and proprioceptors.
2. Describe the location, structure, and function of nociceptors, thermoreceptors, mechanoreceptors, chemoreceptors, and photoreceptors.
3. Explain how sensory transduction takes place at a sensory receptor.
4. Describe the pathway that a sensation takes from its detection in the PNS to its delivery to the CNS.

Build Your Own Glossary

Following is a table listing key terms from Module 13.4. Before you read the module, use the glossary at the back of your book or look through the module to define the following terms.

Key Terms of Module 13.4

Term	Definition
Sensory transduction	
Receptor potential	
Interoceptor	
Exteroceptor	
Mechanoreceptor	
Thermoreceptor	
Chemoreceptor	
Photoreceptor	
Nociceptor	
Proprioceptors	
Posterior root ganglion	
Peripheral process	
Central process	
Receptive field	
Dermatome	
Referred pain	

Survey It: Form Questions

Before you read the module, survey it and form at least three questions for yourself. When you have finished reading the module, return to these questions and answer them.

Question 1: ____________________

Answer: ____________________

Question 2: ____________________

Answer: ____________________

Question 3: ____________________

Answer: ____________________

Complete It: Sensory Transduction

Fill in the blanks to complete the following paragraphs that describe the process of sensory transduction and the types of receptors.

Sensory transduction is the conversion of a __________ into an __________ __________. It begins at the end of an axon in a structure known as a __________ __________. When a stimulus is applied to this structure, __________ ion channels open, causing the membrane potential to __________ __________ __________. This generates a temporary __________ known as a __________ __________. An action potential is generated if the change in potential is strong enough and __________ __________ sodium ion channels in the axon open.

There are many types of receptors. Rapidly adapting receptors respond quickly to the __________ __________ __________, but ignore __________ __________. On the other hand, slowly adapting receptors respond to stimuli with constant __________ __________. Receptors known as __________ detect stimuli originating within the body, whereas __________ detect stimuli originating outside the body.

Key Concept: You walk around your house for 10 minutes looking for your watch, only to realize it has been on your wrist the whole time. Is this the work of rapidly adapting or slowly adapting receptors? Explain.

____

____

____

Team Up

Make a quiz to test the different types of receptors and subclasses of mechanoreceptors. Use a variety of question types, including multiple choice, matching, and fill-in-the-blank. When you have completed your quiz, team up with one or more study partners and trade quizzes. After you and your group have finished taking all the quizzes, discuss the answers to determine places where you need additional study.

Identify It: Somatic Sensory Neuron Structure and Function

Identify and color-code each component of the somatic sensory neuron in Figure 13.7. Then, fill in the steps of sensory transduction next to the numbers.

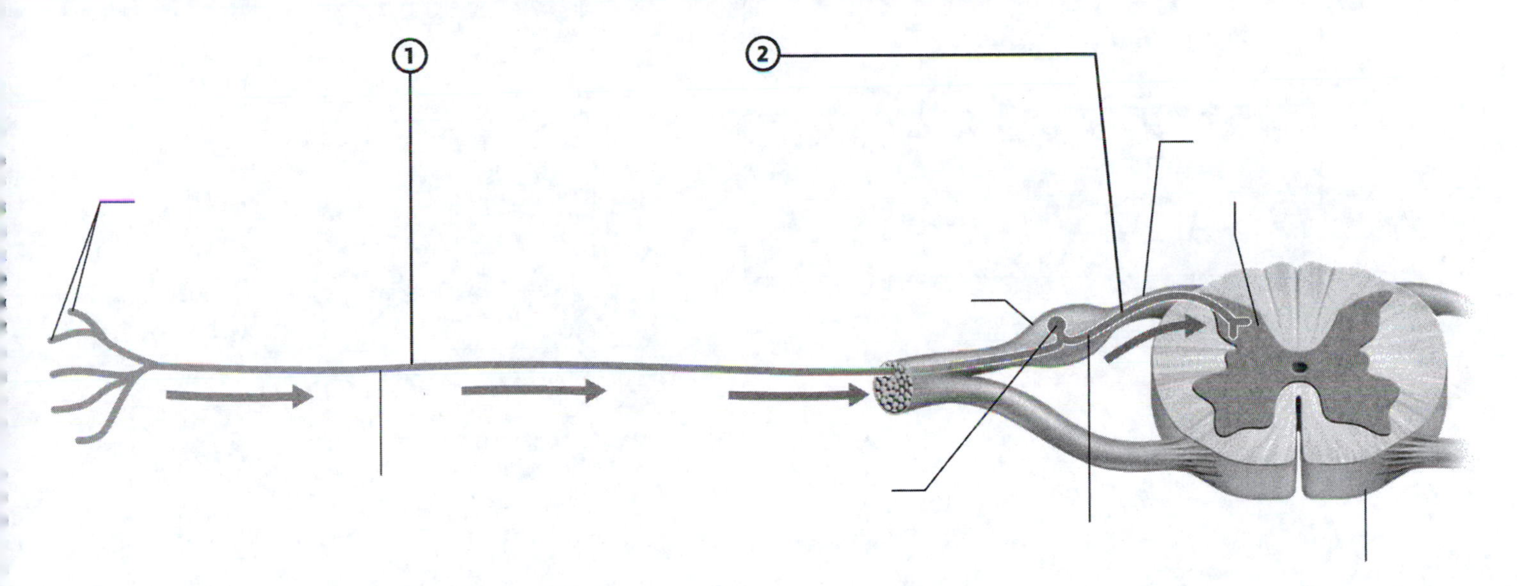

Figure 13.7 Somatic sensory neuron structure and function.

Key Concept: Why are there more numerous, smaller receptive fields in areas like the fingertips than in areas like the back? How does this connect to the somatotopic organization of the primary somatosensory cortex that you saw in the CNS chapter?

____

____

Key Concept: Why is pain originating in the liver often perceived in the right shoulder?

____

____

Describe It: The Big Picture of the Detection and Perception of Somatic Sensation by the Nervous System

Write out the steps of the overall process of sensory detection and perception in Figure 13.8. In addition, label and color-code all key components of this process.

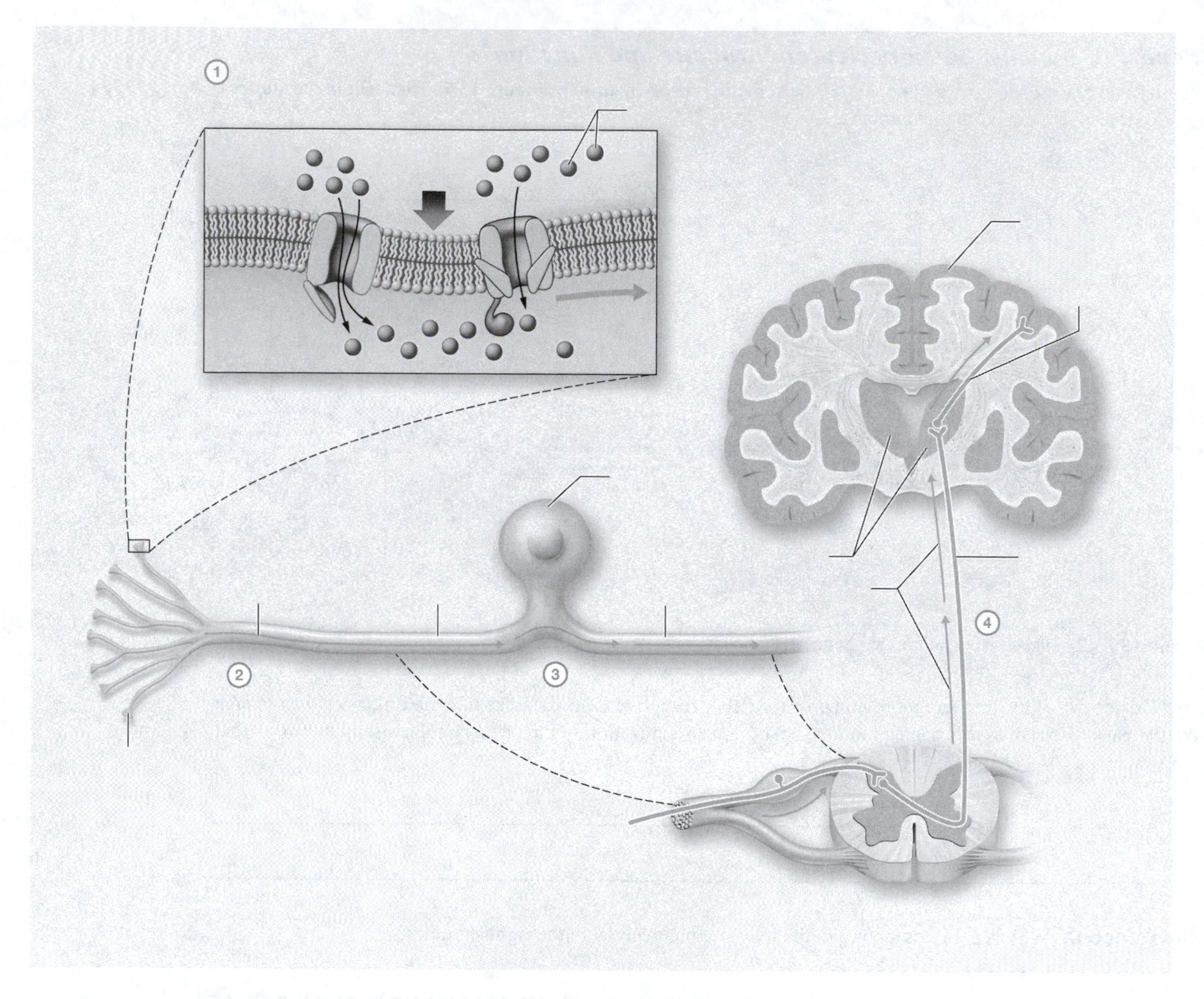

Figure 13.8 The big picture of the detection and perception of somatic sensation by the nervous system.

Module 13.5: Movement Part 2: Role of the PNS in Movement

This module looks at the role of the PNS in movement and ties it together with the role of the CNS for a big picture view of the nervous system control of movement. At the end of this module, you should be able to do the following:

1. Describe the differences between upper motor neurons and lower motor neurons.
2. Describe the overall "big picture" view of how movement occurs.

Build Your Own Glossary

Following is a table listing key terms from Module 13.5. Before you read the module, use the glossary at the back of your book or look through the module to define the following terms.

Key Terms of Module 13.5

Term	Definition
Upper motor neuron	
Lower motor neuron	
α-motor neuron	
γ-motor neuron	

Survey It: Form Questions

Before you read the module, survey it and form at least three questions for yourself. When you have finished reading the module, return to these questions and answer them.

Question 1: ______________________________

Answer: ______________________________

Question 2: ______________________________

Answer: ______________________________

Question 3: ______________________________

Answer: ______________________________

Key Concept: Can an upper motor neuron directly stimulate a skeletal muscle cell to contract? Why or why not?

____

____

____

Describe It: The Big Picture of Control of Movement by the Nervous System

Write out the steps of the overall process of the initiation of movement in Figure 13.9. In addition, label and color-code all key components of this process.

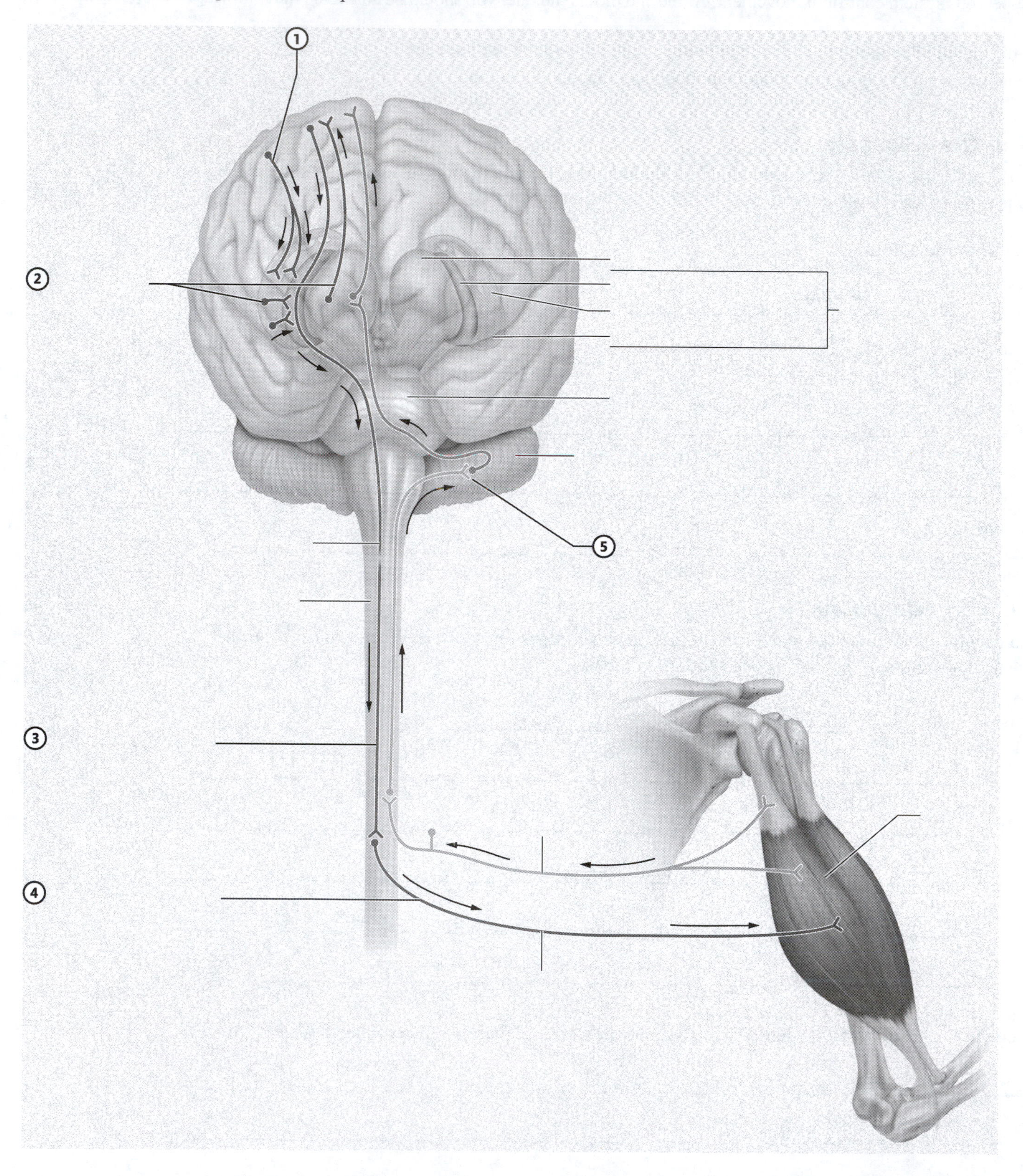

Figure 13.9 The big picture of the control of movement by the nervous system.

Module 13.6: Reflex Arcs: Integration of Sensory and Motor Function

Module 13.6 in your text examines the connection between motor and sensory functions of the nervous system: reflex arcs. By the end of the module, you should be able to do the following:

1. Describe reflex responses in terms of the major structural and functional components of a reflex arc.
2. Distinguish between somatic and visceral reflexes and monosynaptic and polysynaptic reflexes.
3. Describe a simple stretch reflex, a flexion reflex, and a crossed-extension reflex.
4. Describe the role of stretch receptors in skeletal muscles.

Build Your Own Glossary

Following is a table listing key terms from Module 13.6. Before you read the module, use the glossary at the back of your book or look through the module to define the following terms.

Key Terms for Module 13.6

Term	Definition
Reflex	
Muscle spindle	
Extrafusal muscle fiber	
Intrafusal muscle fiber	
Golgi tendon organ	
Monosynaptic reflex	
Polysynaptic reflex	
Simple stretch reflex	
Golgi tendon reflex	
Flexion (withdrawal) reflex	
Crossed-extension reflex	
Peripheral neuropathy	

Survey It: Form Questions

Before you read the module, survey it and form at least two questions for yourself. When you have finished reading the module, return to these questions and answer them.

Question 1: ____________________

Answer: ____________________

Question 2: ____________________

Answer: ____________________

Draw It: The Reflex Arc

Draw, color-code, and label a diagram of a basic reflex. You can use the figure on page 506 in your text as a guide, but make the figure your own and ensure that it makes sense to you.

Key Concept: How does a reflex arc exemplify the cell-cell communication core principle?

____

____

____

Identify It: Muscle Spindle and Golgi Tendon Organ

Identify and color-code the structures of a muscle spindle and Golgi tendon organ in Figure 13.10.

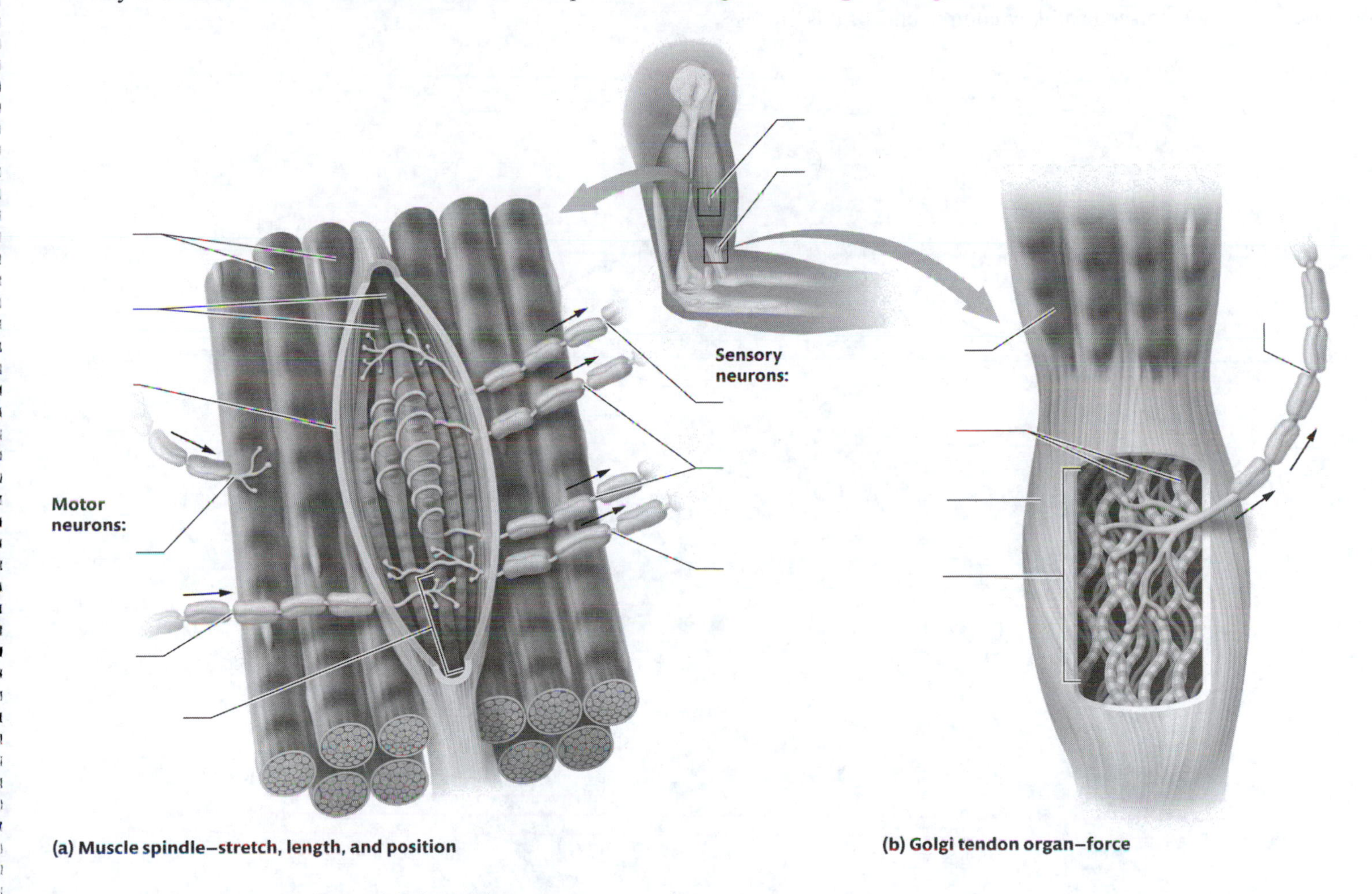

(a) Muscle spindle–stretch, length, and position

(b) Golgi tendon organ–force

Figure 13.10 Muscle spindles and Golgi tendon organs.

Describe It: The Simple Stretch Reflex and Golgi Tendon Reflex

Write a paragraph describing the sequence of events of the simple stretch reflex as if you were writing for a group of students you were teaching. In your paragraph, compare the sequence and purpose of the stretch reflex with that of the Golgi tendon reflex.

Describe It: Flexion and Crossed-Extension Reflexes

Write out the steps of the overall process of the flexion and crossed-extension reflexes in Figure 13.11. In addition, label and color-code all key components of this process.

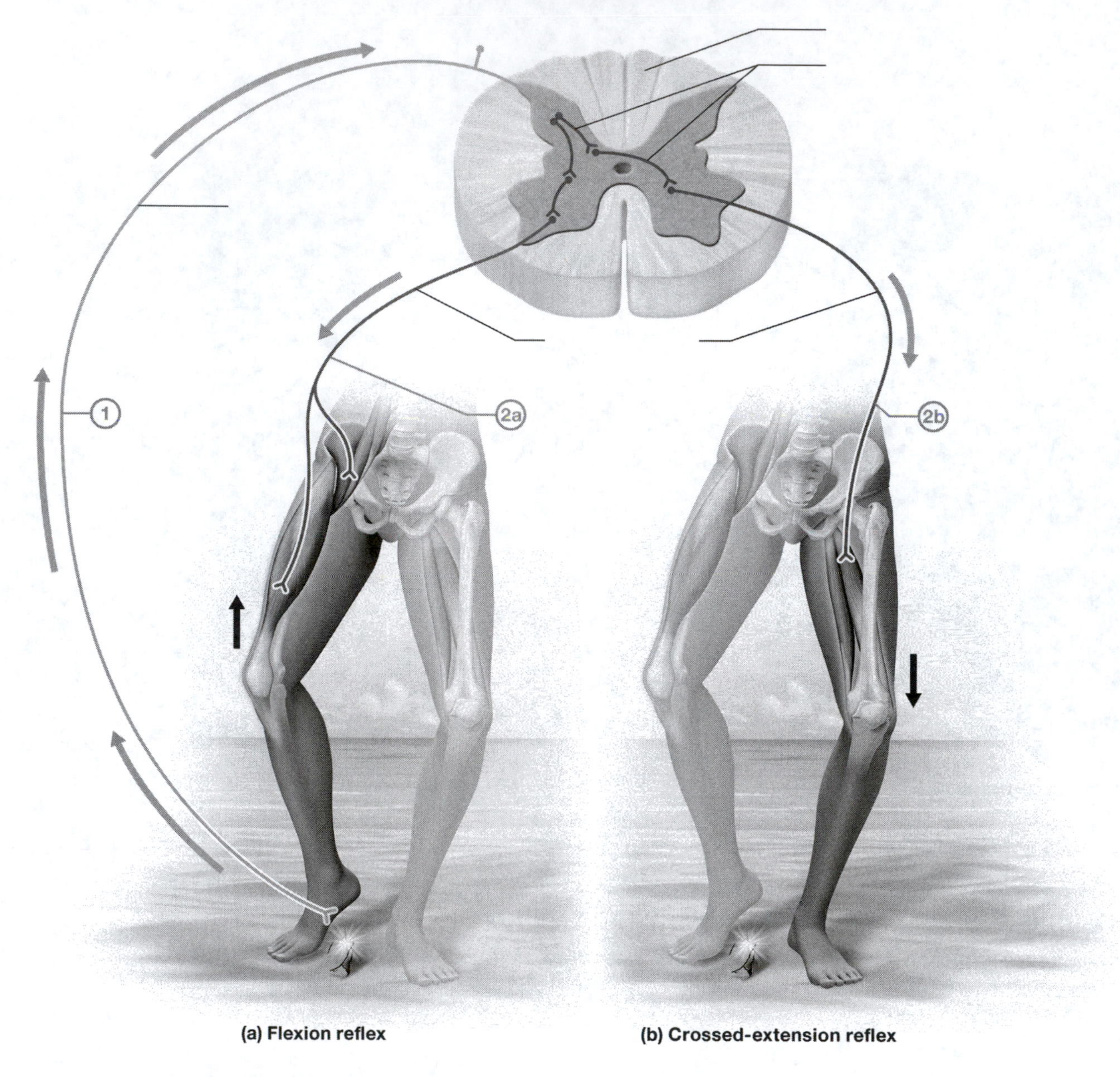

Figure 13.11 The flexion and crossed-extension reflexes.

Key Concept: What might happen if the flexion reflex occurred without the crossed-extension reflex occurring simultaneously?

____

____

Key Concept: Would you expect stretch reflexes to be absent or exaggerated in an upper motor neuron disorder? Why?

What Do You Know Now?

Let's now revisit the questions you answered in the beginning of this chapter. How have your answers changed now that you've worked through the material?

- What is the longest spinal nerve in the body? (*Hint:* Remember that the question asks specifically about *spinal nerves*. How long is an actual spinal nerve before it splits into rami?)
- Why do you blink when something touches your eye?
- What is the purpose of the knee-jerk reflex?

14 The Autonomic Nervous System and Homeostasis

This chapter examines the final arm of the PNS: the autonomic nervous system, or ANS. Here we explore how the divisions of the ANS work with the CNS and the endocrine system to maintain homeostasis in other body systems.

What Do You Already Know?

Try to answer the following questions before proceeding to the next section. If you're unsure of the correct answers, give it your best attempt based on previous courses, previous chapters, or just your general knowledge.

- What are the two divisions of the ANS?

  ____

- What is an "adrenaline rush"?

  ____

- Why does your heart pound when you are nervous?

  ____

Module 14.1: Overview of the Autonomic Nervous System

Module 14.1 in your text introduces you to the divisions of the ANS and compares the somatic and autonomic nervous systems. By the end of the module, you should be able to do the following:

1. Describe the structural and functional details of sensory and motor (autonomic) components of visceral reflex arcs.
2. Distinguish between the target cells of the somatic and autonomic nervous systems.
3. Contrast the cellular components of the somatic and autonomic motor pathways.
4. Discuss the physiological roles of each division of the autonomic nervous system.

Build Your Own Glossary

Following is a table listing key terms from Module 14.1. Before you read the module, use the glossary at the back of your book or look through the module to define the following terms.

Key Terms for Module 14.1

Term	Definition
Visceral reflex arc	
Autonomic ganglion	
Preganglionic neuron	
Postganglionic neuron	
Thoracolumbar division	
Craniosacral division	

Survey It: Form Questions

Before you read the module, survey it and form at least two questions for yourself. When you have finished reading the module, return to these questions and answer them.

Question 1: ______________________________

Answer: ______________________________

Question 2: ______________________________

Answer: ______________________________

Key Concept: What are the steps of a visceral reflex arc?

____

____

Key Concept: What is a key difference between autonomic and somatic neurons?

____

____

Key Concept: Why is the sympathetic nervous system also known as the "thoracolumbar division"? Why is the parasympathetic nervous system also called the "craniosacral division"?

____

____

Draw It: Sympathetic and Parasympathetic Neurons

Draw, color-code, and label a diagram of sympathetic and parasympathetic neurons. Your drawing should include the preganglionic neuron, the ganglion, the postganglionic neuron, and the target cell, and it should emphasize the differences between the parasympathetic and sympathetic anatomy. You can use the figure on page 518 in your text as a guide, but make the figure your own and ensure that it makes sense to you.

Module 14.2: The Sympathetic Nervous System

Now we look at the anatomy and physiology of the sympathetic nervous system. When you finish this module, you should be able to do the following:

1. Explain how the sympathetic nervous system maintains homeostasis.
2. Describe the anatomy of the sympathetic nervous system.
3. Describe the neurotransmitters and neurotransmitter receptors of the sympathetic nervous system.
4. Explain the effects of the sympathetic nervous system on the cells of its target organs.

Build Your Own Glossary

Following is a table listing key terms from Module 14.2. Before you read the module, use the glossary at the back of your book or look through the module to define the following terms.

Key Terms for Module 14.2

Term	Definition
Sympathetic nervous system	
Sympathetic chain ganglia	
Collateral ganglia	
Splanchnic nerves	
Acetylcholine	

Term	Definition
Norepinephrine	
Adrenergic receptor	
Cholinergic receptor	
Adrenal medulla	

Survey It: Form Questions

Before you read the module, survey it and form at least three questions for yourself. When you have finished reading the module, return to these questions and answer them.

Question 1: ______________________________

Answer: ______________________________

Question 2: ______________________________

Answer: ______________________________

Question 3: ______________________________

Answer: ______________________________

Key Concept: How do the axons of splanchnic nerves differ from those of most other preganglionic sympathetic neurons?

____

____

Describe It: The Synapses of Sympathetic Neurons

Describe in Figure 14.1 the three potential places that a preganglionic sympathetic neuron may synapse. Then, trace each potential pathway in a different color.

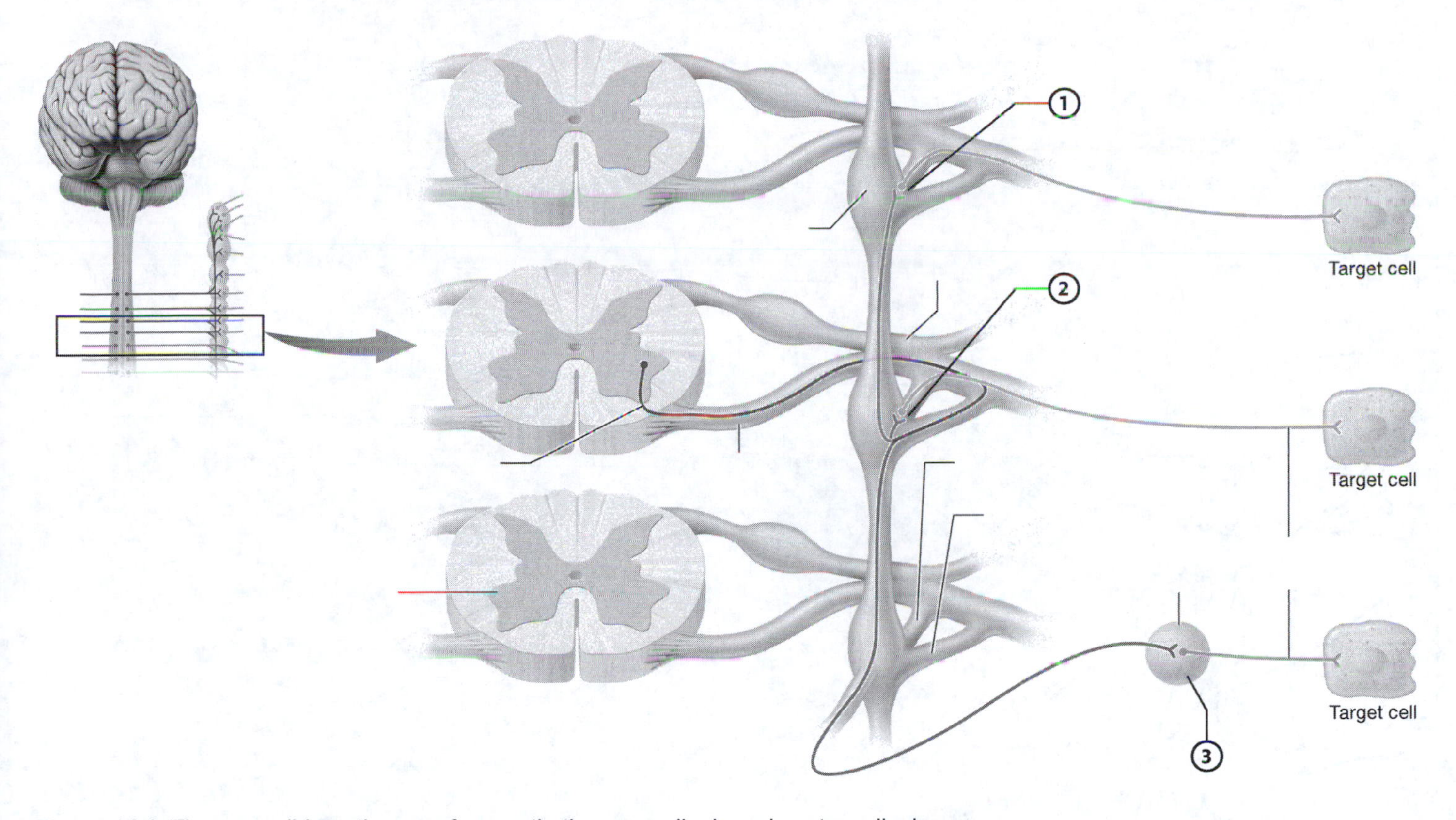

Figure 14.1 Three possible pathways of sympathetic preganglionic and postganglionic neurons.

Key Concept: How do cholinergic and adrenergic receptors differ? Where do we generally find each type of receptor in the sympathetic nervous system?

____

____

Key Concept: How is the alpha-2 receptor different from other adrenergic receptors? What happens when norepinephrine binds these receptors?

____

____

Describe It: Sympathetic Effects on Target Organs

Fill in Figure 14.2 with the neurotransmitter, receptor, and effects of the sympathetic nervous system.

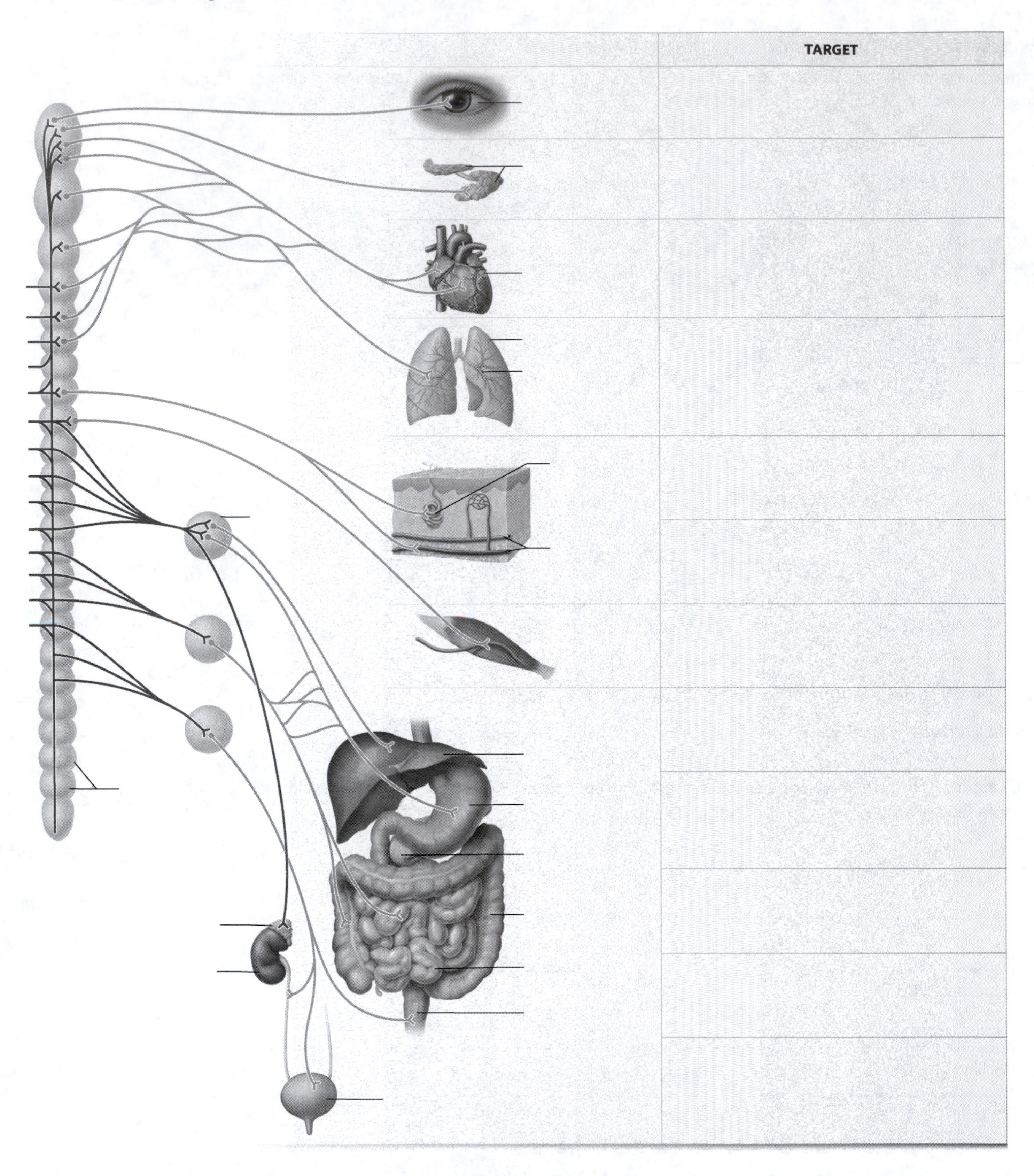

Figure 14.2 The main effects of sympathetic neurons on their target cells.

NT	RECEPTOR	MAIN EFFECTS

Team Up

Make a handout to teach the basics of the sympathetic nervous system, including its anatomy, its receptors, and its effects on target cells. You can use the art in your text as a guide, but the handout should be in your own words and with your own diagrams. At the end of the handout, write a few quiz questions. Once you have completed your handout, team up with one or more study partners and trade handouts. Study your partners' diagrams, and when you have finished, take the quiz at the end. When you and your group have finished taking all the quizzes, discuss the answers to determine places where you need additional study. After you've finished, combine the best elements of each handout to make one "master" diagram for the sympathetic nervous system.

Module 14.3: The Parasympathetic Nervous System

This module explores the structure and function of the parasympathetic nervous system. When you complete this module, you should be able to do the following:

1. Identify the role of the parasympathetic nervous system, and explain how it maintains homeostasis.
2. Describe the anatomy of the parasympathetic nervous system.
3. Describe the neurotransmitter and neurotransmitter receptors of the parasympathetic nervous system.
4. Describe the effects of the parasympathetic nervous system on its target cells.

Build Your Own Glossary

Following is a table listing key terms from Module 14.3. Before you read the module, use the glossary at the back of your book or look through the module to define the following terms.

Key Terms of Module 14.3

Term	Definition
Parasympathetic nervous system	
Pelvic splanchnic nerves	
Nicotinic receptor	
Muscarinic receptor	

Survey It: Form Questions

Before you read the module, survey it and form at least two questions for yourself. When you have finished reading the module, return to these questions and answer them.

Question 1: ______________________________

Answer: ______________________________

Question 2: ______________________________

Answer: ______________________________

Key Concept: Which cranial nerves carry parasympathetic axons? What are the pelvic splanchnic nerves?

____

____

Key Concept: Which neurotransmitter is used by all parasympathetic neurons? To which types of receptors does this neurotransmitter bind?

____

____

Describe It: Parasympathetic Effects on Target Organs

Fill in Figure 14.3 with the target tissue and effects of the parasympathetic nervous system.

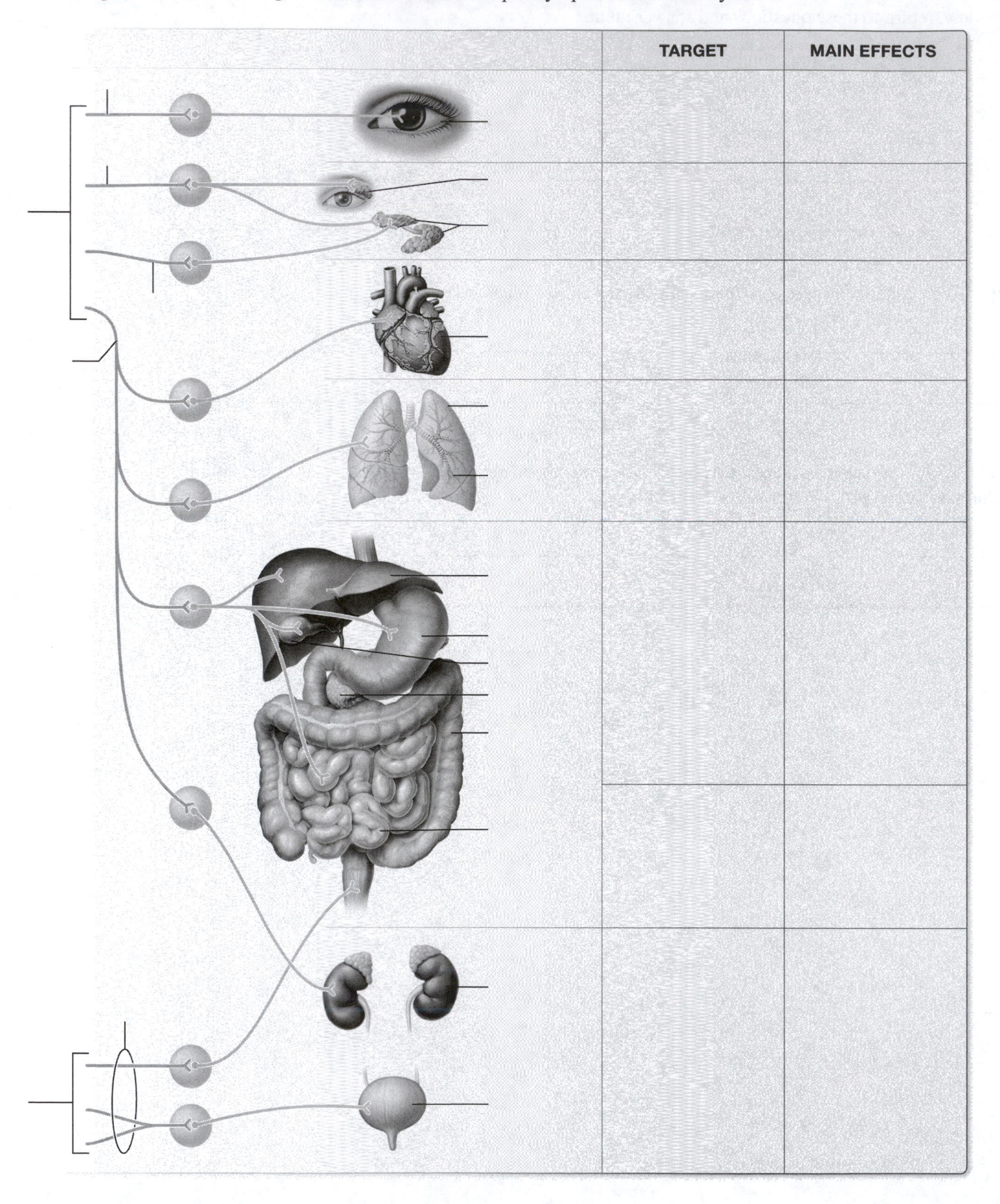

Figure 14.3 The main effects of parasympathetic neurons on target cells.

Team Up

Make a handout to teach basics of the parasympathetic nervous system, including its anatomy, receptors, and effects on target cells. You can use the art in your text as a guide, but the handout should be in your own words and with your own diagrams. At the end of the handout, write a few quiz questions. Once you have completed your handout, team up with one or more study partners and trade handouts. Study your partners' diagrams, and when you have finished, take the quiz at the end. When you and your group have finished taking all the quizzes, discuss the answers to determine places where you need additional study. After you've finished, combine the best elements of each handout to make one "master" diagram for the parasympathetic nervous system.

Module 14.4: Homeostasis Part 2: PNS Maintenance of Homeostasis

In the CNS chapter, we explored the role of the CNS in homeostasis. Now we complete the picture by looking at the role of the PNS. At the end of this module, you should be able to do the following:

1. Define sympathetic and parasympathetic tone.
2. Explain how the nervous system as a whole regulates homeostasis.

Build Your Own Glossary

Following is a table listing key terms from Module 14.4. Before you read the module, use the glossary at the back of your book or look through the module to define the following terms.

Key Terms of Module 14.4

Term	Definition
Autonomic tone	
Autonomic centers	

Survey It: Form Questions

Before you read the module, survey it and form at least two questions for yourself. When you have finished reading the module, return to these questions and answer them.

Question 1: ______________________________

Answer: ______________________________

Question 2: ______________________________

Answer: ______________________________

Describe It: Comparison of the Sympathetic and Parasympathetic Nervous Systems

Fill in Figure 14.4 with the requested information that compares the sympathetic and parasympathetic nervous systems and label the structures involved.

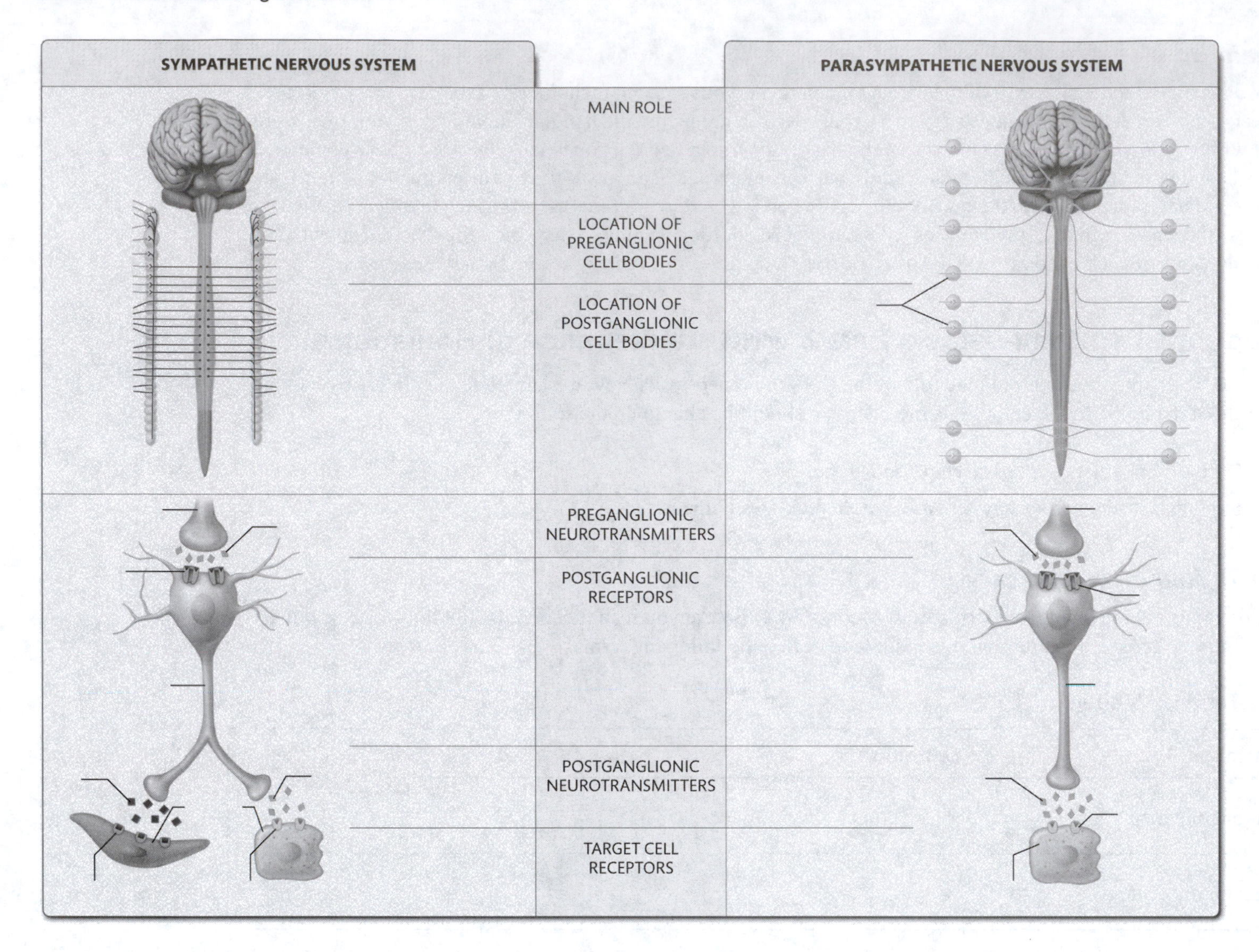

Figure 14.4 Comparison of structure and effects of the sympathetic and parasympathetic nervous systems.

Key Concept: How do the two divisions of the autonomic nervous system interact?

____

____

Build Your Own Summary Table: The Two Divisions of the Autonomic Nervous System

After you have read the chapter, build your own summary table comparing the effects of the sympathetic and parasympathetic nervous systems on their target tissues. If a target tissue is not innervated by one of the divisions, say so in the table.

Summary of the Two Divisions of the Autonomic Nervous System

Target Tissue	Sympathetic Effect	Parasympathetic Effect

What Do You Know Now?

Let's now revisit the questions you answered in the beginning of this chapter. How have your answers changed now that you've worked through the material?

- What are the two divisions of the ANS?

  ____

- What is an "adrenaline rush"?

  ____

- Why does your heart pound when you are nervous?

  ____

15 The Special Senses

In previous chapters we learned that the general senses detect such stimuli as touch, pain, and temperature. We now turn to the special senses, in which specialized sensory organs convey a specific type of information.

What Do You Already Know?

Try to answer the following questions before proceeding to the next section. If you're unsure of the correct answers, give it your best attempt based on previous courses, previous chapters, or just your general knowledge.

- What are the five special senses?

  ____

- What areas of the brain interpret information from the special senses?

  ____

- Which parts of the tongue convey each different taste sensation?

  ____

Module 15.1: Overview of the Special Senses

Module 15.1 in your text provides an overview of the special senses and compares them to the general senses. Although they do have a lot in common, you will see many differences, too. By the end of the module, you should be able to do the following:

1. Describe the basic process of sensory transduction.
2. Compare and contrast the general and special senses.

Build Your Own Glossary

Following is a table listing key terms from Module 15.1. Before you read the module, use the glossary at the back of your book or look through the module to define the following terms.

Key Terms for Module 15.1

Term	Definition
Transduction	
General senses	
Special senses	

Survey It: Form Questions

Before you read the module, survey it and form at least two questions for yourself. When you have finished reading the module, return to these questions and answer them.

Question 1: ______________________________

Answer: ______________________________

Question 2: ______________________________

Answer: ______________________________

Key Concept: How is a physical stimulus transduced into a neural signal?

____

____

Key Concept: How are the special senses different from general sensation?

____

____

Module 15.2: Olfaction

Now we look at how our sense of smell allows us to detect the presence of chemicals in the air and transduces them into signals our brain can interpret. When you finish this module, you should be able to do the following:

1. Describe and identify the location of olfactory receptors.
2. Explain how odorants activate olfactory receptors.
3. Describe the path of action potentials from the olfactory receptors to various parts of the brain.

Build Your Own Glossary

The following table is a list of key terms from Module 15.2. Before you read the module, use the glossary at the back of your book or look through the module to define the following terms.

Key Terms for Module 15.2

Term	Definition
Odorants	
Olfaction	
Chemoreceptors	
Olfactory bulb	

Survey It: Form Questions

Before you read the module, survey it and form at least three questions for yourself. When you have finished reading the module, return to these questions and answer them.

Question 1: ______________________________

Answer: ______________________________

Question 2: ______________________________

Answer: ______________________________

Question 3: ______________________________

Answer: ______________________________

Key Concept: How is binding of an odorant to an olfactory receptor transduced into a neural impulse?

____

____

Identify It: Olfactory Epithelium and Olfactory Neurons

Identify each component of olfactory epithelium and olfactory neurons in Figure 15.1. Then, list the main function and/or a short description of each component.

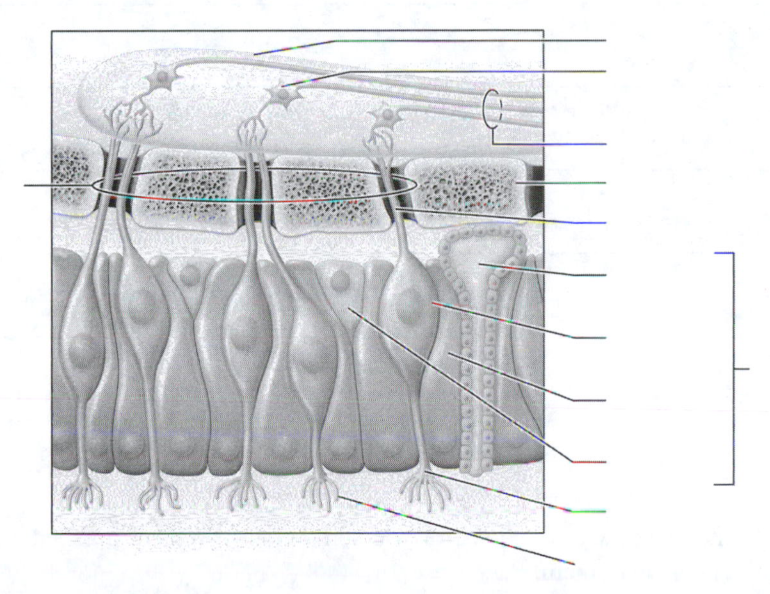

Figure 15.1 Olfactory epithelium and olfactory neurons.

Key Concept: How are olfactory stimuli interpreted by the brain? What is the importance of the connection of the stimuli with the limbic system?

____

____

Module 15.3: Gustation

This module explores taste sensation, from the detection of chemical molecules on the tongue to the processing of neural signals in various regions of the brain. When you complete this module, you should be able to do the following:

1. Describe the location and structure of taste buds.
2. Explain how chemicals dissolved in saliva activate gustatory receptors.
3. Trace the path of action potentials from the gustatory receptors to various parts of the brain.
4. Describe the five primary taste sensations.

Build Your Own Glossary

Below is a table listing key terms from Module 15.3. Before you read the module, use the glossary at the back of your book or look through the module to define the following terms.

Key Terms of Module 15.3

Term	Definition
Gustation	
Taste buds	
Vallate papillae	
Fungiform papillae	
Foliate papillae	
Filiform papillae	
Gustatory (taste) cells	
Primary gustatory cortex	

Survey It: Form Questions

Before you read the module, survey it and form at least two questions for yourself. When you have finished reading the module, return to these questions and answer them.

Question 1: ______________________________

Answer: ______________________________

Question 2: ______________________________

Answer: ______________________________

Key Concept: How does a gustatory cell transduce a chemical taste into a neural signal?

____

____

Describe It: The Gustatory Pathway

Write in the steps of the gustatory pathway in Figure 15.2. In addition, label and color-code important components of this process.

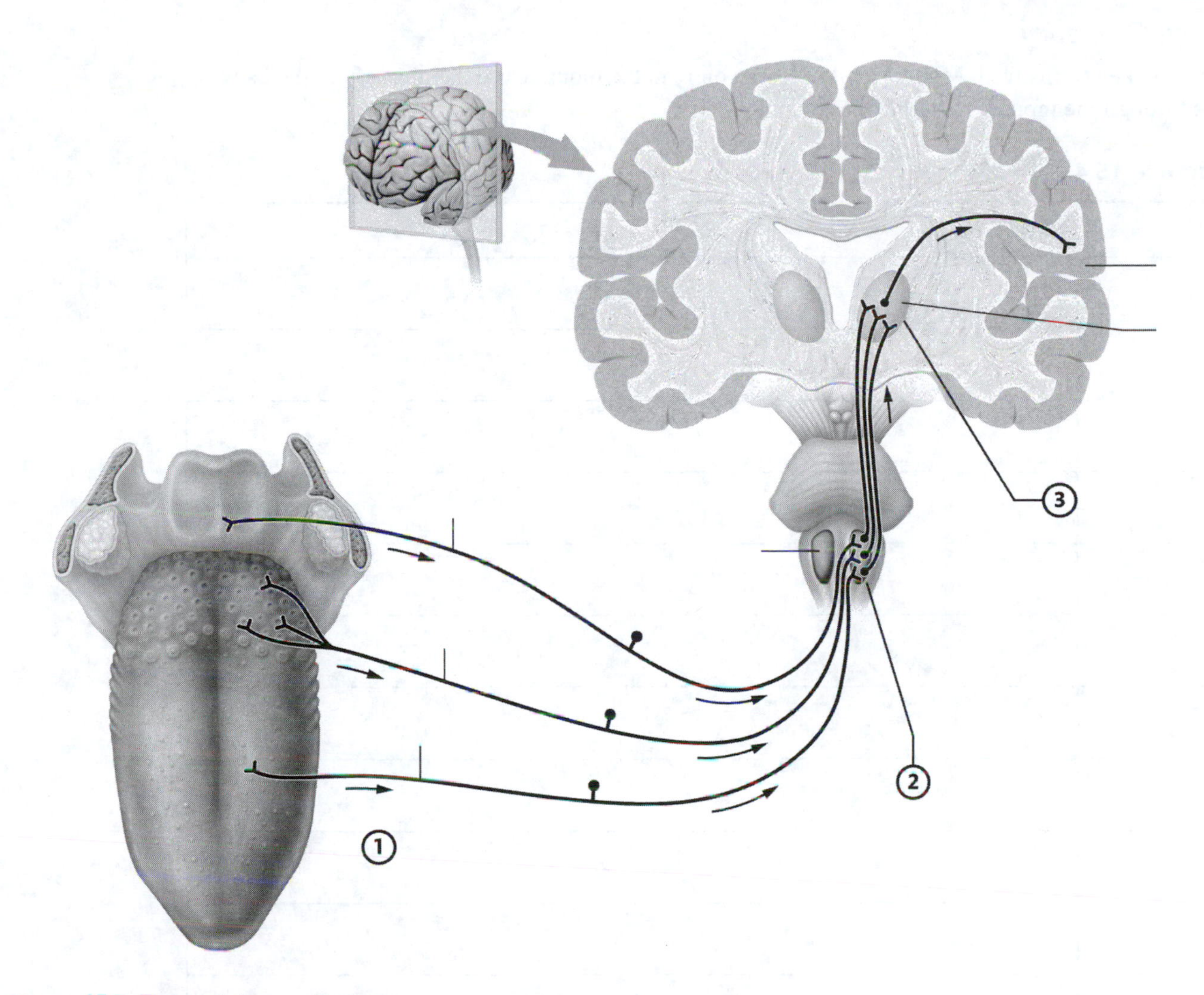

Figure 15.2 The gustatory pathway.

Key Concept: Which parts of the brain are involved in taste interpretation?

____

____

Module 15.4: Anatomy of the Eye

Module 15.4 in your text explores the location and anatomy of the eye and its accessory structures. At the end of this module, you should be able to do the following:

1. Discuss the structure and functions of the accessory structures of the eye.
2. Describe the innervation and actions of the extrinsic eye muscles.
3. Identify and describe the three layers of the eyeball.
4. Describe the structure of the retina.

Build Your Own Glossary

Below is a table listing key terms from Module 15.4. Before you read the module, use the glossary at the back of your book or look through the module to define the following terms.

Key Terms of Module 15.4

Term	Definition
Palpebrae	
Conjunctiva	
Lacrimal gland	
Nasolacrimal duct	
Sclera	
Cornea	
Choroid	
Ciliary body	
Iris	
Retina	
Macula lutea	
Optic disc	
Lens	
Vitreous humor	
Aqueous humor	

Survey It: Form Questions

Before you read the module, survey it and form at least three questions for yourself. When you have finished reading the module, return to these questions and answer them.

Question 1: ____________________

Answer: ____________________

Question 2: ____________________

Answer: ____________________

Question 3: ____________________

Answer: ____________________

Key Concept: What are the accessory structures of the eye, and what are their functions?

____

____

Identify It: The Lacrimal Apparatus

Identify and color-code each component of the lacrimal apparatus in Figure 15.3. Then, list the main function(s) of each component.

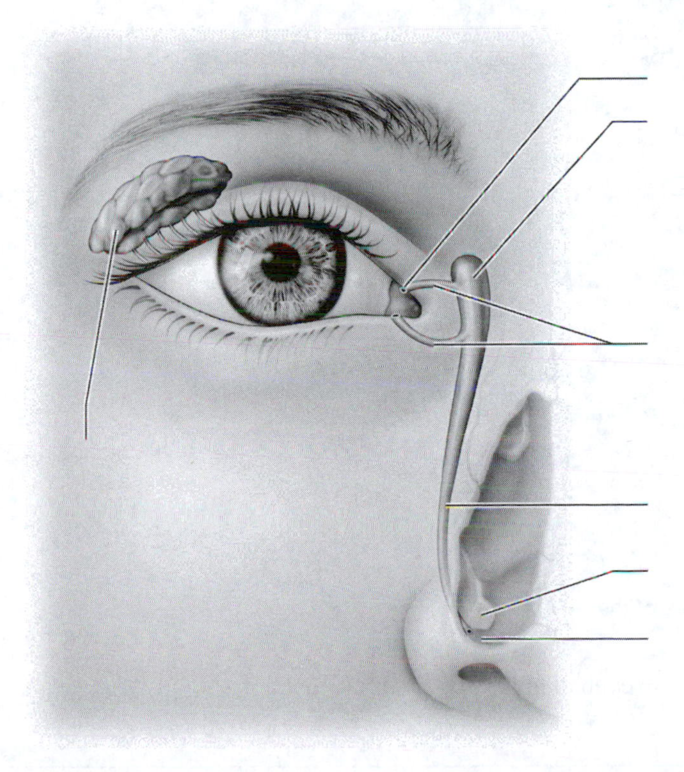

Figure 15.3 The lacrimal apparatus.

Describe It: The Extrinsic Eye Muscles

Write a paragraph describing the way the extrinsic eye muscles are attached to the eye. You may refer to Figure 9.9 in your text, if you get stuck.

Identify It: The Internal Structures of the Eye

Identify and color-code each component of the internal structures of the eye in Figure 15.4. Then, list the main function(s) of each component.

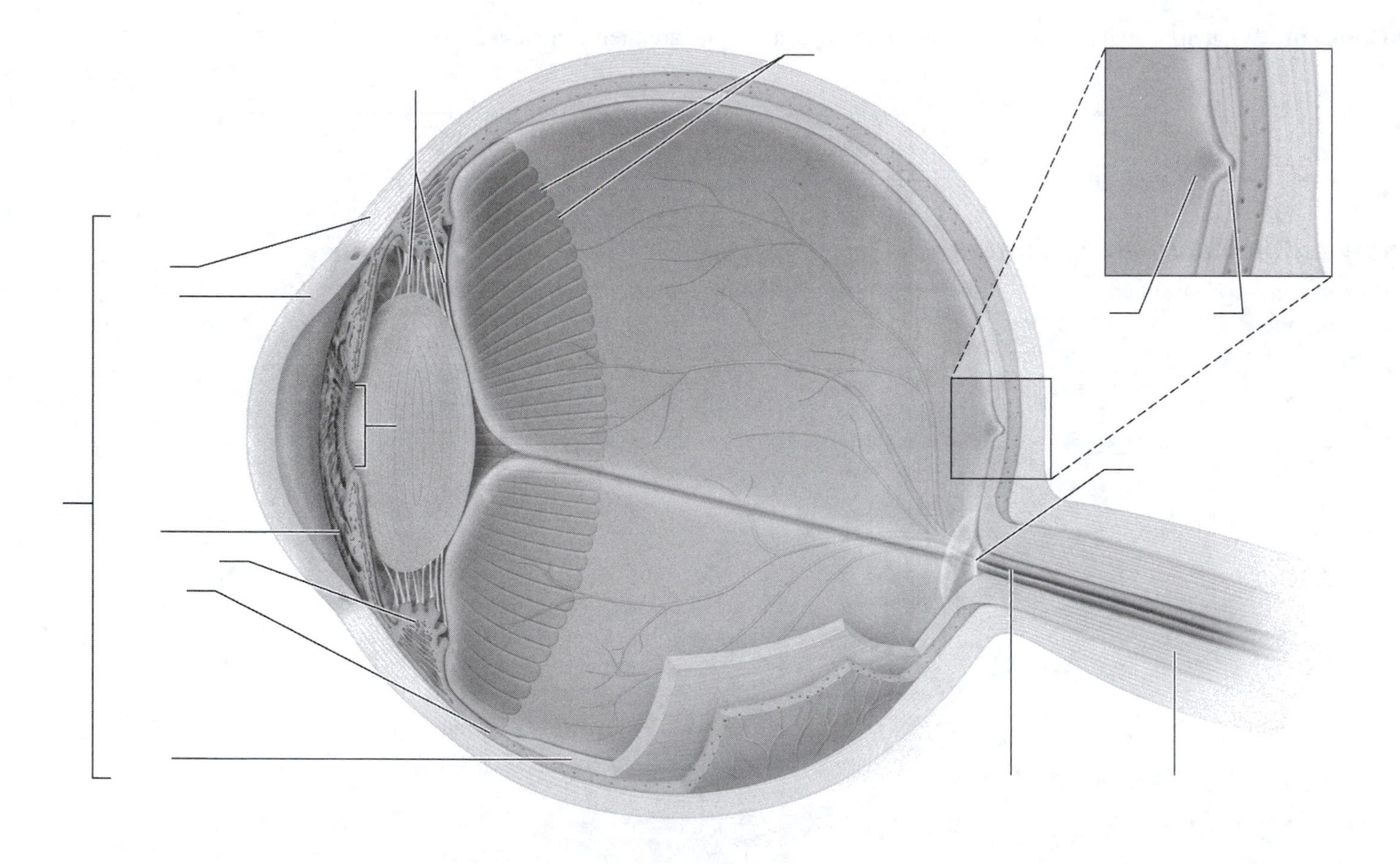

Figure 15.4 Sagittal section of internal structures of the eye.

Key Concept: What are the three layers of the eyeball, and what are their functions?

____

____

Key Concept: What are the main features of the retina? Describe each of these features.

____

____

Module 15.5: Physiology of Vision

This module turns to how the eye works: the physiology of vision. At the end of this module, you should be able to do the following:

1. Describe how light activates photoreceptors.
2. Explain how the optical system of the eye creates an image on the retina.
3. Compare the functions of rods and cones in vision.
4. Trace the path of light as it passes through the eye to the retina and the path of action potentials from the retina to various parts of the brain.
5. Explain the processes of light and dark adaptation.

Build Your Own Glossary

Following is a table listing key terms from Module 15.5. Before you read the module, use the glossary at the back of your book or look through the module to define the following terms.

Key Terms of Module 15.5

Term	Definition
Vision	
Photon	
Accommodation	
Emmetropia	
Hyperopia	
Myopia	
Bipolar cells	
Retinal ganglion cells	
Cones	
Rods	
Rhodopsin	

Term	Definition
Iodopsin	
Visual field	
Optic chiasma	

Survey It: Form Questions

Before you read the module, survey it and form at least three questions for yourself. When you have finished reading the module, return to these questions and answer them.

Question 1: ______________________________

Answer: ______________________________

Question 2: ______________________________

Answer: ______________________________

Question 3: ______________________________

Answer: ______________________________

Key Concept: How does the eye bend light to focus on objects at different distances?

____

____

Draw It: Refraction of Light

In the space below, draw convex and concave surfaces like the ones on page 551 in your text, and add line arrows to show how they bend light. Label your drawings with the following terms: diverging rays, converging rays, and focal point.

Key Concept: How does the shape of the eyeball affect focusing of light on the retina?

____

____

Identify It: Layers of the Retina

Identify the cells and structural components in the layers of the retina in Figure 15.5. Then, list the main function(s) and/or description of each component.

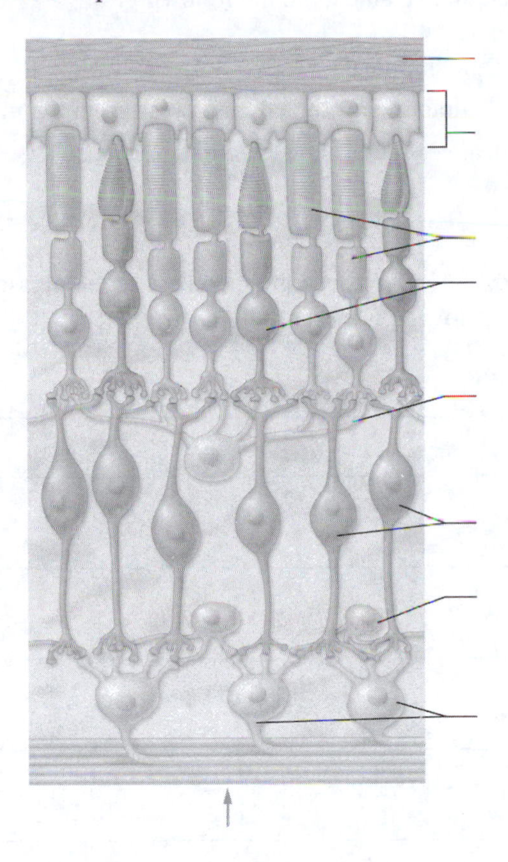

Figure 15.5 Layers of the retina.

Complete It: The Big Picture of Vision

Fill in the blanks to complete the following paragraph that describes the overview of vision.

The cornea and lens produce ____________ to focus light rays on the ____________. Light causes the pigment molecule ____________ to separate from opsin. This triggers chemical reactions that cause photoreceptors to ____________, and retinal ganglion cells send action potentials through the ____________ ____________. Selected axons cross at the ____________ ____________ so that stimuli from each half of the visual field are sent to the opposite hemisphere. The ____________ ____________ ____________ of the cerebrum provides conscious awareness and initial processing of shape, color, and movement.

Key Concept: What changes occur in the retina during dark adaptation and light adaptation?

____

____

Module 15.6: Anatomy of the Ear

Module 15.6 in your text explores the structure of the regions of the ear: outer (external) ear, middle ear, and inner (internal) ear. By the end of the module, you should be able to do the following:

1. Describe the structure and function of the outer and middle ear.
2. Discuss the role of the pharyngotympanic tube in draining and equalizing pressure in the middle ear.
3. Describe the structure and function of the cochlea, vestibule, and semicircular canals.

Build Your Own Glossary

Below is a table listing key terms from Module 15.6. Before you read the module, use the glossary at the back of your book or look through the module to define the following terms.

Key Terms for Module 15.6

Term	Definition
Auricle	
External auditory canal	
Ceruminous glands	
Tympanic membrane	
Pharyngotympanic tube	
Auditory ossicles	
Bony labyrinth	
Endolymph	
Perilymph	
Vestibule	
Semicircular canals	
Cochlea	
Spiral organ	

Survey It: Form Questions

Before you read the module, survey it and form at least two questions for yourself. When you have finished reading the module, return to these questions and answer them.

Question 1: ______________________________

Answer: ______________________________

Question 2: ______________________________

Answer: ______________________________

Key Concept: What are the three main regions of the ear, and what are the main parts of each region?

____

____

Identify It: Regions of the Ear

Identify the structural components in the regions of the ear in Figure 15.6. Then, describe each component.

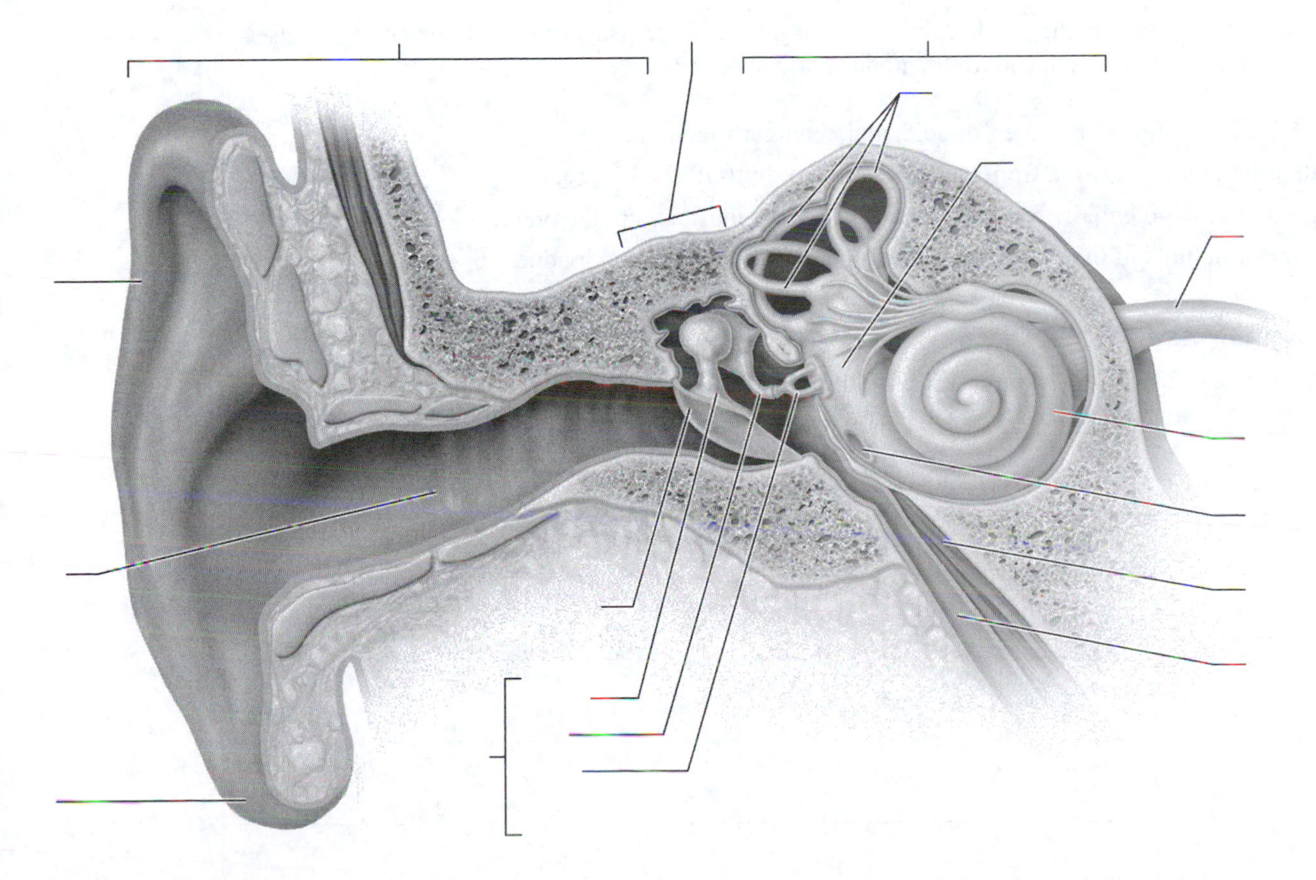

Figure 15.6 Regions of the ear.

Key Concept: What are the three main parts of the inner ear, and what are their main functions?

____

____

Draw It: Structure of the Cochlea

Draw and color-code a diagram of the cross section of the cochlea. You can use Figure 15.26 in your text as a guide, but make the figure your own and ensure that it makes sense to you. Make sure to label the scala media, scala vestibuli, scala tympani, basilar membrane, and spiral organ (organ of Corti).

Module 15.7: Physiology of Hearing

Module 15.7 in your text covers how the ear detects sound vibrations and transduces them into neural impulses that the brain interprets as sound. By the end of the module, you should be able to do the following:

1. Describe how the structures of the outer, middle, and inner ear function in hearing.
2. Trace the path of sound conduction from the auricle to the fluids of the inner ear.
3. Trace the path of action potentials from the spiral organ to various parts of the brain.
4. Explain how the structures of the ear enable differentiation of the pitch and loudness of sounds.

Build Your Own Glossary

Below is a table listing key terms from Module 15.7. Before you read the module, use the glossary at the back of your book or look through the module to define the following terms.

Key Terms for Module 15.7

Term	Definition
Audition	
Pitch	
Hair cells	
Stereocilia	
Tectorial membrane	

Helicotrema	
Vestibulocochlear nerve	

Survey It: Form Questions

Before you read the module, survey it and form at least three questions for yourself. When you have finished reading the module, return to these questions and answer them.

Question 1: ______________________________

Answer: ______________________________

Question 2: ______________________________

Answer: ______________________________

Question 3: ______________________________

Answer: ______________________________

Key Concept: How do the structures of the ear deliver sound vibrations to the correct part of the cochlea for transduction?

____

____

Draw It: Processing of Sound in the Inner Ear

Draw yourself a diagram of the processing of sound in the inner ear and describe each step. You may use Figure 15.28 in your text as a reference, but make sure that the drawing is your own diagram and words so that it makes the most sense to you.

Key Concept: How do hair cells convert mechanical energy into nervous stimuli?

____

____

Describe It: The Auditory Pathway

Write in the steps of the auditory pathway in Figure 15.7, and label and color-code important components of the process. You may use text Figure 15.31 as a reference, but write the steps in your own words.

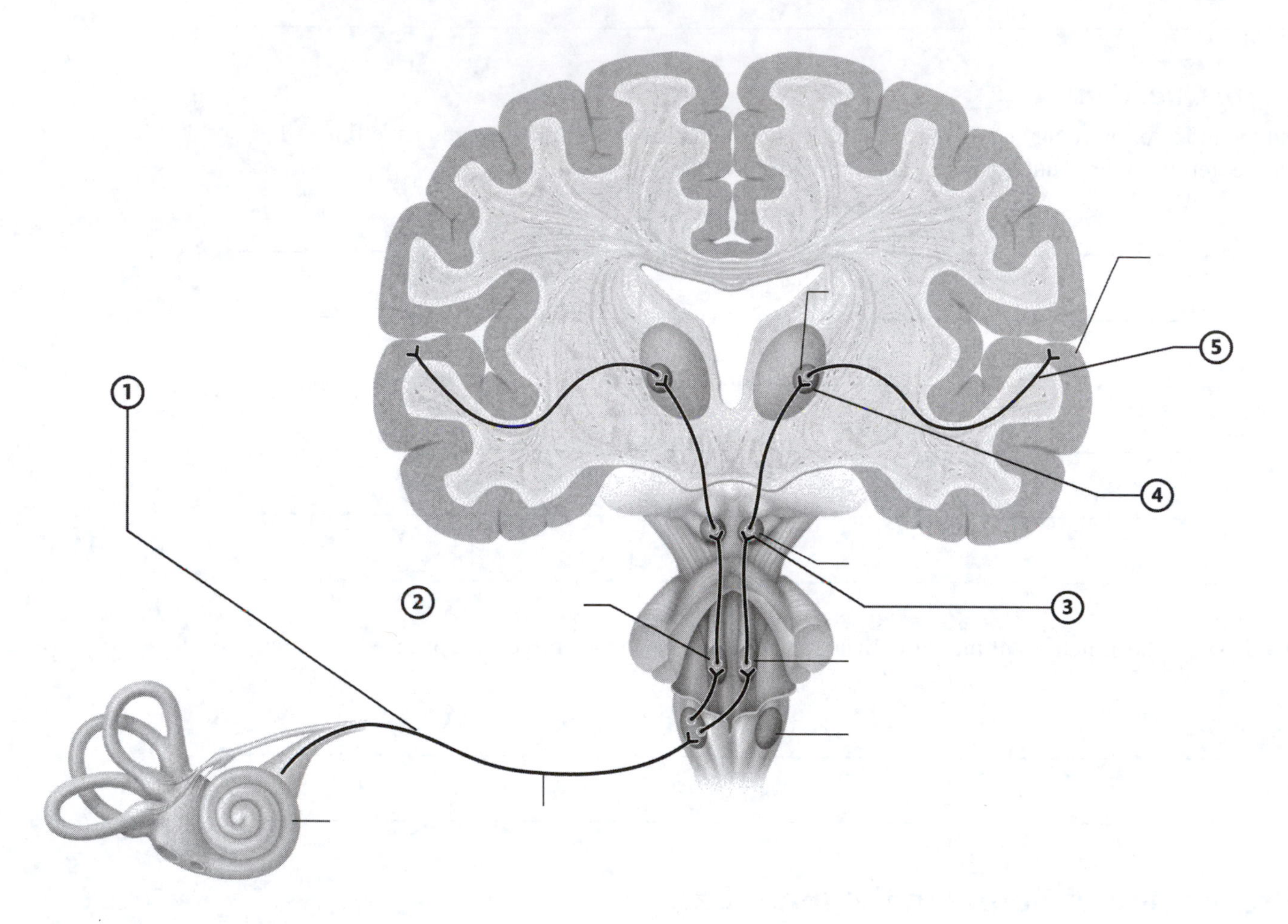

Figure 15.7 The auditory pathway.

Key Concept: How are sensory hearing loss and neural hearing loss different? Which parts of the hearing pathway are affected in each?

____

____

Module 15.8: Vestibular Sensation

Module 15.8 in your text discusses the role of the inner ear in your sense of equilibrium or balance. By the end of the module, you should be able to do the following:

1. Distinguish between static and dynamic equilibrium.
2. Describe the structure of the maculae, and explain their function in static equilibrium.
3. Describe the structure of the crista ampullaris, and explain its function in dynamic equilibrium.

Build Your Own Glossary

Below is a table listing key terms from Module 15.8. Before you read the module, use the glossary at the back of your book or look through the module to define the following terms.

Key Terms for Module 15.8

Term	Definition
Vestibular system	
Static equilibrium	
Dynamic equilibrium	
Utricle	
Saccule	
Macula	
Kinocilium	
Otolithic membrane	
Ampulla	
Crista ampullaris	
Cupula	

Survey It: Form Questions

Before you read the module, survey it and form at least three questions for yourself. When you have finished reading the module, return to these questions and answer them.

Question 1: ____________________

Answer: ____________________

Question 2: ____________________

Answer: ____________________

Question 3: ____________________

Answer: ____________________

Identify It: Maculae of the Utricle and Saccule

Identify and color-code the cells and structural components in the maculae of the utricle and saccule in Figure 15.8. Then, briefly describe each component.

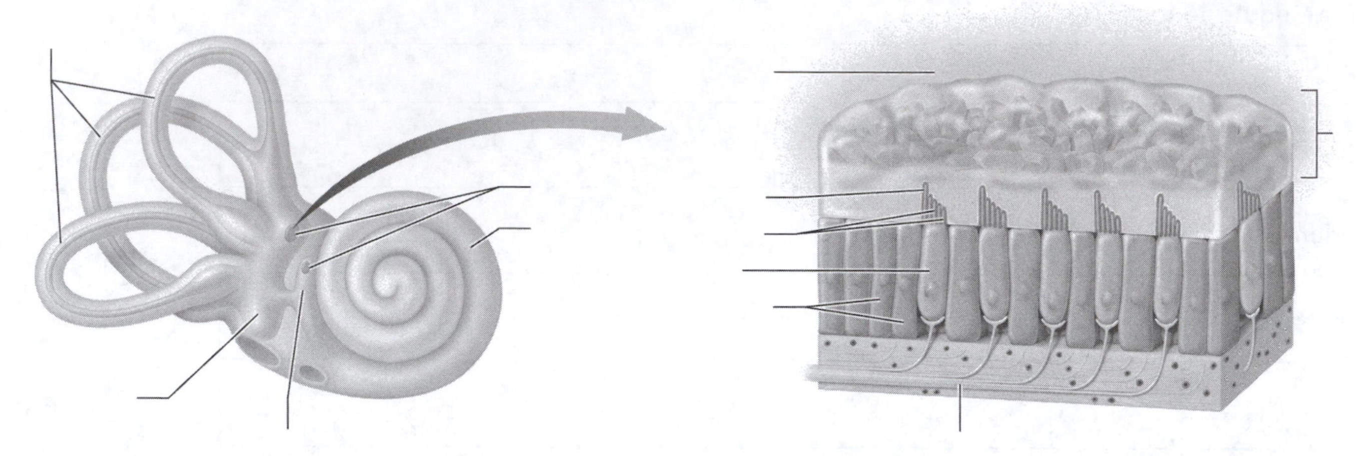

Figure 15.8 Maculae of the utricle and saccule.

Key Concept: How do static equilibrium and dynamic equilibrium contribute to your sense of balance in different ways?

____

____

Describe It: The Vestibular Sensation Pathway

Write in the steps of the auditory pathway in Figure 15.9, and label and color-code important components of the process. You may use text Figure 15.36 as a reference, but write the steps in your own words.

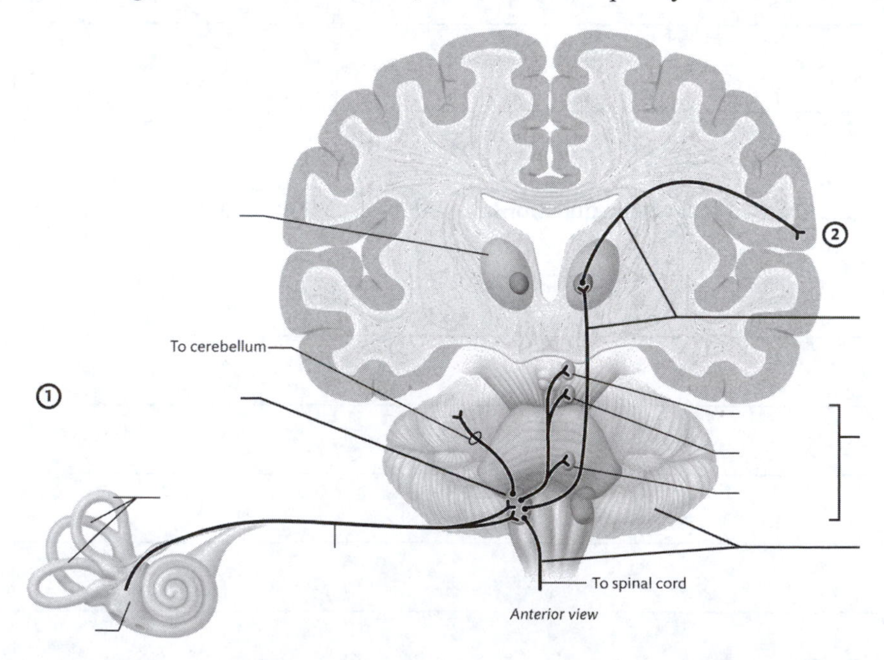

Figure 15.9 The vestibular sensation pathway.

Key Concept: What is the relationship between the movement of fluid in the semicircular canals and bending of the cupula in the crista ampullaris?

____

____

Module 15.9: How the Special Senses Work Together

Module 15.9 in your text discusses how the special senses work together to provide a complete understanding of the world around us. By the end of the module, you should be able to do the following:

1. Summarize the pathways for each of the special senses.
2. Describe how the frontal lobe and limbic system integrate the signals from the special senses into a meaningful picture of a situation.

Survey It: Form Questions

Before you read the module, survey it and form at least three questions for yourself. When you have finished reading the module, return to these questions and answer them.

Question 1: ____

Answer: ____

Question 2: ____

Answer: ____

Question 3: ____

Answer: ____

Key Concept: How do the frontal lobe and limbic system work together to integrate the signals from the special senses into a meaningful understanding of the world we are experiencing?

____

____

Describe It: How the Special Senses Work Together

Write in the steps summarizing how the special senses are processed and how their information is integrated. You may use text Figure 15.37 as a reference, but write the steps in your own words.

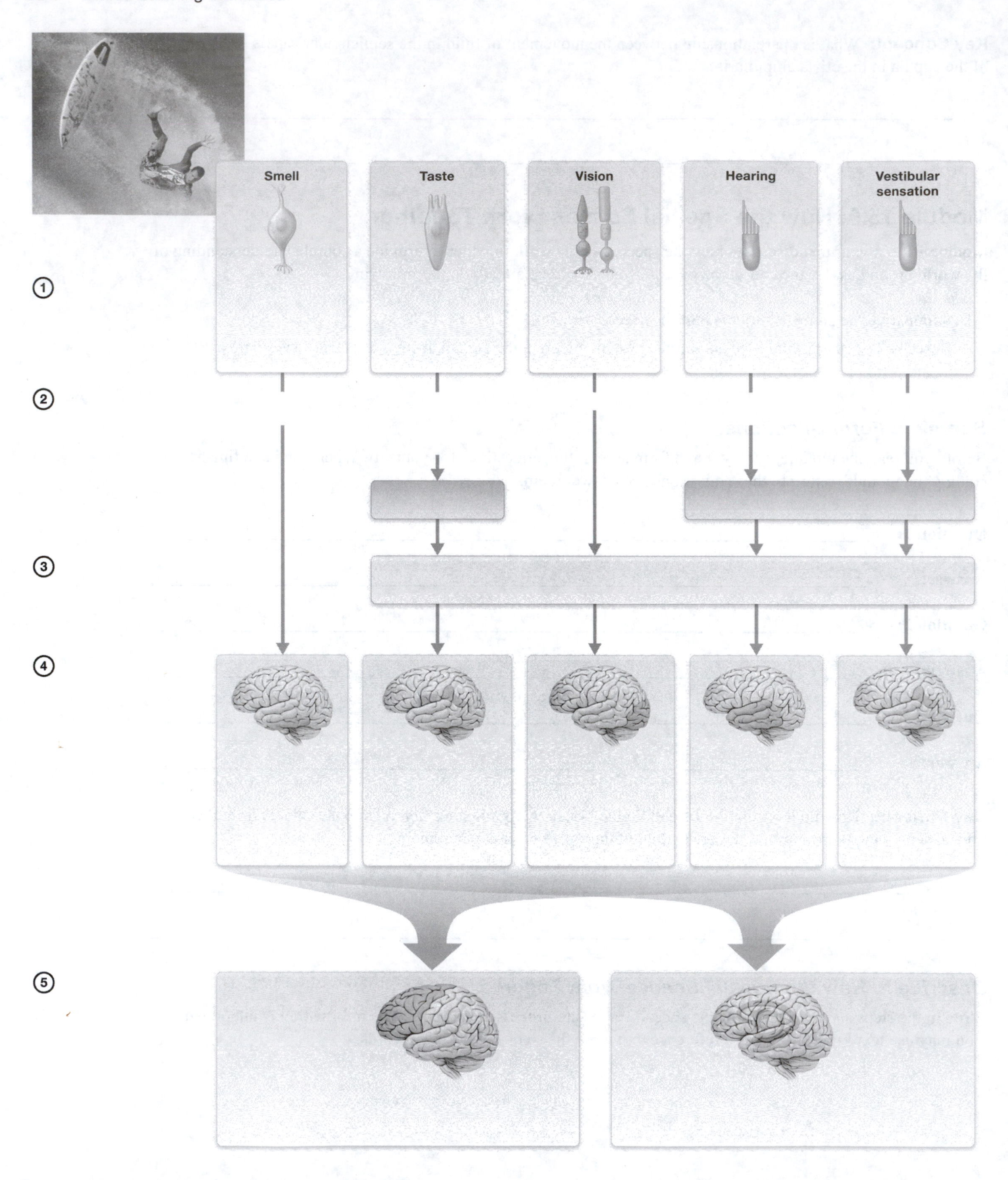

Figure 15.10 How the special senses work together.

What Do You Know Now?

Let's now revisit the questions you answered in the beginning of this chapter. How have your answers changed now that you've worked through the material?

- What are the five special senses?

  ____

  ____

- Which areas of the brain interpret information from the special senses?

  ____

  ____

- Which parts of the tongue convey each different taste sensation?

  ____

  ____

16 The Endocrine System

What Do You Already Know?

Try to answer the following questions before proceeding to the next section. If you're unsure of the correct answers, give it your best attempt based on previous courses, previous chapters, or just your general knowledge.

- What does it mean when someone has a "hormonal" or "glandular" disorder?

  ____

- Can a person actually become overweight due to a hormonal problem?

  ____

- Which of this book's Core Principles are exemplified by the endocrine system?

  ____

Module 16.1: Overview of the Endocrine System

Module 16.1 in your text introduces you to the endocrine system and the mechanism of action of its chemical messengers: hormones. By the end of the module, you should be able to do the following:

1. Compare and contrast how the endocrine and nervous systems control body functions.
2. Describe the major structures and functions of the endocrine system.
3. Explain the different types of chemical signaling used by the body.
4. Describe the major chemical classes of hormones found in the human body, and compare and contrast the types of receptors to which each class of hormone binds.
5. Describe several types of stimuli that control production and secretion of hormones, including the roles of negative and positive feedback.

Build Your Own Glossary

Following is a table listing key terms from Module 16.1. Before you read the module, use the glossary at the back of your book or look through the module to define the following terms.

Key Terms for Module 16.1

Term	Definition
Paracrine	
Autocrine	

Term	Definition
Endocrine gland	
Neuroendocrine organ	
Target cell	
Receptor	
Upregulation	
Downregulation	
Amino acid–based hormones	
Steroid hormones	
G protein	

Survey It: Form Questions

Before you read the module, survey it and form at least three questions for yourself. When you have finished reading the module, return to these questions and answer them.

Question 1: ____________________

Answer: ____________________

Question 2: ____________________

Answer: ____________________

Question 3: ____________________

Answer: ____________________

Key Concept: What is the difference between a neurotransmitter and a hormone?

____

____

Identify It: Structures of the Endocrine System

Identify and color-code each component of the endocrine system in Figure 16.1. Also, identify which of the organs are primary endocrine organs and which are neuroendocrine organs.

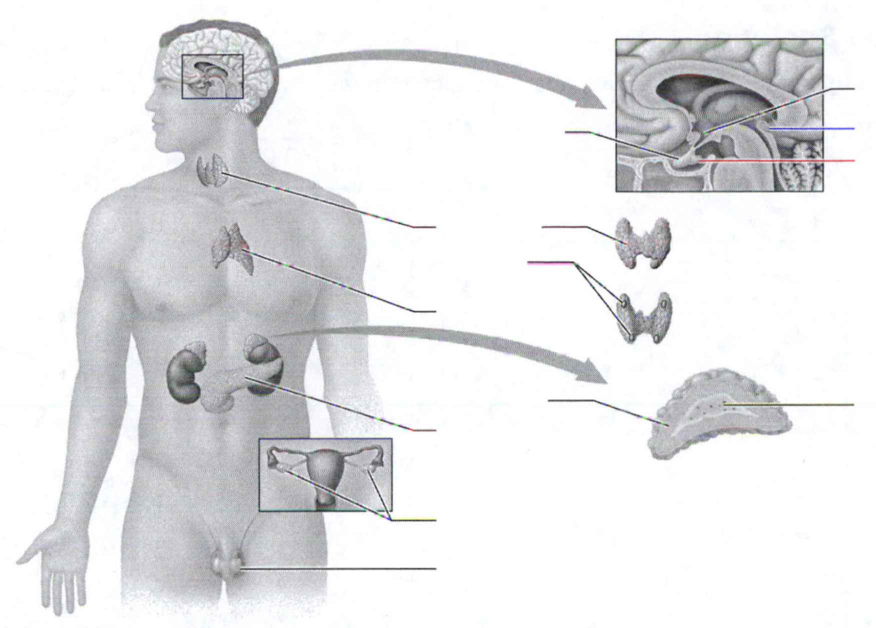

Figure 16.1 Overview of the endocrine system.

Key Concept: Why does a hydrophilic hormone generally have difficulty crossing the plasma membrane? Why does a hydrophobic hormone generally cross the plasma membrane easily?

____

____

Describe It: Mechanisms of Action of Hydrophilic and Hydrophobic Hormones

Write out the steps of the processes by which hydrophobic and hydrophilic hormones exert their effects in Figure 16.2. In addition, label and color-code all key components of these processes.

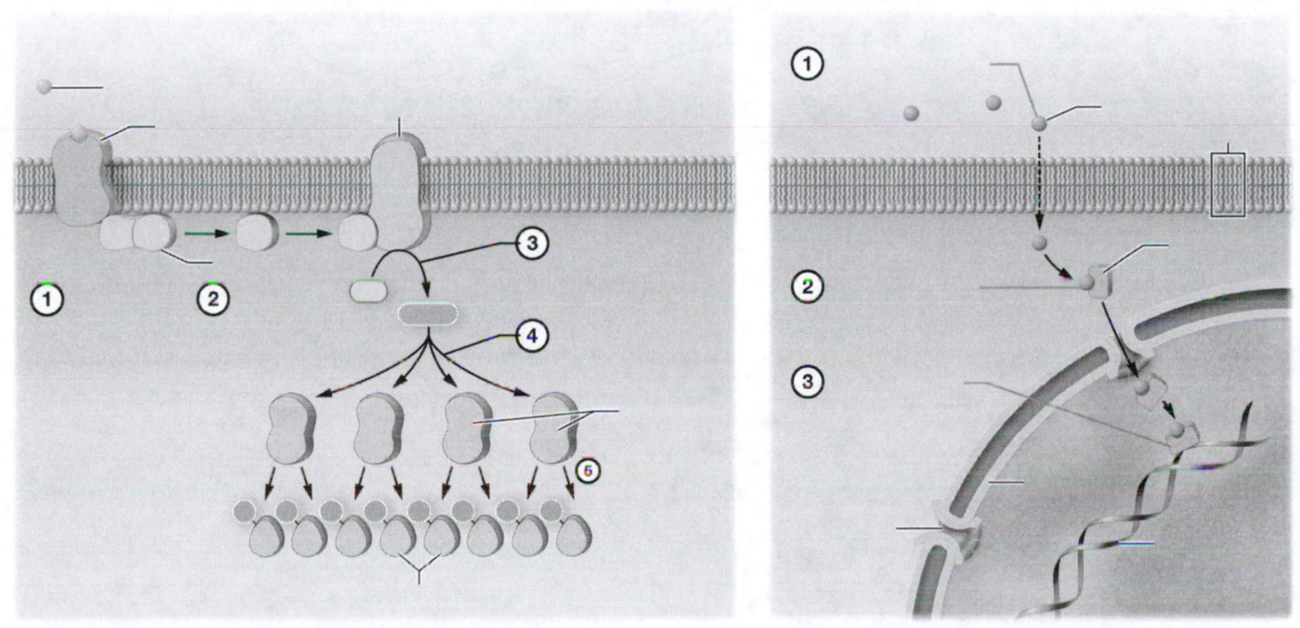

(a) Hydrophilic hormone and second-messenger system

(b) Hydrophobic hormone and intracellular receptor mechanism

Figure 16.2 Mechanisms of action of hydrophilic and hydrophobic hormones.

Draw It: Stimuli for Hormone Secretion

In the boxes below, draw diagrams of the three ways in which hormone secretion may be stimulated. Label and color-code important components of each process.

Describe It: Regulation of Hormone Secretion

Write out the steps of the negative feedback loop by which hormone secretion is regulated in Figure 16.3. In addition, label and color-code all key components of these processes.

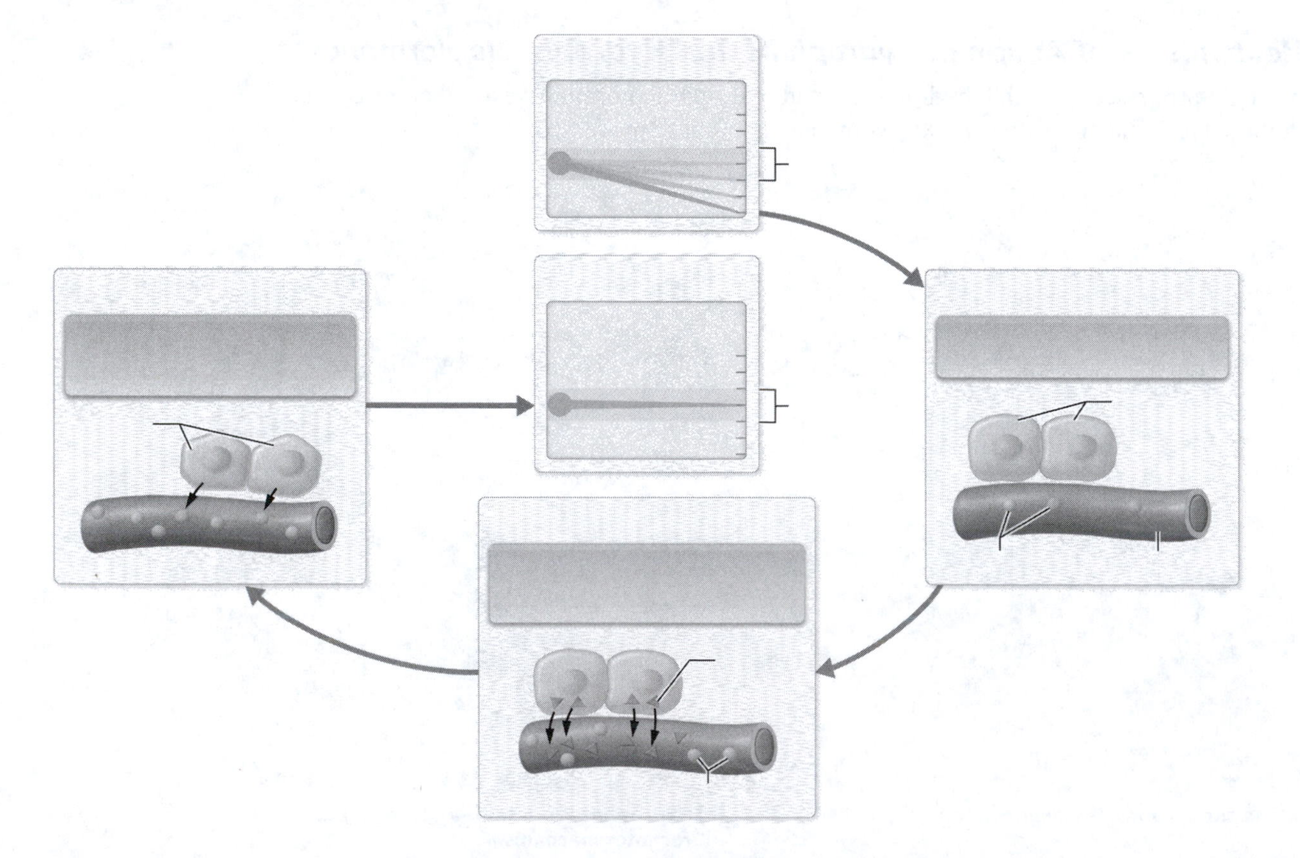

Figure 16.3 Maintaining homeostasis: regulation of hormone secretion by negative feedback loops.

Module 16.2: The Hypothalamus and the Pituitary Gland

Now we look at the anatomy and physiology of two closely structurally and functionally related endocrine organs, the hypothalamus and the pituitary gland. When you finish this module, you should be able to do the following:

1. Describe the anatomical and functional relationships between the hypothalamus and the anterior and posterior pituitary glands.
2. Explain how tropic hormones affect the secretion of other hormones.
3. Describe the stimulus for release, the target tissue, and the functional effect of each hormone released from the posterior pituitary.
4. Describe the stimulus for release, the target tissue, and the functional effect of each hormone produced and secreted by the anterior pituitary.
5. Explain the negative feedback loops that regulate the production and release of anterior pituitary hormones.

Build Your Own Glossary

Following is a table listing key terms from Module 16.2. Before you read the module, use the glossary at the back of your book or look through the module to define the following terms.

Key Terms for Module 16.2

Term	Definition
Hypothalamus	
Anterior pituitary	
Posterior pituitary	
Hypothalamic-hypophyseal portal system	
Tropic hormone	
Releasing hormone	
Inhibiting hormone	
Gigantism	
Acromegaly	
Pituitary dwarfism	

Survey It: Form Questions

Before you read the module, survey it and form at least three questions for yourself. When you have finished reading the module, return to these questions and answer them.

Question 1: ______________________________

Answer: ______________________________

Question 2: ______________________________

Answer: ______________________________

Question 3: ______________________________

Answer: ______________________________

Key Concept: Why is the anterior pituitary also called the adenohypophysis? Why is the posterior pituitary also called the neurohypophysis?

____

____

Describe It: Relationship between the Hypothalamus and Pituitary

Write out the steps by which the hypothalamus interacts with the anterior and posterior pituitary glands in Figure 16.4. In addition, label and color-code all key components of these processes.

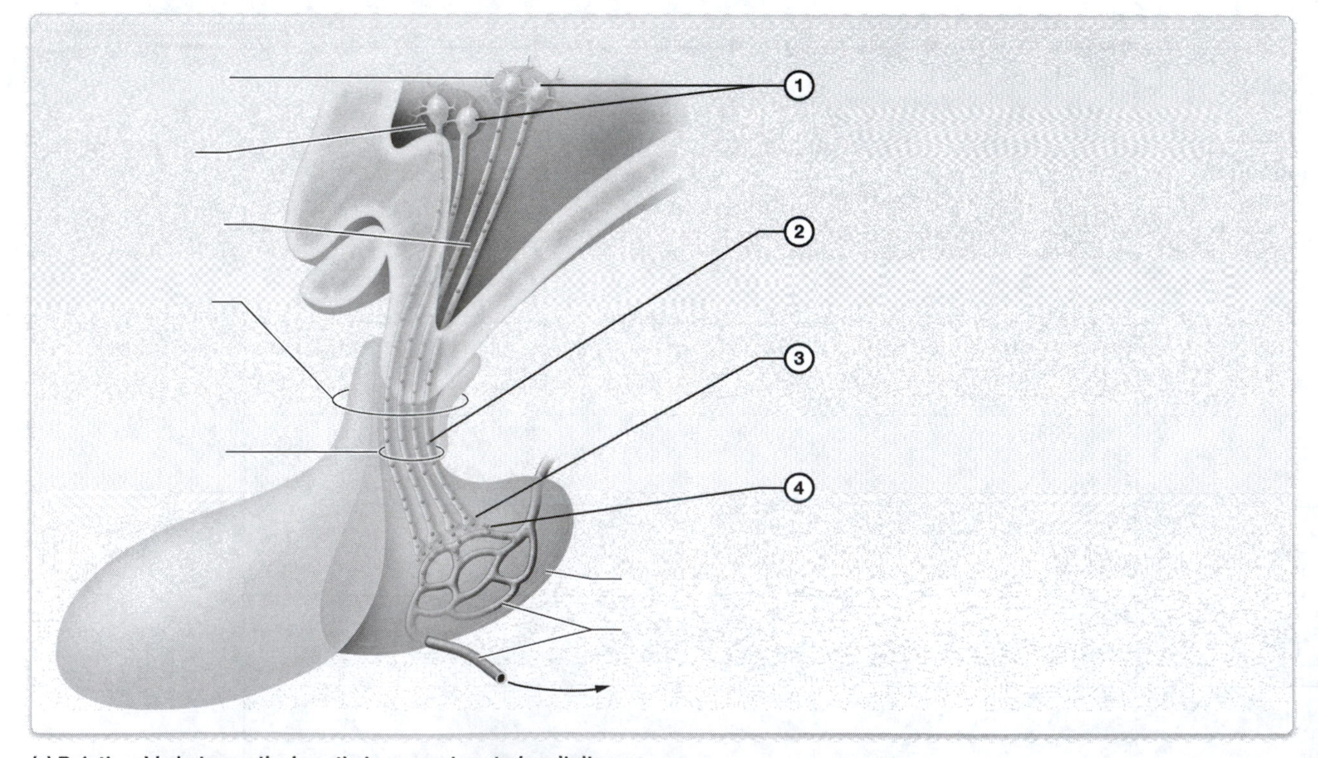

(a) Relationship between the hypothalamus and posterior pituitary

Figure 16.4 Functional relationships between the hypothalamus and pituitary glands.

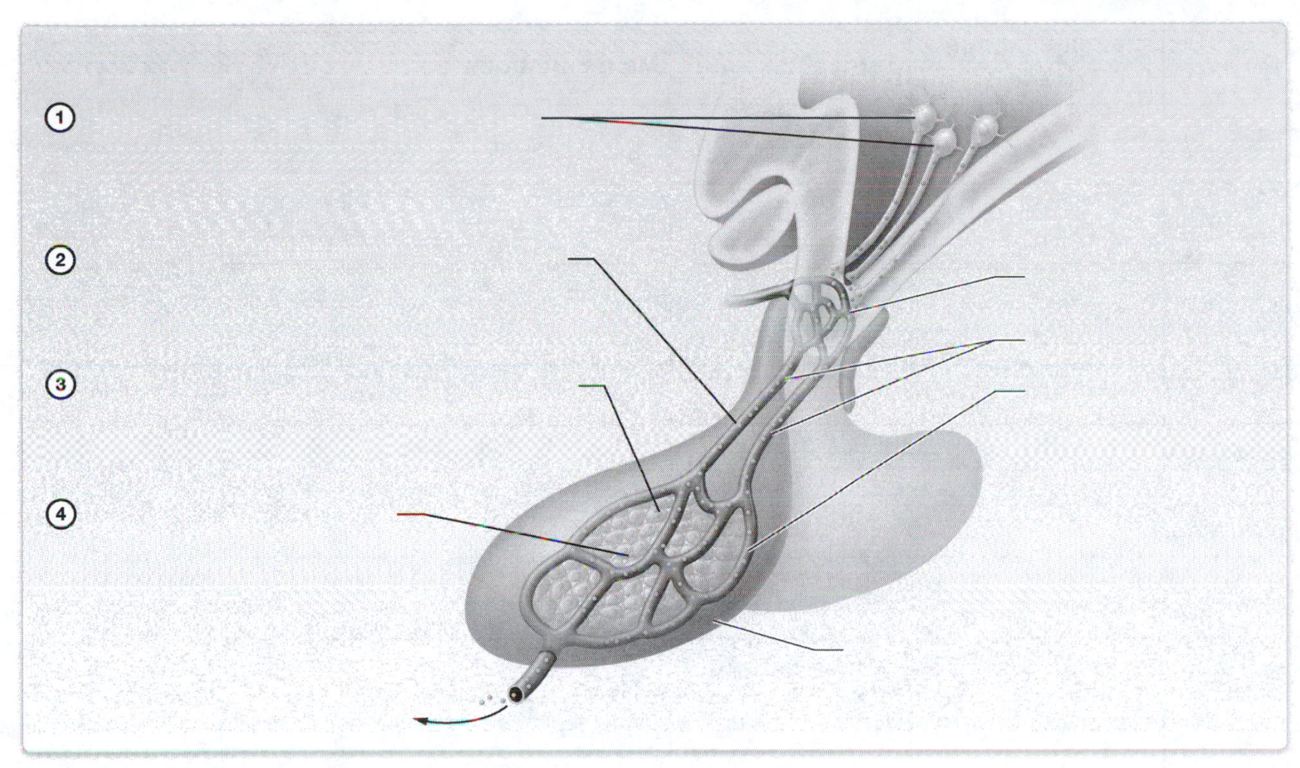

(b) Relationship between the hypothalamus and anterior pituitary

Figure 16.4 *(continued)*

Key Concept: What are the advantages of multiple tiers of control in hormone secretion?

Build Your Own Summary Table: Hormones of the Hypothalamus and Anterior Pituitary

As you read Module 16.2, build your own summary table about the hormones of the hypothalamus and anterior pituitary by filling in the template provided.

Summary of the Hypothalamic and Anterior Pituitary Hormones

Hormone	Stimulus/Inhibitor for Release	Target Tissue(s)	Main Functions
Hypothalamic Hormones			
Antidiuretic hormone			
Oxytocin			
Thyrotropin-releasing hormone			
Somatostatin			

Hormone	Stimulus/Inhibitor for Release	Target Tissue(s)	Main Functions
Corticotropin-releasing hormone			
Prolactin-releasing hormone			
Prolactin-inhibiting factor			
Gonadotropin-releasing hormone			
Growth hormone–releasing hormone			
Anterior Pituitary Hormones			
Thyroid-stimulating hormone			
Adrenocorticotropic hormone			
Prolactin			
Follicle-stimulating hormone			
Luteinizing hormone			
Growth hormone			

Key Concept: How are the effects of growth hormones different from the other anterior pituitary hormones? What is the difference between growth hormone's short-term and long-term effects?

___

___

Team Up

Team up with one or more partners and have each person write a 10-question quiz on the hormones of the hypothalamus and anterior pituitary. Once you have completed writing your quiz questions, combine them and take the whole quiz. When you and your group have finished taking the quiz, discuss the answers to determine places where you need additional study.

Module 16.3: The Thyroid and Parathyroid Glands

This module explores the structure and function of the thyroid and parathyroid glands and their hormones. When you complete this module, you should be able to do the following:

1. Describe the gross and microscopic anatomy of the thyroid and parathyroid glands.
2. Identify and describe the types of cells within the thyroid gland that produce thyroid hormone and calcitonin.

3. Describe the stimulus for release, the target tissue, and the effects of thyroid hormone.
4. Explain how negative feedback loops regulate the production of thyroid hormones.
5. Describe the stimulus for release, the target tissue, and the effects of parathyroid hormone.

Build Your Own Glossary

Following is a table listing key terms from Module 16.3. Before you read the module, use the glossary at the back of your book or look through the module to define the following terms.

Key Terms of Module 16.3

Term	Definition
Thyroid gland	
Thyroid follicle	
Colloid	
Triiodothyronine	
Thyroxine	
Parathyroid gland	
Parathyroid hormone	
Calcitonin	

Survey It: Form Questions

Before you read the module, survey it and form at least three questions for yourself. When you have finished reading the module, return to these questions and answer them.

Question 1: ______________________________

Answer: ______________________________

Question 2: ______________________________

Answer: ______________________________

Question 3: ______________________________

Answer: ______________________________

Identify It: Structure of the Thyroid Gland

Identify and color-code the macroscopic and microscopic structures of the thyroid gland in Figure 16.5.

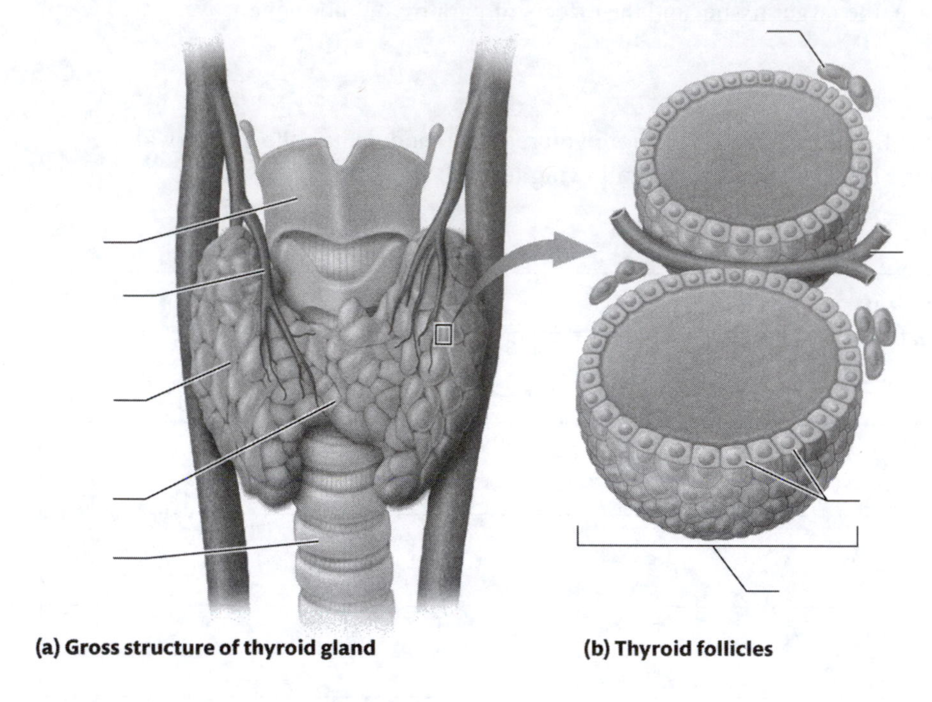

(a) Gross structure of thyroid gland (b) Thyroid follicles

Figure 16.5 Anatomy and histology of the thyroid gland.

Complete It: Functions of Thyroid Hormone

Fill in the blanks to complete the following paragraph that describes the functions of thyroid hormones.

Thyroid hormones are major regulators of the ____________ ____________ and the process of heat generation, known as ____________. One way they accomplish this is to cause the ____________ ____________ pumps to function at a faster rate and increasing the rate of ____________ reactions. Thyroid hormones also promote ____________ and ____________, and have synergistic actions with the ____________ ____________ ____________.

Key Concept: What would you expect to happen to the metabolic rate if thyroid hormone synthesis decreased? Why? Be specific.

____

____

Describe It: Thyroid Hormone Production

Write out the steps by which thyroid hormone is produced in Figure 16.6. In addition, label and color-code all key components of these processes.

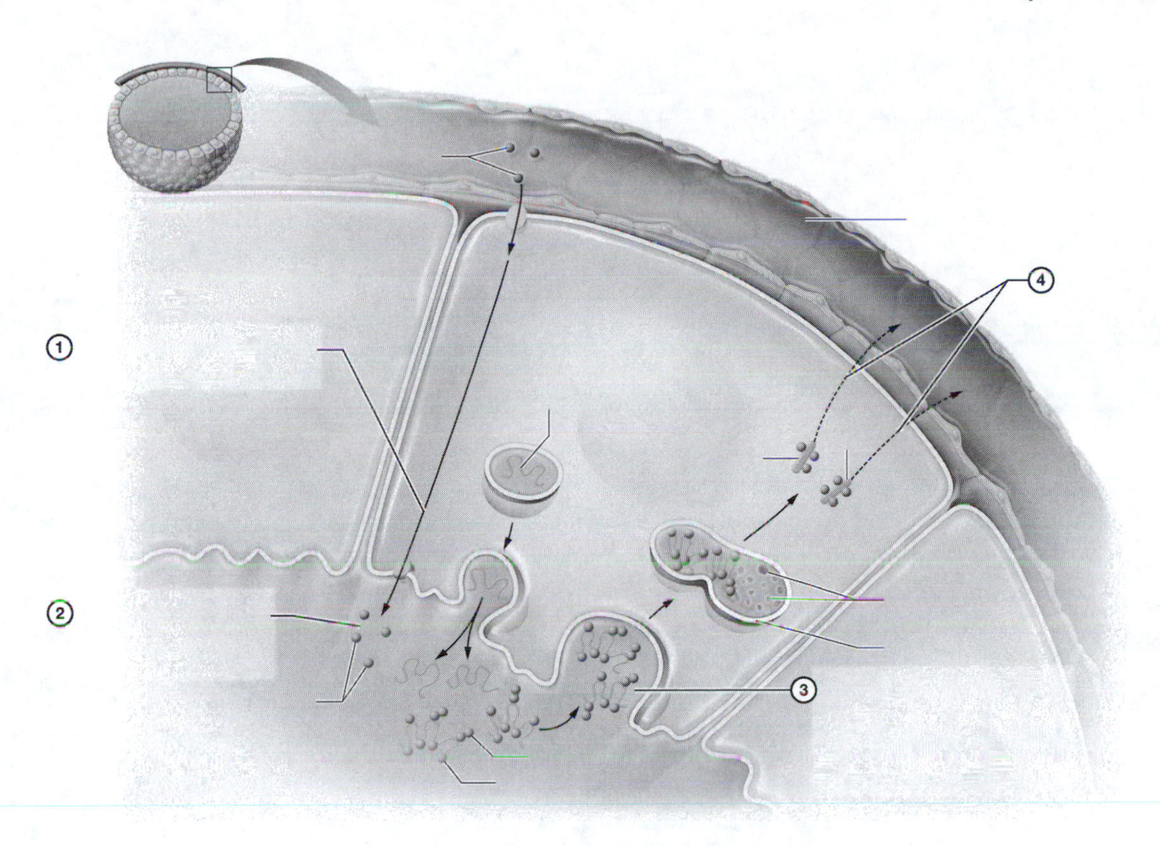

Figure 16.6 Production of thyroid hormone.

Key Concept: What is the importance of iodine to thyroid hormone synthesis? What do you think would happen if insufficient iodine were available?

____

____

Key Concept: Why does the level of TSH increase in hypothyroidism? What happens to the level of TSH in hyperthyroidism? Explain.

____

____

Describe It: Regulation of Calcium Ion Concentration by Parathyroid Hormone

Write out the steps of the negative feedback loop by which calcium ion concentration is regulated in Figure 16.7. In addition, label and color-code all key components of these processes.

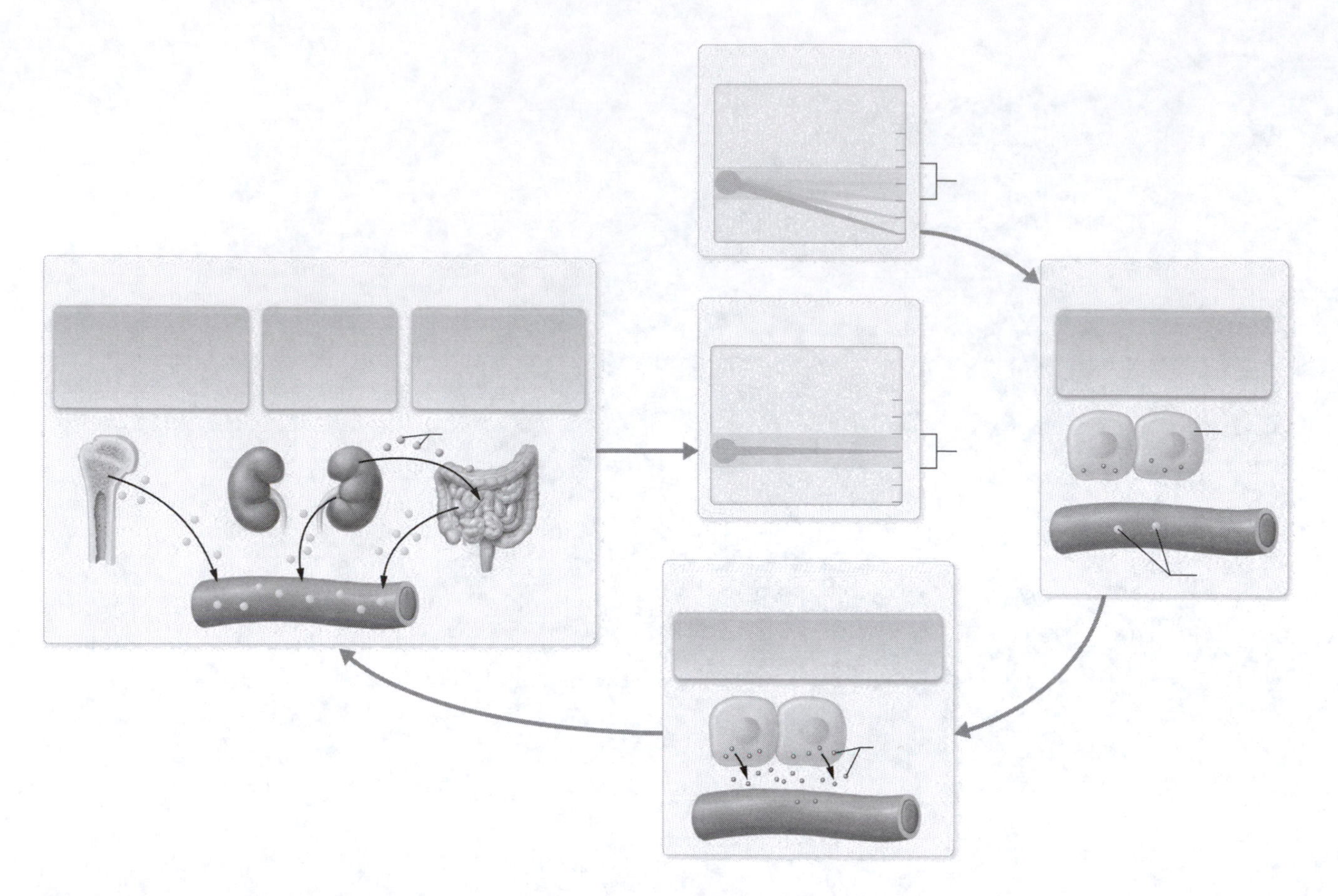

Figure 16.7 Maintaining homeostasis: regulation of blood calcium concentration by a negative feedback loop.

Key Concept: How are parathyroid hormone and calcitonin antagonists?

____

____

Module 16.4: The Adrenal Glands

This module examines the structure and function of the adrenal cortex and adrenal medulla. At the end of this module, you should be able to do the following:

1. Describe the gross and microscopic anatomy of the cortex and medulla of the adrenal gland.
2. Describe the stimulus for release, the target tissue, and the effects of the mineralocorticoids secreted by the adrenal cortex.
3. Describe the stimulus for release, the target tissue, and the effects of the glucocorticoids secreted by the adrenal cortex.
4. Explain the relationship of the adrenal medulla to the sympathetic nervous system.
5. Describe the stimulus for release, the target tissue, and the effect of catecholamines.

Build Your Own Glossary

Following is a table listing key terms from Module 16.4. Before you read the module, use the glossary at the back of your book or look through the module to define the following terms.

Key Terms of Module 16.4

Term	Definition
Adrenal cortex	
Adrenal medulla	
Mineralocorticoids	
Glucocorticoids	
Stress response	
Cushing's disease	
Androgenic steroids	
Chromaffin cells	
Catecholamines	

Survey It: Form Questions

Before you read the module, survey it and form at least two questions for yourself. When you have finished reading the module, return to these questions and answer them.

Question 1: ______________________________

Answer: ______________________________

Question 2: ______________________________

Answer: ______________________________

Complete It: Hormones of the Adrenal Cortex

Fill in the blanks to complete the following paragraphs that describe the hormones of the adrenal cortex.

The outermost zone of the adrenal cortex is the ____________ ____________, which produces hormones called ____________, the main one of which is ____________. This hormone impacts water balance by causing the retention of ____________ ____________, which causes water ____________. It also increases secretion of ____________ ____________ and ____________ ____________. This helps to maintain ____________ pressure, the concentration of ____________ ____________ and ____________ ____________ in the extracellular fluid, and ____________ ____________ homeostasis.

The next two zones of the adrenal cortex are the ____________ ____________ and the ____________ ____________. These zones produce hormones called ____________, one of the main ones of which is ____________. This hormone helps the body adapt to ____________ by causing metabolic changes, including stimulating ____________ in the liver, stimulating the release of ____________ ____________ from muscle tissue, and stimulating the release of ____________ ____________ from adipose tissue. This hormone also has effects on the inflammatory response, ____________ inflammation.

Key Concept: Are aldosterone and cortisol hydrophobic or hydrophilic hormones? What does this mean about how they likely interact with their target cells?

____

____

Describe It: The Hypothalamic-Pituitary-Adrenal Axis

Write out the steps by which the hypothalamus interacts with the anterior pituitary and adrenal glands in Figure 16.8. In addition, label and color-code all key components of these processes.

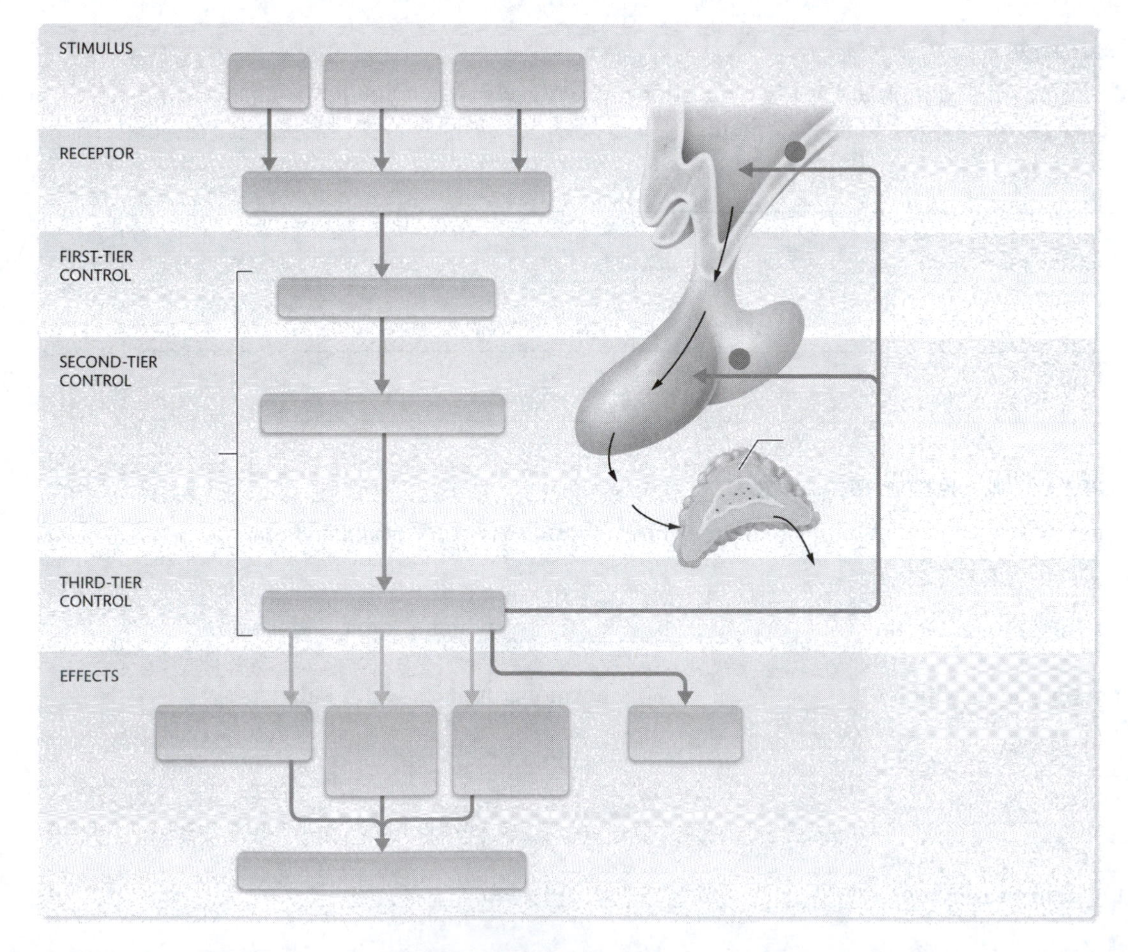

Figure 16.8 Regulation of cortisol production by the HPA axis.

Key Concept: How are the cells of the adrenal cortex and medulla different, structurally and functionally?

____

____

Key Concept: What role does the adrenal medulla play in the sympathetic nervous system response? What hormones does it release, and what are their effects on target organs?

____

____

Module 16.5: The Endocrine Pancreas

This module looks at the structure, function, and hormones of the endocrine portion of the pancreas. At the end of this module, you should be able to do the following:

1. Describe the structure of the endocrine pancreas and its hormone-secreting cells.
2. Describe the stimulus for release, the target tissue, and the effect of glucagon.
3. Describe the stimulus for release, the target tissue, and the effect of insulin.
4. Explain how insulin and glucagon work together to maintain the blood glucose level within the normal range.
5. Describe the causes, symptoms, and treatments for the two types of diabetes mellitus.

Build Your Own Glossary

Following is a table listing key terms from Module 16.5. Before you read the module, use the glossary at the back of your book or look through the module to define the following terms.

Key Terms of Module 16.5

Term	Definition
Pancreas	
Pancreatic islet	
Glucagon	
Alpha cells	
Insulin	
Beta cells	
Hypoglycemia	
Hyperglycemia	

Term	Definition
Type 1 diabetes mellitus	
Type 2 diabetes mellitus	

Survey It: Form Questions

Before you read the module, survey it and form at least three questions for yourself. When you have finished reading the module, return to these questions and answer them.

Question 1: ____________________

Answer: ____________________

Question 2: ____________________

Answer: ____________________

Question 3: ____________________

Answer: ____________________

Identify It: Structure of the Pancreas

Identify and color-code the gross and microscopic structures of the pancreas in Figure 16.9.

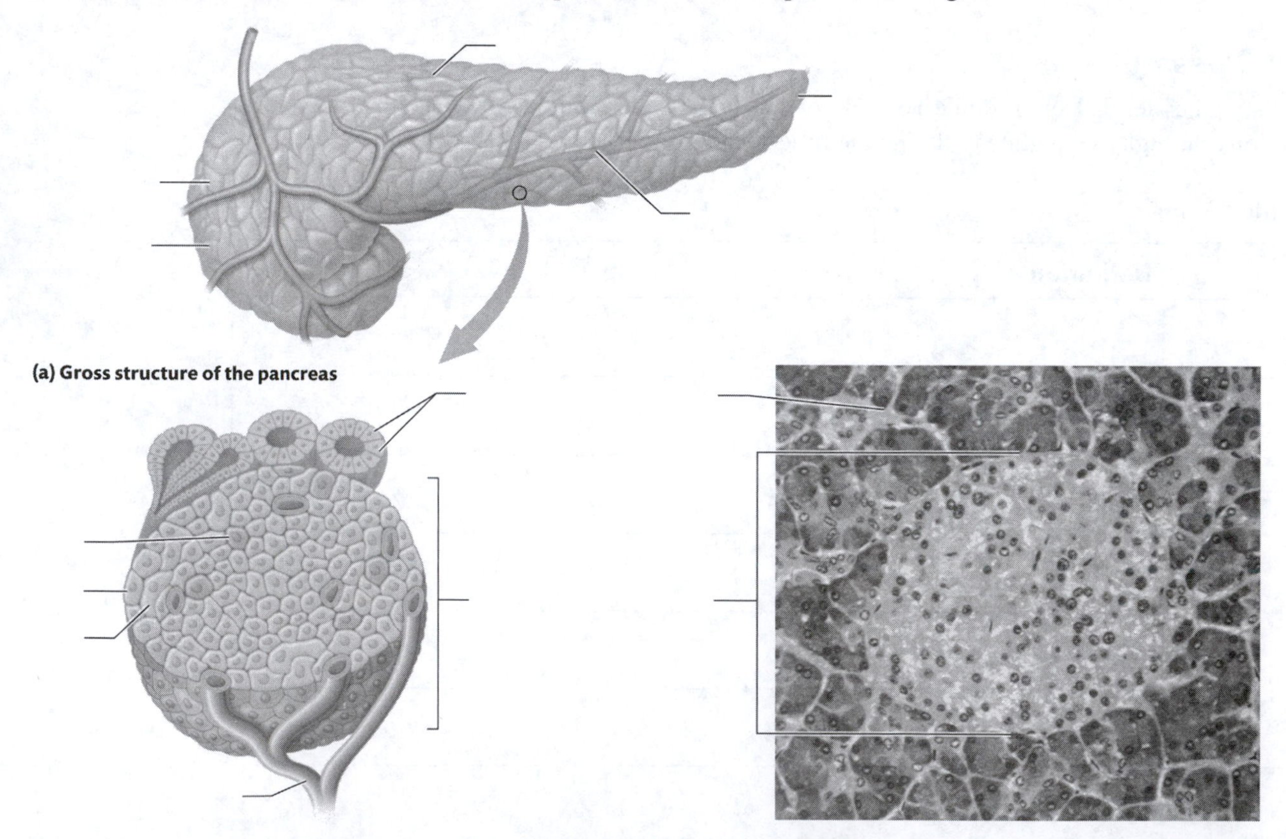

Figure 16.9 Anatomy and histology of the pancreas.

Build Your Own Summary Table: Glucagon and Insulin

As you read Module 16.5, build your own summary table about insulin and glucagon by filling in the information in the template provided.

Summary of the Hypothalamic and Anterior Pituitary Hormones

Property	Glucagon	Insulin
Cell type		
Stimulus/stimuli for release		
Main target tissues		
Overall effect on blood glucose		
Effects on target tissues		

Key Concept: How are glucagon and insulin antagonists?

____

____

Key Concept: Mr. Sorgi is a type I diabetic patient who forgot to take his insulin and is now feeling faint. Will his blood glucose be high or low? Why? Will giving him glucose help his condition?

____

____

Draw It: Feedback Loops for Glucose Control

Draw, color-code, and label a diagram of a negative feedback loop for increased and decreased blood glucose levels. You can use the figure on page 618 in your text as a guide, but make the figure your own and ensure that it makes sense to you.

a. Increased blood glucose

b. Decreased blood glucose

Module 16.6: Other Endocrine Glands and Hormone-Secreting Tissues

Module 16.6 in your text examines the other organs and tissues that are either primary or secondary endocrine organs. By the end of the module, you should be able to do the following:

1. Describe the stimulus for release, the target tissue, and the effect of the hormones produced by the pineal and thymus glands.
2. Describe the stimulus for release, the target tissue, and the effect of the hormones produced by the gonads, adipose tissue, the heart, and the kidneys.

Build Your Own Glossary

Following is a table listing key terms from Module 16.6. Before you read the module, use the glossary at the back of your book or look through the module to define the following terms.

Key Terms for Module 16.6

Term	Definition
Pineal gland	
Melatonin	
Thymus	
Testes	
Ovaries	
Leptin	
Atrial natriuretic peptide	
Erythropoietin	

Survey It: Form Questions

Before you read the module, survey it and form at least three questions for yourself. When you have finished reading the module, return to these questions and answer them.

Question 1: ______________________________

Answer: ______________________________

Question 2: ______________________________

Answer: ______________________________

Question 3: ______________________________

Answer: ______________________________

Complete It: Other Hormone-Secreting Tissues

Complete the following statements describing the properties of the other hormone-secreting tissues discussed in Module 16.6.

- The pineal gland is located in the ____________ and secretes the hormone ____________, which is involved in regulating ____________.
- The thymus gland is located in the ____________ and secretes the hormones ____________ and ____________, which are involved in ____________ ____________ ____________.
- The testes are organs known as ____________ that produce the hormone ____________, which has both ____________ effects and ____________ effects.
- The ovaries are the female ____________ that produce the hormones ____________ and ____________, which are involved in: ____

  ____

  ____
- Adipose tissue produces the hormone ____________, which is involved in promoting ____________.
- The cells of the atrial muscle of the heart produce ____________ ____________ ____________, which causes sodium ion and water ____________ to the urine.
- The kidneys produce the hormone ____________, which stimulates the development of ____________. They also produce ____________ and the active form of ____________.

Module 16.7: Three Examples of Endocrine Control of Physiological Variables

The final module takes a big picture look at how the endocrine system controls different variables in the body in order to maintain homeostasis. By the end of the module, you should be able to do the following:

1. Provide specific examples to demonstrate how hormones maintain homeostasis in the body.
2. Describe how hormones work together with the nervous system to mediate the stress response.

Survey It: Form Questions

Before you read the module, survey it and form at least two questions for yourself. When you have finished reading the module, return to these questions and answer them.

Question 1: ______________________________

Answer: ______________________________

Question 2: ______________________________

Answer: ______________________________

Predict It: Hormones and Homeostasis

Predict the hormonal response to each of the following situations. State which hormones will be secreted in response and the effects each hormone will trigger to restore homeostasis.

- A person drinks 2 gallons of water in a short period of time, increasing plasma volume significantly:
- A person drinks nothing all day except a whole bottle of soy sauce, which has an extremely high sodium ion concentration.
- A person eats a meal that is low in carbohydrates but rich in proteins.
- A person accidentally stepped in an alligator's nest and is now being chased by a very unhappy mother alligator.

What Do You Know Now?

Let's now revisit the questions you answered in the beginning of this chapter. How have your answers changed now that you've worked through the material?

- What does it mean when someone has a "hormonal" or "glandular" disorder?

  ____

- Can a person actually become overweight due to a hormonal problem?

  ____

- Which of this book's Core Principles are exemplified by the endocrine system?

  ____

17 The Cardiovascular System I: The Heart

What Do You Already Know?

Try to answer the following questions before proceeding to the next section. If you're unsure of the correct answers, give it your best attempt based on previous courses, previous chapters, or just your general knowledge.

- What is/are the functions of the heart?

  ____

- How are the two sides of the heart different?

  ____

- Do arteries carry oxygenated or deoxygenated blood? What about veins?

  ____

Module 17.1: Overview of the Heart

Module 17.1 in your text introduces you to the basic structure, location, and functions of the heart. By the end of the module, you should be able to do the following:

1. Describe the position of the heart in the thoracic cavity.
2. Describe the basic surface anatomy of the chambers of the heart.
3. Explain how the heart functions as a double pump and why this is significant.

Build Your Own Glossary

Following is a table listing key terms from Module 17.1. Before you read the module, use the glossary at the back of your book or look through the module to define the following terms.

Key Terms for Module 17.1

Term	Definition
Atria	
Ventricles	
Atrioventricular sulcus	
Arteries	
Veins	

Term	Definition
Pulmonary circuit	
Capillaries	
Systemic circuit	

Survey It: Form Questions

Before you read the module, survey it and form at least three questions for yourself. When you have finished reading the module, return to these questions and answer them.

Question 1: ______________________________

Answer: ______________________________

Question 2: ______________________________

Answer: ______________________________

Question 3: ______________________________

Answer: ______________________________

Identify It: Basic Anatomy of the Heart

Identify and color-code the structures in Figure 17.1. Also, identify which side of the heart is the pulmonary pump and which side is the systemic pump.

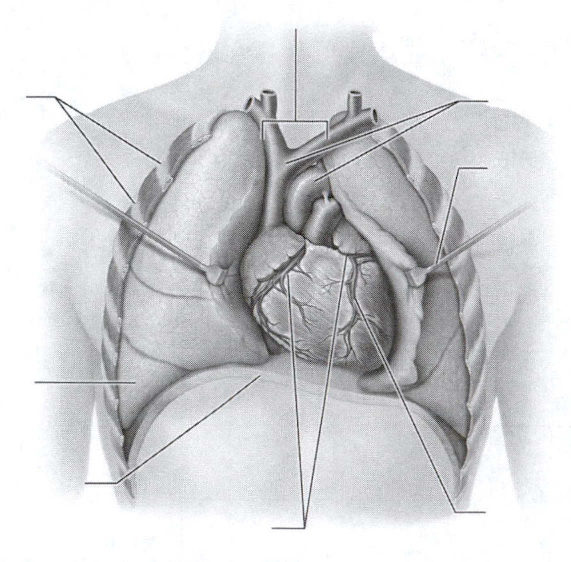

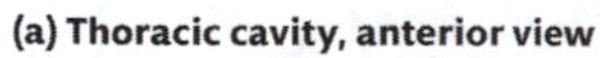

(a) Thoracic cavity, anterior view

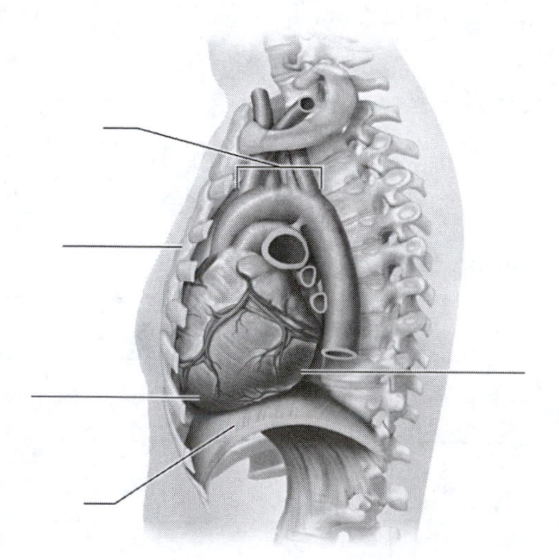

(b) Thoracic cavity, left lateral view

Figure 17.1 Basic anatomy of the heart.

Key Concept: What are the key differences between the right side and left side of the heart?

____

____

Describe It: The Pulmonary and Systemic Circuits

Write out the steps of the blood flow through the pulmonary and systemic circuits in Figure 17.2. In addition, label and color-code the blood in the vessels and heart to indicate if it is oxygenated (red) or deoxygenated (blue).

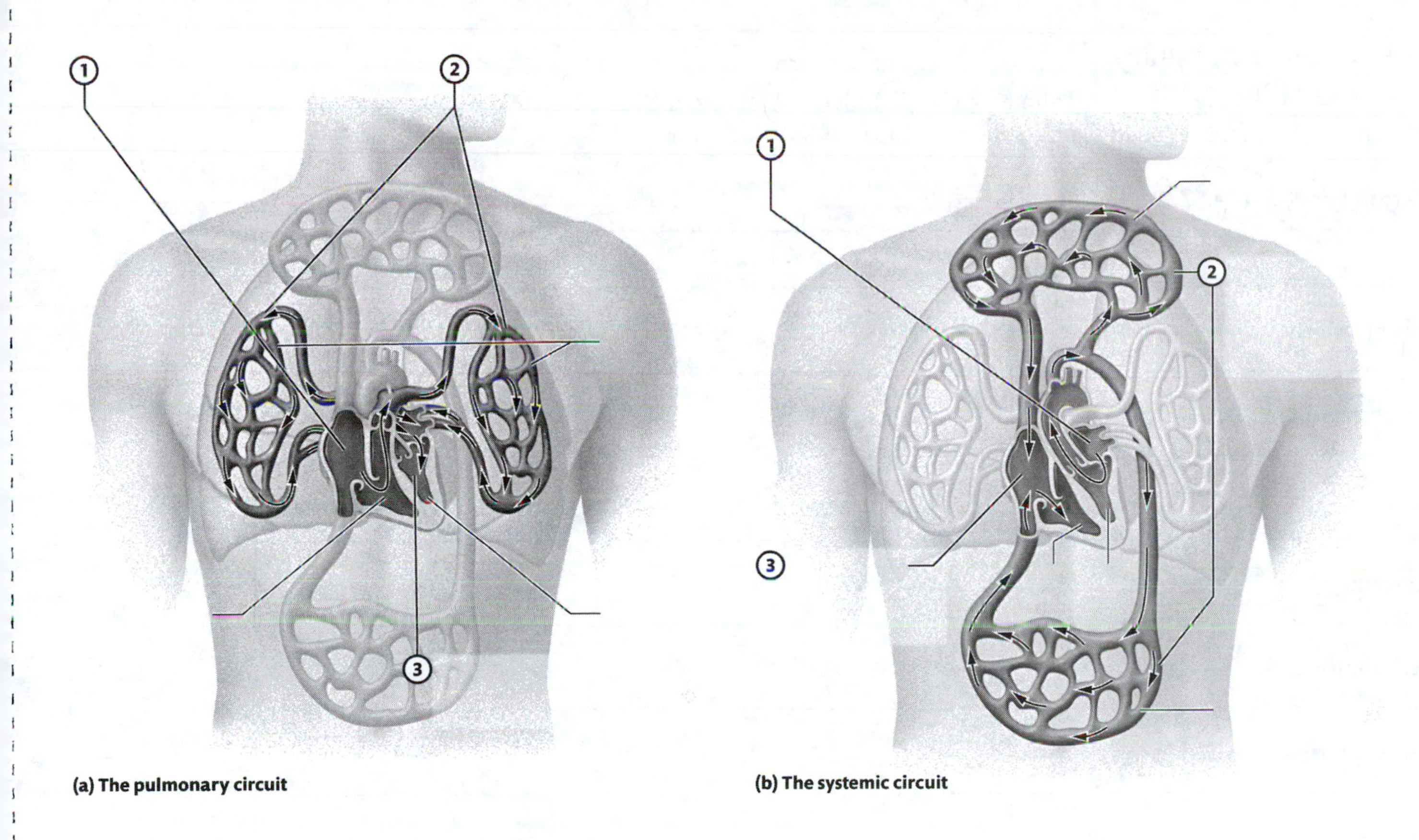

Figure 17.2 The pulmonary and systemic circuits.

Module 17.2: Heart Anatomy and Blood Flow Pathway

Now we look more closely at the anatomy of the heart and the pathway that blood takes as it passes through the heart. When you finish this module, you should be able to do the following:

1. Describe the layers of the pericardium and heart wall.
2. Describe the location and function of the coronary circulation and great vessels.
3. Describe the structure and function of the chambers, septa, valves, and other structural features of the heart.
4. Trace the pathway of blood flow through the heart, and explain how structures of the heart ensure that blood flows in a single direction.

Build Your Own Glossary

Following is a table listing key terms from Module 17.2. Before you read the module, use the glossary at the back of your book or look through the module to define the following terms.

Key Terms for Module 17.2

Term	Definition
Fibrous pericardium	
Serous pericardium	
Pericardial cavity	
Myocardium	
Endocardium	
Coronary arteries	
Coronary veins	
Great vessels	
Papillary muscles	
Atrioventricular valves	
Chordae tendineae	
Semilunar valves	

Survey It: Form Questions

Before you read the module, survey it and form at least four questions for yourself. When you have finished reading the module, return to these questions and answer them.

Question 1: ______________________________

Answer: ______________________________

Question 2: ______________________________

Answer: ______________________________

Question 3: ______________________________

Answer: ______________________________

Question 4: ______________________________

Answer: ______________________________

Identify It: The Pericardium and Layers of the Heart Wall

Identify and color-code the structures in Figure 17.3. Also, describe the functions of each structure that you label.

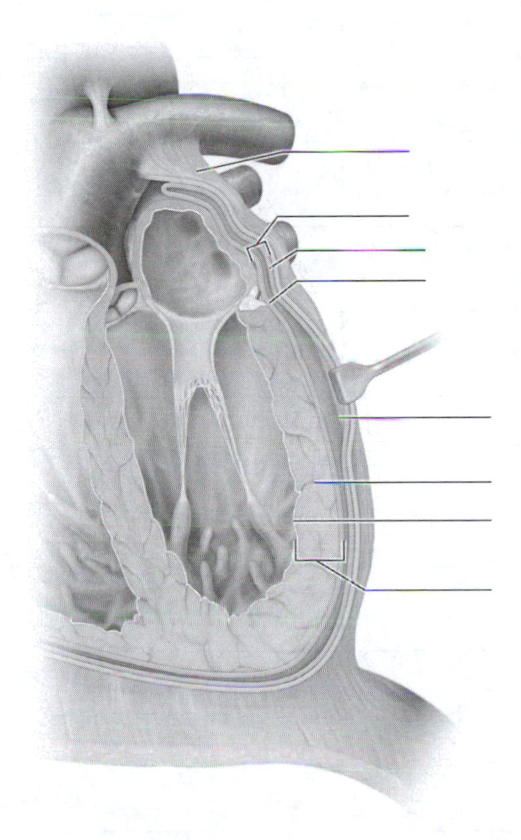

Figure 17.3 The pericardium and layers of the heart wall.

Key Concept: How does the serous pericardium envelop the heart? What is its function?

____

____

Build Your Own Summary Table: Coronary Vessels

As you read Module 17.2, build your own summary table about the coronary vessels by filling in the information in the following table.

Summary of the Hypothalamic and Anterior Pituitary Hormones

Vessel	Location	Area Supplied or Drained
Coronary Arteries		
Right coronary artery		
Marginal artery		
Posterior interventricular artery		
Left coronary artery		
Anterior interventricular artery		
Circumflex artery		
Coronary Veins		
Great cardiac vein		
Middle cardiac vein		
Small cardiac vein		
Coronary sinus		

Key Concept: What happens when blood flow through one or more of the coronary arteries is blocked? What can cause blockage of coronary arteries?

____

____

Identify It: Anatomy of the Heart

Identify and color-code the structures of the heart in Figure 17.4.

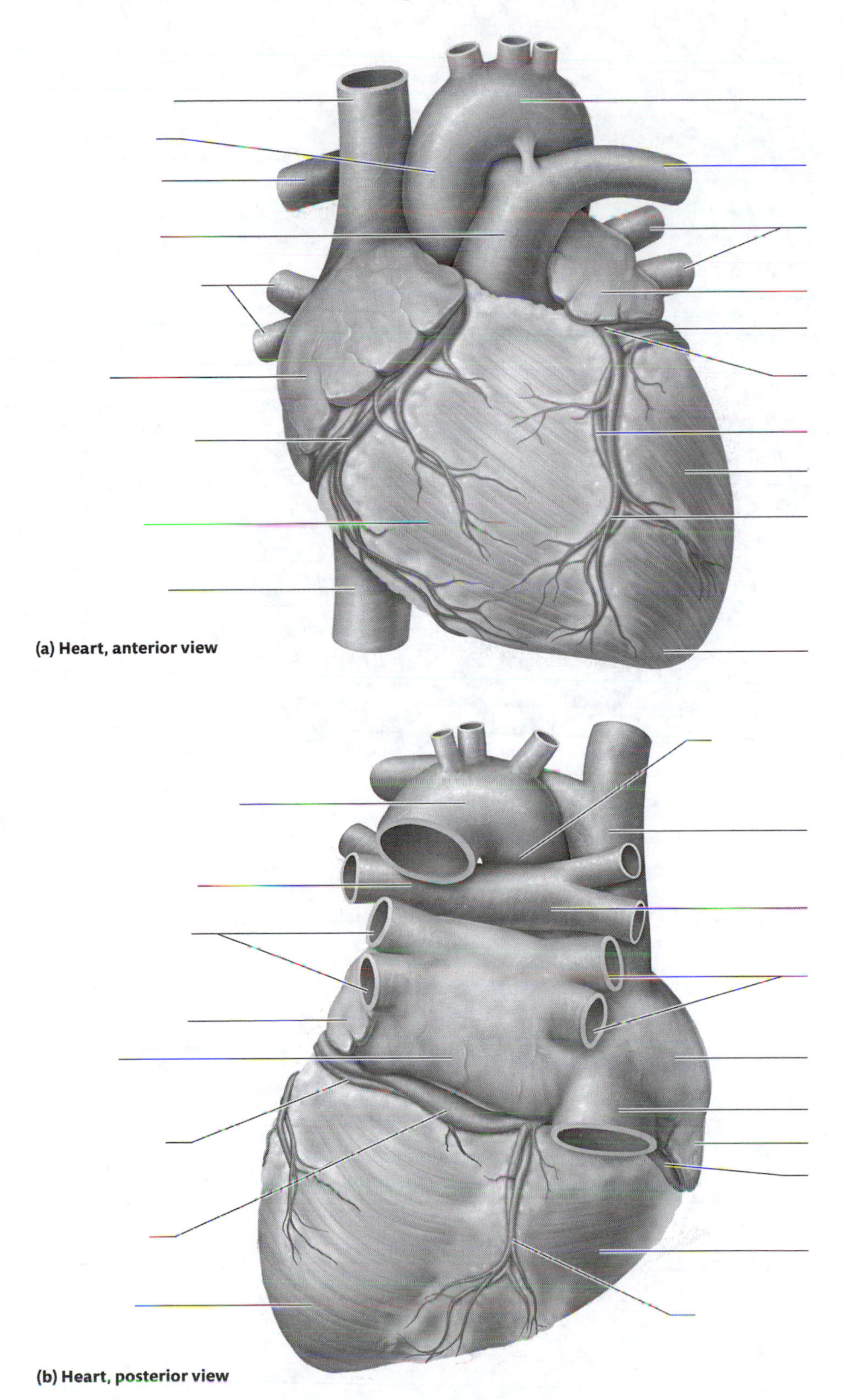

Figure 17.4 The heart.

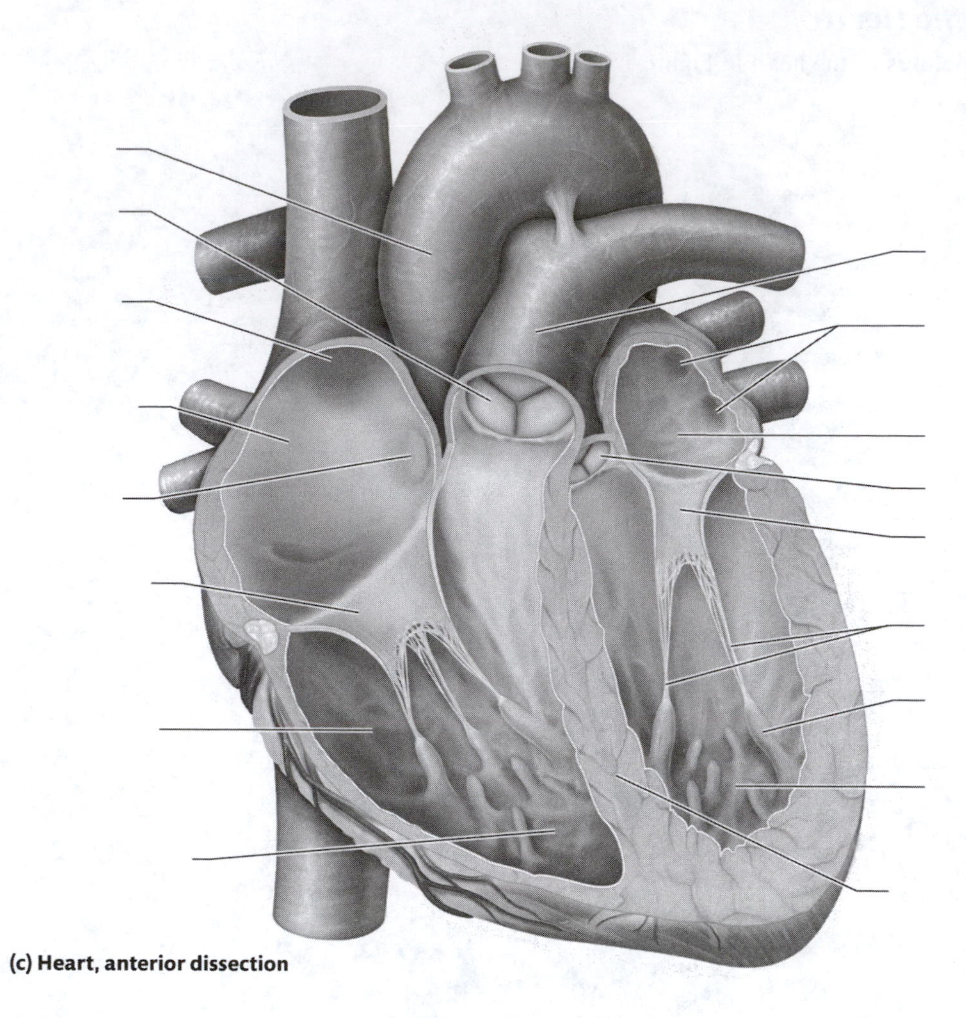

(c) Heart, anterior dissection

Figure 17.4 (*continued*)

Key Concept: How is the right side of the heart different structurally from the left side? Why?

____

____

Key Concept: Why is it important to prevent backflow of blood in the heart? Which structures prevent backflow, and how do they function?

____

____

Trace It: Pathway of Blood Flow through the Heart

In the space below, write out the overall pathway of blood flow through the heart, starting with the superior and inferior venae cavae and ending with the aorta. Be sure to include the valves and the pulmonary circulation, and indicate where blood is oxygenated or deoxygenated.

Module 17.3: Cardiac Muscle Tissue Anatomy and Electrophysiology

This module explores the structure and function of the cardiac muscle tissue and examines its functional properties by visiting a familiar concept: electrophysiology. When you complete this module, you should be able to do the following:

1. Describe the histology of cardiac muscle tissue, and differentiate it from that of skeletal muscle.
2. Describe the phases of the cardiac muscle action potential, including the ion movements that occur in each phase, and explain the importance of the plateau phase.
3. Contrast the way action potentials are generated in cardiac pacemaker cells, cardiac contractile cells, and skeletal muscle cells.
4. Describe the parts of the cardiac conduction system, and explain how the system functions.
5. Identify the waveforms in a normal electrocardiogram (ECG), and relate the ECG waveforms to electrical activity in the heart.

Build Your Own Glossary

Following is a table listing key terms from Module 17.3. Before you read the module, use the glossary at the back of your book or look through the module to define the following terms.

Key Terms of Module 17.3

Term	Definition
Pacemaker cell	
Contractile cells	
Autorhythmicity	
Intercalated discs	
Plateau phase	
Cardiac conduction system	
Pacemaker potential	
Sinoatrial node	
Atrioventricular node	
Purkinje fiber system	
Sinus rhythm	
Electrocardiogram	

Survey It: Form Questions

Before you read the module, survey it and form at least four questions for yourself. When you have finished reading the module, return to these questions and answer them.

Question 1: ________________________________

Answer: ________________________________

Question 2: ________________________________

Answer: ________________________________

Question 3: ________________________________

Answer: ________________________________

Question 4: ________________________________

Answer: ________________________________

Key Concept: Why can the heart continue beating even when the brain has ceased most of its functions?

Key Concept: What is the importance of the intercalated discs to heart function? Why do skeletal muscle fibers lack intercalated discs?

Draw It: Cardiac Action Potentials

Draw two graphs: one of a contractile cell cardiac action potential and one of a pacemaker cell action potential. Label each part of the graph and indicate the ion channels that are opening or closing and the ions that are moving in or out of the cell. Then, study your graphs, and make a list of the key differences between the action potentials in the two cell types.

Contractile cell action potential:

Pacemaker cell action potential:

Key differences:

Key Concept: What causes the plateau phase of the contractile cardiac cell action potential? Why is the plateau phase so important to the electrophysiology of the heart?

Identify It: The Cardiac Conduction System

Identify and color-code the structures of the cardiac conduction system in Figure 17.5. Then, trace the spread of the action potential starting from the sinoatrial node and ending with the ventricular contractile cells.

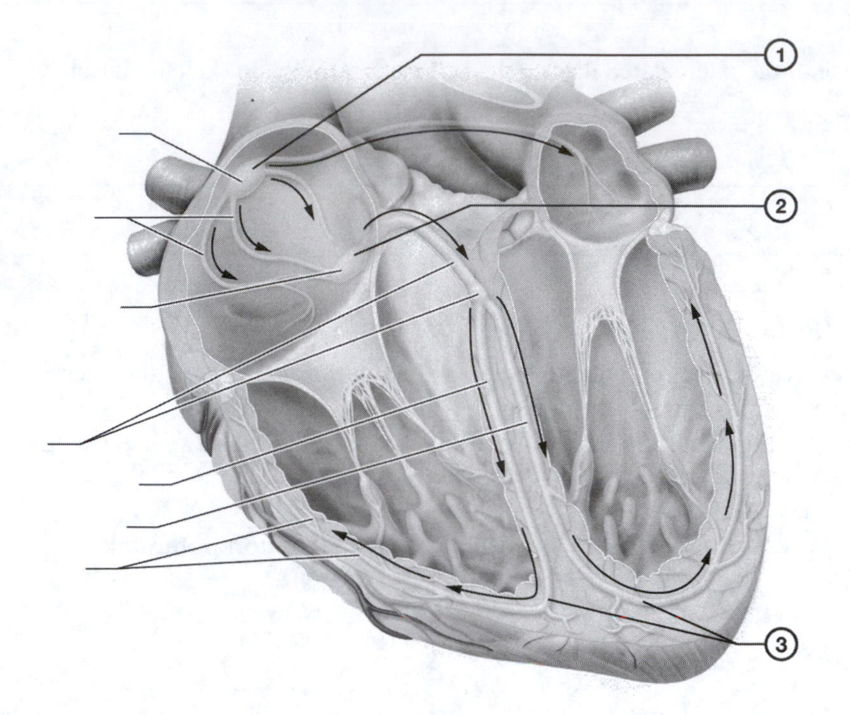

Figure 17.5 The cardiac conduction system.

Key Concept: Why is there a delay at the AV node? What would happen if the delay were too short? What would happen if it were too long?

____

____

Identify It and Describe It: The Electrocardiogram

Identify each wave and interval on the ECG in Figure 17.6. Then, draw the wave or interval in the following table and describe what it represents on the ECG.

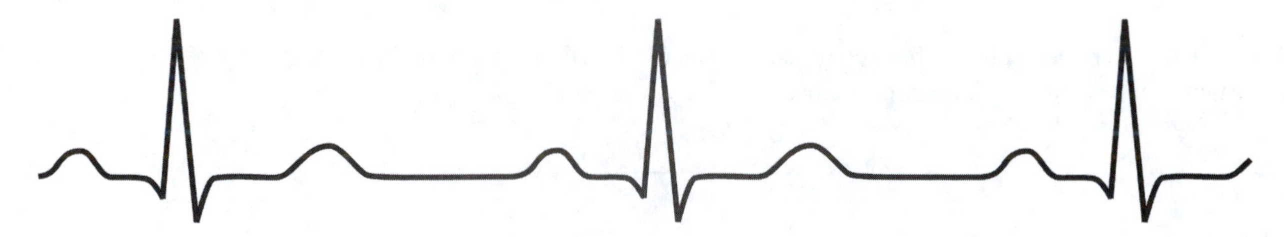

Figure 17.6 The electrocardiogram.

Waves and Intervals of the ECG

Portion of the ECG	Drawing	Definition/Meaning
P wave		
QRS complex		
T wave		
R-R interval		
P-R interval		
Q-T interval		
S-T segment		

Key Concept: What part of the heart has an anomaly if the P wave is abnormal on an ECG? Explain.

____

____

Module 17.4: Mechanical Physiology of the Heart: The Cardiac Cycle

We now examine the other facet of cardiac physiology: mechanical physiology, or the actual pumping of the heart. At the end of this module, you should be able to do the following:

1. Describe the phases of the cardiac cycle.
2. Relate the opening and closing of specific heart valves in each phase of the cardiac cycle to pressure changes in the heart chambers.
3. Relate the heart sounds and ECG waveforms to the normal mechanical events of the cardiac cycle.
4. Compare and contrast pressure and volume changes of the left and right ventricles and the aorta during one cardiac cycle.

Build Your Own Glossary

Following is a table listing key terms from Module 17.4. Before you read the module, use the glossary at the back of your book or look through the module to define the following terms.

Key Terms of Module 17.4

Term	Definition
Mechanical physiology	
Heartbeat	
Cardiac cycle	
Heart sounds	
Systole	
Diastole	
End-diastolic volume	
End-systolic volume	

Survey It: Form Questions

Before you read the module, survey it and form at least three questions for yourself. When you have finished reading the module, return to these questions and answer them.

Question 1: ______________________________

Answer: ______________________________

Question 2: ______________________________

Answer: ______________________________

Question 3: ______________________________

Answer: ______________________________

Key Concept: How do pressure gradients drive blood flow through the heart? How do pressure gradients influence the functioning of the heart valves?

____

____

Describe It: Heart Sounds

Describe the location and timing of each heart sound in the following table.

Location and Timing of Heart Sounds

Area	Pulmonic	Tricuspid	Mitral	Aortic
Location of sound				
Timing of sound				

Describe It: The Cardiac Cycle

Write out the steps of the cardiac cycle in Figure 17.7. In addition, label and color-code all key components of the cycle, and be sure to color deoxygenated blood blue and oxygenated blood red.

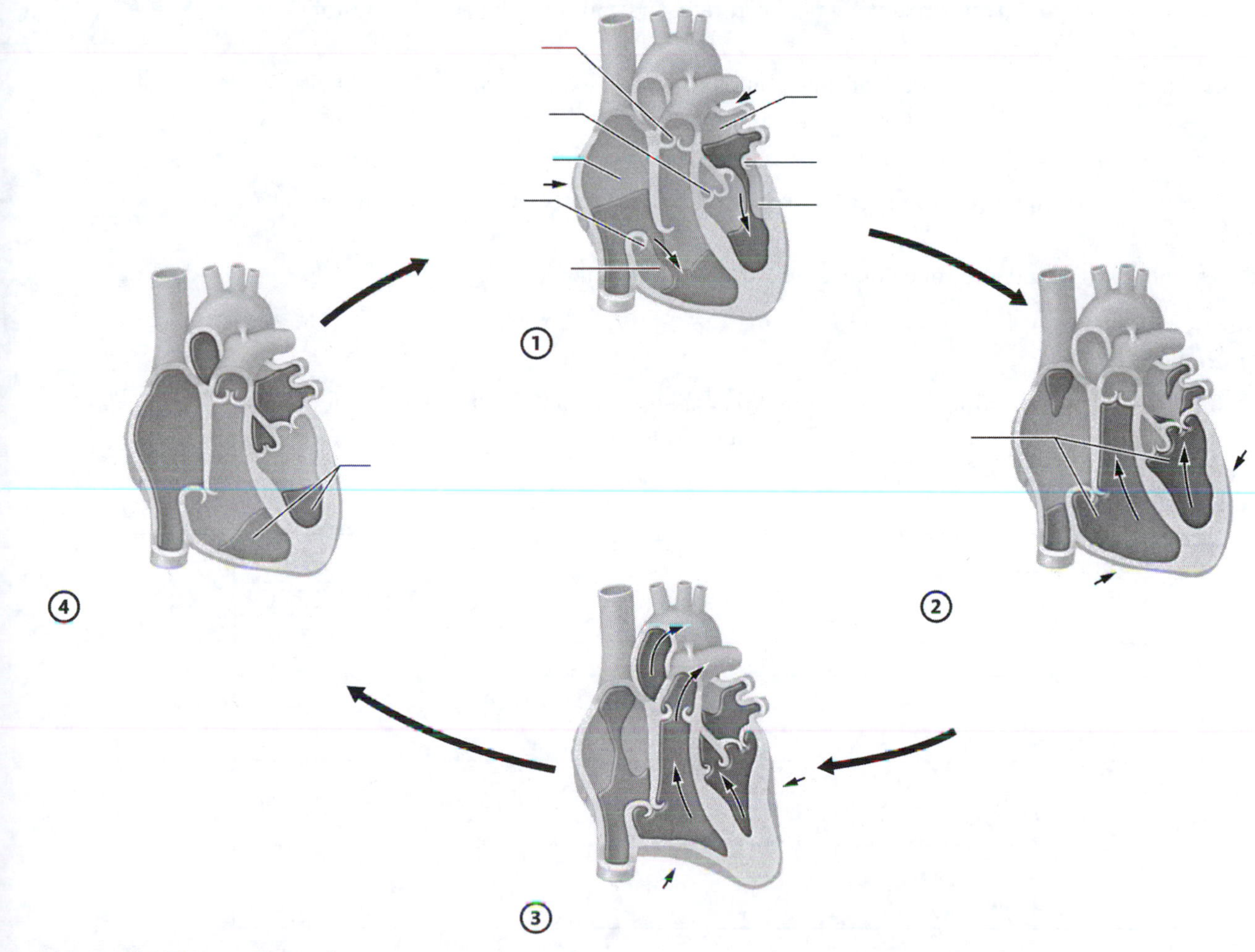

Figure 17.7 The cardiac cycle.

Team Up

Team up with one or more partners and have each person develop a handout to teach the Wiggers diagram, which connects the mechanical and electrical events of the heart. At the end of the handout, write a 10-question quiz on the mechanical and electrical physiology of the heart. Once you have completed your diagrams, exchange them with your partners, study them, and then take the quiz at the end. When you and your group have finished taking the quizzes, discuss the answers to determine places where you need additional study. Finally, combine the best elements of each handout to make a master diagram teaching the correlation between the heart's electrical and mechanical events.

Key Concept: You notice that a patient has QRS waveforms on his ECG, indicating that his ventricles are undergoing electrical activity. Does this finding always mean that a heart is actually pumping blood? Explain.

____

____

Module 17.5: Cardiac Output and Regulation

This module looks at the amount of blood pumped by the heart and the factors that influence this value. At the end of this module, you should be able to do the following:

1. Define and calculate cardiac output, given stroke volume, heart rate, and end-diastolic and end-systolic volumes.
2. Describe the factors that influence preload, afterload, and contractility, and explain how they affect cardiac output.
3. Explain the significance of the Frank-Starling law for the heart.
4. Discuss the influence of positive and negative inotropic and chronotropic agents on stroke volume and heart rate, respectively.
5. Predict how changes in heart rate and/or stroke volume will affect cardiac output.

Build Your Own Glossary

Following is a table listing key terms from Module 17.5. Before you read the module, use the glossary at the back of your book or look through the module to define the following terms.

Key Terms of Module 17.5

Term	Definition
Heart rate	
Stroke volume	
Cardiac output	
Preload	

Term	Definition
Frank-Starling law	
Contractility	
Afterload	
Heart failure	

Survey It: Form Questions

Before you read the module, survey it and form at least two questions for yourself. When you have finished reading the module, return to these questions and answer them.

Question 1: ____________________

Answer: ____________________

Question 2: ____________________

Answer: ____________________

Key Concept: If a person has a stroke volume of 55 ml and a heart rate of 80 beats per minute, what is his or her cardiac output? Is this a normal value? If not, how is it abnormal?

____

____

Complete It: Factors That Influence Stroke Volume

Fill in the blanks to complete the following paragraphs that describe the factors influencing stroke volume.

The ____________ refers to the amount a ventricular cell is stretched before contracting. Generally, as it increases, stroke volume also ____________ due to a phenomenon known as the ____________ ____________ ____________. This law states that cells contract ____________ when their sarcomeres are stretched prior to contraction. One of the major determinants of preload is ____________ ____________, or the amount of blood veins deliver to the heart.

The heart's ____________ is its inherent ability to pump. As this increases, the stroke volume also

____________ and the ____________ ____________ volume.

A final factor that influences stroke volume is ____________, which is the force against which the heart must

pump. It is largely determined by the ____________ of the circuit. When afterload increases, stroke volume

generally ____________ and ____________ ____________ volume ____________.

Key Concept: Does a high ESV signify a high stroke volume or a low stroke volume? What could cause a high ESV, and what would this do to cardiac output?

____

____

Predict It: Effect of Cardiac Output

Following are several scenarios in which one or more variables that affect cardiac output are altered. Predict in each scenario which factor is affected (preload, contractility, or afterload) and whether cardiac output is likely to increase or decrease, and explain why.

- The heart muscle is damaged after a heart attack. ____

  ____

- A person is taking an angiotensin-receptor blocker, which causes vasodilation. ____________________

  ____

- A person is taking the medication digitalis, which increases the strength of the heart's contraction.

  ____

  ____

- Venous return decreases due to significant edema. ____

  ____

- A person is running from an angry badger, and the sympathetic nervous system has activated

  in full force. ____

  ____

Key Concept: Why does left ventricular failure cause pulmonary congestion and edema? Why do both right ventricular failure and left ventricular failure cause peripheral edema?

____

____

What Do You Know Now?

Let's now revisit the questions you answered in the beginning of this chapter. How have your answers changed now that you've worked through the material?

- What is/are the functions of the heart?

  ____

- How are the two sides of the heart different?

  ____

- Do arteries carry oxygenated or deoxygenated blood? What about veins?

  ____

18 The Cardiovascular System II: The Blood Vessels

We now turn to the next part of the cardiovascular system with the organs that transport blood to and from the heart: blood vessels. This chapter explores the structure and function of blood vessels and the anatomy of the arteries and veins of the body.

What Do You Already Know?

Try to answer the following questions before proceeding to the next section. If you're unsure of the correct answers, give it your best attempt based on previous courses, previous chapters, or just your general knowledge.

- What effect does the sympathetic nervous system have on blood pressure?

  ____

- What are systole and diastole?

  ____

- Why can a person lose consciousness when placed in a choke hold?

  ____

Module 18.1: Overview of Arteries and Veins

Module 18.1 in your text introduces you to the structure and function of arteries and veins. By the end of the module, you should be able to do the following:

1. Compare and contrast the structures of arteries and veins, and of arterioles and venules.
2. Define vascular anastomosis, and explain the significance of anastomoses.

Build Your Own Glossary

Following is a table listing key terms from Module 18.1. Before you read the module, use the glossary at the back of your book or look through the module to define the following terms.

Key Terms for Module 18.1

Term	Definition
Artery	
Capillary	

Term	Definition
Vein	
Tunica intima	
Tunica media	
Tunica externa	
Venous valve	
Vascular anastomosis	

Survey It: Form Questions

Before you read the module, survey it and form at least two questions for yourself. When you have finished reading the module, return to these questions and answer them.

Question 1: ______________________________

Answer: ______________________________

Question 2: ______________________________

Answer: ______________________________

Identify It: Layers of the Blood Vessel Wall

Identify and color-code each component of the blood vessel wall in Figure 18.1. Then, list the main function(s) of each tunic (layer).

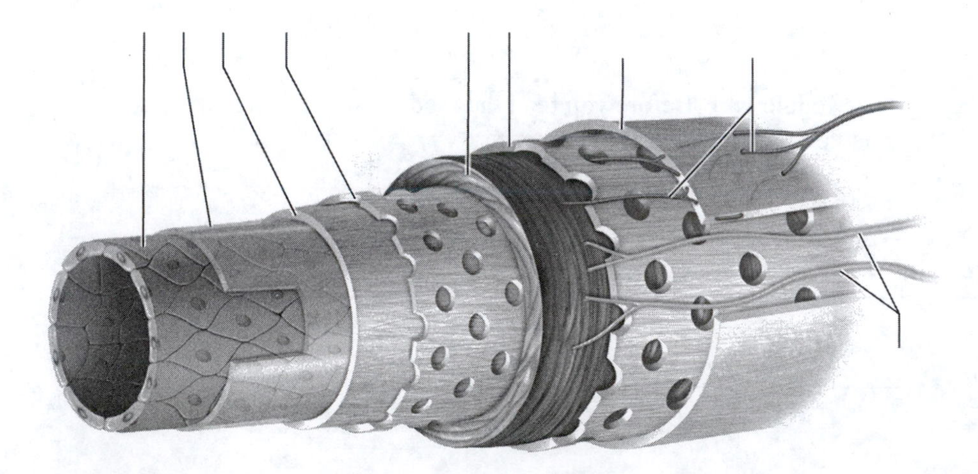

Figure 18.1 The tunics (layers) of the blood vessel wall.

Key Concept: Why is it important that a vessel be able to stretch? Which structures allow a vessel to stretch? What prevents it from overstretching?

___

___

Build Your Own Summary Table: Properties of Blood Vessels

As you read Module 18.1, build your own summary table about the different types of blood vessels by filling in the information in the following table. You can use text Table 18.1 on page 674 for reference, but the descriptions should be in your own words so that they make sense to you.

Properties of Blood Vessels and Blood Distribution

	Elastic Artery	**Muscular Artery**	**Arteriole**	**Venule**	**Vein**
Structure					
Description					
Diameter					
Functions					

Key Concept: Why is a greater percentage of blood found in veins? Why is this important?

___

___

Module 18.2: Physiology of Blood Flow

Now we look at some of the fundamental principles of blood flow, or hemodynamics. When you finish this module, you should be able to do the following:

1. Describe the factors that influence blood flow, blood pressure, and peripheral resistance.
2. Explain the relationships between vessel diameter, cross-sectional area, blood pressure, and blood velocity.
3. Explain how blood pressure varies in different parts of the systemic and pulmonary circuits.
4. Describe how blood pressure changes in the arteries, capillaries, and veins.
5. Explain how mean arterial pressure is calculated.
6. Describe the mechanisms that assist in the return of venous blood to the heart.

Build Your Own Glossary

Following is a table listing key terms from Module 18.2. Before you read the module, use the glossary at the back of your book or look through the module to define the following terms.

Key Terms for Module 18.2

Term	Definition
Blood pressure	
Blood flow	
Peripheral resistance	
Viscosity	
Vessel compliance	
Mean arterial pressure	
Systolic pressure	
Diastolic pressure	
Pulse pressure	

Survey It: Form Questions

Before you read the module, survey it and form at least two questions for yourself. When you have finished reading the module, return to these questions and answer them.

Question 1: ______________________________

Answer: ______________________________

Question 2: ______________________________

Answer: ______________________________

Key Concept: How does the heart drive blood through the blood vessels?

____

____

Complete It: Blood Flow

Fill in the blanks to complete the following paragraphs that describe properties of blood flow.

____________ ____________ is the outward force the blood exerts on the wall of the ____________ ____________. The magnitude of this gradient is one factor that determines ____________ ____________, which is the ____________ of blood that flows per minute. In general, this value matches the ____________ ____________ of about 5–6 liters/min. The second factor that determines blood flow is ____________, which is any ____________ to blood flow. Generally, as resistance increases, blood flow ____________.

The velocity with which blood flows is largely determined by the ____________ ____________ ____________ of the blood vessel. As this area increases, the velocity of blood flow ____________.

Predict It: Factors That Influence Blood Pressure

Predict whether blood pressure will increase or decrease given each of the following conditions. Justify each of your responses.

- A person does hot yoga and forgets to drink any water, becoming dehydrated.

  ____

- A patient is given a drug that causes vasoconstriction.

  ____

- The blood vessels contain numerous large plaques throughout the systemic circuit.

____

- A person meditates, lowering her heart rate.

____

Key Concept: What three factors determine blood pressure? How does each factor influence blood pressure?

____

____

Build Your Own Summary Table: Pressures in the Systemic Circuit

As you read Module 18.2, build your own summary table about the blood pressure in different parts of the systemic circuit by filling in the information in the following table.

Blood Pressure in Different Parts of the Systemic Circuit

Vessel	Pressure Range	Reason for Pressure Value
Arteries during systole		
Arteries during diastole		
Arterioles		
Capillaries: arteriolar end		
Capillaries: venular end		
Venules		
Veins		

Key Concept: Why does venous blood need assistance in returning to the heart? What mechanisms are in place to assist in venous return?

____

____

Module 18.3: Maintenance of Blood Pressure

Blood pressure must be within a certain range to maintain life, but it must also be able to adjust to the changing demands of the body. This module explores how blood pressure is maintained and modified to ensure the needs of the body are met at all times. When you complete this module, you should be able to do the following:

1. Describe the role of arterioles in regulating tissue blood flow and systemic arterial blood pressure.
2. Describe the local, hormonal, and neural factors that affect and regulate blood pressure.
3. Explain the main effects and importance of the baroreceptor reflex.
4. Explain how the respiratory and cardiovascular systems maintain blood flow to tissues via the chemoreceptor reflex.
5. Describe common causes of and common treatments for hypertension.

Build Your Own Glossary

Following is a table listing key terms from Module 18.3. Before you read the module, use the glossary at the back of your book or look through the module to define the following terms.

Key Terms of Module 18.3

Term	Definition
Carotid and aortic sinuses	
Baroreceptor reflex	
Valsalva maneuver	
Peripheral chemoreceptor	
Hypertension	
Hypotension	

Survey It: Form Questions

Before you read the module, survey it and form at least two questions for yourself. When you have finished reading the module, return to these questions and answer them.

Question 1: ______________________________

Answer: ______________________________

Question 2: ______________________________

Answer: ______________________________

Key Concept: Which body systems are responsible for short-term maintenance of blood pressure? Which factors do these systems regulate?

____

____

Describe It: Effects of the Autonomic Nervous System on Blood Pressure

Describe the effects of the sympathetic and parasympathetic nervous systems in Figure 18.2. In addition, draw in the ECG change (tachycardia or bradycardia) and the arrows on the blood vessels to indicate vasoconstriction or vasodilation.

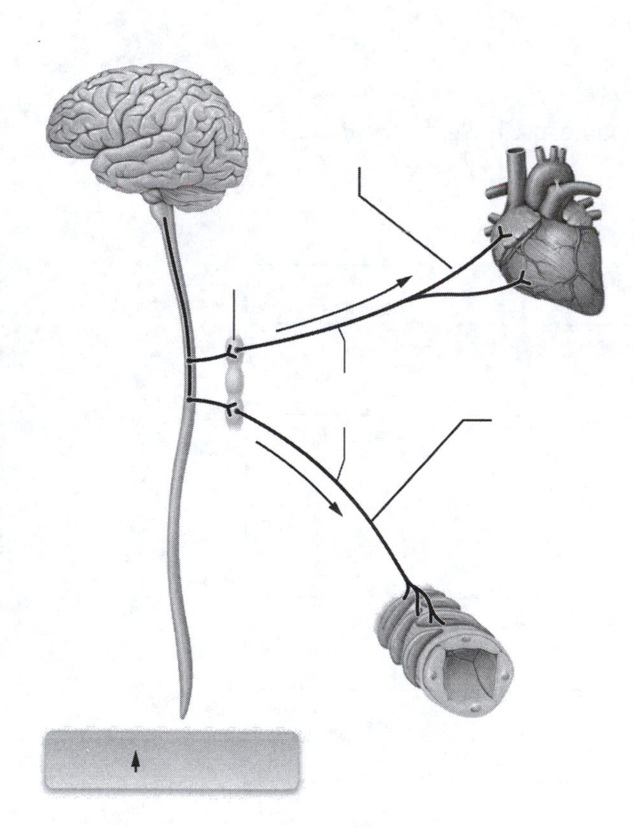

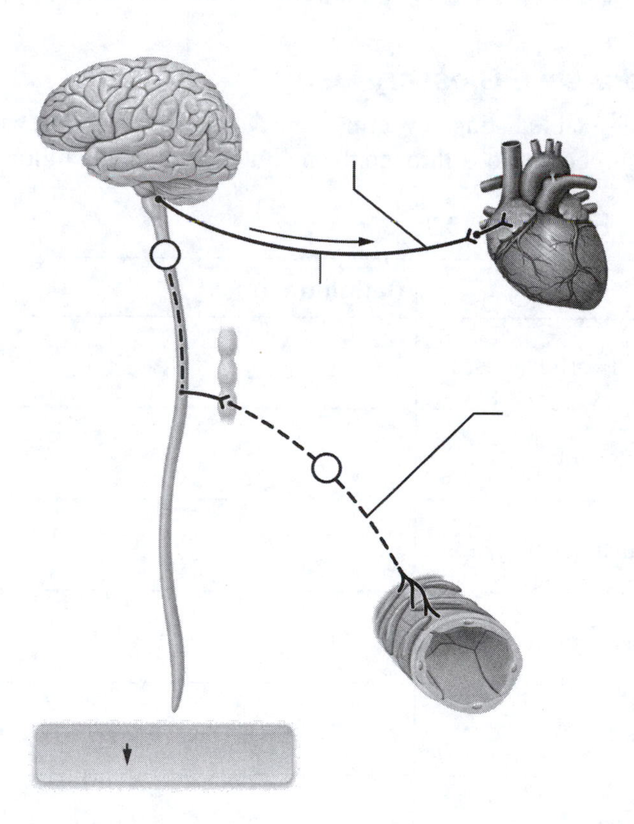

Figure 18.2 Effects of the ANS on blood pressure.

Key Concept: Why may someone temporarily lose consciousness when held in a choke hold?

____

____

Describe It: Effects of the Autonomic Nervous System on Blood Pressure

Describe the steps of the baroreceptor reflex feedback loops in Figure 18.3. In addition, label and color-code all relevant structures of the processes.

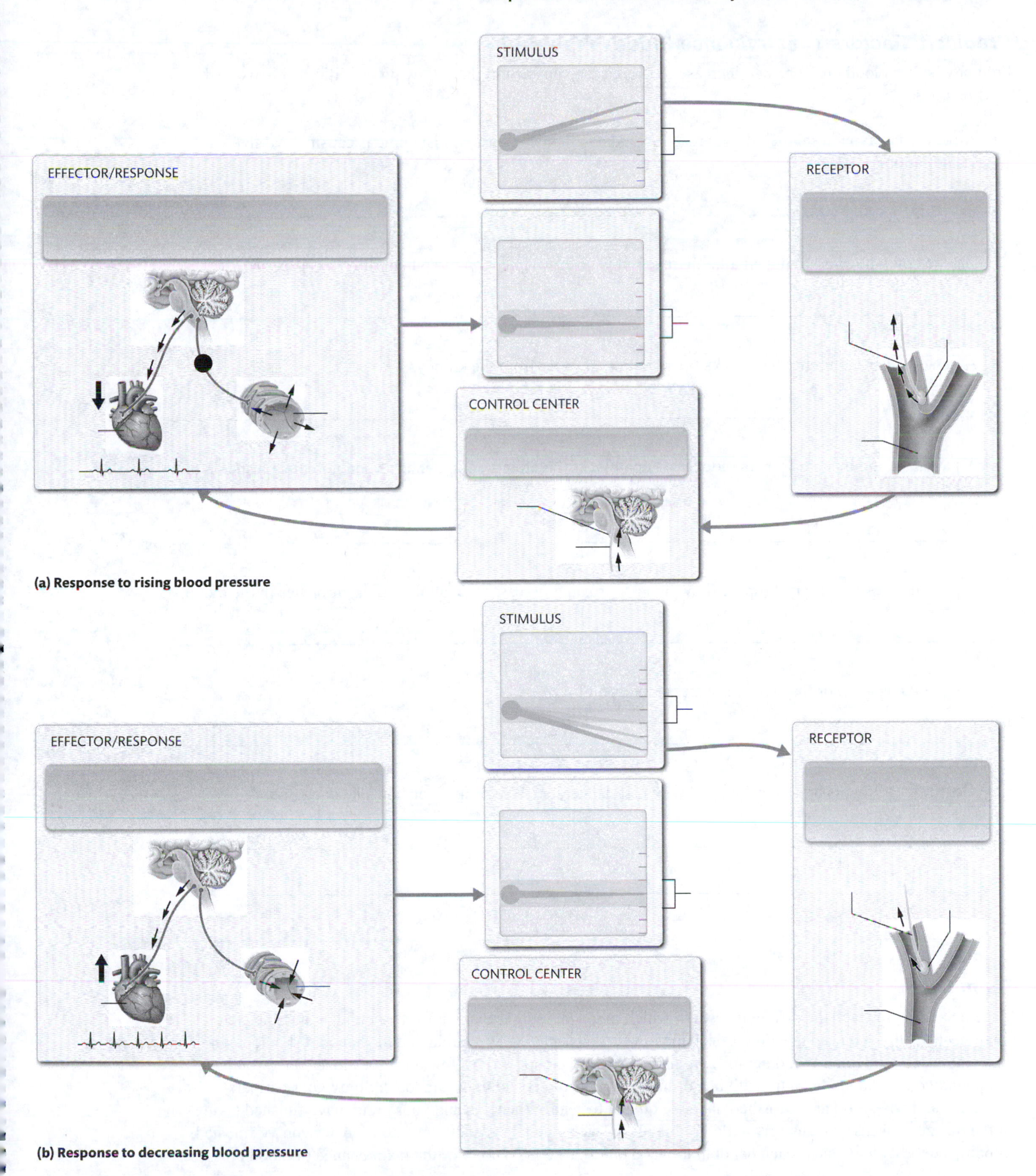

Figure 18.3 Regulation of blood pressure by the baroreceptor reflex.

Predict It: Factors That Influence Blood Pressure II

Predict whether blood pressure will increase or decrease given each of the following conditions. Justify each of your responses.

- A person is having a severe allergic reaction and a systemic release of histamine is causing massive vasodilation.

  ____

- The person who was having an allergic reaction is given an injection of epinephrine.

  ____

- A person takes a diuretic that blocks the retention of water from the kidneys.

  ____

- A patient is administered a beta blocker, which blocks the effects of epinephrine and norepinephrine on the heart.

  ____

- A person takes excess erythropoietin, which abnormally increases the number of erythrocytes in the blood.

  ____

- A patient has a tumor that secretes excess amounts of ADH.

  ____

Key Concept: What is hypertension, and why is it dangerous? Why is hypotension also dangerous?

____

____

____

Team Up

Make a handout to teach the different ways in which blood pressure is maintained by neural, endocrine, and renal mechanisms. You can use Figure 18.9 in your text on page 685 as a guide, but the handout should be in your own words and with your own diagrams. At the end of the handout, write a few quiz questions. Once you have completed your handout, team up with one or more study partners and trade handouts. Study your partners' diagrams, and when you have finished, take the quiz at the end. When you and your group have finished taking all the quizzes, discuss the answers to determine places where you need additional study. After you've finished, combine the best elements of each handout to make one "master" diagram for the maintenance of blood pressure.

Module 18.4: Capillaries and Tissue Perfusion

Module 18.4 in your text explores the ways in which materials are exchanged in capillary beds. At the end of this module, you should be able to do the following:

1. Describe the different types of capillaries, and explain how their structure relates to their function.
2. Explain the roles of diffusion, filtration, and osmosis in capillary exchange.
3. Describe how autoregulation controls blood flow to tissues.

Build Your Own Glossary

Following is a table listing key terms from Module 18.4. Before you read the module, use the glossary at the back of your book or look through the module to define the following terms.

Key Terms of Module 18.4

Term	Definition
Tissue perfusion	
Capillary exchange	
Continuous capillary	
Fenestrated capillary	
Sinusoid	
Autoregulation	
Myogenic mechanism	

Survey It: Form Questions

Before you read the module, survey it and form at least three questions for yourself. When you have finished reading the module, return to these questions and answer them.

Question 1: ____________________

Answer: ____________________

Question 2: ____________________

Answer: ____________________

Question 3: ____________________

Answer: ____________________

Key Concept: Why is tissue perfusion tightly regulated?

____

____

Identify It: Capillary Structure and Function

Identify and color-code each component of the capillary in part (a) of Figure 18.4. Then, describe the ways in which substances cross capillary walls in part (b).

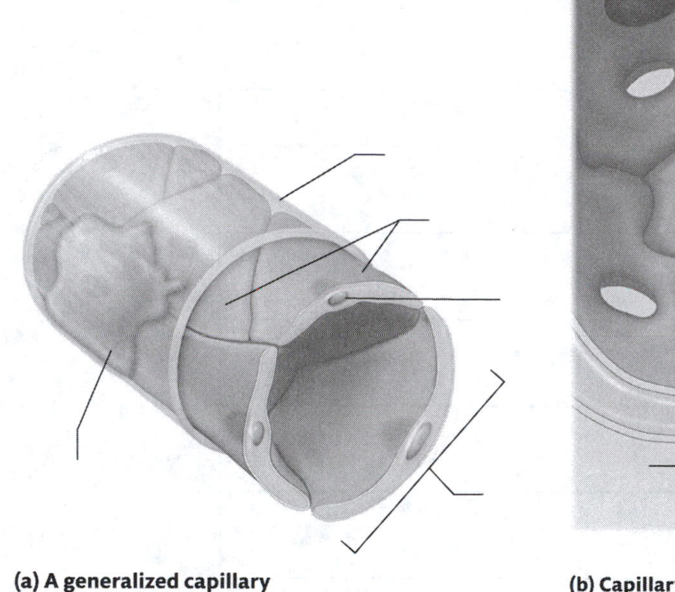

(a) A generalized capillary

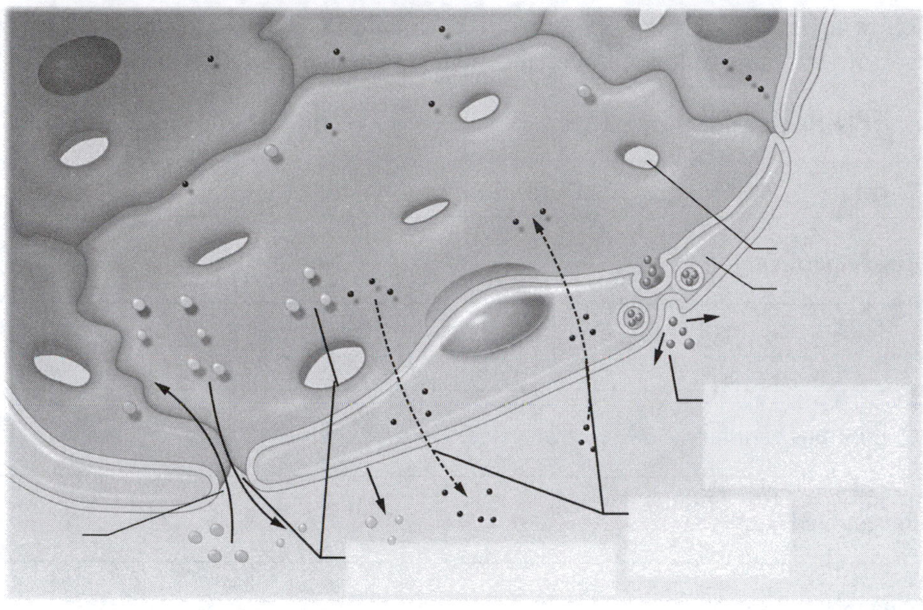

(b) Capillary exchange mechanisms

Figure 18.4 Structure and function of a capillary.

Identify It: Types of Capillaries

Identify each of the following statements as being properties of continuous, fenestrated, or sinusoidal capillaries.

- Located in the skin, muscle tissue, and most nervous and connective tissue ____________
- Have irregular basal laminae with very large pores; discontinuous sheets of endothelium ____________
- Endothelial cells joined by many tight junctions ____________
- Located in the kidneys, endocrine glands, and small intestine ____________
- Extremely leaky, allow large substances to cross the capillary walls ____________
- Contain fenestrations in their endothelial cells ____________
- Least leaky, allow a narrow range of substances to cross the capillary wall ____________

Key Concept: How does the myogenic mechanism regulate local tissue perfusion?

____

____

Key Concept: A cell is actively metabolizing glucose and producing carbon dioxide and consuming oxygen at a high rate. What effect will this have on local arterioles?

____

____

Complete It: Tissue Perfusion in Special Circuits

Fill in the blanks to complete the following paragraphs that describe tissue perfusion in different circuits.

Perfusion to the tissues of the heart ____________ during systole and ____________ during diastole. Perfusion to the heart increases dramatically during strenuous activity due to a low level of ____________ in the interstitial fluid, which triggers ____________.

Blood flow to the brain is maintained nearly constantly at ____________; however, perfusion to areas of the brain varies with ____________.

The increase in blood flow to skeletal muscle during exercise is called ____________. When exercise begins, the ____________ arterioles dilate, which triggers the ____________ to dilate, which finally triggers the ____________ ____________ to dilate.

Perfusion to the skin is regulated by the ____________ ____________ ____________ as part of the body's ____________ ____________ physiology.

Module 18.5: Capillary Pressures and Water Movement

This module examines the pressures that drive water movement across capillaries. At the end of this module, you should be able to do the following:

1. Describe hydrostatic pressure and colloid osmotic pressure.
2. Explain how net filtration pressure across the capillary wall determines movement of fluid across that wall.
3. Explain how changes in hydrostatic and colloid osmotic pressure may cause edema.

Build Your Own Glossary

Following is a table listing key terms from Module 18.5. Before you read the module, use the glossary at the back of your book or look through the module to define the following terms.

Key Terms of Module 18.5

Term	Definition
Hydrostatic pressure	
Filtration	
Osmotic pressure	
Colloid osmotic pressure	
Net filtration pressure	
Edema	

Survey It: Form Questions

Before you read the module, survey it and form at least two questions for yourself. When you have finished reading the module, return to these questions and answer them.

Question 1: ____________________

Answer: ____________________

Question 2: ____________________

Answer: ____________________

Key Concept: What is hydrostatic pressure? In which direction does hydrostatic pressure push fluid in a capillary bed?

____

____

Key Concept: What is osmotic pressure? In which direction does osmotic pressure push or pull fluid in a capillary bed?

____

____

Draw It: Hydrostatic and Osmotic Pressure Gradients

Draw a blood vessel diagram similar to that in Figure 18.13 in your text, with a capillary connected by an arteriole and a venule. Write in the normal hydrostatic and colloid osmotic pressures and indicate with arrows whether each end will have net filtration or net absorption.

Calculate It: Net Filtration Pressure

Calculate the net filtration pressure for each of the following sets of values, and determine if there will be net filtration or net absorption.

1. Example 1:
    - Arteriolar end HP: 45 mm Hg
    - Venular end HP: 15 mm Hg
    - OP (at both ends): 19 mm Hg

2. Example 1:
    - Arteriolar end HP: 25 mm Hg
    - Venular end HP: 15 mm Hg
    - OP (at both ends): 22 mm Hg

Module 18.6: Anatomy of the Systemic Arteries

Module 18.6 in your text explores the anatomy of the systemic arteries. By the end of the module, you should be able to do the following:

1. Describe the patterns of arterial blood flow for the head and neck, the thoracic cavity, the abdominopelvic cavity, and the upper and lower limbs.
2. Identify major arteries of the systemic circuit.
3. Identify the major pulse points.

Build Your Own Glossary

Following is a table listing key terms from Module 18.6. Before you read the module, use the glossary at the back of your book or look through the module to define the following terms.

Key Terms for Module 18.6

Term	Definition
Aorta	
Common carotid artery	
Cerebral arterial circle	
Celiac trunk	
Common iliac artery	
Subclavian artery	
Pulse	
Pulse point	

Survey It: Form Questions

Before you read the module, survey it and form at least two questions for yourself. When you have finished reading the module, return to these questions and answer them.

Question 1: ____________________

Answer: ____________________

Question 2: ____________________

Answer: ____________________

Key Concept: What are the four divisions of the aorta? How does the last division of the aorta end, and what structures do these vessels supply?

____

____

Identify It: Arteries of the Head and Neck

Identify and color-code each of the arteries of the head and neck in Figure 18.5. Then, list the main structures that each artery supplies.

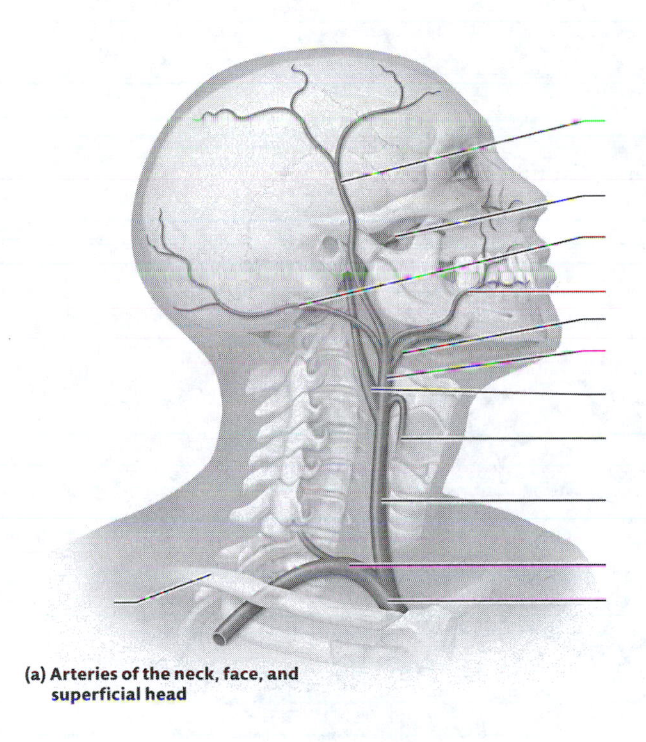

(a) Arteries of the neck, face, and superficial head

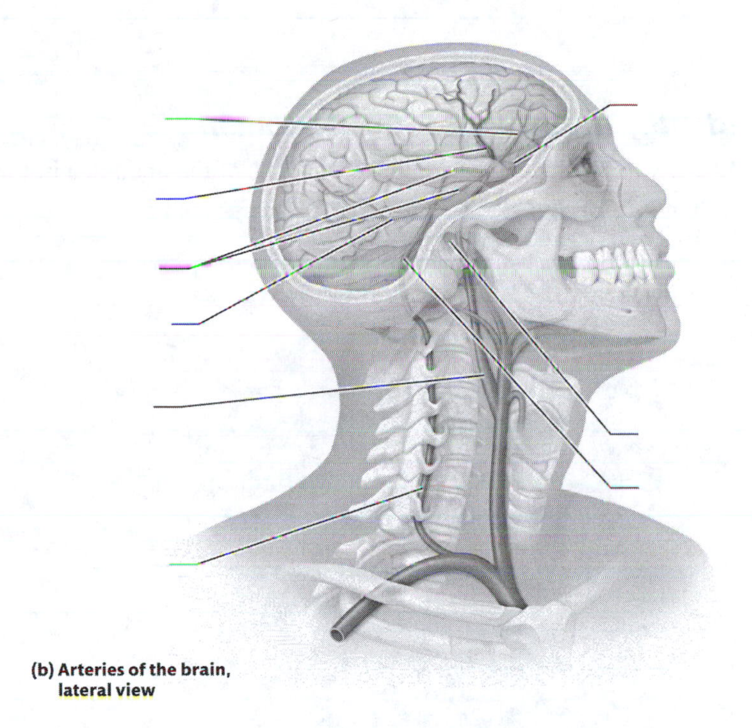

(b) Arteries of the brain, lateral view

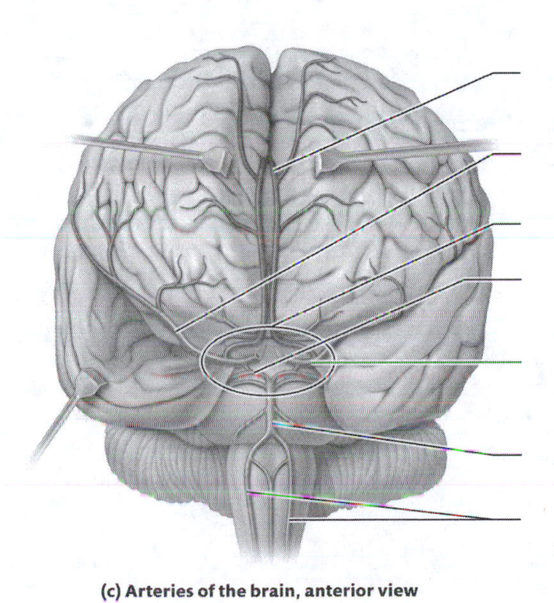

(c) Arteries of the brain, anterior view

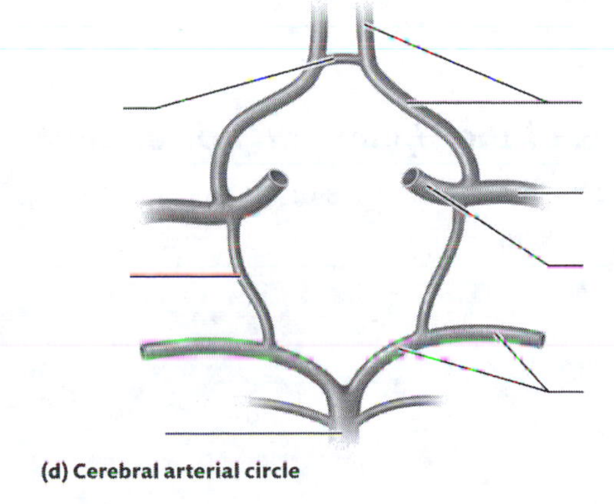

(d) Cerebral arterial circle

Figure 18.5 Arteries of the head and neck.

Key Concept: What is the importance of the cerebral arterial circle?

____

____

Identify It: Arteries of the Abdomen

Identify and color-code each of the arteries of the abdomen in Figure 18.6. Then, list the main structures that each artery supplies.

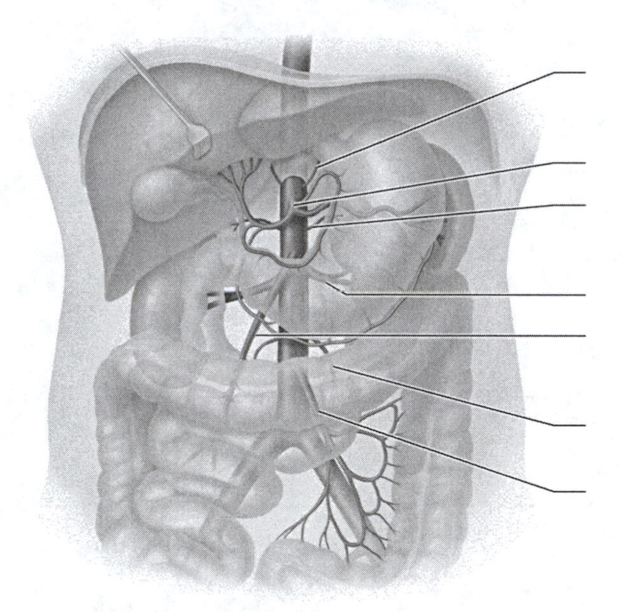

(a) Overview of abdominal arteries

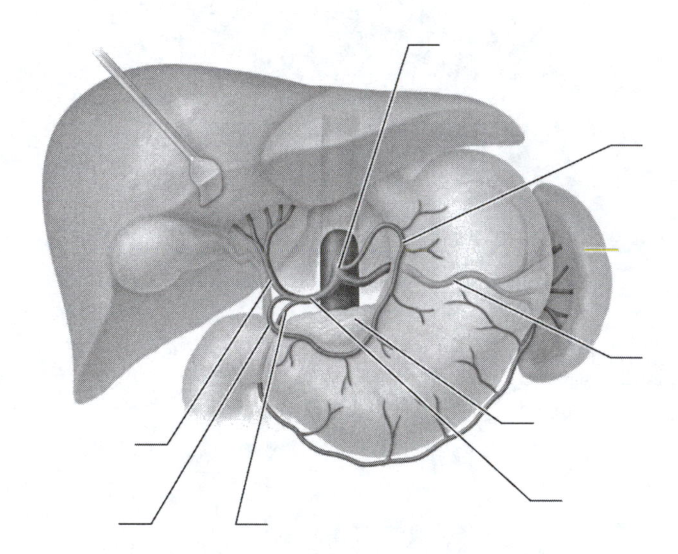

(b) Distribution of the celiac trunk

Figure 18.6 Arteries of the abdomen.

Identify It: Arteries of the Upper and Lower Limbs

Identify and color-code each of the arteries of the upper and lower limbs in Figure 18.7. Then, list the main structures that each artery supplies.

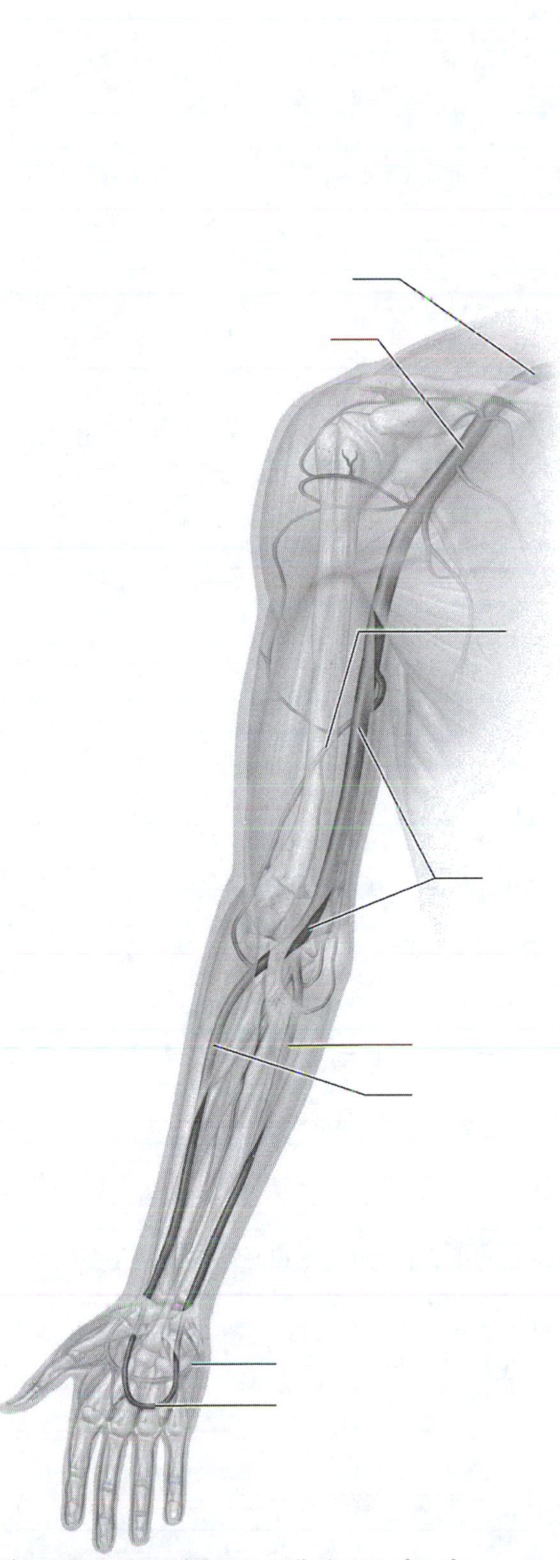

(a) Arteries of the right upper limb, anterior view

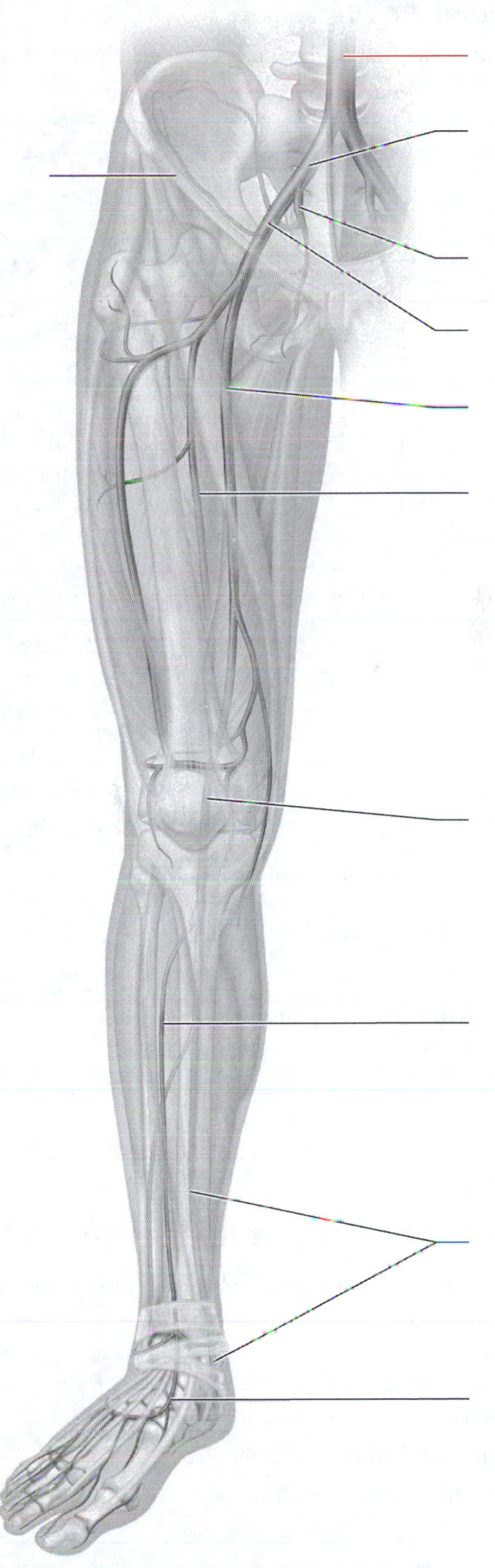

(b) Arteries of the right lower limb, anterior view

Figure 18.7 Arteries of the upper and lower limbs.

Identify It: Pulse Points

Identify and color-code each of the pulse points in Figure 18.8.

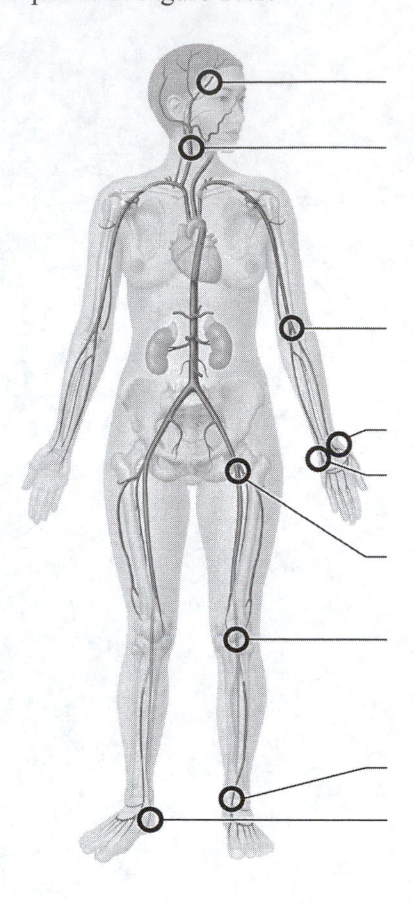

Figure 18.8 Common pulse points.

Key Concept: Why do arteries pulsate?

____

____

Module 18.7: Anatomy of the Systemic Veins

Module 18.7 in your text discusses the anatomy of the systemic veins. By the end of the module, you should be able to do the following:

1. Describe the patterns of venous blood drainage for the head and neck, the thoracic cavity, the abdominopelvic cavity, and the upper and lower limbs.
2. Identify major veins of the systemic circuit.
3. Describe the structure and function of the hepatic portal system.

Build Your Own Glossary

Following is a table listing key terms from Module 18.7. Before you read the module, use the glossary at the back of your book or look through the module to define the following terms.

Key Terms for Module 18.7

Term	Definition
Superior vena cava	
Brachiocephalic vein	
Common iliac vein	
Inferior vena cava	
Dural sinuses	
Azygos system	
Hepatic portal system	

Survey It: Form Questions

Before you read the module, survey it and form at least two questions for yourself. When you have finished reading the module, return to these questions and answer them.

Question 1: ______________________________

Answer: ______________________________

Question 2: ______________________________

Answer: ______________________________

Identify It: Venous Drainage of the Head and Neck

Identify and color-code each of the veins of the head and neck in Figure 18.9. Then, list the main structures that each vein drains.

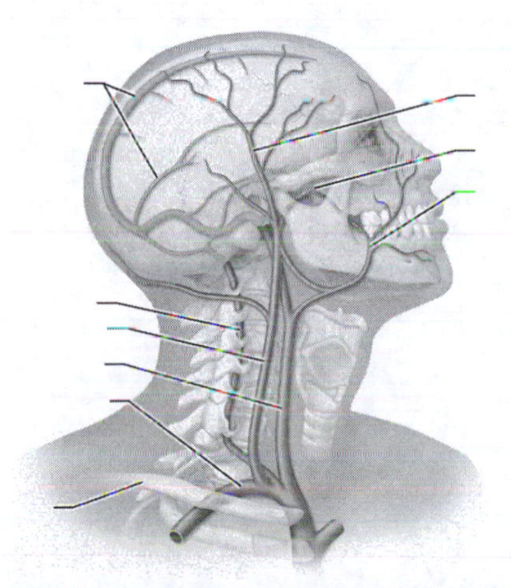

(a) Veins of the neck and superficial head

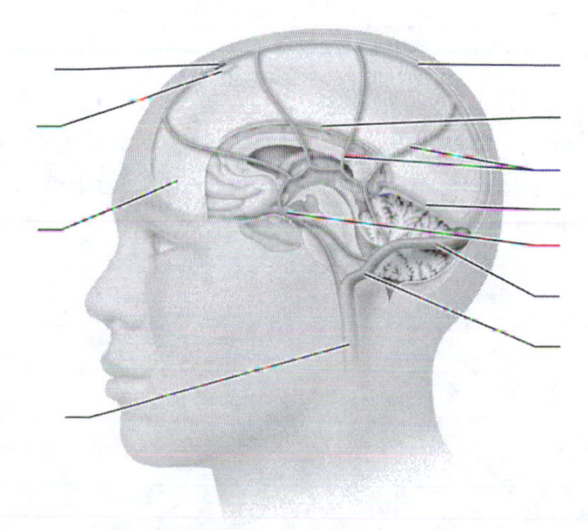

(b) Dural sinuses and other veins of the brain

Figure 18.9 Venous drainage of the head and neck.

Key Concept: How is the venous drainage of the brain different from other structures in the body?

____

____

Identify It: Venous Drainage of the Thorax and Abdomen

Identify and color-code each of the veins of the thorax and abdomen in Figure 18.10. Then, list the main structures that each vein drains.

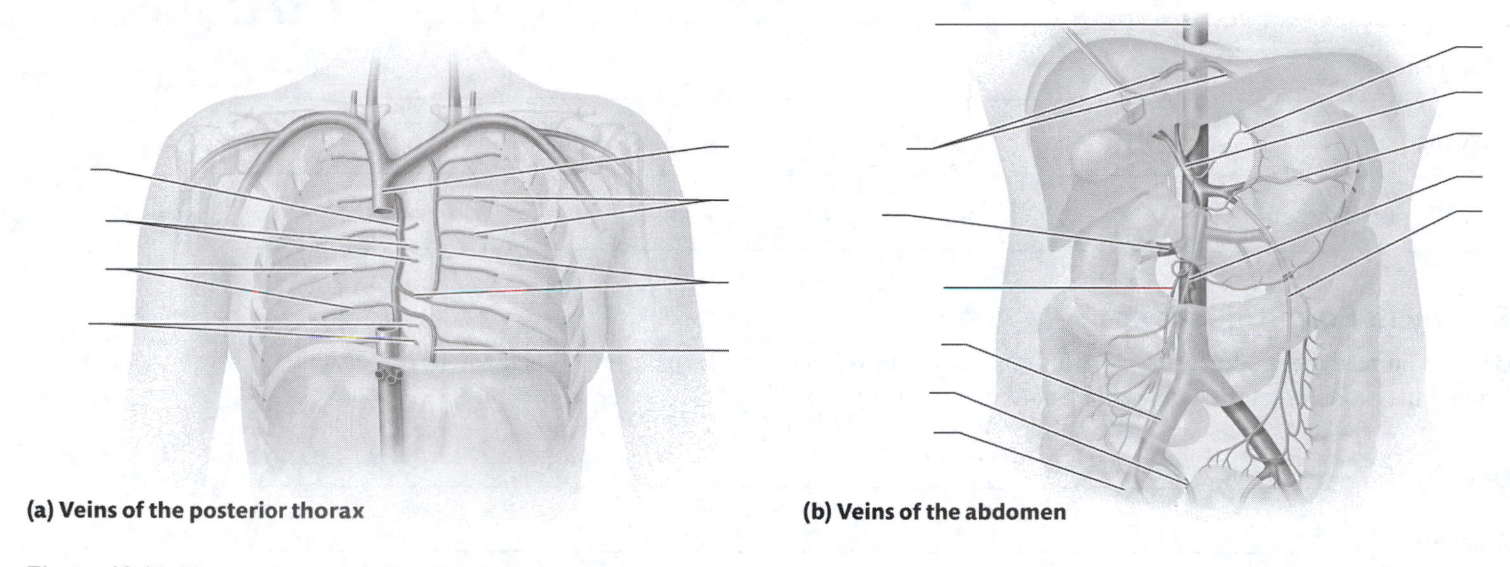

(a) Veins of the posterior thorax **(b) Veins of the abdomen**

Figure 18.10 Venous drainage of the thorax and abdomen.

Key Concept: Where does the blood from most of the abdominal organs go before it enters the inferior vena cava? Why?

____

____

Identify It: Venous Drainage of the Upper and Lower Limbs

Identify and color-code each of the veins of the upper and lower limbs in Figure 18.11. Then, list the main structures that each vein drains.

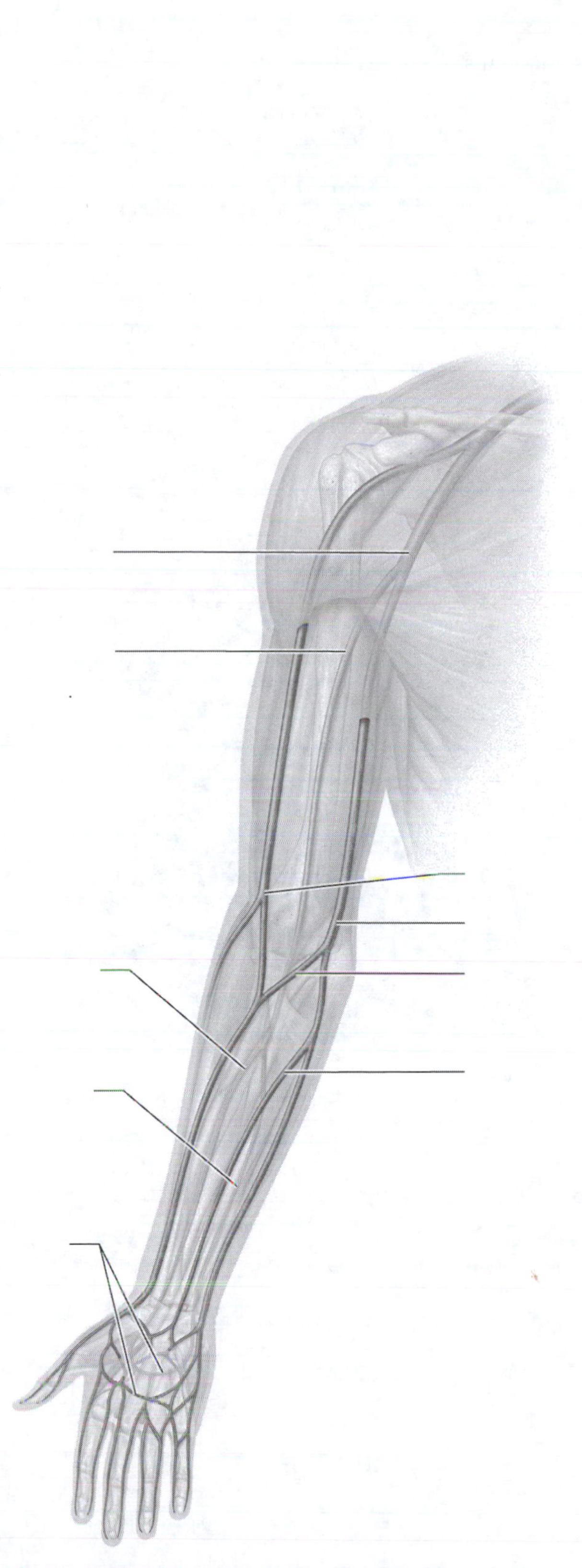

(a) Veins of the right upper limb, anterior view

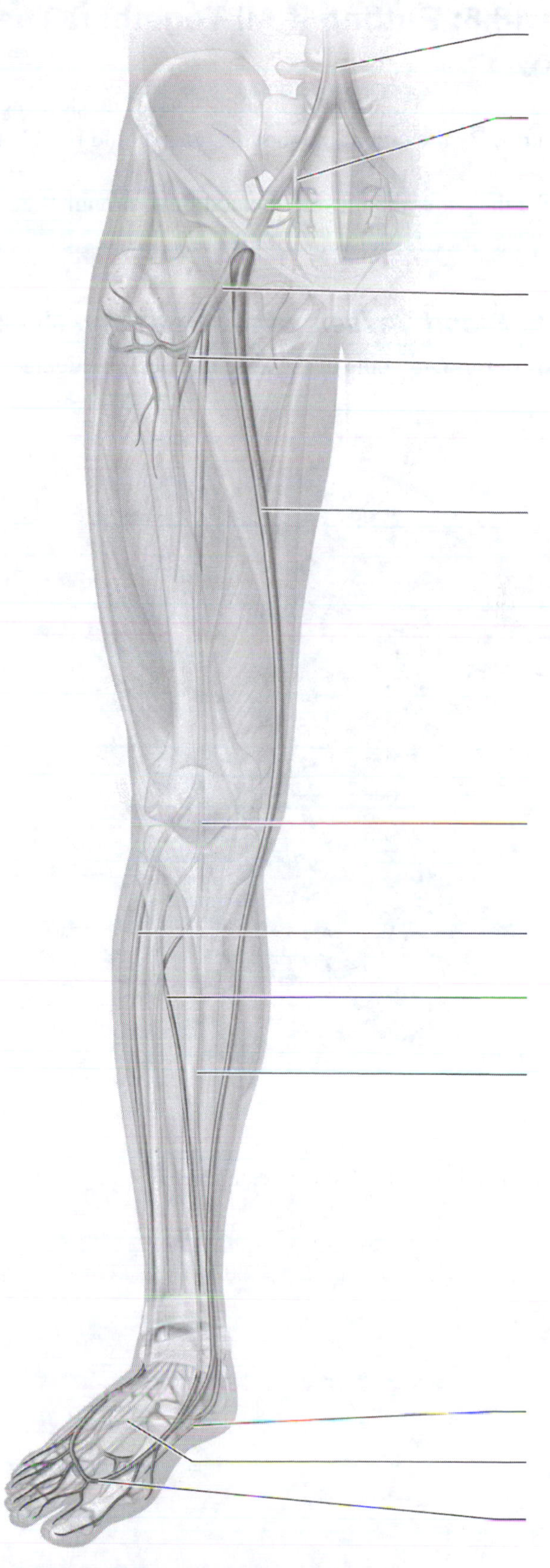

(b) Veins of the right lower limb, anterior view

Figure 18.11 Venous drainage of the upper and lower limbs.

Module 18.8: Putting It All Together: The Big Picture of Blood Vessel Anatomy

Module 18.8 gives you a larger view of the vessels by showing you the arteries and veins in context with other body structures. By the end of the module, you should be able to do the following:

1. Describe the general pathway of blood flow through the body.
2. Identify the arteries and veins of the body.

Identify It: Blood Vessels of the Head and Neck

Identify the arteries and veins of the head and neck in Figure 18.12.

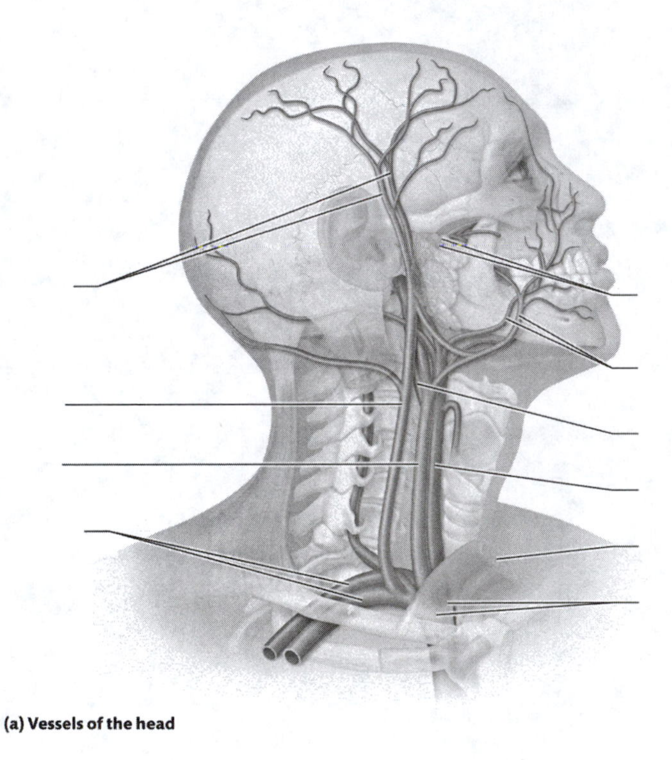

(a) Vessels of the head

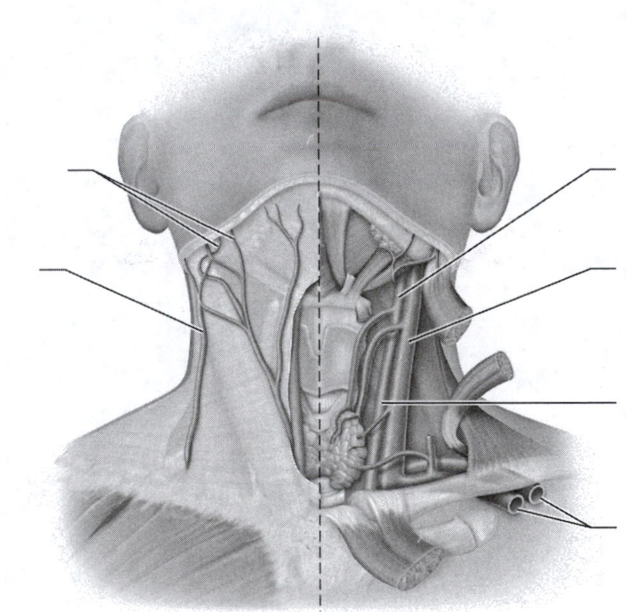

(b) Vessels of the neck

Figure 18.12 Blood vessels of the head and neck.

Identify It: Blood Vessels of the Abdomen

Identify the arteries and veins of the abdomen in Figure 18.13.

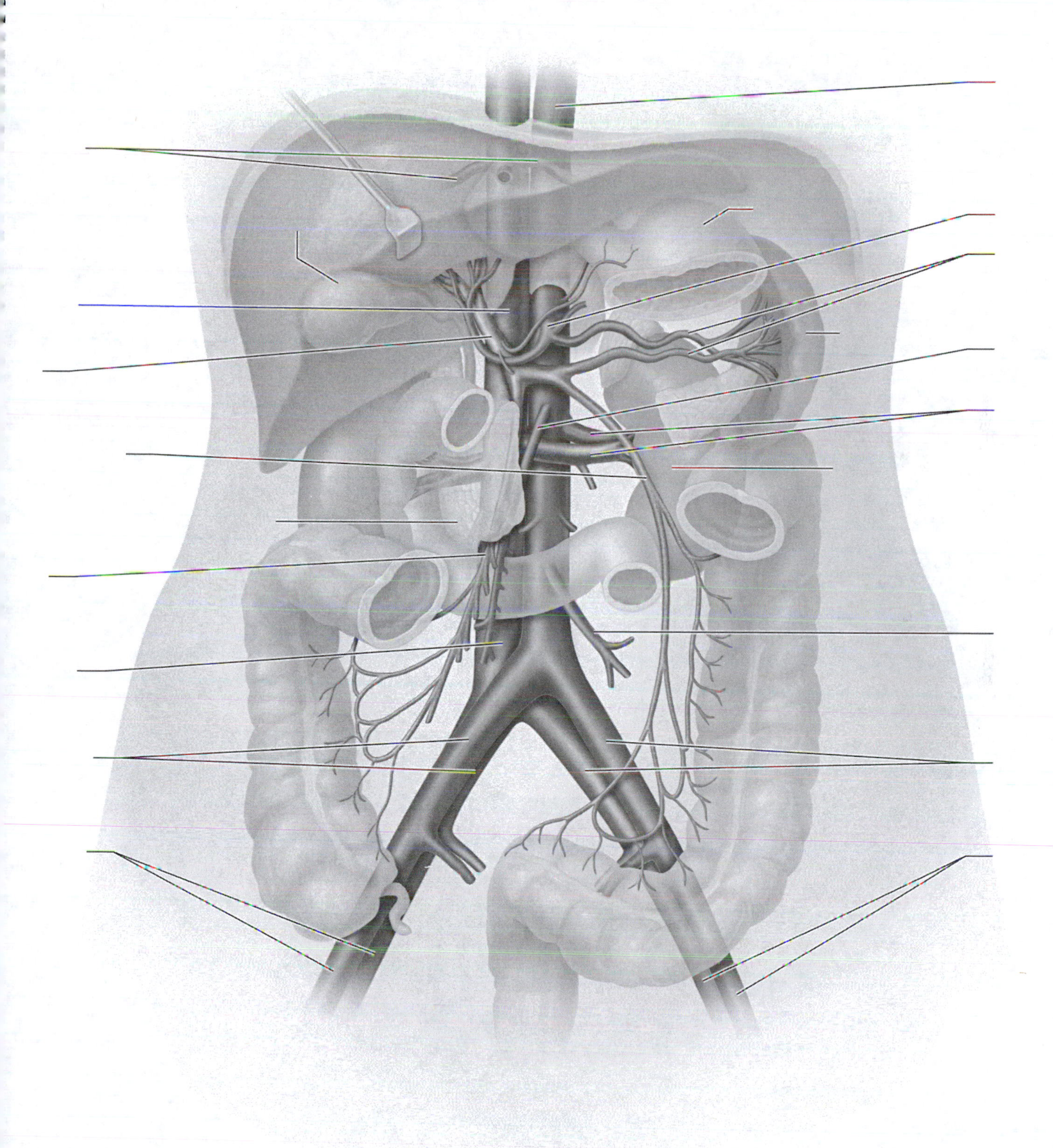

Figure 18.13 Blood vessels of the abdomen.

Identify It: Blood Vessels of the Upper and Lower Limbs

Identify the arteries and veins of the upper and lower limbs in Figure 18.14.

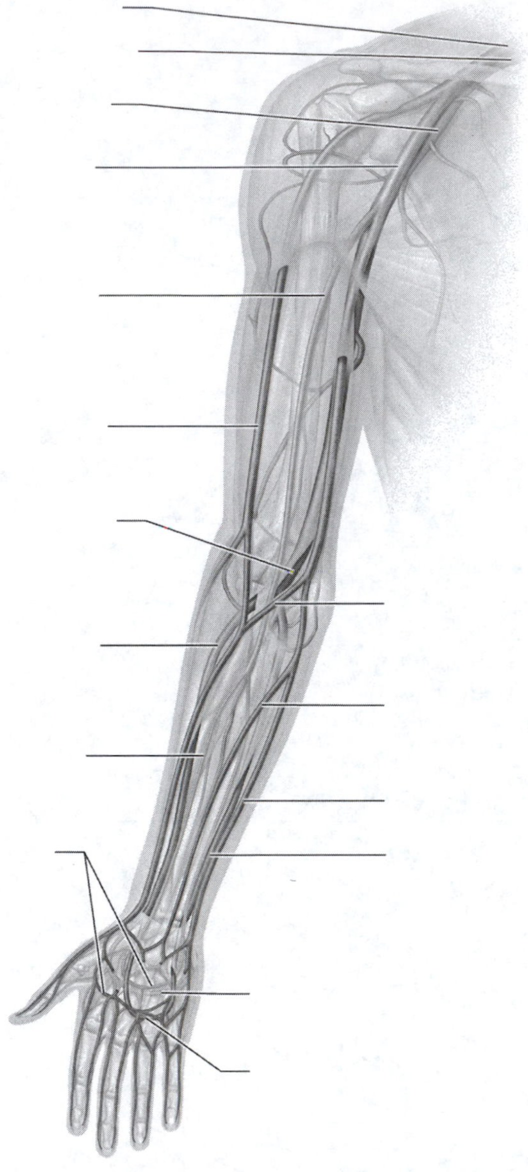

(a) Arteries and veins of the right upper limb, anterior view

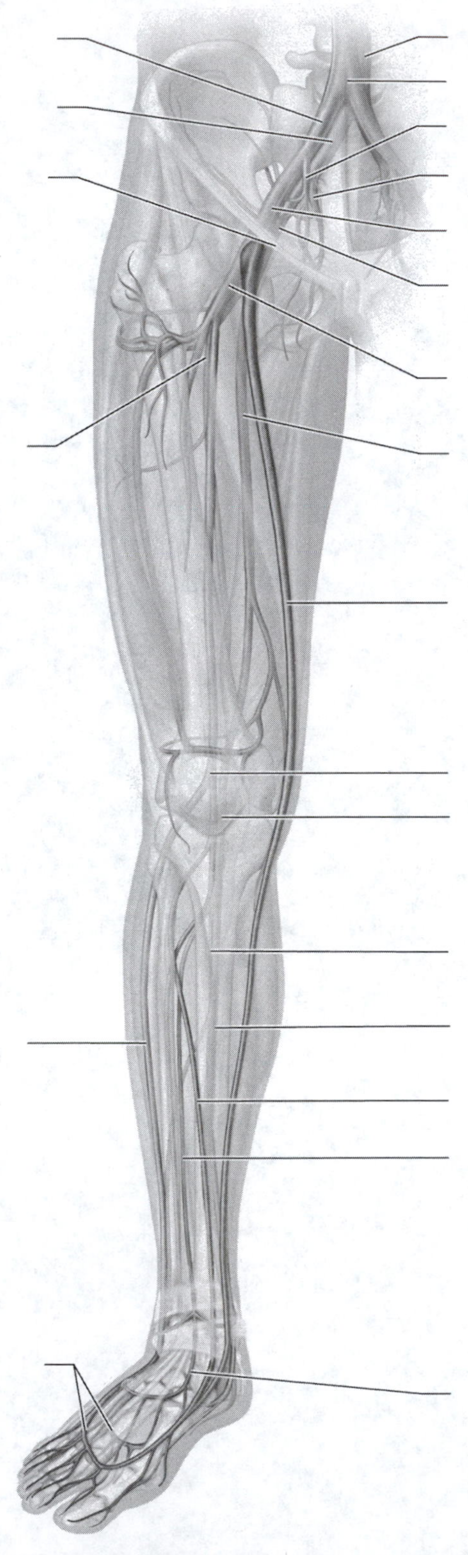

(b) Arteries and veins of the right lower limb, anterior view

Figure 18.14 Blood vessels of the upper and lower limbs.

What Do You Know Now?

Let's now revisit the questions you answered in the beginning of this chapter. How have your answers changed now that you've worked through the material?

- What effect does the sympathetic nervous system have on blood pressure?

  ____

- What are systole and diastole?

  ____

- Why can a person lose consciousness when placed in a choke hold?

  ____

19 Blood

Now that we've discussed the organs that pump and transport blood, let's talk about blood itself. This chapter explores the structure and function of the tissue blood, including blood clotting and blood typing.

What Do You Already Know?

Try to answer the following questions before proceeding to the next section. If you're unsure of the correct answers, give it your best attempt based on previous courses, previous chapters, or just your general knowledge.

- How much blood does an average person have in his or her body?

  ____

- Why is blood red? What color is venous blood?

  ____

- Which blood type is the universal donor? Why?

  ____

Module 19.1: Overview of Blood

Module 19.1 in your text introduces you to the structure and function of blood. By the end of the module, you should be able to do the following:

1. Describe the major components of blood.
2. Describe the basic functions of blood.
3. Describe the overall composition of plasma, including the major types of plasma proteins, their functions, and where in the body they are produced.

Build Your Own Glossary

Following is a table listing key terms from Module 19.1. Before you read the module, use the glossary at the back of your book or look through the module to define the following terms.

Key Terms for Module 19.1

Term	Definition
Blood	
Plasma	
Formed elements	
Hematocrit	
Plasma proteins	
Albumin	

Survey It: Form Questions

Before you read the module, survey it and form at least two questions for yourself. When you have finished reading the module, return to these questions and answer them.

Question 1: ______________________________

Answer: ______________________________

Question 2: ______________________________

Answer: ______________________________

Identify It: Components of Blood

Identify and color-code each component of blood in Figure 19.1.

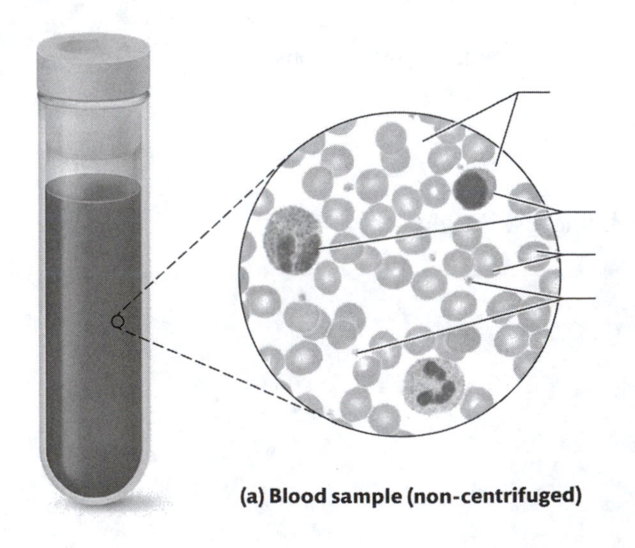

(a) Blood sample (non-centrifuged)

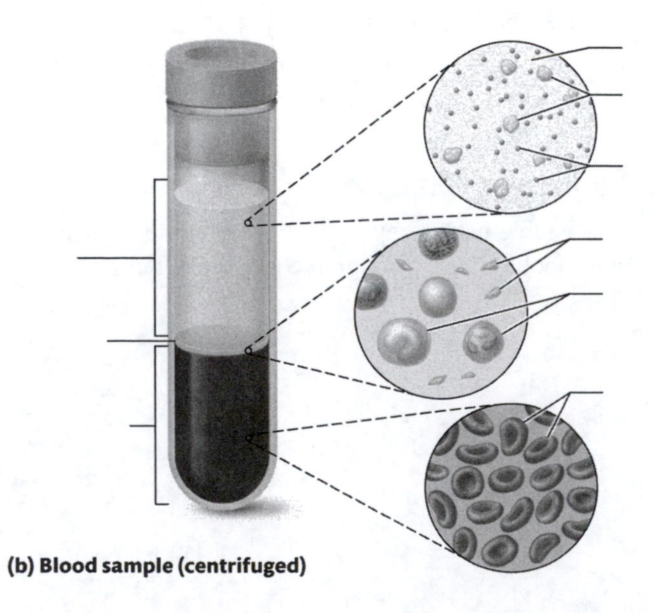

(b) Blood sample (centrifuged)

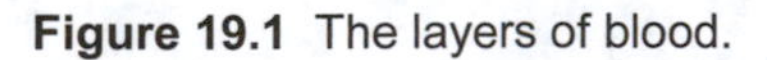

Figure 19.1 The layers of blood.

Key Concept: What are the main functions of blood?

____

____

Key Concept: What are the four types of plasma proteins, and what are their functions?

____

____

____

Module 19.2: Erythrocytes and Oxygen Transport

Now we look at the structure and development of the most common formed elements—the erythrocytes—and their role in the transport of oxygen through the body. When you finish this module, you should be able to do the following:

1. Describe the structure and functions of erythrocytes.
2. Discuss the structure and function of hemoglobin, as well as its breakdown products.
3. Explain the basic process of erythropoiesis and its regulation through erythropoietin.
4. Describe the causes and symptoms of anemia.

Build Your Own Glossary

Following is a table listing key terms from Module 19.2. Before you read the module, use the glossary at the back of your book or look through the module to define the following terms.

Key Terms for Module 19.2

Term	Definition
Erythrocyte	
Hemoglobin	
Hematopoiesis	
Hematopoietic stem cell	
Erythropoietin	
Bilirubin	
Anemia	

Survey It: Form Questions

Before you read the module, survey it and form at least two questions for yourself. When you have finished reading the module, return to these questions and answer them.

Question 1: ______________________________

Answer: ______________________________

Question 2: ______________________________

Answer: ______________________________

Key Concept: How does the structure of an erythrocyte allow it to carry out its functions?

____

____

Complete It: Erythrocyte Structure

Fill in the blanks to complete the following paragraph that describes the structure and function of erythrocytes.

A mature erythrocyte consists of a plasma membrane surrounding cytosol filled with the protein ____________. This large protein consists of four polypeptide subunits: two ____________ chains and two ____________ chains. Each polypeptide chain has a ____________ group, which contains an ____________ ion. When this ion binds to oxygen, the overall protein is called ____________. Binding to oxygen causes the ion to become ____________, which is what gives blood its ____________ color. When this ion is not bound to oxygen, it is called ____________.

Key Concept: Why is blood red? What makes venous blood darker than arterial blood?

____

____

Key Concept: What is the stimulus for erythropoietin production and release? What action does erythropoietin trigger, and how does this return the variable to the homeostatic range?

____

____

Draw It: Hematopoiesis

Draw the stages that an erythrocyte goes through during the process of hematopoiesis. Under each drawing, name the cell type and explain what is happening in the cell at that stage.

Name: Description:	Name: Description:	Name: Description:
Name: Description:	Name: Description:	Name: Description:

Describe It: Processing of Old Erythrocytes

Describe the process by which old erythrocytes are processed and destroyed in Figure 19.2. In addition, label and color-code important structures involved in the process.

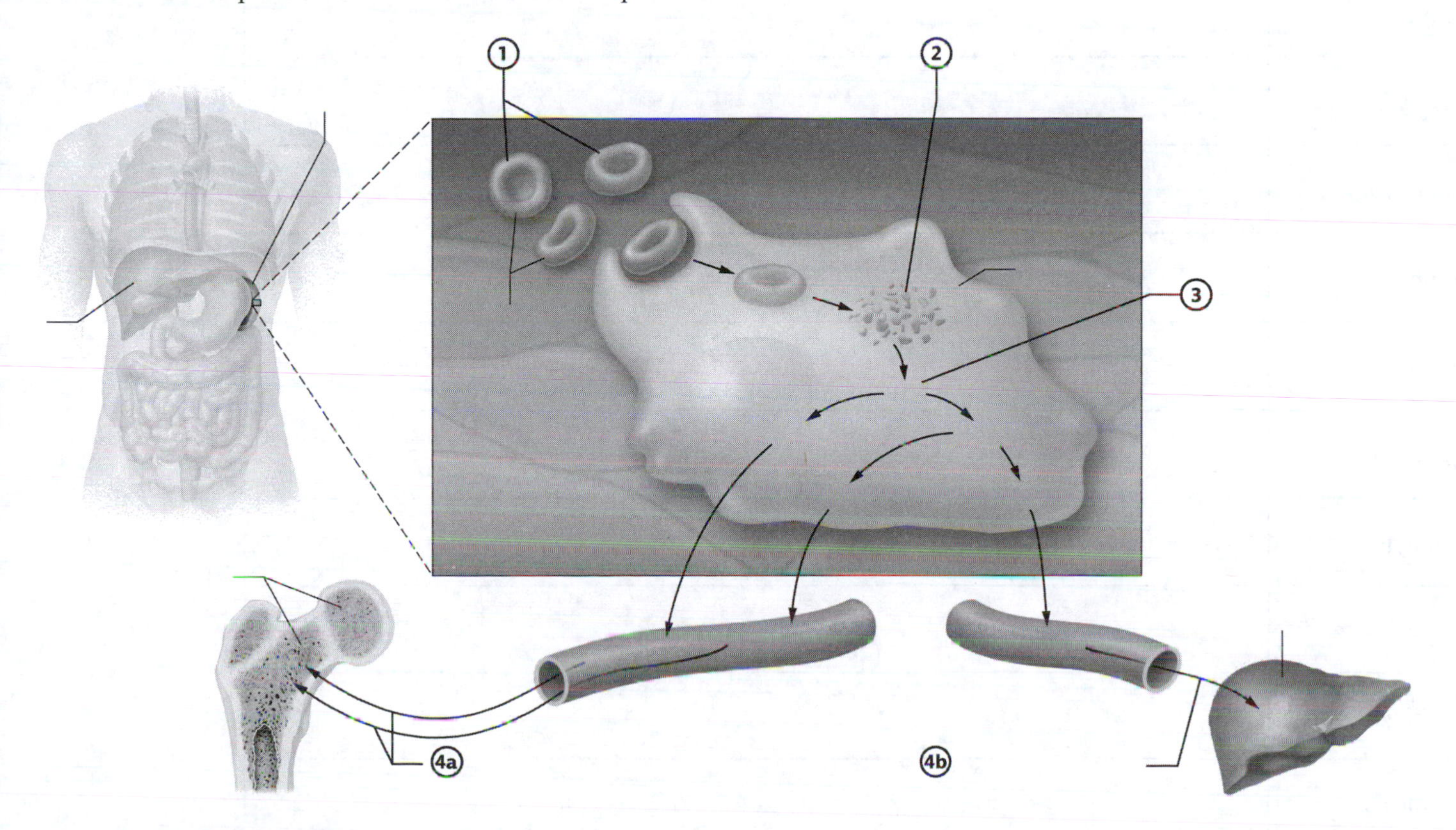

Figure 19.2 Erythrocyte death.

Key Concept: What are the three causes of anemia? How does each cause decrease the oxygen-carrying capacity of the blood?

____

____

Module 19.3: Leukocytes and Immune Function

The next major type of formed element we examine is the leukocyte. This module explores the different types of leukocytes, their development, and their functions. When you complete this module, you should be able to do the following:

1. Compare and contrast the relative prevalence and morphological features of the five types of leukocytes.
2. Describe functions for each of the five major types of leukocytes.
3. Discuss the difference in leukopoiesis of granulocytes and agranulocytes.

Build Your Own Glossary

Following is a table listing key terms from Module 19.3. Before you read the module, use the glossary at the back of your book or look through the module to define the following terms.

Key Terms of Module 19.3

Term	Definition
Leukocytes	
Granulocyte	
Agranulocyte	
Neutrophil	
Eosinophil	
Basophil	
Lymphocyte	
Monocyte	
Macrophage	
Leukopoiesis	

Survey It: Form Questions

Before you read the module, survey it and form at least three questions for yourself. When you have finished reading the module, return to these questions and answer them.

Question 1: ______________________________

Answer: ______________________________

Question 2: ______________________________

Answer: ______________________________

Question 3: ______________________________

Answer: ______________________________

Key Concept: What are the key differences between erythrocytes and leukocytes?

____

____

Build Your Own Summary Table: Leukocytes

As you read Module 19.3, build your own summary table about the different types of leukocytes by filling in the information in the following table. Under the "Appearance" column, draw the cell with colored pencils and point out the unique structural features of each type.

Summary of Leukocytes

Leukocyte	Appearance	Functions
Granulocytes		
Neutrophils		
Eosinophils		

Leukocyte	Appearance	Functions
Basophils		
Agranulocytes		
Lymphocytes		
Monocytes		

Key Concept: How do lymphocytes differ from all other cells in terms of their development in the bone marrow?

____

____

Module 19.4: Platelets

Module 19.4 in your text explores the final type of formed element: the platelet. At the end of this module, you should be able to do the following:

1. Explain how platelets differ structurally from the other formed elements of blood.
2. Discuss the role of the megakaryocyte in the formation of platelets.

Build Your Own Glossary

Following is a table listing key terms from Module 19.4. Before you read the module, use the glossary at the back of your book or look through the module to define the terms.

Key Terms of Module 19.4

Term	Definition
Platelets	
Megakaryocyte	

Survey It: Form Questions

Before you read the module, survey it and form at least two questions for yourself. When you have finished reading the module, return to these questions and answer them.

Question 1: ____________________

Answer: ____________________

Question 2: ____________________

Answer: ____________________

Key Concept: Are platelets cells? Explain your answer.

____

____

Describe It: Platelet Formation

Describe the process by which platelets are formed in Figure 19.3. In addition, label and color-code important structures involved in the process.

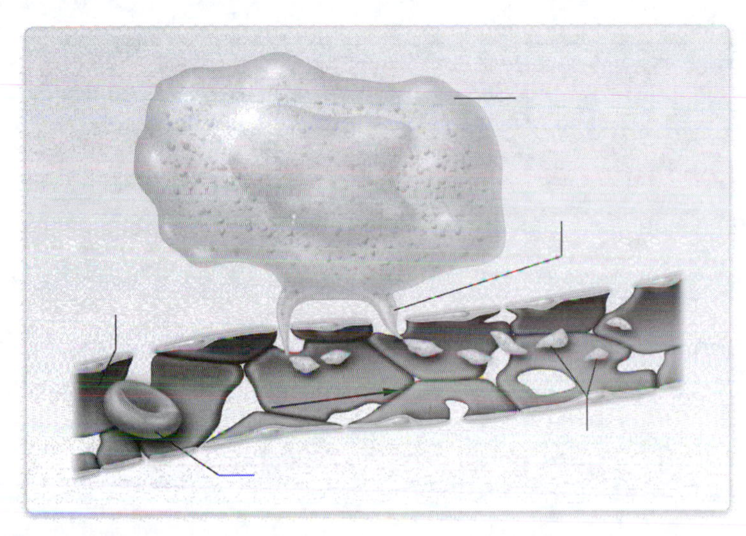

Figure 19.3 Platelet formation.

Description of process:

Module 19.5: Hemostasis

This module examines how blood loss from an injured vessel is minimized by a process called hemostasis. At the end of this module, you should be able to do the following:

1. Distinguish between the terms hemostasis and coagulation.
2. Describe the process of hemostasis, including the vascular phase, the formation of the platelet plug, and the formation of fibrin.
3. Explain the differences between the intrinsic and extrinsic clotting cascades.
4. Describe the role of calcium ions and vitamin K in blood clotting.
5. Explain how the positive feedback loops in the platelet and coagulation phases promote hemostasis.
6. Discuss the process of thrombolysis.

Build Your Own Glossary

Following is a table listing key terms from Module 19.5. Before you read the module, use the glossary at the back of your book or look through the module to define the following terms.

Key Terms of Module 19.5

Term	Definition
Hemostasis	
Blood clot	
Platelet plug	
Coagulation	
Coagulation cascade	
Thrombin	
Fibrin	
Clot retraction	
Fibrinolysis	
Plasmin	
Anticoagulant	
Thrombosis	

Survey It: Form Questions

Before you read the module, survey it and form at least two questions for yourself. When you have finished reading the module, return to these questions and answer them.

Question 1: ______________________________

Answer: ______________________________

Question 2: ______________________________

Answer: ______________________________

Key Concept: What is hemostasis? What are the five steps of hemostasis?

____

____

Key Concept: How and why does vascular spasm occur?

____

____

Describe It: Platelet Plug Formation

Describe the process by which platelet plug formation occurs in Figure 19.4. In addition, label and color-code important structures involved in the process.

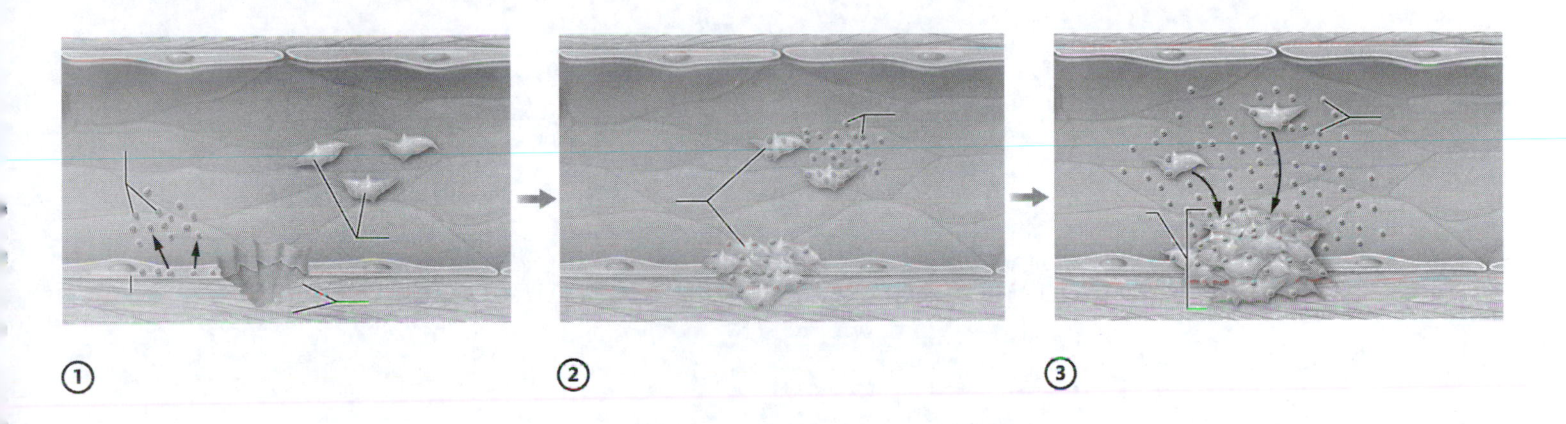

Figure 19.4 Hemostasis: Platelet plug formation.

Describe It: The Coagulation Cascade

Describe the process by which coagulation takes place in the intrinsic/contact activation pathway, extrinsic/tissue factor pathway, and common pathway in Figure 19.5. In addition, label and color-code important structures involved in the process.

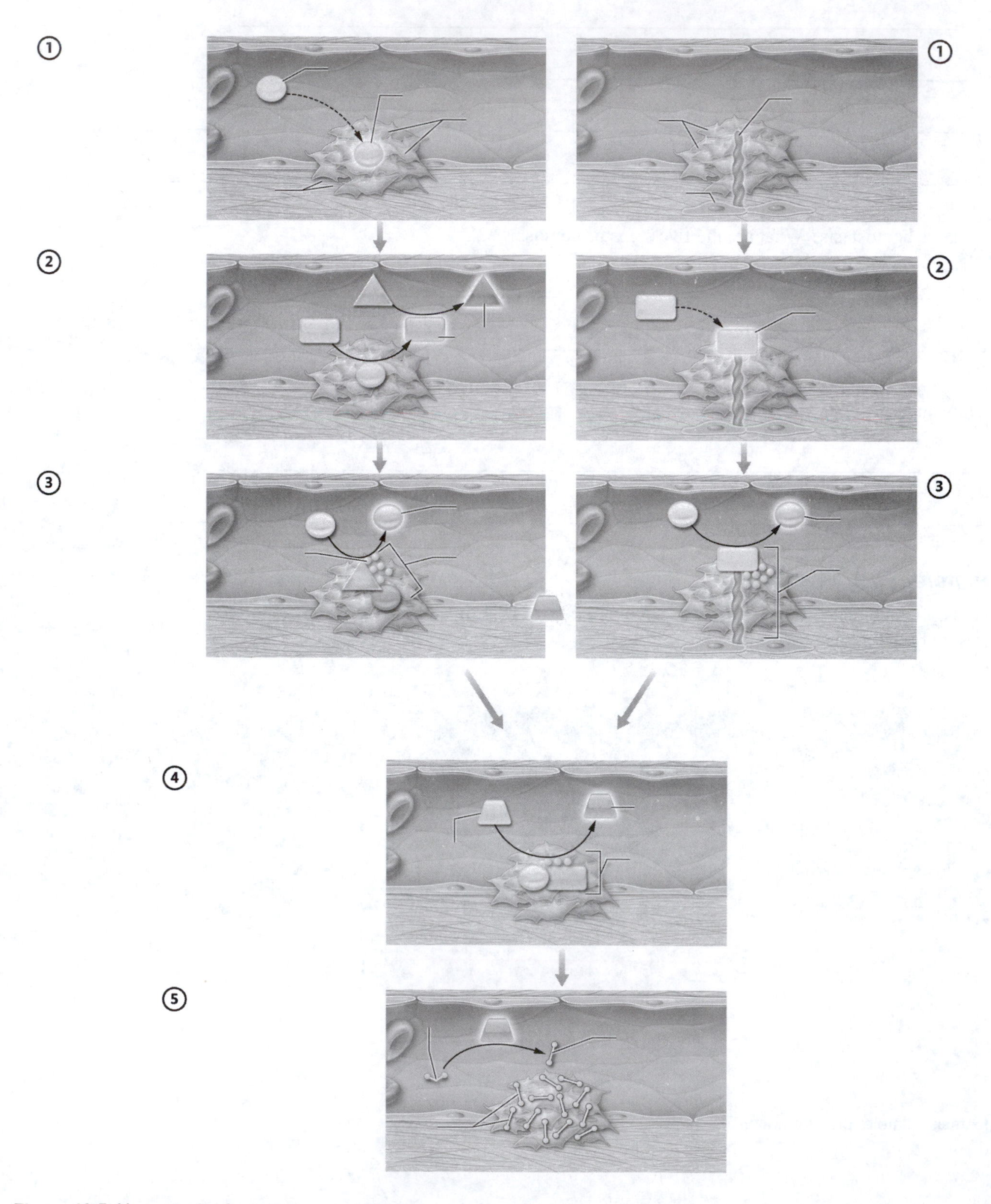

Figure 19.5 Hemostasis: Coagulation cascade.

Key Concept: What are the fundamental differences between the intrinsic and extrinsic pathways?

____

____

Key Concept: What is the overall purpose of coagulation?

____

____

Complete It: Clot Retraction and Thrombolysis

Fill in the blanks to complete the following paragraphs that describe clot retraction and fibrinolysis.

At the end of the coagulation cascade when ____________ threads "glue" the platelet plug together, the ____________ and ____________ fibers in the platelets contract. This action, called ____________ ____________, brings the edges of the vessel closer together. It also squeezes ____________ out of the clot, which contains plasma without ____________ ____________.

When the wounded vessel has healed, the clot is removed by the process of ____________. The first step of this process is to break down ____________ by the process of ____________. It begins when endothelial cells release ____________ ____________ ____________, which catalyzes the conversion of plasminogen to the active enzyme ____________. This enzyme then degrades ____________, which causes the clot to dissolve.

Key Concept: It is obviously important to clot a broken blood vessel. Why is it equally important to eventually break down that clot?

____

____

Team Up

Hemostasis is complicated, without a doubt, but working with your classmates can make it easier. Form a group and have each person make a handout to teach the process of hemostasis, including fibrinolysis. You can use Figure 19.17 in your text on page 745 as a guide, but the handout should be in your own words and with your own diagrams. At the end of the handout, write a few quiz questions. Once you have completed your handout, trade handouts with your group members. Study your partners' diagrams, and when you have finished, take the quiz at the end. When you and your group have finished taking all the quizzes, discuss the answers to determine places where you need additional study. After you've finished, combine the best elements of each handout to make one "master" diagram for the process of hemostasis.

Key Concept: What is an anticoagulant? Why is it important to regulate positive feedback loops like coagulation?

____

____

Module 19.6: Blood Typing and Matching

Now we look at the glycoproteins on the surface of cells, called antigens, which give rise to different blood types. By the end of the module, you should be able to do the following:

1. Explain the role of surface antigens on erythrocytes in determining blood groups.
2. List the type of antigen and the type of antibodies present in each ABO and Rh blood type.
3. Explain the differences between the development of anti-Rh antibodies and the development of anti-A and anti-B antibodies.
4. Predict which blood types are compatible, and explain what happens when the incorrect ABO or Rh blood type is transfused.
5. Explain why blood type O– is the universal donor and type AB+ is the universal recipient.

Build Your Own Glossary

Following is a table listing key terms from Module 19.6. Before you read the module, use the glossary at the back of your book or look through the module to define the following terms.

Key Terms for Module 19.6

Term	Definition
Blood transfusion	
Recipient	
Donor	
Antigen	
Blood groups	
ABO blood group	
Rh blood group	
Antibody	
Universal donor	
Universal recipient	

Survey It: Form Questions

Before you read the module, survey it and form at least three questions for yourself. When you have finished reading the module, return to these questions and answer them.

Question 1: ______________________________

Answer: ______________________________

Question 2: ______________________________

Answer: ______________________________

Question 3: ______________________________

Answer: ______________________________

Key Concept: What are antigens? What are the key antigens on erythrocytes?

____

____

Draw It: Blood Types

Draw each cell and its antigens indicated here by the blood type. Then, list the antibodies this blood type will have in the blood (assume previous exposure to the Rh antigen).

Type A+ Antibodies:	Type B− Antibodies:	Type O− Antibodies:
Type AB− Antibodies:	Type O+ Antibodies:	Type AB+ Antibodies:

Key Concept: What happens when antibodies bind their specific antigens?

Practice It: Blood Donation

Following are people who want to donate blood. Determine to which blood types they could safely donate. Explain in each case why the person could or could not donate to the recipient. Assume each recipient has been exposed to the Rh antigen and so has anti-Rh antibodies.

1. Donor Jane, blood type: B+
   - Recipient 1: O+
   - Recipient 2: AB–
   - Recipient 3: AB+
2. Donor Jerome, blood type: O+
   - Recipient 1: B+
   - Recipient 2: AB–
   - Recipient 3: A–
3. Donor Ngoc, blood type: A–
   - Recipient 1: O–
   - Recipient 2: B–
   - Recipient 3: AB–

Key Concept: When considering blood donation, do we consider the antigens on the donor's erythrocytes, the antibodies in the donor's blood, or both? Explain.

What Do You Know Now?

Let's now revisit the questions you answered in the beginning of this chapter. How have your answers changed now that you've worked through the material?

- How much blood does an average person have in his or her body?
- Why is blood red? What color is venous blood?
- Which blood type is the universal donor? Why?

20 The Lymphatic System and Immunity

In the last chapter, we looked at the structure and function of blood, including leukocytes, which function in immunity. Now we look at these cells more closely as we explore the lymphatic system and the organs, cells, and proteins that function in immunity.

What Do You Already Know?

Try to answer the following questions before proceeding to the next section. If you're unsure of the correct answers, give it your best attempt based on previous courses, previous chapters, or just your general knowledge.

- Why does a healthcare provider palpate (feel) your neck when you are sick?

  ____

- What causes flu-like symptoms when we get sick?

  ____

- What is the purpose of a fever?

  ____

Module 20.1: Structure and Function of the Lymphatic System

Module 20.1 in your text introduces you to the structure and function of the cells, tissues, and organs of the lymphatic system. By the end of the module, you should be able to do the following:

1. Describe the major functions of the lymphatic system.
2. Compare and contrast lymphatic vessels and blood vessels in terms of structure and function.
3. Explain the mechanisms of lymph formation, and trace the pathway of lymph circulation through the body.
4. Describe the basic structure and cellular composition of lymphatic tissue, and relate them to the overall functions of the lymphatic system.
5. Describe the structures and functions of the lymphoid organs.

Build Your Own Glossary

Following is a table listing key terms from Module 20.1. Before you read the module, use the glossary at the back of your book or look through the module to define the following terms.

Key Terms for Module 20.1

Term	Definition
Lymph	
Lymphatic vessels	
Lymphatic capillaries	
Lymphoid tissue	
Dendritic cells	
Reticular cells	
Mucosa-associated lymphatic tissue (MALT)	
Lymphoid follicles	
Tonsils (pharyngeal, lingual, and palatine)	
Lymph node	
Spleen	
Thymus	

Survey It: Form Questions

Before you read the module, survey it and form at least three questions for yourself. When you have finished reading the module, return to these questions and answer them.

Question 1: ______________________________

Answer: ______________________________

Question 2: ______________________________

Answer: ______________________________

Question 3: ______________________________

Answer: ______________________________

Identify It: Structures of the Lymphatic System

Identify and color-code each component of the lymphatic system in Figure 20.1.

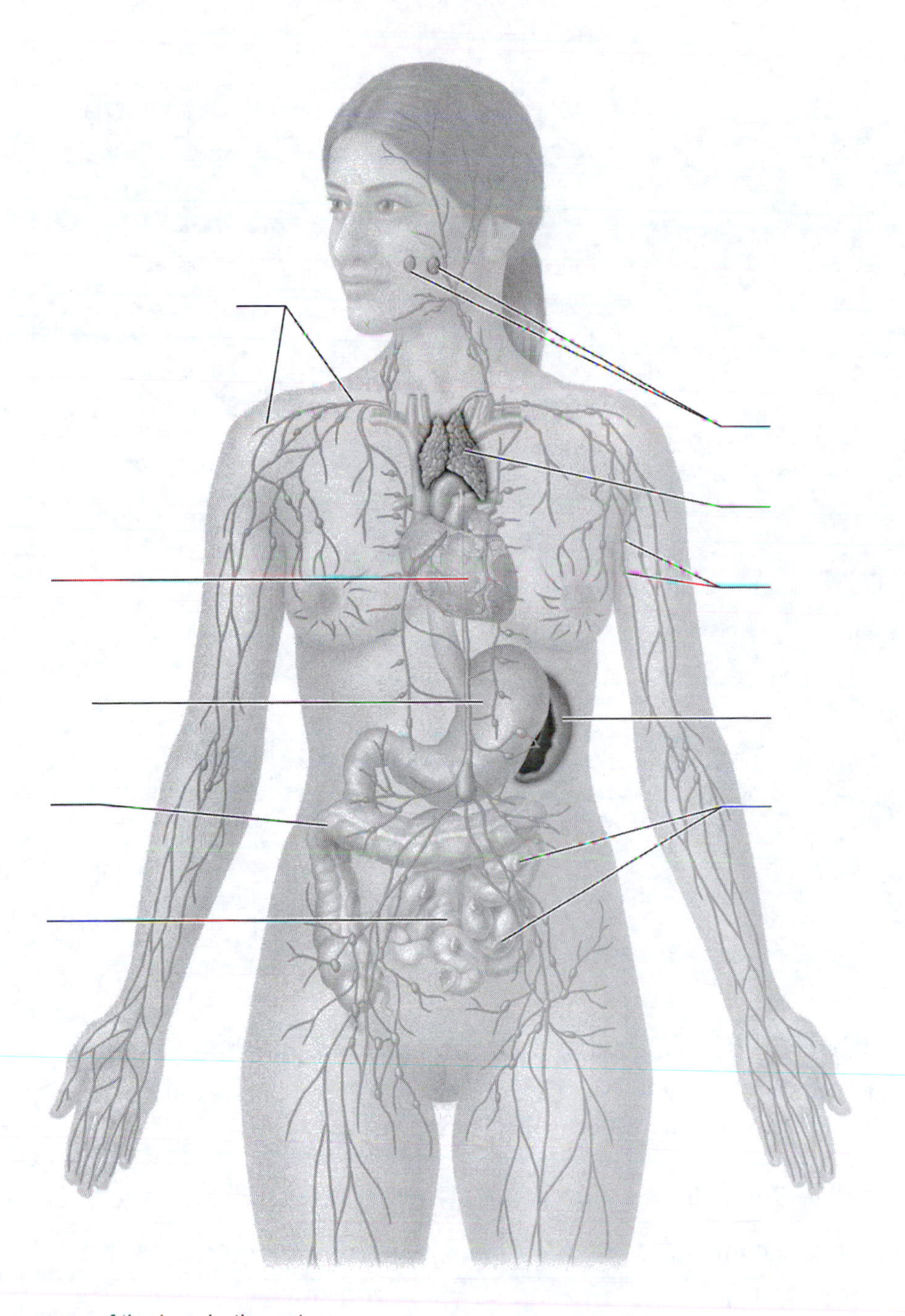

Figure 20.1 The organs of the lymphatic system.

Key Concept: Why does severe swelling result when lymphatic vessels are blocked or damaged?

____

____

Identify It: Main Lymph Ducts and Trunks

Identify and color-code the main lymph ducts and trunks in Figure 20.2. Then, list the body regions that each duct and trunk drains.

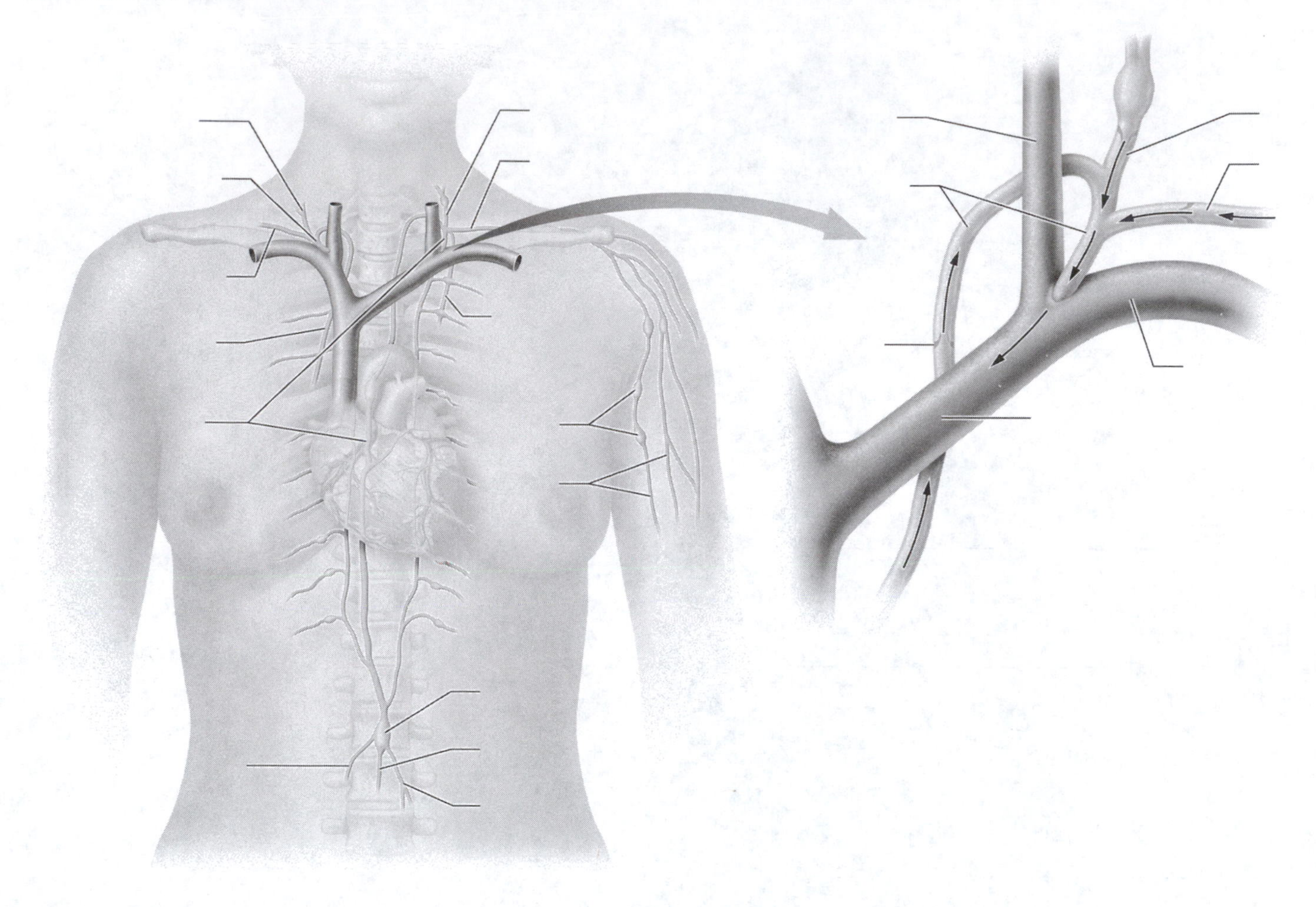

Figure 20.2 Main lymph ducts and trunks.

Build Your Own Summary Table: Cells, Tissues, and Organs of the Lymphatic System

As you read Module 20.1, build your own summary table about the different types of cells, tissues, and organs that make up the lymphatic system by filling in the information in the following table.

Summary of Lymphatic Cells, Tissues, and Organs

Lymphatic Structure	Description/Location	Function
Cells		
Leukocytes		
Dendritic cells		

Reticular cells		
Mucosa-Associated Lymphatic Tissue		
Peyer's patches		
Appendix		
Pharyngeal tonsil		
Lingual tonsil		
Palatine tonsils		
Organs		
Lymph nodes		
Spleen		
Thymus		

Key Concept: Why do lymph nodes get swollen when you are sick?

____

____

Module 20.2: Overview of the Immune System

Now we look at some of the fundamental principles of immunity. When you finish this module, you should be able to do the following:

1. Identify the differences between innate and adaptive immunity, and explain how the two types of immunity work together.
2. Describe the basic differences between antibody-mediated and cell-mediated immunity.
3. Describe the roles that surface barriers play in immunity.
4. Describe the cells and proteins that make up the immune system.
5. Explain how the immune and lymphatic systems are connected structurally and functionally.

Build Your Own Glossary

Following is a table listing key terms from Module 20.2. Before you read the module, use the glossary at the back of your book or look through the module to define the following terms.

Key Terms for Module 20.2

Term	Definition
Innate immunity	
Adaptive immunity	
Antigen	
Cell-mediated immunity	
Antibody-mediated immunity	
Surface barriers	
Phagocyte	
Natural killer cell	
Cytokine	

Survey It: Form Questions

Before you read the module, survey it and form at least three questions for yourself. When you have finished reading the module, return to these questions and answer them.

Question 1: ______________________________

Answer: ______________________________

Question 2: ____________________

Answer: ____________________

Question 3: ____________________

Answer: ____________________

Key Concept: When you cut your finger, which line of defense is compromised? Why do we cover a wound with a bandage?

____

____

Practice It: Types of Immunity

Identify each of the following statements as being a property of either innate immunity or adaptive immunity.

- Requires 3 to 5 days to mount a response ____________
- Brought about by T and B lymphocytes ____________
- Requires no stimulus to produce; already present in the blood ____________
- Responds to unique antigens ____________
- Responds quickly to cellular injury; the dominant response in the first 12 hours ____________
- Has the capacity for immunological memory ____________

Complete It: Overview of Immunity

Fill in the blanks to complete the following paragraphs that describe the basic principles of the immune system.

The first line of defense consists of ____________ ____________ that prevent the entry of pathogens into the body. Certain surface barriers secrete substances that deter the growth of, kill, or trap pathogens, such as ____________ in the skin, ____________ in mucous membranes, and ____________ in the stomach.

The second line of defense is also known as ____________ ____________ and the third line as ____________ ____________. The main cells of both lines are ____________. Many cells of the second line ingest pathogens or damaged cells by the process of ____________. The other components of both lines are ____________, such as ____________, ____________, and ____________.

Key Concept: Why does a problem with the lymphatic system potentially lead to a problem with the immune system?

____

____

Module 20.3: Innate Immunity: Internal Defenses

We now take a look at the defenses provided by the cells and proteins of innate immunity. When you complete this module, you should be able to do the following:

1. Describe the roles that phagocytic and nonphagocytic cells and plasma proteins such as complement and interferon play in innate immunity.
2. Walk through the stages of the inflammatory response and describe its purpose.
3. Describe the process by which fever is generated and explain its purpose.

Build Your Own Glossary

Following is a table listing key terms from Module 20.3. Before you read the module, use the glossary at the back of your book or look through the module to define the following terms.

Key Terms of Module 20.3

Term	Definition
Antigen-presenting cell	
Neutrophil	
Eosinophil	
Basophil	
Mast cell	
Complement system	
Interferons	
Inflammatory response	
Fever	
Pyrogen	

Survey It: Form Questions

Before you read the module, survey it and form at least three questions for yourself. When you have finished reading the module, return to these questions and answer them.

Question 1: ____________________

Answer: ____________________

Question 2: ______________________________

Answer: ______________________________

Question 3: ______________________________

Answer: ______________________________

Build Your Own Summary Table: Cells and Proteins of Innate Immunity

As you read Module 20.3, build your own summary table about the cells and proteins of innate immunity by filling in the information in the following table.

Cells and Proteins of Innate Immunity

Cell or Protein	Description	Functions
Cells		
Macrophages		
Neutrophils		
Eosinophils		
NK cells		
Dendritic cells		
Basophils		
Mast cells		
Proteins		
Complement		
Tumor necrosis factor		

Cell or Protein	Description	Functions
Interferons		
Interleukins		

Key Concept: What often causes flu-like symptoms when one is infected with a pathogen?

____

____

Describe It: The Inflammatory Response, Part 1

Describe the sequence of events of the first part of the inflammatory response in Figure 20.3. In addition, label and color-code each important component of the process.

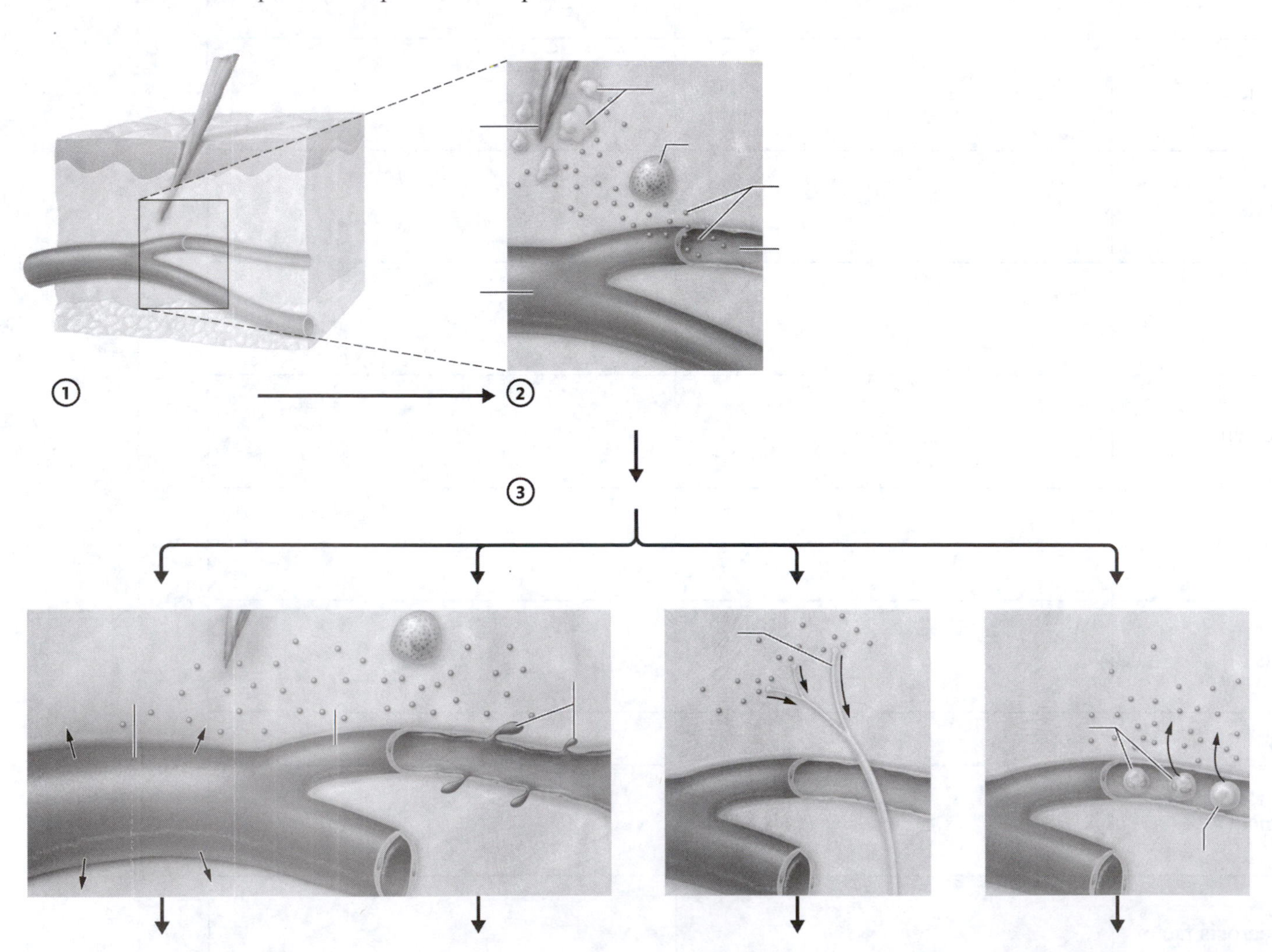

Figure 20.3 The inflammatory response, part 1.

Describe It: The Inflammatory Response, Part 2

Describe the sequence of events of the second part of the inflammatory response in Figure 20.4. In addition, label and color-code each important component of the process.

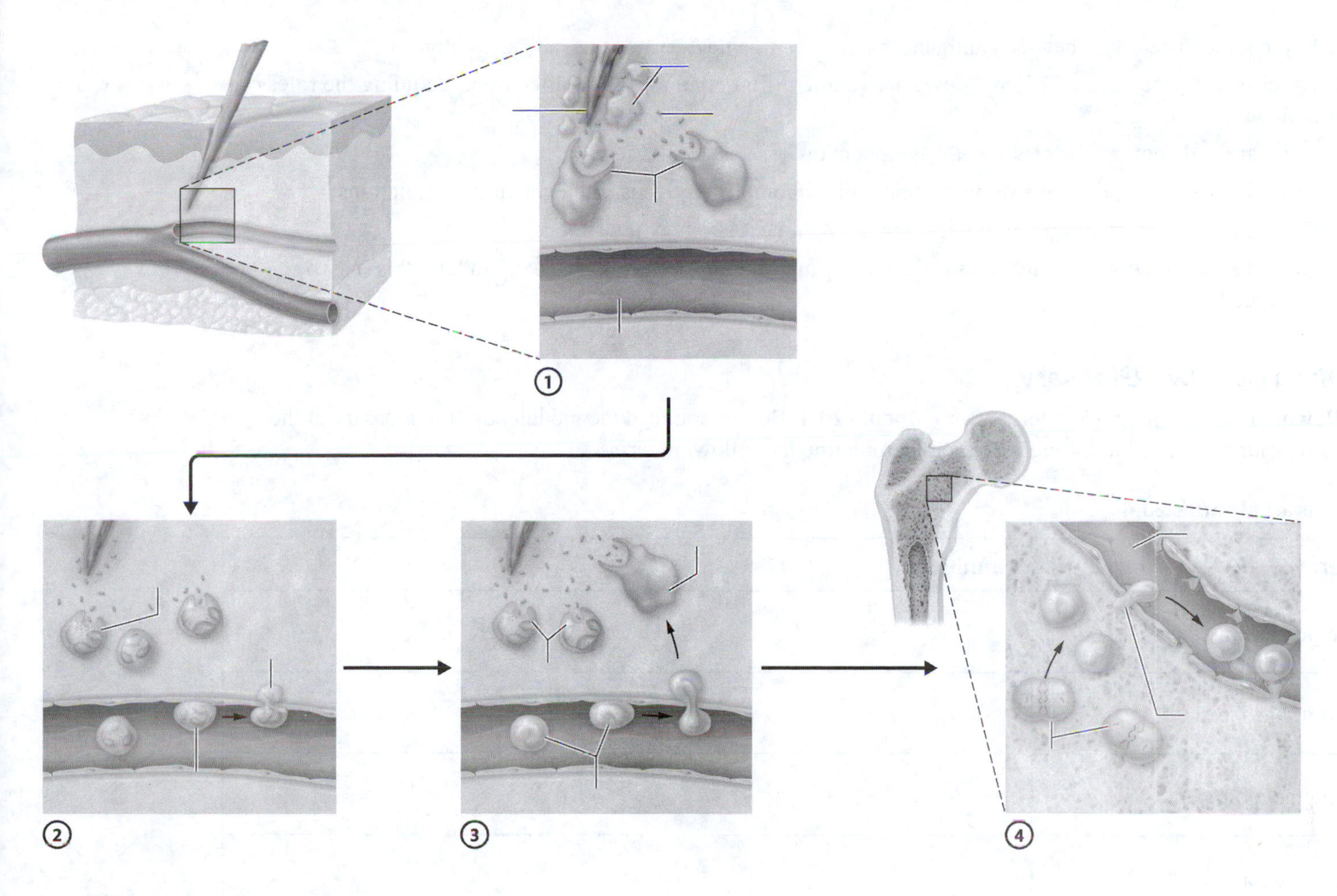

Figure 20.4 The inflammatory response, part 2.

Key Concept: What causes a fever?

____

____

Key Concept: Your classmate insists that since your patient has a fever, she must have an infection because the elevated temperature of fever kills the bacteria. Explain to your classmate why this is wrong.

____

____

Module 20.4: Adaptive Immunity: Cell-Mediated Immunity

Module 20.4 in your text explores the functions of the cells of the first arm of adaptive immunity: cell-mediated immunity. At the end of this module, you should be able to do the following:

1. Explain the differences between antigens, haptens, antigenic determinants, and self antigens.
2. Describe the processes of T lymphocyte activation, differentiation, and proliferation, including the roles of antigen-presenting cells.
3. Compare and contrast the classes of T lymphocytes.
4. Describe the two types of major histocompatibility complex antigens, and explain their functions.
5. Describe the purpose of immunological memory, and explain how it develops.
6. Describe the process by which a transplanted organ or tissue is rejected, and explain how this may be prevented.

Build Your Own Glossary

Following is a table listing key terms from Module 20.4. Before you read the module, use the glossary at the back of your book or look through the module to define the following terms.

Terms of Module 20.4

Term	Definition
Helper T cell	
Cytotoxic T cell	
Immunocompetent	
Self antigen	
Class I MHC molecule	
Class II MHC molecule	
Immunological memory	

Survey It: Form Questions

Before you read the module, survey it and form at least three questions for yourself. When you have finished reading the module, return to these questions and answer them.

Question 1: ____

Answer: ____

Question 2: ____

Answer: ____

Question 3: ____

Answer: ____

Key Concept: What is a T cell clone? How is each T cell clone genetically unique? Which clones are destroyed in the thymus, and why?

____

____

Complete It: T Cells and MHC Molecules

Fill in the blanks to complete the following paragraph that describes the properties of MHC molecules and their relationship to cytotoxic and helper T cells.

The T cell receptor can only bind to portions of antigen bound to ___________ ___________ ___________ on the surface of cells. Nearly all cells have ___________ ___________ ___________ molecules on their cell surface that display ___________ antigens. ___________ ___________ ___________ molecules are found only on antigen-presenting cells and display ___________ antigens. T_H cells bind to ___________ ___________ ___________ molecules only, whereas T_C cells bind to ___________ ___________ ___________.

Describe It: Display of MHC Molecules

Describe the sequence of events by which class I and class II MHC molecules are displayed in Figure 20.5. In addition, label and color-code each important component of the process.

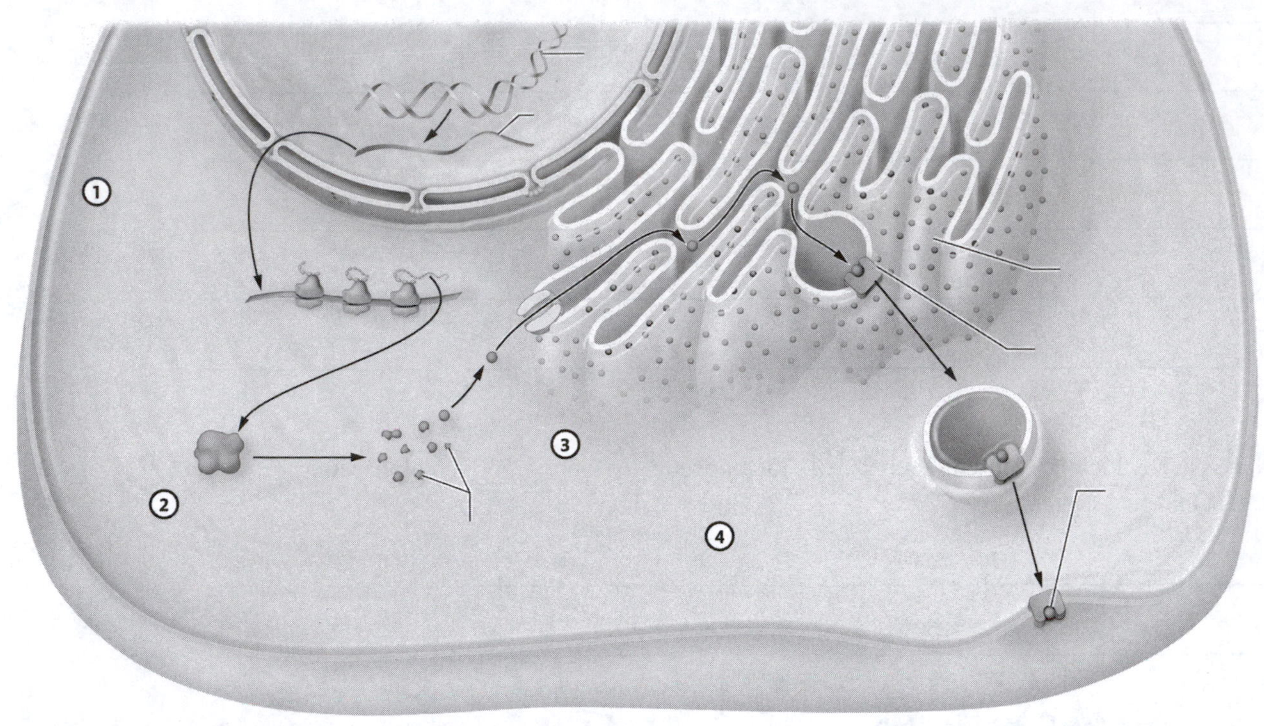

(a) Class I MHC molecules process and display endogenous antigens.

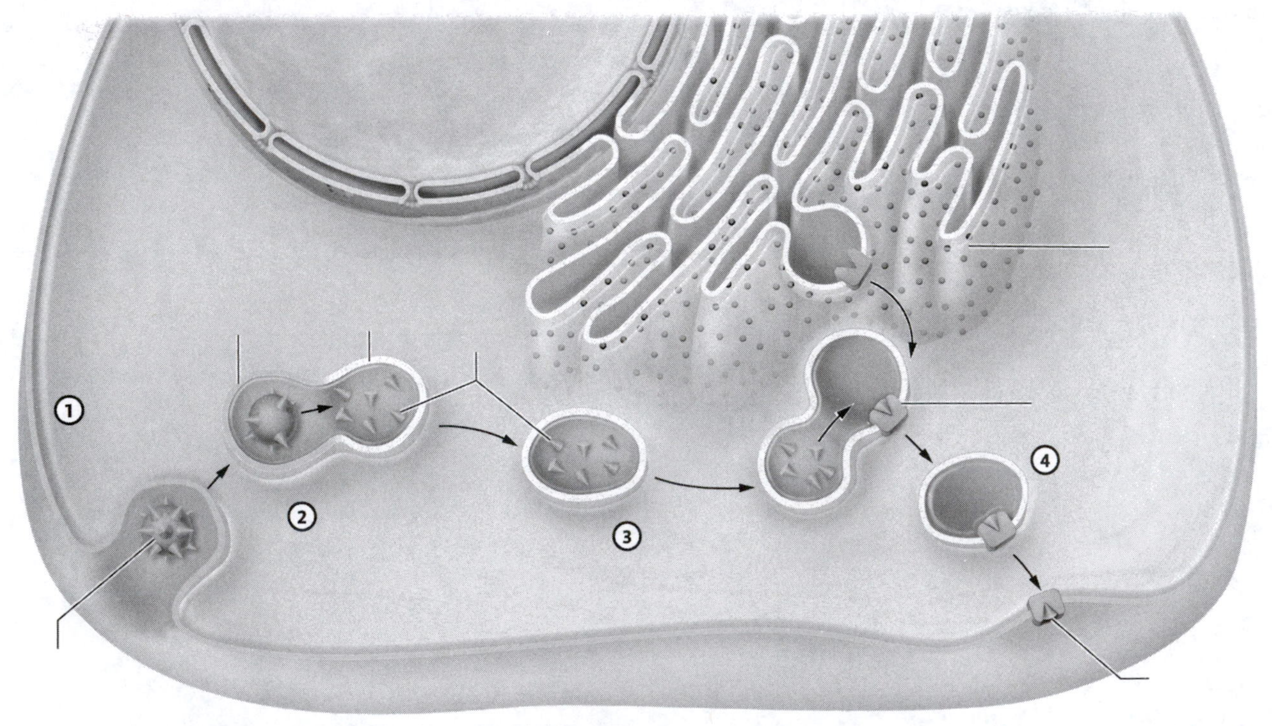

(b) Class II MHC molecules process and display exogenous antigens.

Figure 20.5 Antigen processing and display by MHC molecules.

Describe It: T Cell Activation

Describe the sequence of events by which T_C and T_H cells are activated in Figure 20.6. In addition, label and color-code each component of the process.

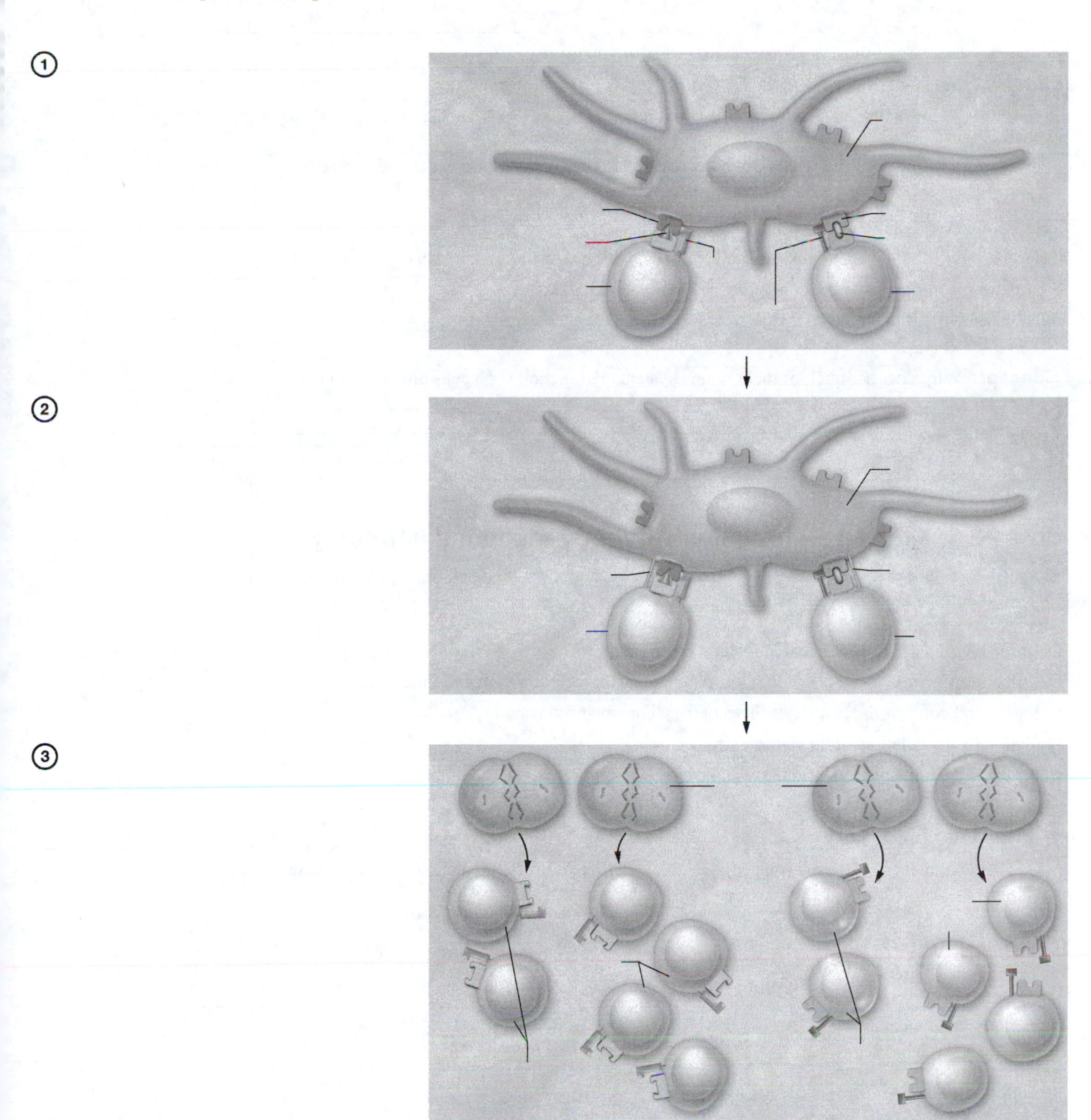

Figure 20.6 T cell activation.

Key Concept: Why are both class I and class II MHC molecules needed?

____

____

Practice It: Functions of T_H and T_C Cells

Identify each of the following statements as being a function of T_H cells or T_C cells.

- Releases the enzyme perforin, which catalyzes reactions to perforate the target cell membrane ____________
- Secretes interleukin-2 to activate T_C cells ____________
- Directly binds and stimulates B cells ____________
- Detects cancer cells, foreign cells, and cells infected with intracellular pathogens ____________
- Stimulates macrophages to become more efficient phagocytes ____________
- Releases enzymes that catalyze the destruction of a target cell's DNA ____________

Key Concept: Why does so much of the immune system malfunction if T_H cells are not working properly?

____

____

Module 20.5: Adaptive Immunity: Antibody-Mediated Immunity

This module examines the second arm of adaptive immunity: antibody-mediated immunity. At the end of this module, you should be able to do the following:

1. Describe the process of B cell activation and proliferation.
2. Describe the five major classes of antibodies, and explain their structure and functions.
3. Compare and contrast the primary and secondary immune responses.
4. Explain how vaccinations induce immunity.
5. Compare and contrast active immunity and passive immunity.

Build Your Own Glossary

Following is a table listing key terms from Module 20.5. Before you read the module, use the glossary at the back of your book or look through the module to define the following terms.

Key Terms of Module 20.5

Term	Definition
Antibody	
Plasma cell	
Memory B cell	

Term	Definition
Agglutination	
Precipitation	
Opsonization	
Neutralization (of antigen)	
Primary immune response	
Secondary immune response	
Vaccination	
Active immunity	
Passive immunity	

Survey It: Form Questions

Before you read the module, survey it and form at least three questions for yourself. When you have finished reading the module, return to these questions and answer them.

Question 1: ______________________________

Answer: ______________________________

Question 2: ______________________________

Answer: ______________________________

Question 3: ______________________________

Answer: ______________________________

Key Concept: Where do B cells mature? Why do only about 10% of the cells complete their maturation process?

____

____

Describe It: B Cell Activation

Describe the sequence of events by which B cells are activated in Figure 20.7. In addition, label and color-code each important component of the process.

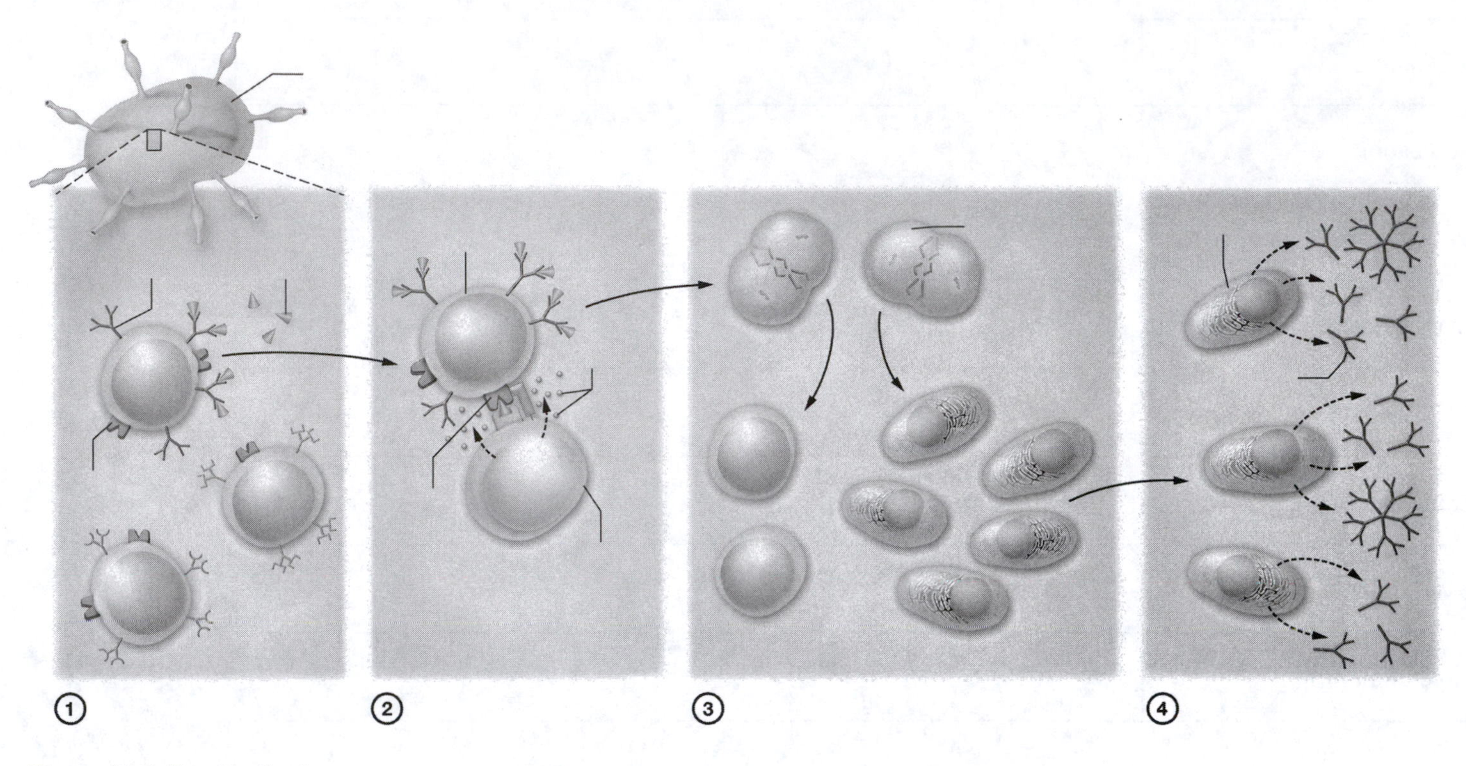

Figure 20.7 B cell activation.

Key Concept: How do plasma cells and memory B cells differ?

____

____

Draw It: Functions of Antibodies

Draw and describe the functions of antibodies. Label and color any important structures of the processes.

Build Your Own Summary Table: Antibody Classes

Build your own summary table about the different classes of antibodies by filling in the information in the following table.

Antibody Classes

Class	Structure	Functions/Properties
IgG		
IgA		
IgM		
IgE		
IgD		

Team Up

Form a group and have each member make a diagram to teach the differences between the primary and secondary immune responses and active versus passive immunity. At the end of your handout, write at least 10 quiz questions. When you have finished with your handouts, trade them with other group members and study them, taking the quiz at the end. Check your answers to determine areas where you need further study, then combine the best elements of each handout to make a "master" diagram teaching these concepts.

Key Concept: You have just been envenomated by a spider and the physician gives you pre-formed antibodies to the venom. Is this active or passive immunity? Six months later, you find that your antibody titer to the venom is still high. Is this due to active or passive immunity? Explain.

____

____

Key Concept: Your friend insists that she gets the flu every time she gets a flu shot. However, the flu vaccine contains killed viruses. What do you tell her?

____

____

Module 20.6: Putting It All Together: The Big Picture of the Immune Response

None of the parts of the immune system can function in isolation—all components must work with one another to protect the body from pathogens. So now we put the pieces together and see the big picture of the immune response. By the end of the module, you should be able to do the following:

1. Describe how the immune and lymphatic systems work together to respond to internal and external threats.
2. Explain how the immune response differs for different types of threats.
3. Describe the immune response to cancerous cells.
4. Explain how certain pathogens can evade the immune response.

Survey It: Form Questions

Before you read the module, survey it and form at least two questions for yourself. When you have finished reading the module, return to these questions and answer them.

Question 1: ____

Answer: ____

Question 2: ____

Answer: ____

Describe It: Response to a Viral Infection

Describe the sequence of events by which the immune system responds to a viral common cold in Figure 20.8. In addition, label and color-code each component of the process.

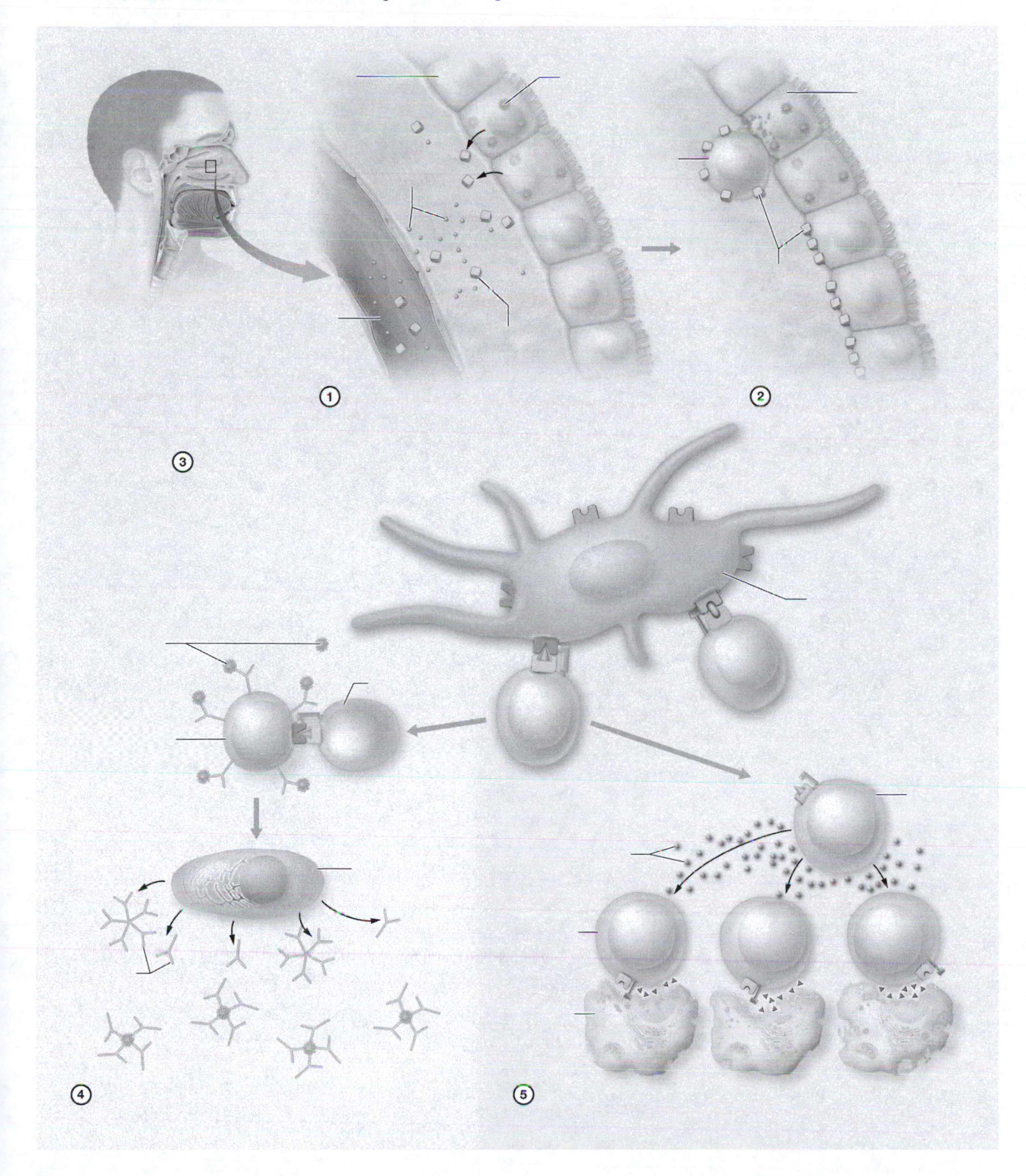

Figure 20.8 Immune response to the common cold.

Describe It: Response to a Bacterial Infection

Describe the sequence of events by which the immune system responds to a bacterial infection in Figure 20.9. In addition, label and color-code each component of the process.

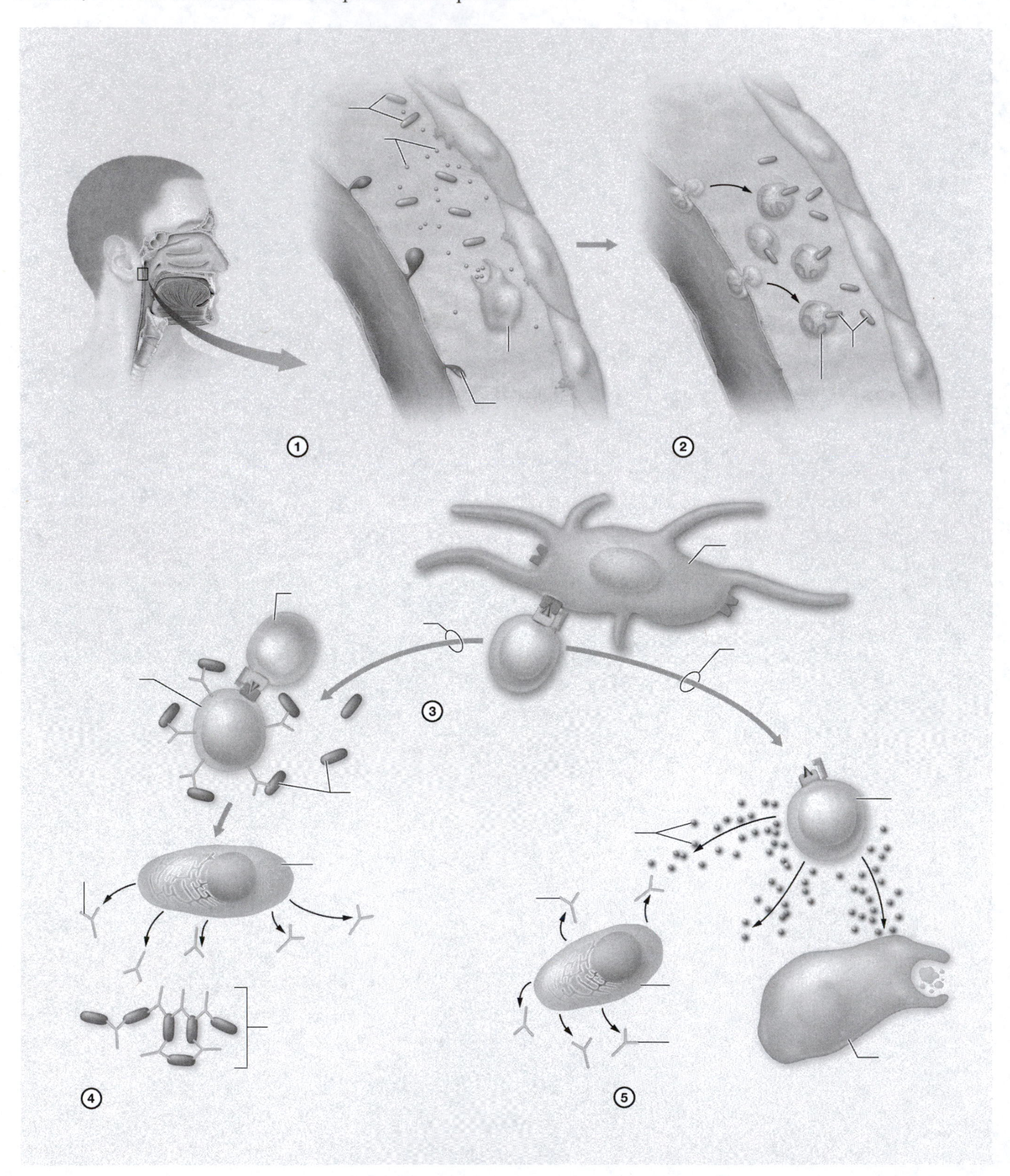

Figure 20.9 Immune response to a bacterial infection.

Key Concept: How does the response to a bacterial infection differ from the response to a viral infection?

____

____

Describe It: Immune Response to Cancer

Describe the sequence of events by which the immune system responds to cancer cells in Figure 20.10. In addition, label and color-code each component of the process.

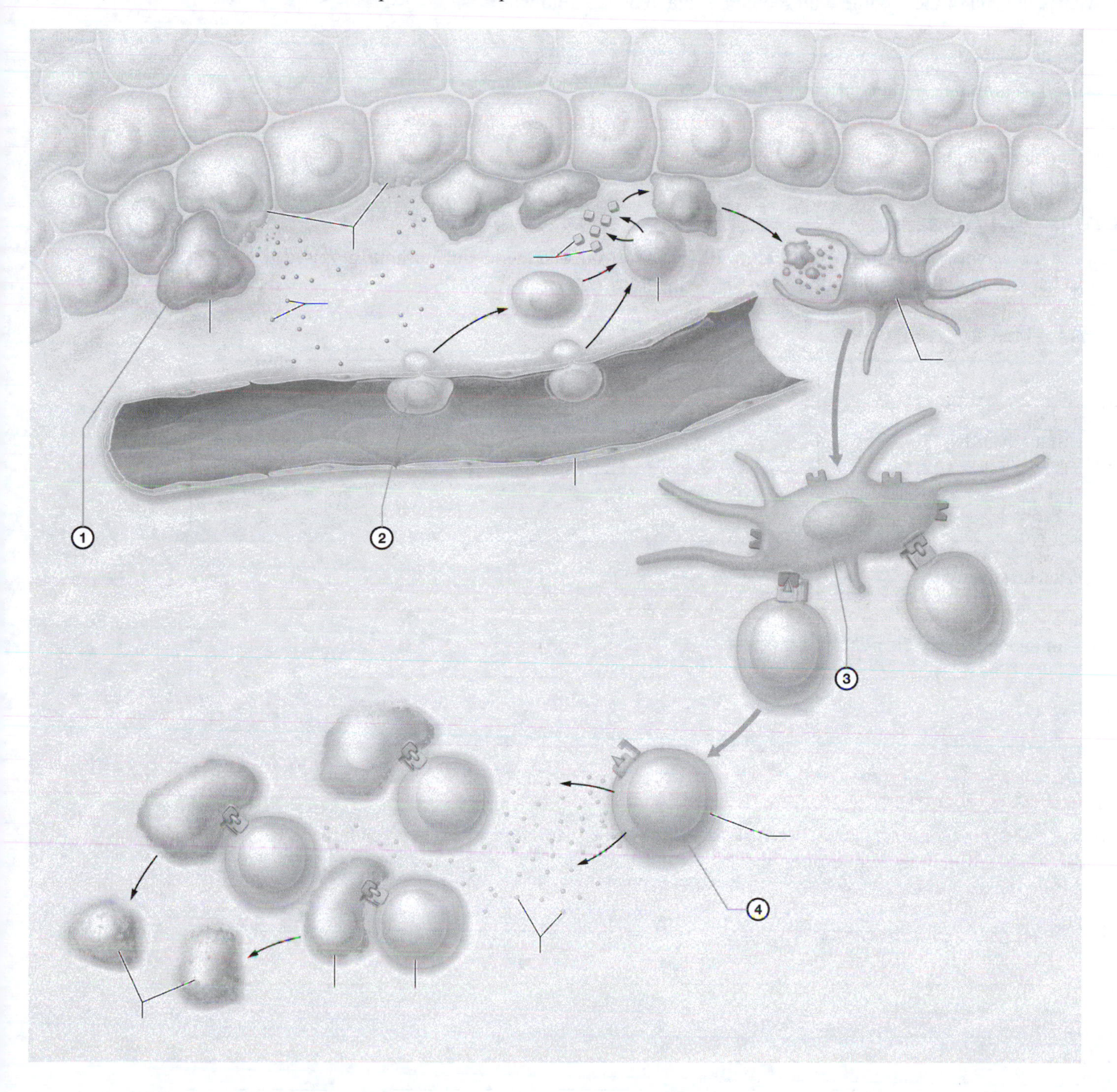

Figure 20.10 Immune response to cancer cells.

Key Concept: How does the response to cancer cells differ from the responses to most infectious pathogens?

____

____

Module 20.7: Disorders of the Immune System

Module 20.7 in your text discusses what happens when parts of the immune system are either overactive or underactive. By the end of the module, you should be able to do the following:

1. Describe the characteristics of the types of hypersensitivity disorders.
2. Describe the common immunodeficiency disorders.
3. Explain why HIV targets certain cell types, and describe the effects this virus has on the immune system.
4. Describe the characteristics of common autoimmune disorders.

Build Your Own Glossary

Following is a table listing key terms from Module 20.7. Before you read the module, use the glossary at the back of your book or look through the module to define the following terms.

Key Terms for Module 20.7

Term	Definition
Hypersensitivity disorder	
Anaphylactic shock	
Immunodeficiency disorder	
Acquired immunodeficiency disorder	
Autoimmune disorder	

Survey It: Form Questions

Before you read the module, survey it and form at least two questions for yourself. When you have finished reading the module, return to these questions and answer them.

Question 1: ____

Answer: ____

Question 2: ____

Answer: ____

Key Concept: Why is a type I hypersensitivity reaction much more severe upon a second exposure to the allergen?

____

____

Build Your Own Summary Table: Immune Disorders

As you read Module 20.7, build your own summary table about the different types of immune disorders by filling in the information in the following table.

Types of Immune Disorders

Type of Immune Disorder	Description/Causes	Example
Hypersensitivity Disorders		
Type I hypersensitivity		
Type II hypersensitivity		
Type III hypersensitivity		
Type IV hypersensitivity		
Immunodeficiency Disorders		
Primary immunodeficiency disorders		
Secondary immunodeficiency disorders		
Autoimmune Disorders		
Autoimmune disorders		

Key Concept: Why does the loss of functional T_H cells in HIV/AIDS cause problems for the entire immune system?

____

____

What Do You Know Now?

Let's now revisit the questions you answered in the beginning of this chapter. How have your answers changed now that you've worked through the material?

- Why does a healthcare provider palpate (feel) your neck when you are sick?

  ____

- What causes flu-like symptoms when we get sick?

  ____

- What is the purpose of a fever?

  ____

21 The Respiratory System

This chapter discusses the respiratory system. We begin by taking a tour of its organs and then explore its functions in greater detail. The chapter concludes with the two types of diseases that affect the respiratory system and their effects on homeostasis.

What Do You Already Know?

Try to answer the following questions before proceeding to the next section. If you're unsure of the correct answers, give it your best attempt based on previous courses, previous chapters, or just your general knowledge.

- What is/are the main function(s) of the respiratory system?

  ____

- How do we breathe?

  ____

- Why do we inhale oxygen and exhale carbon dioxide?

  ____

Module 21.1: Overview of the Respiratory System

This module introduces the basic structures of the respiratory system and its many roles in maintaining homeostasis. By the end of the module, you should be able to do the following:

1. Describe and distinguish between the upper and lower respiratory tracts.
2. Describe and distinguish between the conducting and respiratory zones of the respiratory tract.
3. Describe the major functions of the respiratory system.
4. Define and describe the four respiratory processes—pulmonary ventilation, pulmonary gas exchange, gas transport, and tissue gas exchange.

Build Your Own Glossary

Following is a table listing key terms from Module 21.1. Before you read the module, use the glossary at the back of your book or look through the module to define the following terms.

Key Terms for Module 21.1

Term	Definition
Respiratory tract	
Conducting zone	

Term	Definition
Respiratory zone	
Respiration	

Survey It: Form Questions

Before you read the module, survey it and form at least two questions for yourself. When you have finished reading the module, return to these questions and answer them.

Question 1: ______________________________

Answer: ______________________________

Question 2: ______________________________

Answer: ______________________________

Identify It: Structures of the Respiratory System

Identify and color-code the structures of the respiratory system shown in Figure 21.1. In addition, identify each part of the respiratory tract as belonging to the conducting zone or respiratory zone.

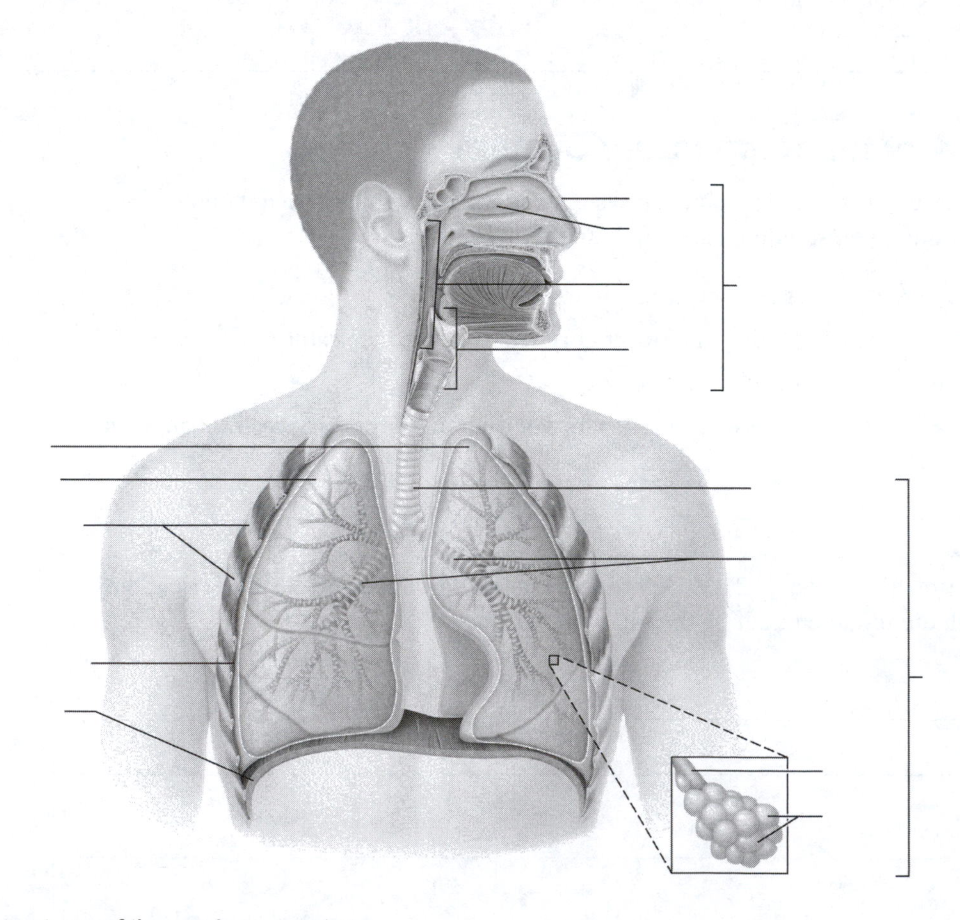

Figure 21.1 Structures of the respiratory system.

Key Concept: What is respiration? Which processes make up respiration?

____

____

Key Concept: What are the other functions of the respiratory system, and how do they help maintain homeostasis?

____

____

Module 21.2: Anatomy of the Respiratory System

In this module, we explore the organs of the respiratory system and follow the pathway that air takes through the group of structures collectively called the respiratory tract. When you finish this module, you should be able to do the following:

1. Trace the pathway through which air passes during inspiration.
2. Describe the gross anatomical features and function of each region of the respiratory tract, the pleural and thoracic cavities, and the pulmonary blood vessels and nerves.
3. Describe the histology of the different regions of the respiratory tract, the types of cells present in alveoli, and the structure of the respiratory membrane.
4. Explain how the changes in epithelial and connective tissue in air passageways relate to their function.
5. Describe the structure of the lungs and pleural cavities.

Build Your Own Glossary

Following is a table listing key terms from Module 21.2. Before you read the module, use the glossary at the back of your book or look through the module to define the following terms.

Key Terms for Module 21.2

Term	Definition
Nasal cavity	
Paranasal sinuses	
Respiratory mucosa	
Pharynx	
Larynx	
Epiglottis	
Vocal folds	
Trachea	

Term	Definition
Bronchial tree	
Alveoli	
Respiratory membrane	
Lung	
Pleural cavity	

Survey It: Form Questions

Before you read the module, survey it and form at least three questions for yourself. When you have finished reading the module, return to these questions and answer them.

Question 1: ______________________

Answer: ______________________

Question 2: ______________________

Answer: ______________________

Question 3: ______________________

Answer: ______________________

Identify It: The Nose and Nasal Cavity

Identify and color-code the structures of the nose and nasal cavity in Figure 21.2.

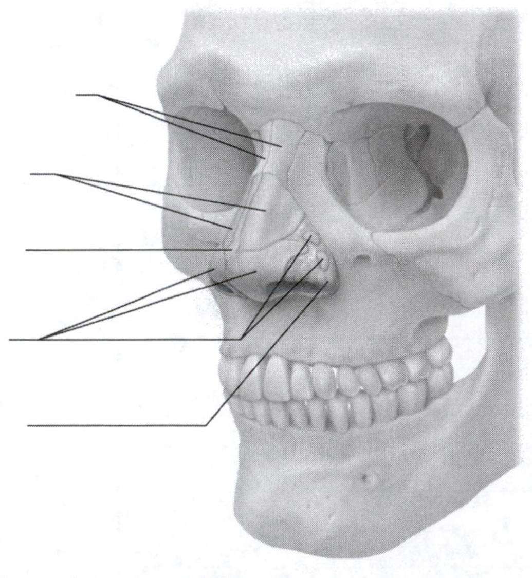

(a) Internal structures of the nose

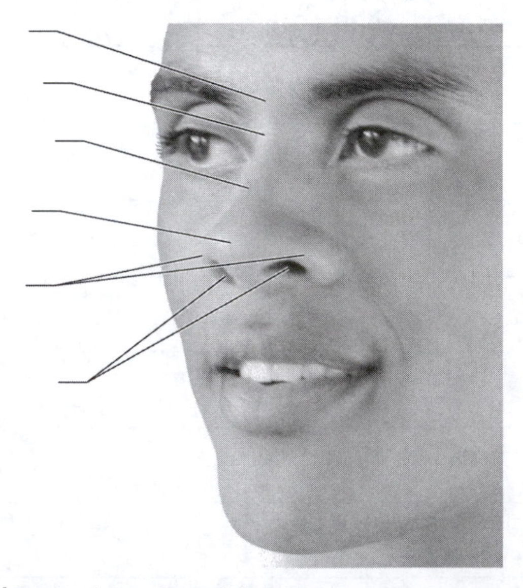

(b) External structures of the nose

Figure 21.2 Structures of the nose and nasal cavity.

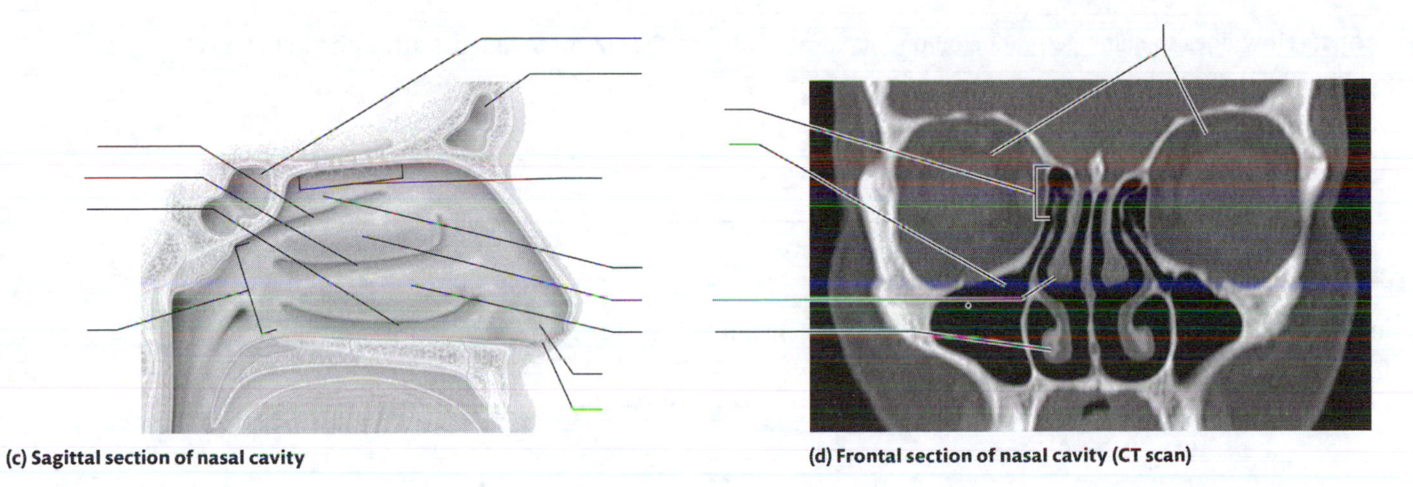

(c) Sagittal section of nasal cavity

(d) Frontal section of nasal cavity (CT scan)

Figure 21.2 *(continued)*

Key Concept: Where are the paranasal sinuses located? What are their functions? Could they perform these functions if they were lined with stratified squamous epithelium instead of respiratory epithelium?

____

____

Key Concept: What delineates the three different regions of the pharynx? How do they differ, structurally and functionally?

____

____

Identify It: The Larynx

Identify and color-code the structures of the larynx in Figure 21.3.

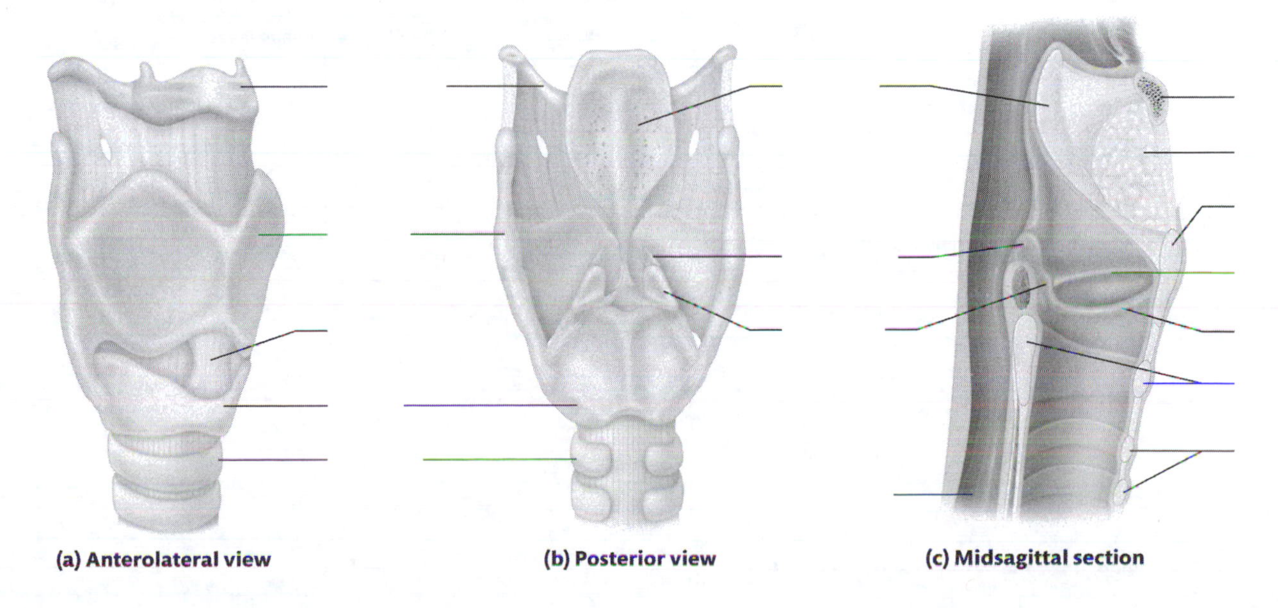

(a) Anterolateral view

(b) Posterior view

(c) Midsagittal section

Figure 21.3 Anatomy of the larynx.

Key Concept: How does the change in tension on the vocal cords and the size of the glottis affect the loudness and pitch of the sound produced?

____

____

Identify It: The Trachea and Bronchial Tree

Identify and color-code the structures of the trachea and bronchial tree in Figure 21.4.

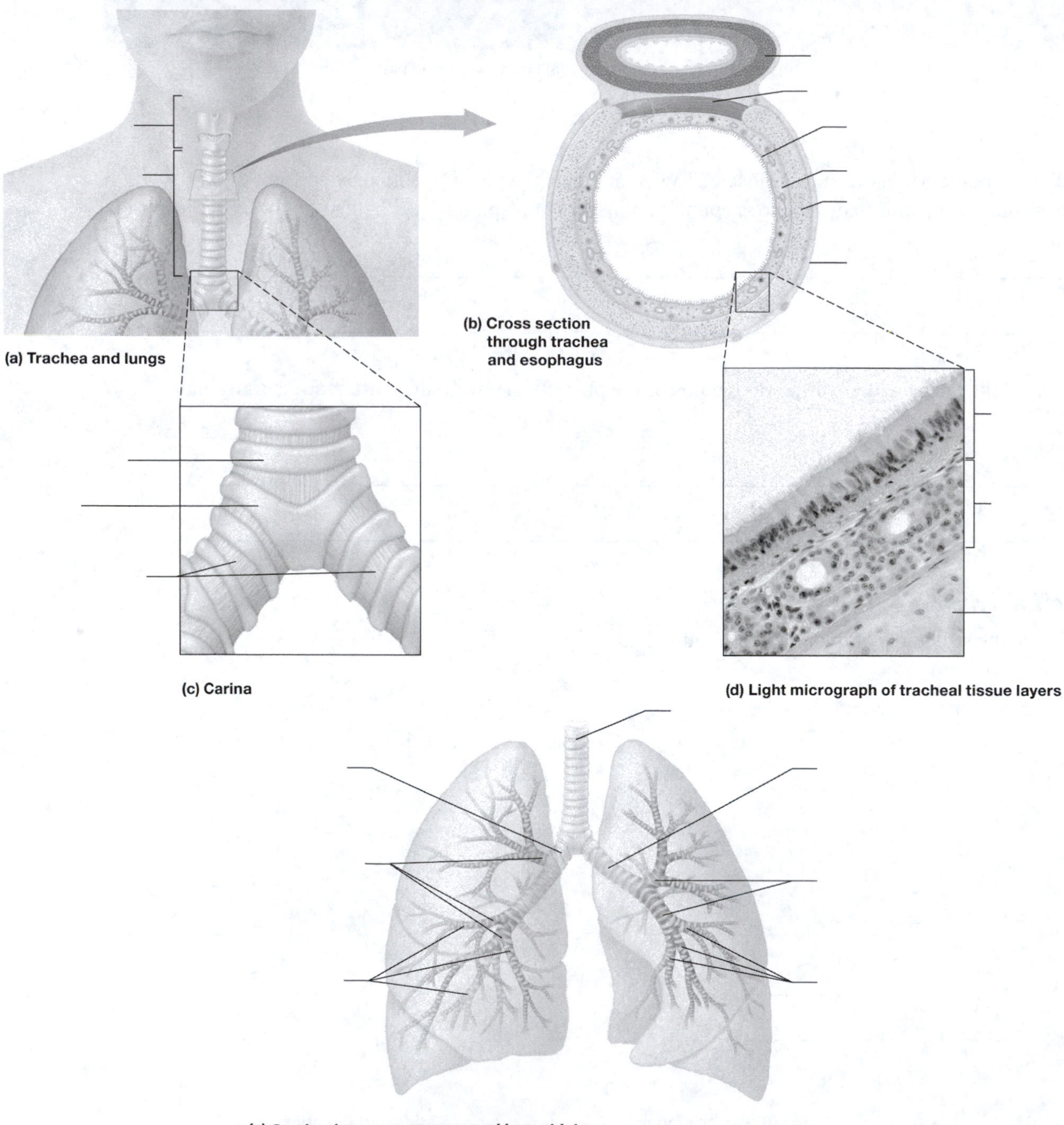

Figure 21.4 The trachea and bronchial tree.

Identify It: Anatomy of the Respiratory Zone

Identify and color-code the structures of the respiratory zone in Figure 21.5.

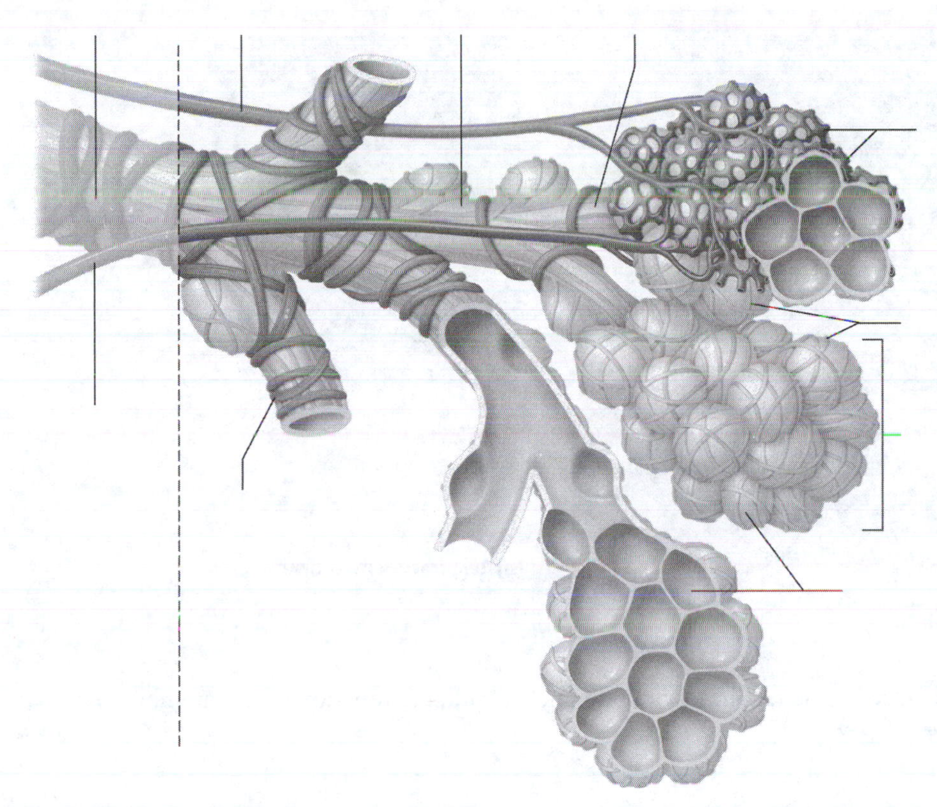

Figure 21.5 Anatomy of the respiratory zone.

Trace It: Pathway of Air Flow through the Respiratory System

Trace the pathway of air as it flows from the nares through the bronchial tree to the alveoli.

Start: Nares →

→ End: Alveoli

Identify It: The Respiratory Membrane

Identify and color-code the structures of the alveoli and respiratory membrane in Figure 21.6.

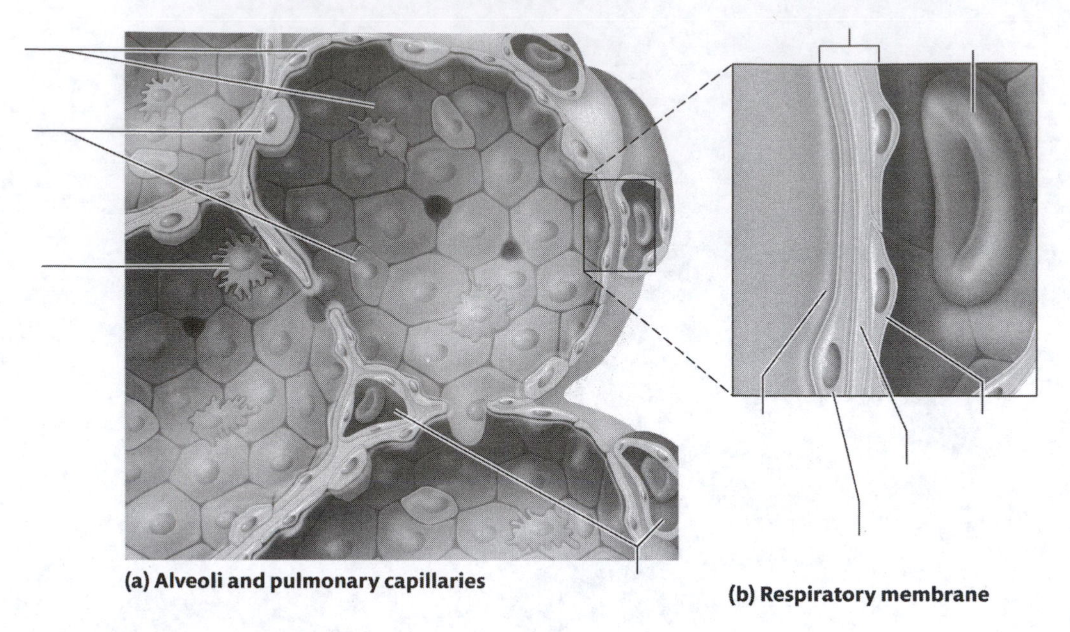

(a) Alveoli and pulmonary capillaries

(b) Respiratory membrane

Figure 21.6 The respiratory membrane.

Key Concept: Which structures make up the respiratory membrane? Would the function of the membrane change if it were to become thicker? How?

____

____

Identify It: The Lungs and Pleural Cavities

Identify and color-code the structures of the lungs and pleural cavities in Figure 21.7.

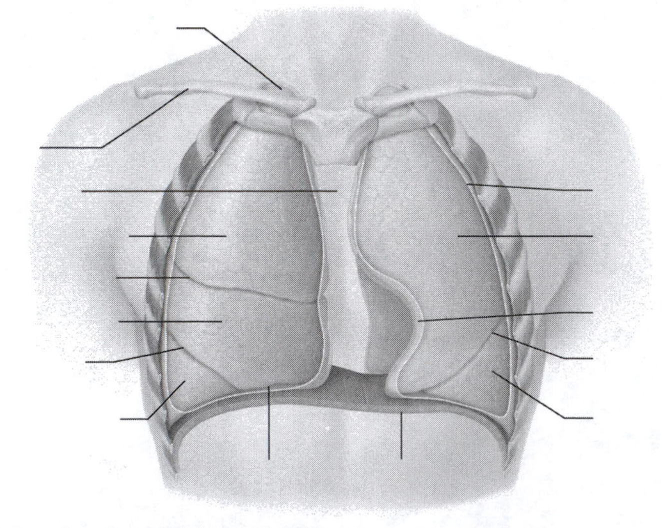

(a) Anterior view of right and left lungs

(b) Mediastinal surface of right lung

Figure 21.7 The lungs and pleural cavities.

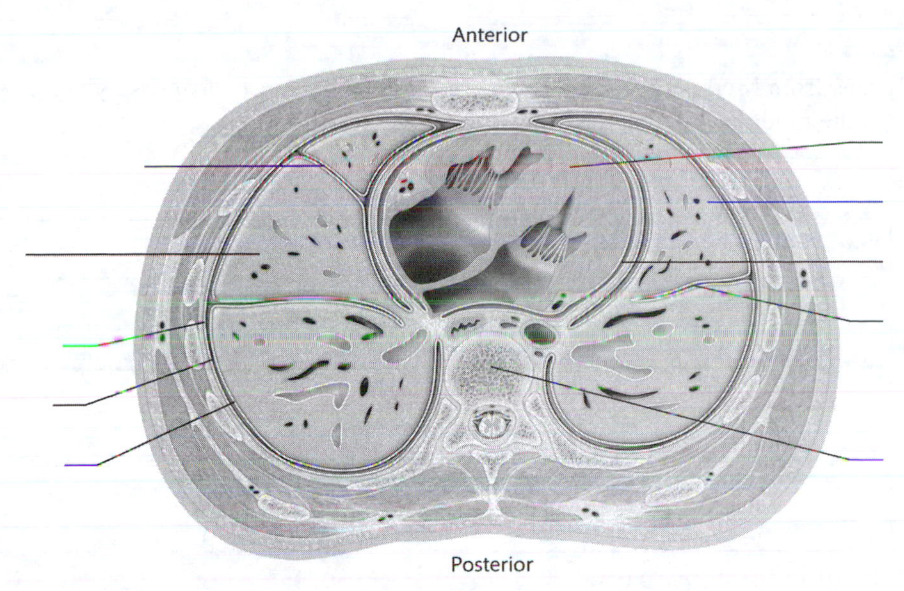

(c) Transverse section through the thoracic cavity

Figure 21.7 The lungs and pleural cavities.

Try It

We call the pleural cavities potential spaces because the two membranes are held together by the pleural fluid. This is quite easy to see in action. For this simple demonstration, obtain either four pieces of paper or four glass slides. Take two of the pieces of dry paper (or two dry slides) and place them on top of one another. Now, take the two other pieces of paper, line them up so they are the same size, and get them soaking wet. If you are using slides, place a small drop of water on one slide and put the other slide on top. Now, try to separate the dry pieces of paper (or slides). What happens? Then, try to separate the wet pieces of paper (or slides). What happens? Why? (*Hint:* Think about the hydrogen bonds that form between water molecules.)

Module 21.3: Pulmonary Ventilation

This module begins our venture into respiratory physiology. In it, we will discuss the mechanics of breathing: how we actually get air into and out of our lungs. When you have completed the module, you should be able to do the following:

1. Describe how pressure and volume are related, and explain how this relationship applies to pulmonary ventilation.
2. Explain how the inspiratory muscles, accessory muscles of inspiration, and accessory muscles of expiration change the volume of the thoracic cavity.
3. Explain how the values for atmospheric pressure, intrapulmonary pressure, and intrapleural pressure change with inspiration and expiration.
4. Explain how each of the following factors affects pulmonary ventilation: airway resistance, pulmonary compliance, and alveolar surface tension.
5. Describe and identify the values for the respiratory volumes and the respiratory capacities.

Build Your Own Glossary

Following is a table listing key terms from Module 21.3. Before you read the module, use the glossary at the back of your book or look through the module to define the following terms.

Key Terms for Module 21.3

Term	Definition
Pulmonary ventilation	
Inspiration	
Expiration	
Pressure gradient	
Boyle's law	
Inspiratory muscles	
Atmospheric pressure	
Intrapulmonary pressure	
Intrapleural pressure	
Airway resistance	
Alveolar surface tension	
Pulmonary compliance	
Spirometer	
Pulmonary capacities	

Survey It: Form Questions

Before you read the module, survey it and form at least three questions for yourself. When you have finished reading the module, return to these questions and answer them.

Question 1: ____

Answer: ____

Question 2: ____

Answer: ____

Question 3: ____

Answer: ____

Key Concept: What is a pressure gradient? How do pressure gradients drive the movement of gases into and out of the lungs?

____

____

Complete It: Overview of Pulmonary Ventilation

Fill in the blanks to complete the following paragraphs that describe the basic principles of pulmonary ventilation.

The main inspiratory muscle is the ____________. It creates a pressure gradient when it contracts by ____________ the volume of the lungs, which ____________ the pressure in the lungs. When the pressure in the lungs, or the ____________ ____________, falls below ____________ ____________, air enters the lungs via ____________. During forced inspiration, other muscles called ____________ ____________ assist in the process.

The process of expiration is largely ____________ due to the ____________ ____________ ____________. This causes the volume of the lungs to ____________, which ____________ intrapulmonary pressure. When intrapulmonary pressure is ____________ ____________ atmospheric pressure, ____________ occurs.

Key Concept: Why is intrapleural pressure slightly lower than intrapulmonary pressure? What happens if it rises above intrapulmonary pressure?

____

____

Describe It: The Big Picture of Pulmonary Ventilation

Describe the changes in lung volume and pressure during inspiration and expiration in Figure 21.8. Also, draw arrows to indicate muscle movement and the changes in lung volume.

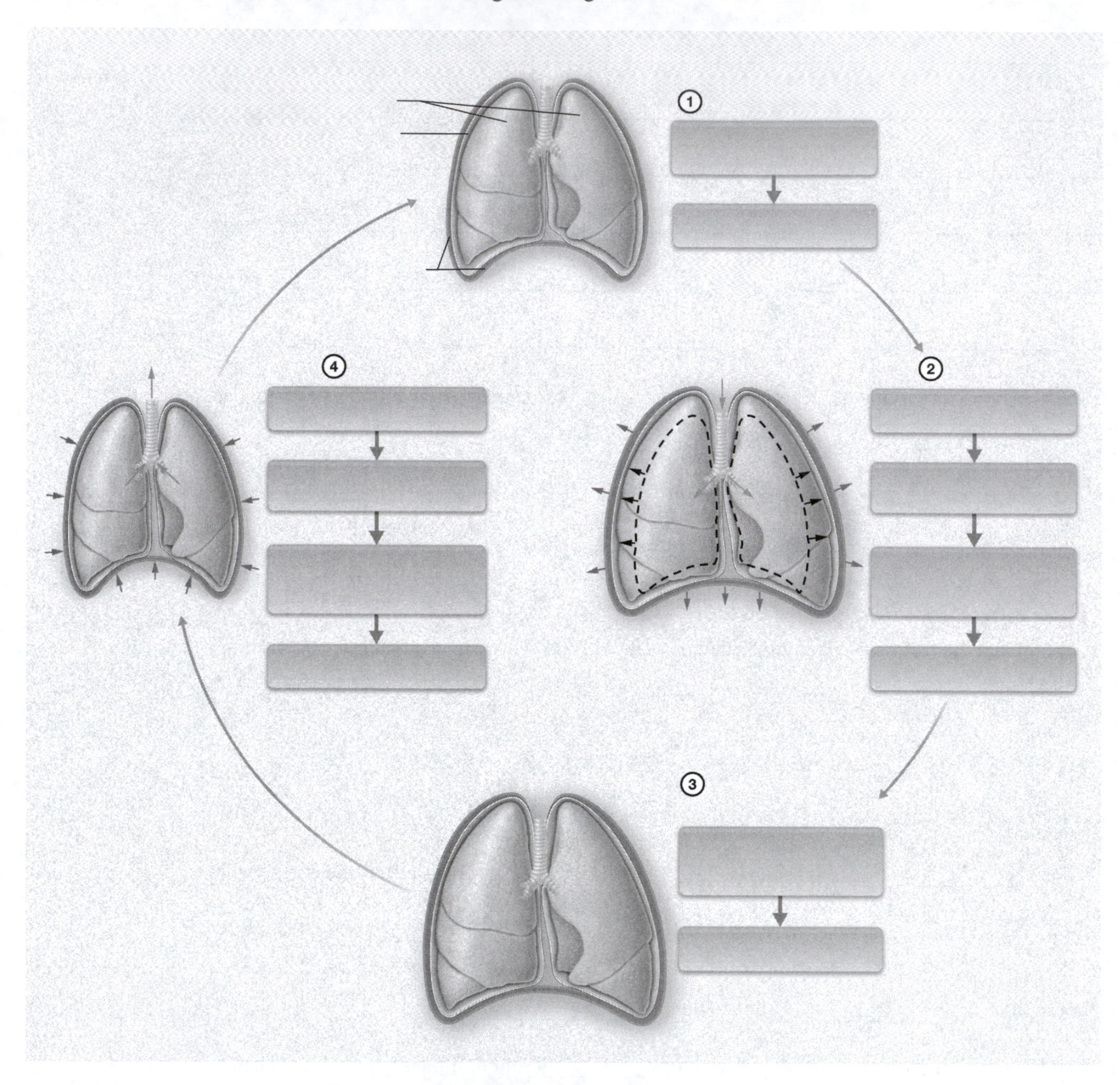

Figure 21.8 The big picture of pulmonary ventilation.

Predict It: Factors That Influence Pulmonary Ventilation

Predict whether the efficiency of pulmonary ventilation will increase or decrease given each of the following conditions. Justify each of your responses.

- A person inhales methacholine, which causes bronchoconstriction.

___

- An infant is born premature and does not produce surfactant.

____

- The premature infant is given inhalable surfactant as a treatment.

____

- A person develops emphysema, causing the destruction of alveolar walls.

____

- A person is administered epinephrine, which causes bronchodilation.

____

- A chest wall deformity prevents normal expansion of the lungs.

____

Key Concept: What are the three physical factors that influence ventilation? How does each influence the efficiency of ventilation?

____

____

Build Your Own Summary Table: Pulmonary Volumes and Capacities

As you read Module 21.3, build your own summary table about the pulmonary volumes and capacities by filling in the information in the following table. When you have finished, check your answers with text Table 21.4 on page 827.

Pulmonary Volumes and Capacities

Measurement	Average Value (Female/Male)	Definition
Pulmonary Volumes		
Tidal volume		
Inspiratory reserve volume		
Expiratory reserve volume		
Residual volume		

Measurement	Average Value (Female/Male)	Definition
Pulmonary Capacities		
Inspiratory capacity		
Functional residual capacity		
Vital capacity		
Total lung capacity		

Module 21.4: Gas Exchange

This module examines how gases diffuse across the respiratory membrane and the walls of the systemic capillaries. When you have finished the module, you should be able to do the following:

1. Describe the relationship of Dalton's law and Henry's law to pulmonary and tissue gas exchange and to the amounts of oxygen and carbon dioxide dissolved in plasma.
2. Describe oxygen and carbon dioxide pressure gradients and net gas movements in pulmonary and tissue gas exchange.
3. Explain how oxygen and carbon dioxide movements are affected by changes in partial pressure gradients.
4. Describe the mechanisms of ventilation-perfusion matching.
5. Explain the factors that maintain oxygen and carbon dioxide gradients between blood and tissue cells.

Build Your Own Glossary

Following is a table listing key terms from Module 21.4. Before you read the module, use the glossary at the back of your book or look through the module to define the following terms.

Key Terms for Module 21.4

Term	Definition
Pulmonary gas exchange	
Tissue gas exchange	
Dalton's law of partial pressures	
Henry's law	
Ventilation-perfusion matching	

Survey It: Form Questions

Before you read the module, survey it and form at least two questions for yourself. When you have finished reading the module, return to these questions and answer them.

Question 1: ____________________

Answer: ____________________

Question 2: ____________________

Answer: ____________________

Complete It: The Behavior of Gases

Fill in the blanks to complete the following paragraphs that describe the behavior of gases and relevant gas laws.

Each gas in a mixture exerts its own pressure on the mixture, which is known as ____________ ____________ ____________ ____________. The partial pressures of gases in a mixture determine whether the gases will move by ____________—gases move from an area of ____________ pressure to an area of ____________ pressure, following a pressure gradient. A second relevant gas law is Henry's law, which states that a gas's ability to dissolve in a liquid is proportional to its ____________ ____________ and its ____________ in the liquid. Henry's law helps to explain why we find very little ____________ ____________ in plasma in spite of its high partial pressure in the air we breathe.

During pulmonary gas exchange, oxygen moves from an area of ____________ partial pressure in the ____________ to an area of ____________ partial pressure in the ____________ ____________. Carbon dioxide moves in the opposite direction, going from an area of ____________ partial pressure in the ____________ ____________ to an area of ____________ partial pressure in the ____________. In the tissues, the opposite occurs, with oxygen moving from a high partial pressure in the ____________ ____________ to an area of low partial pressure in the ____________, and carbon dioxide moving from an area of high partial pressure in the ____________ to an area of low partial pressure in the ____________ ____________.

Key Concept: In which direction would oxygen diffuse if its partial pressure in the blood were 40 mm Hg and its partial pressure in the alveoli were 35 mm Hg? How would this affect homeostasis overall?

____

____

Practice It

For each of the following scenarios, state whether pulmonary and/or tissue gas exchange will increase or decrease.

1. Emphysema causes destruction of alveolar walls. _______________
2. Long-standing pulmonary hypertension thickens the pulmonary capillary walls. _______________
3. Vasodilators increase the perfusion to a systemic capillary bed. _______________
4. A pulmonary embolus (clot) blocks blood flow to a certain part of the lung. _______________
5. A person inhales 100% oxygen. _______________
6. Systemic capillary beds are damaged by hyperglycemia due to unmanaged diabetes mellitus. _______________

Module 21.5: Gas Transport through the Blood

Gases will not dissolve to a great extent in plasma, which means that our bodies must transport them in other ways. This module explores these transport mechanisms and the significant ways in which gas transport contributes to the body's general homeostasis. When you have finished the module, you should be able to do the following:

1. Describe the ways in which oxygen is transported in blood, including the reversible reaction for oxygen binding to hemoglobin.
2. Interpret the oxygen-hemoglobin dissociation curve, and describe the factors that affect the curve.
3. Describe the ways in which carbon dioxide is transported in blood, including the reversible reaction that converts carbon dioxide and water to carbonic acid.
4. Predict how changing the partial pressure of carbon dioxide will affect the pH of plasma.
5. Describe the conditions hyperventilation and hypoventilation.

Build Your Own Glossary

Following is a table listing key terms from Module 21.5. Before you read the module, use the glossary at the back of your book or look through the module to define the following terms.

Key Terms for Module 21.5

Term	Definition
Gas transport	
Deoxyhemoglobin	
Oxyhemoglobin	
Percent saturation of hemoglobin	
Oxygen-hemoglobin dissociation curve	

Bohr effect	
Bicarbonate ion	
Carbonic anhydrase	
Carbonic acid–bicarbonate buffer system	
Hyperventilation	
Hypoventilation	

Survey It: Form Questions

Before you read the module, survey it and form at least three questions for yourself. When you have finished reading the module, return to these questions and answer them.

Question 1: ______________________________

Answer: ______________________________

Question 2: ______________________________

Answer: ______________________________

Question 3: ______________________________

Answer: ______________________________

Key Concept: Why is most oxygen transported on hemoglobin?

____

____

Describe It: Hemoglobin Loading and Unloading

Describe, label, and color-code the overall process of hemoglobin loading and unloading in Figure 21.9. In addition, draw in the arrows to indicate the direction in which oxygen is diffusing in each part.

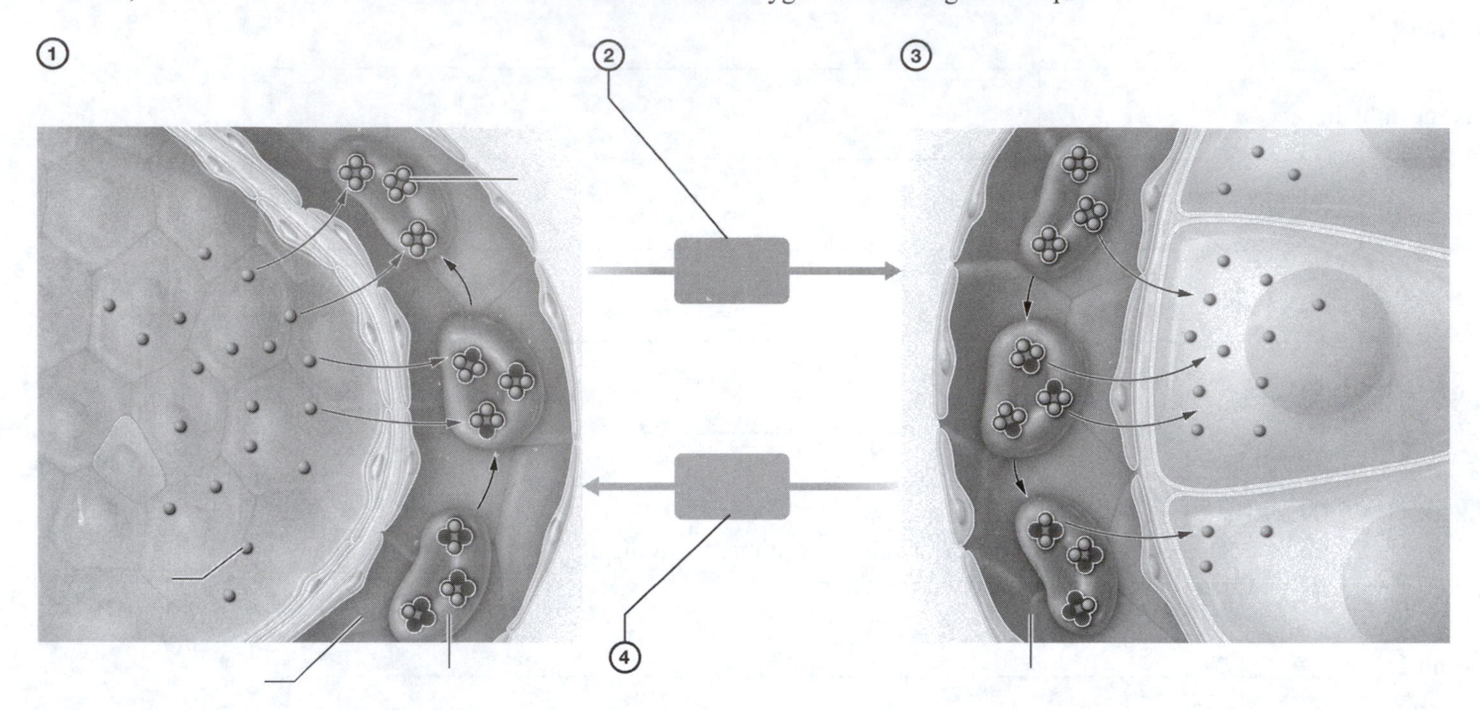

Figure 21.9 Hemoglobin loading and unloading.

Team Up

Make a handout to teach the oxygen-hemoglobin dissociation curve and the main factors that influence this curve (temperature, hydrogen ion concentration, BPG, and P_{CO_2}). You can use Figure 21.22 in your text on page 833 and the Concept Boost on page 834 as guides, but the handout should be in your own words and with your own diagrams. At the end of the handout, write a few quiz questions. Once you have completed your handout, team up with one or more study partners and trade handouts. Study your partners' diagrams, and when you have finished, take the quiz at the end. When you and your group have finished taking all the quizzes, discuss the answers to determine places where you need additional study. After you've finished, combine the best elements of each handout to make one "master" diagram for the oxygen-hemoglobin dissociation curve.

Key Concept: How does the percent saturation of hemoglobin affect hemoglobin's ability to unload oxygen? Why is this important to maintaining homeostasis?

____

____

Practice It

For each of the following scenarios, state whether the situation would cause more or less oxygen to be unloaded from Hb in the tissues.

1. pH increase ______________
2. Temperature decrease ______________
3. P_{O_2} increase ______________
4. P_{CO_2} decrease ______________

5. Acidity increase _______________
6. BPG concentration increase _______________
7. P_{CO_2} increase _______________
8. P_{O_2} decrease _______________
9. Temperature increase _______________
10. Vigorous exercise _______________

Describe It: Transport of Carbon Dioxide as Bicarbonate Ions

Describe, label, and color-code the overall process of carbon dioxide transport as bicarbonate ions in Figure 21.10.

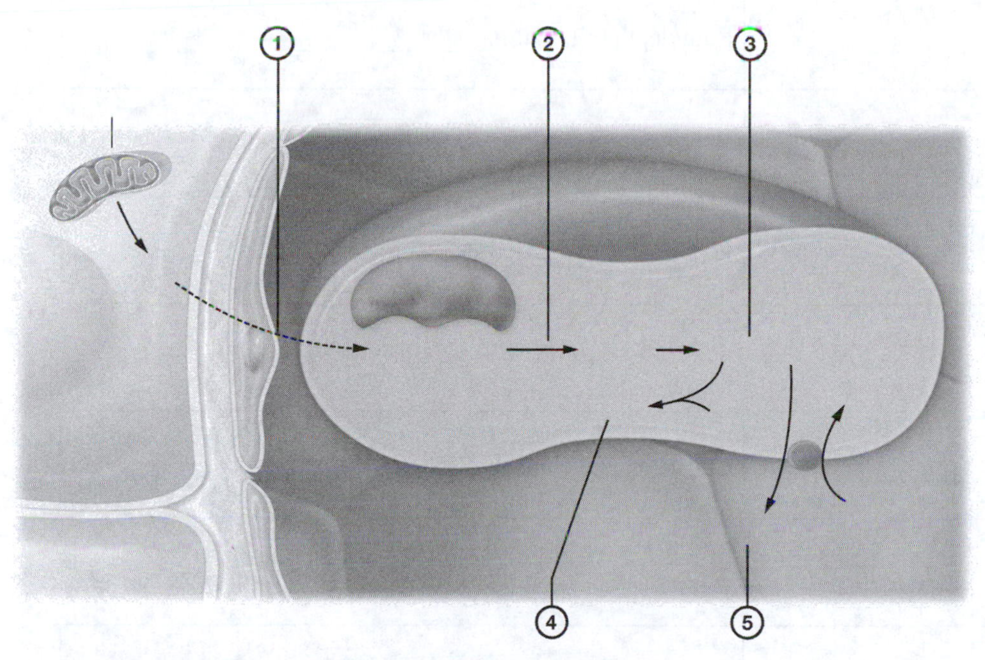

(a) Bicarbonate formation in an erythrocyte in a systemic capillary

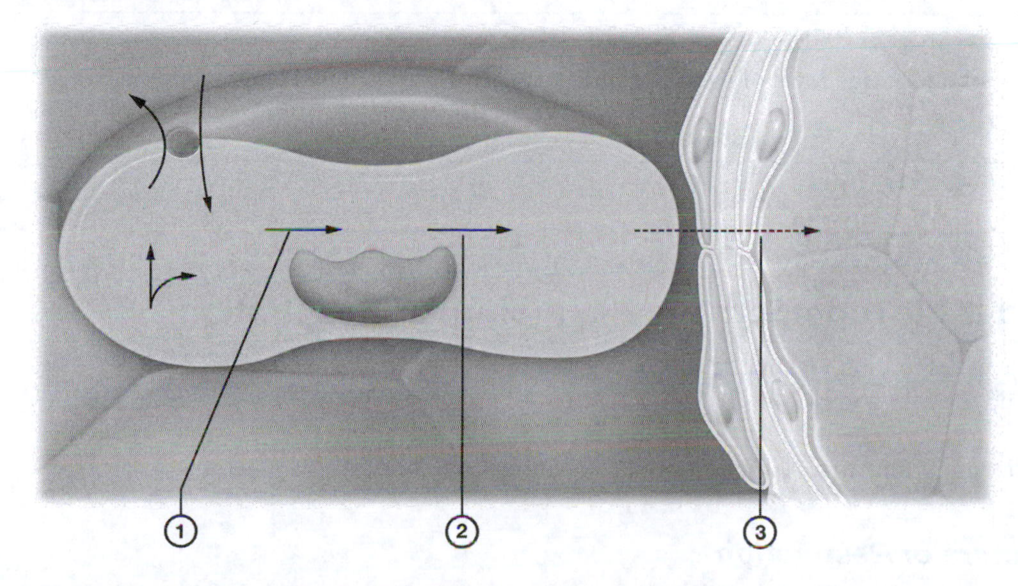

(b) Carbon dioxide formation in an erythrocyte in a pulmonary capillary

Figure 21.10 Transport of carbon dioxide as bicarbonate ions.

Trace It: Effect of Ventilation Changes on pH

In the following boxes, trace each step of the changes that take place according to the rate and depth of ventilation, along with the resulting pH change.

Hypoventilation	Hyperventilation
⇩	⇩
Rate/depth of breathing ____________	Rate/depth of breathing ____________
⇩	⇩
P_{CO_2} in blood ____________	P_{CO_2} in blood ____________
⇩	⇩
Carbonic acid concentration ____________	Carbonic acid concentration ____________
⇩	⇩
Body fluid pH ____________	Body fluid pH ____________

Key Concept: Why does the carbon dioxide level of the blood influence the pH of the blood and body fluids?

____

____

Module 21.6: Putting It All Together: The Big Picture of Respiration

Let's now put all four processes of respiration together to see the big picture. When you have completed this module, you should be able to do the following:

1. Describe the overall big picture of the processes involved in respiration.

Describe It: The Big Picture of Respiration

Describe, label, and color-code the overall process of respiration in Figure 21.11. In addition, draw in the arrows to indicate the direction in which oxygen and carbon dioxide are moving in each step.

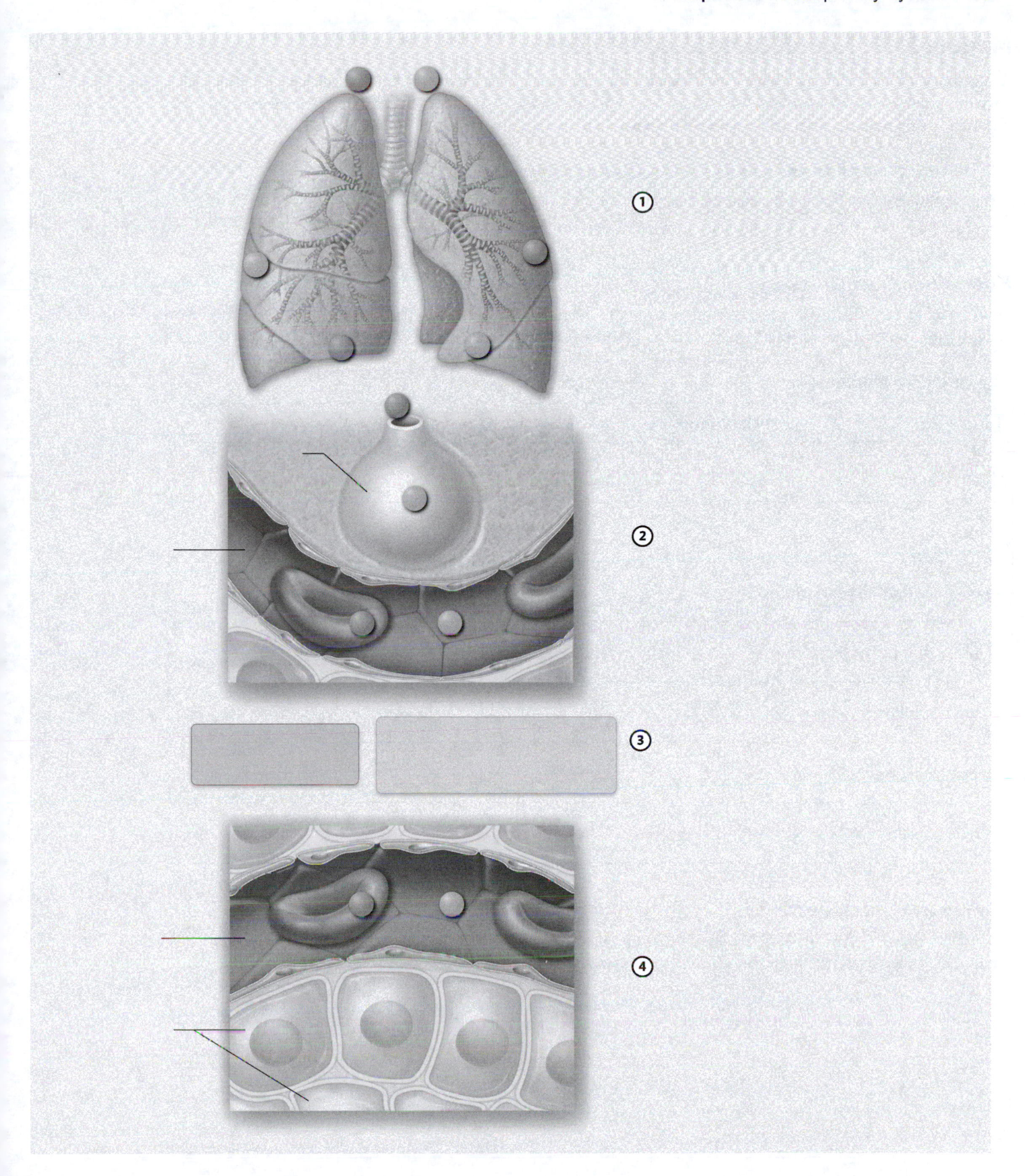

Figure 21.11 The big picture of respiration.

Module 21.7: Neural Control of Ventilation

At this point in the chapter we have examined all four processes of gas exchange, but we still have one final but very important topic left to cover: What causes us to breathe in the first place? We explore the answer to this question in this module, after which you should be able to do the following:

1. Describe the locations and functions of the brainstem respiratory centers.
2. List and describe the major chemical and neural stimuli to the respiratory centers.
3. Compare and contrast the central and peripheral chemoreceptors.

Build Your Own Glossary

Following is a table listing key terms from Module 21.7. Before you read the module, use the glossary at the back of your book or look through the module to define the following terms.

Key Terms for Module 21.7

Term	Definition
Dyspnea	
Eupnea	
Respiratory rhythm generator	
Ventral respiratory group	
Dorsal respiratory group	
Central chemoreceptors	
Peripheral chemoreceptors	

Survey It: Form Questions

Before you read the module, survey it and form at least two questions for yourself. When you have finished reading the module, return to these questions and answer them.

Question 1: ______________________________

Answer: ______________________________

Question 2: ______________________________

Answer: ______________________________

Identify It: Neural Control of Ventilation

Identify and color-code each structure involved in the neural control of ventilation in Figure 21.12. Then, list the main function(s) of each component.

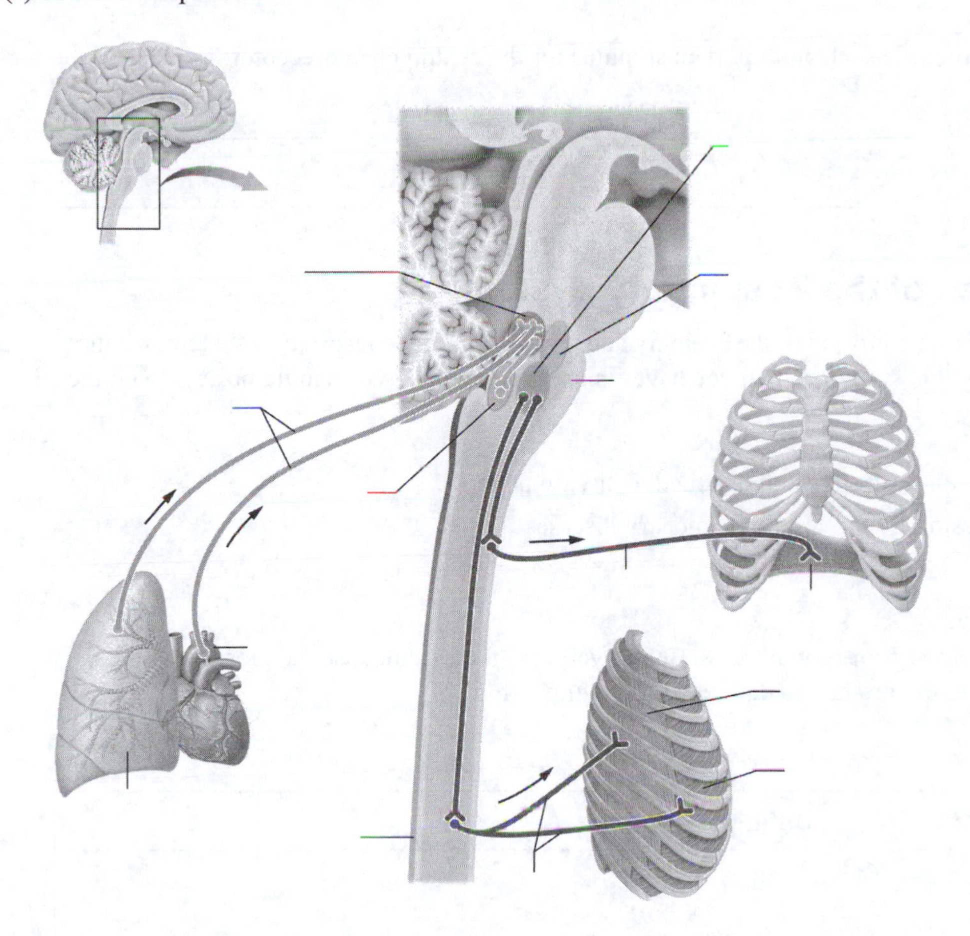

Figure 21.12 Neural control of the basic pattern of ventilation.

Key Concept: What is the respiratory rhythm generator? Which nuclei work with the RRG to maintain eupnea?

____

____

Practice It: Central and Peripheral Chemoreceptors

For each of the following scenarios, determine if the stimulus would be detected by the central chemoreceptors, peripheral chemoreceptors, or both. Then, determine the change in ventilation that would be triggered by each stimulus.

- Decreased P_{O_2} ____
- Increased P_{CO_2} ____
- Decreased pH ____
- Increased P_{O_2} ____

- Increased pH ________________________________
- Decreased P_{CO_2} ________________________________

Key Concept: Why is carbon dioxide such an important stimulus for the central chemoreceptors?

____

____

Module 21.8: Diseases of the Respiratory System

Now that we have explored the basic concepts of the anatomy and physiology of the respiratory system, we turn to the common diseases affecting this system. When you have finished the module, you should be able to do the following:

1. Explain the difference between restrictive and obstructive disease patterns.
2. Describe the basic pathophysiology for certain pulmonary diseases.

Build Your Own Glossary

Following is a table listing key terms from Module 21.8. Before you read the module, use the glossary at the back of your book or look through the module to define the following terms.

Key Terms for Module 21.8

Term	Definition
Restrictive lung disease	
Obstructive lung disease	
Chronic obstructive pulmonary disease	
Asthma	

Survey It: Form Questions

Before you read the module, survey it and form at least two questions for yourself. When you have finished reading the module, return to these questions and answer them.

Question 1: ________________________________

Answer: ________________________________

Question 2: ________________________________

Answer: ________________________________

Try It: Obstructive and Restrictive Diseases

To experience the feeling of both a restrictive and an obstructive lung disease, obtain a regular-sized drinking straw and put one end in your mouth. To mimic the feeling of a restrictive disease, inhale as deeply as you can through the straw (pinch your nose closed so you don't cheat). To mimic the feeling of an obstructive disease, take in a normal inhalation (not through the straw), then put the straw in your mouth and *exhale* through it (again, pinch your nose closed).

Practice It: Obstructive and Restrictive Lung Diseases

Would the following pulmonary volumes and capacities be expected to be abnormal in an obstructive lung disease, a restrictive lung disease, or both?

- Vital capacity ______________________________
- Functional residual capacity ______________________________
- Residual volume ______________________________
- Inspiratory capacity ______________________________
- Total lung capacity ______________________________

Key Concept: What is the key difference between a restrictive and an obstructive respiratory disease?

____

____

What Do You Know Now?

Let's now revisit the questions you answered in the beginning of this chapter. How have your answers changed now that you've worked through the material?

- What is/are the main function(s) of the respiratory system?

  ____

- How do we breathe?

  ____

- Why do we inhale oxygen and exhale carbon dioxide?

  ____

22 The Digestive System

We now turn to the digestive system—the set of organs that take the food we eat and break it down so its nutrients can be absorbed and used by the body.

What Do You Already Know?

Try to answer the following questions before proceeding to the next section. If you're unsure of the correct answers, give it your best attempt based on previous courses, previous chapters, or just your general knowledge.

- What are the main organs of the digestive system?

  ____

- What are building-block molecules of the four classes of macromolecules (carbohydrates, lipids, proteins, and nucleic acids)?

  ____

- What are enzymes, and how do they make chemical reactions occur more easily?

  ____

Module 22.1: Overview of the Digestive System

Module 22.1 in your text introduces you to the basic structure and tissue composition of the organs of the digestive tract and the accessory organs of the digestive system. By the end of the module, you should be able to do the following:

1. Describe the major functions of the digestive system.
2. Describe the histological structure and function of each of the four layers of the alimentary canal wall.
3. Explain the basic anatomy, organization, and regulation of the digestive system.

Build Your Own Glossary

Below is a table listing key terms from Module 22.1. Before you read the module, use the glossary at the back of your book or look through the module to define the following terms.

Key Terms for Module 22.1

Term	Definition
Gastrointestinal (GI) tract	
Alimentary canal	

Term	Definition
Accessory organs	
Ingestion	
Secretion	
Propulsion	
Mechanical digestion	
Chemical digestion	
Absorption	
Defecation	
Mucosa	
Submucosa	
Muscularis externa	
Serosa	
Peritoneum	
Mesenteries	

Survey It: Form Questions

Before you read the module, survey it and form at least two questions for yourself. When you have finished reading the module, return to these questions and answer them.

Question 1: ______________________________

Answer: ______________________________

Question 2: ______________________________

Answer: ______________________________

Key Concept: What are the accessory organs of the digestive system? How do their functions differ from those of the alimentary canal?

____

____

Describe It: The Alimentary Canal

Starting with the mouth, list the organs through which an indigestible molecule (like those found in dietary fiber) would pass during its trip through the alimentary canal. You may use text Figure 22.1 as a reference, but try it from memory first.

Draw It: The Basic Tissue Organization of the Alimentary Canal

In the space below, you will find concentric circles that separate the cross section of the alimentary canal into its four main tissue areas (as well as the hollow space in the middle). Label each space for the appropriate alimentary canal tissue layer or space (mucosa, submucosa, muscularis externa, serosa, and lumen). Then, color each with colored pencils according to this scheme: muscle = brown; connective tissue = gray; epithelium = pink. Note that one of the layers will require the use of all three colors. You may refer to Figure 22.3 in your text.

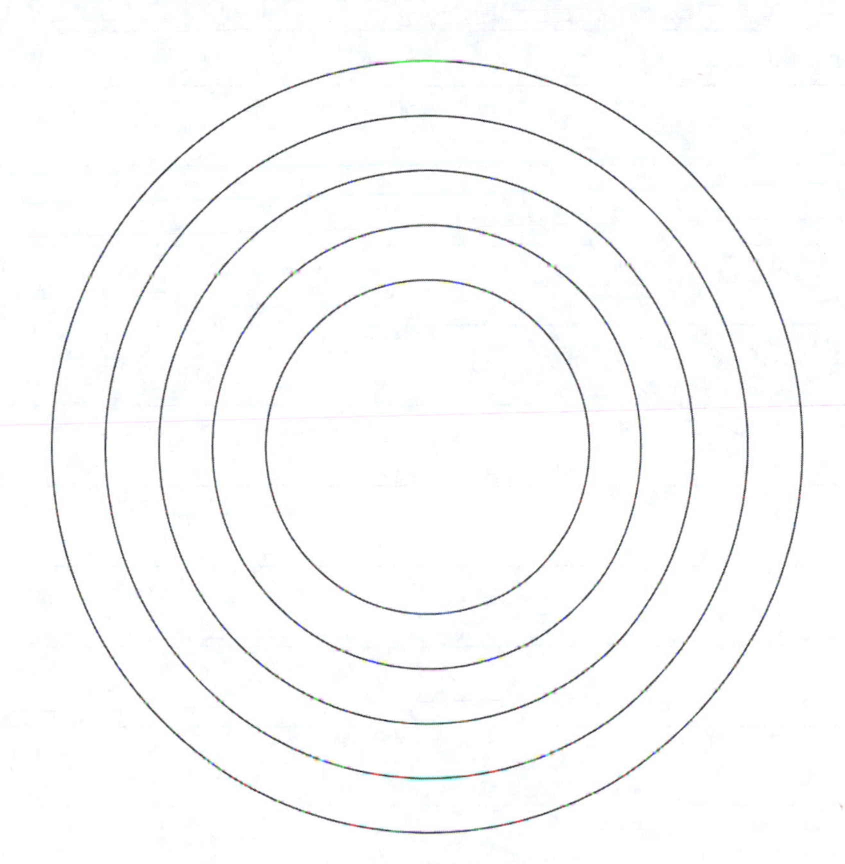

Key Concept: How are the peritoneal membranes similar to other serous membranes we have covered? How are they different?

____

____

Module 22.2: The Oral Cavity, Pharynx, and Esophagus

Now we look at the portion of the digestive tract where food is chewed and swallowed. When you finish this module, you should be able to do the following:

1. Discuss the structure and basic functions of the oral cavity, the different types of teeth, and the tongue.
2. Describe the structure and function of the salivary glands, their respective ducts, and the products secreted by their cells.
3. Describe and classify the regions of the pharynx with respect to passage of food and/or air through them.
4. Describe the structure and function of the esophagus, including the locations of skeletal and smooth muscle within its wall.
5. Explain the process of deglutition, including the changes in position of the glottis and larynx.

Build Your Own Glossary

Following is a table listing key terms from Module 22.2. Before you read the module, use the glossary at the back of your book or look through the module to define the following terms.

Key Terms for Module 22.2

Term	Definition
Palate	
Masticate	
Deciduous teeth	
Enamel	
Dentin	
Tongue	
Salivary glands	
Serous cells	
Salivary amylase	
Lysozyme	
Pharynx	
Esophagus	
Deglutition	

Survey It: Form Questions

Before you read the module, survey it and form at least three questions for yourself. When you have finished reading the module, return to these questions and answer them.

Question 1: ______________________________

Answer: ______________________________

Question 2: ______________________________

Answer: ______________________________

Question 3: ______________________________

Answer: ______________________________

Key Concept: What are the main differences between primary and secondary dentition?

____

____

Identify It: Oral Cavity and Pharynx

Identify each component of the oral cavity and pharynx in Figure 22.1. Then, list the main function(s) of each component.

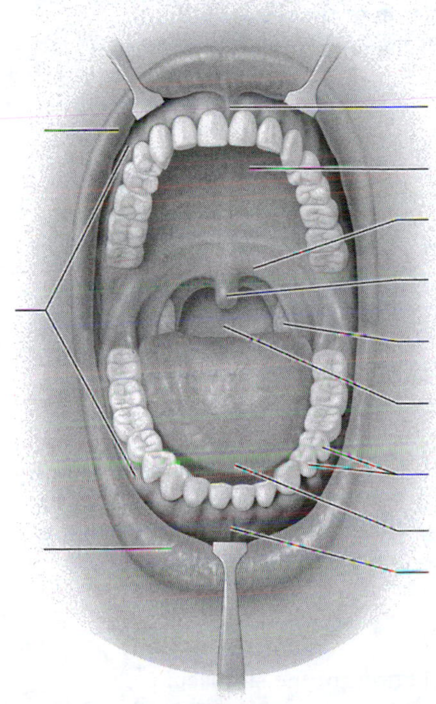

(a) Anterior view of the oral cavity

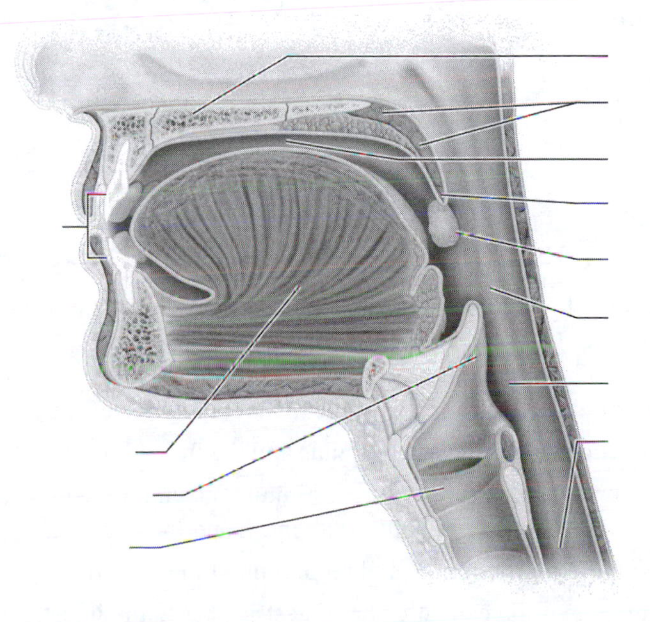

(b) Oral cavity and pharynx, sagittal section

Figure 22.1 The oral cavity and pharynx.

Key Concept: What are the three salivary glands, and what does each one contribute to saliva?

___

___

Describe It: The Process of Swallowing (Deglutition)

Write in the steps of the process of swallowing (deglutition) in Figure 22.2, and label and color-code important components of this process. You may use text Figure 22.9 as a reference, but write the steps in your own words.

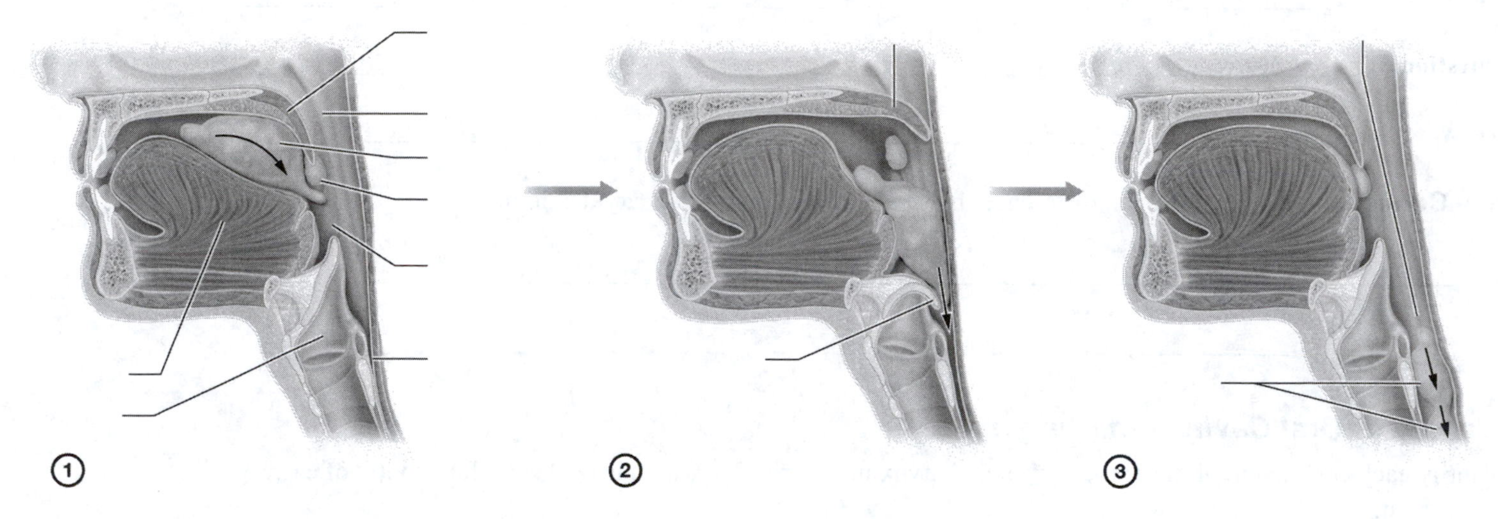

Figure 22.2 The process of swallowing (deglutition).

Key Concept: How is the esophagus structurally different from the basic tissue organization of the rest of the alimentary canal? Why is the esophagus composed of these different tissues? (*Hint:* Recall the core principle of structure-function.)

___

___

Module 22.3: The Stomach

Now we will examine the gross and microscopic anatomy of the stomach, as well as its role in digestion. When you complete this module, you should be able to do the following:

1. Describe the structure and function of the different regions of the stomach.
2. Describe the structure of the gastric glands and the functions of the types of cells they contain.
3. Explain how hormones, nervous system stimulation, and the volume, chemical composition, and osmolarity of the chyme affect motility in both the stomach and the duodenum.
4. Discuss the function, production, and regulation of the secretion of hydrochloric acid.
5. Explain the effects of the cephalic phase, gastric phase, and intestinal phase on the functions of the stomach and small intestine, and give examples for each phase.

Build Your Own Glossary

Below is a table listing key terms from Module 22.3. Before you read the module, use the glossary at the back of your book or look through the module to define the following terms.

Key Terms of Module 22.3

Term	Definition
Cardia	
Fundus	
Body (of the stomach)	
Pyloric antrum	
Pylorus	
Rugae	
Chyme	
Gastric pits	
Gastric glands	
Parietal cells	
Chief cells	
Pepsinogen	
Gastrin	
Cephalic phase	
Gastric phase	
Enterogastric reflex	

Survey It: Form Questions

Before you read the module, survey it and form at least three questions for yourself. When you have finished reading the module, return to these questions and answer them.

Question 1: ______________________________

Answer: ______________________________

Question 2: ______________________________

Answer: ______________________________

Question 3: ______________________________

Answer: ______________________________

Key Concept: What is the importance of the extra oblique layer of smooth muscle in the stomach?

____

____

Key Concept: What are the three types of gastric gland cells, and what do they contribute to the function of the stomach?

____

____

Identify It: Gross Anatomy of the Stomach

Identify each component of the stomach in Figure 22.3. Then, list the main function(s) of each component.

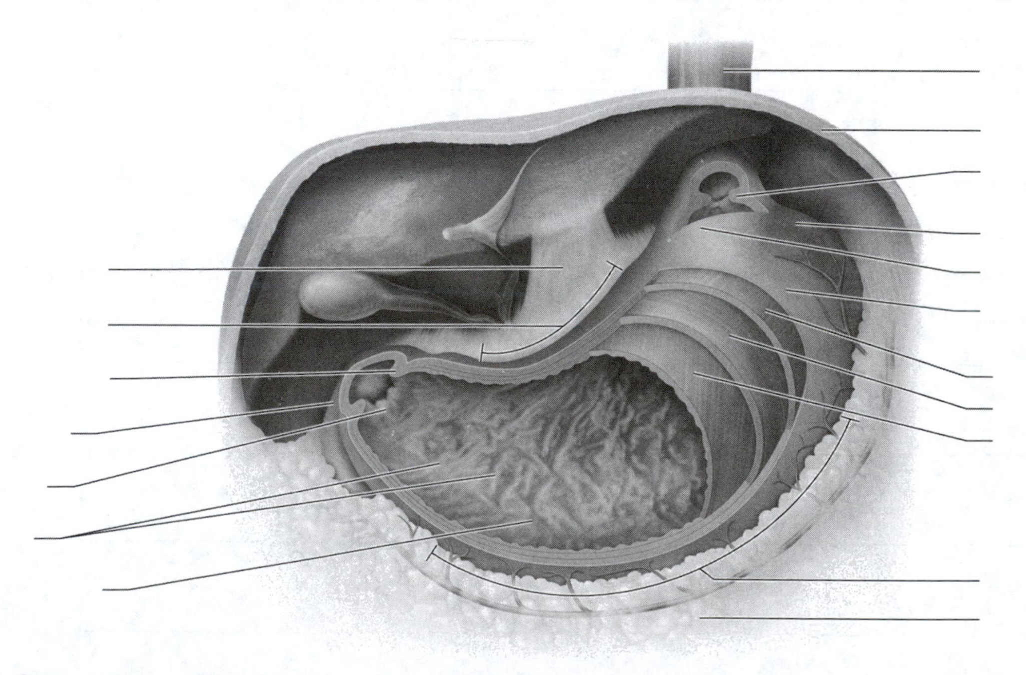

Figure 22.3 Gross anatomy of the stomach.

Complete It: The Stimulation of Stomach Acid Secretion

Fill in the blanks to complete the following paragraphs that describe the cephalic and gastric phases of stomach secretion.

The ____________ phase is mediated by the sight, smell, taste, or ____________ of food. In this phase, those stimuli trigger the ____________ nerve to stimulate ____________ cells to release ____________ ions, and G cells to release ____________. This ____________ nervous system stimulation also stimulates enteroendocrine cells to release ____________ while inhibiting ____________ release, both of which enhance H^+ secretion and ____________ stomach pH.

The ____________ phase begins when food enters the ____________. The presence of food and stretching or ____________ of the stomach stimulates ____________ neurons and vagus nerve ____________. The neurotransmitter ____________ released by these neurons stimulates parietal cells to produce ____________. The presence of partially digested ____________ in gastric juice stimulates ____________ release, which in turn stimulates acid secretion. This results in a ____________ feedback loop that lowers stomach pH to about 2.0.

Key Concept: What factors assist with turning off stomach acid production?

____

____

Describe It: The Process of Churning/Emptying in the Stomach

Describe the steps in churning/emptying in the stomach in Figure 22.4. (*Hint:* You may refer to Figure 22.13 in the text, but do not just copy the descriptions—make sure to put your descriptions in your own words.)

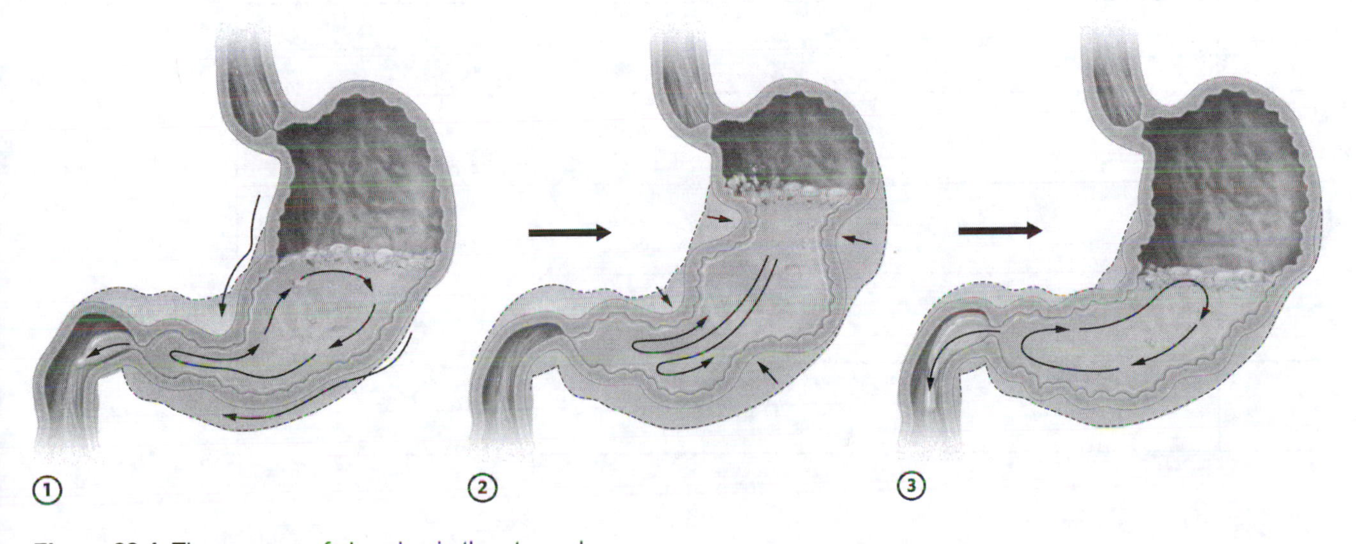

Figure 22.4 The process of churning in the stomach.

Key Concept: Why must stomach emptying be a slow, limited process?

____

____

Module 22.4: The Small Intestine

Module 22.4 in your text covers the small intestine—the longest section of the digestive tract, where most of the digestive system's secretion, digestion, and absorption occur. At the end of this module, you should be able to do the following:

1. Describe the structure and functions of the duodenum, jejunum, and ileum.
2. Discuss the histology and functions of the circular folds, villi, and microvilli of the small intestine.
3. Describe the functions and regulation of motility in the small intestine.

Build Your Own Glossary

Below is a table listing key terms from Module 22.4. Before you read the module, use the glossary at the back of your book or look through the module to define the following terms.

Key Terms of Module 22.4

Term	Definition
Duodenum	
Jejunum	
Ileum	
Ileocecal valve	
Villi	
Microvilli	
Brush border	
Intestinal crypts	
Segmentation	

Survey It: Form Questions

Before you read the module, survey it and form at least two questions for yourself. When you have finished reading the module, return to these questions and answer them.

Question 1: ______________________________

Answer: ______________________________

Question 2: ______________________________

Answer: ______________________________

Key Concept: Is it accurate to state that the majority of digestion and absorption occurs in the small intestine? Explain your answer.

____

____

Identify It: Gross Anatomy of the Small Intestine

Identify each component of the small intestine in Figure 22.5. Then, list the main function(s) and/or description of each component.

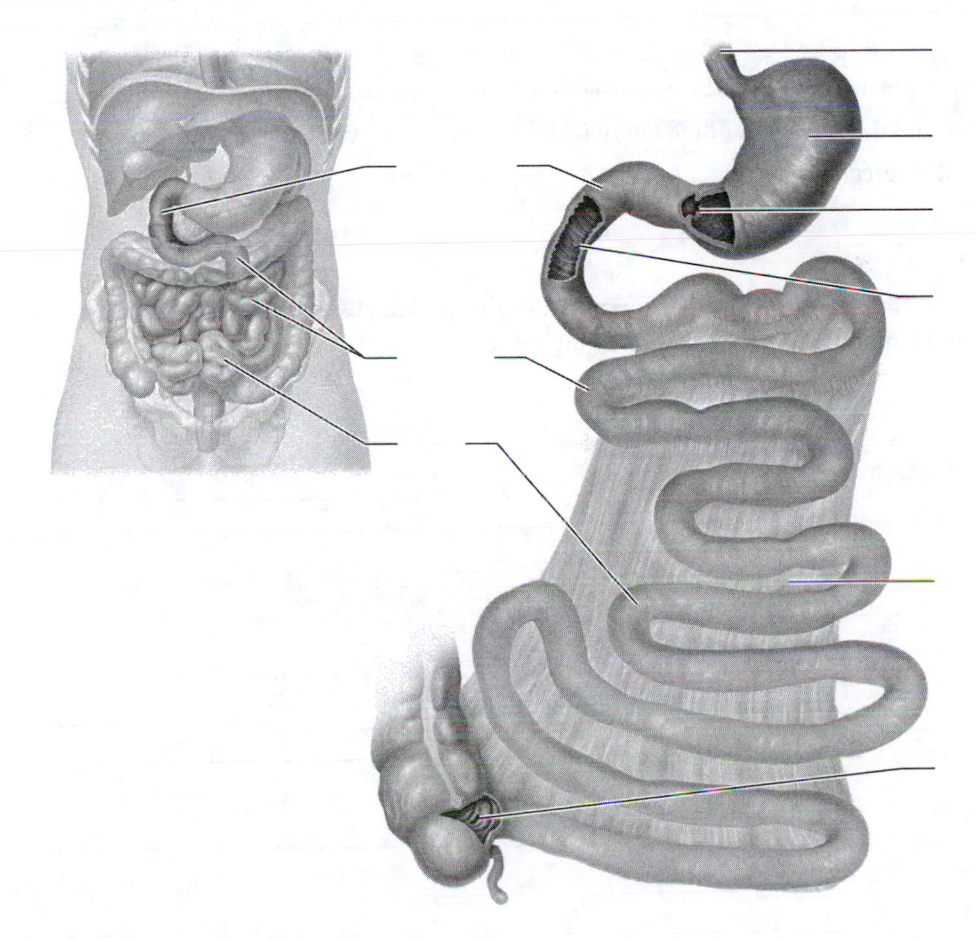

Figure 22.5 Gross anatomy of the small intestine.

Describe It: Structure and Functions of the Small Intestine

Write a paragraph describing how the small intestine is modified to increase surface area for absorption.

Key Concept: How do the circular and longitudinal muscle layers work to move material through the small intestine? Is this the same during fasting and after eating?

____

____

Module 22.5: The Large Intestine

Although the small intestine absorbs nearly all the nutrients from the food we eat, there is still much left to do. Water, electrolytes, and vitamins are absorbed by the large intestine. At the end of this module, you should be able to do the following:

1. Describe the gross and microscopic anatomy of the divisions of the large intestine.
2. Describe the defecation reflex and the functions of the internal and external anal sphincters.
3. Discuss conscious control of the defecation reflex.

Build Your Own Glossary

Following is a table listing key terms from Module 22.5. Before you read the module, use the glossary at the back of your book or look through the module to define the following terms.

Key Terms of Module 22.5

Term	Definition
Feces	
Cecum	
Vermiform appendix	
Ascending colon	
Transverse colon	

Term	Definition
Descending colon	
Sigmoid colon	
Rectum	
Anal canal	
Taeniae coli	
Haustra	
Mass movement	
Defecation	

Survey It: Form Questions

Before you read the module, survey it and form at least three questions for yourself. When you have finished reading the module, return to these questions and answer them.

Question 1: ______________________________

Answer: ______________________________

Question 2: ______________________________

Answer: ______________________________

Question 3: ______________________________

Answer: ______________________________

Key Concept: How do the internal and external anal sphincters work together to control defecation?

____

____

Key Concept: What features of the tissues of the large intestine produce its "pocketed" appearance?

____

____

Key Concept: What is the importance of the normal bacterial flora of the large intestine? What effect would overuse of antibiotics have on large intestine function?

____

____

Practice Labeling: Gross Anatomy of the Large Intestine

Identify each component of the large intestine in Figure 22.6. Then, list the main function(s) of each component.

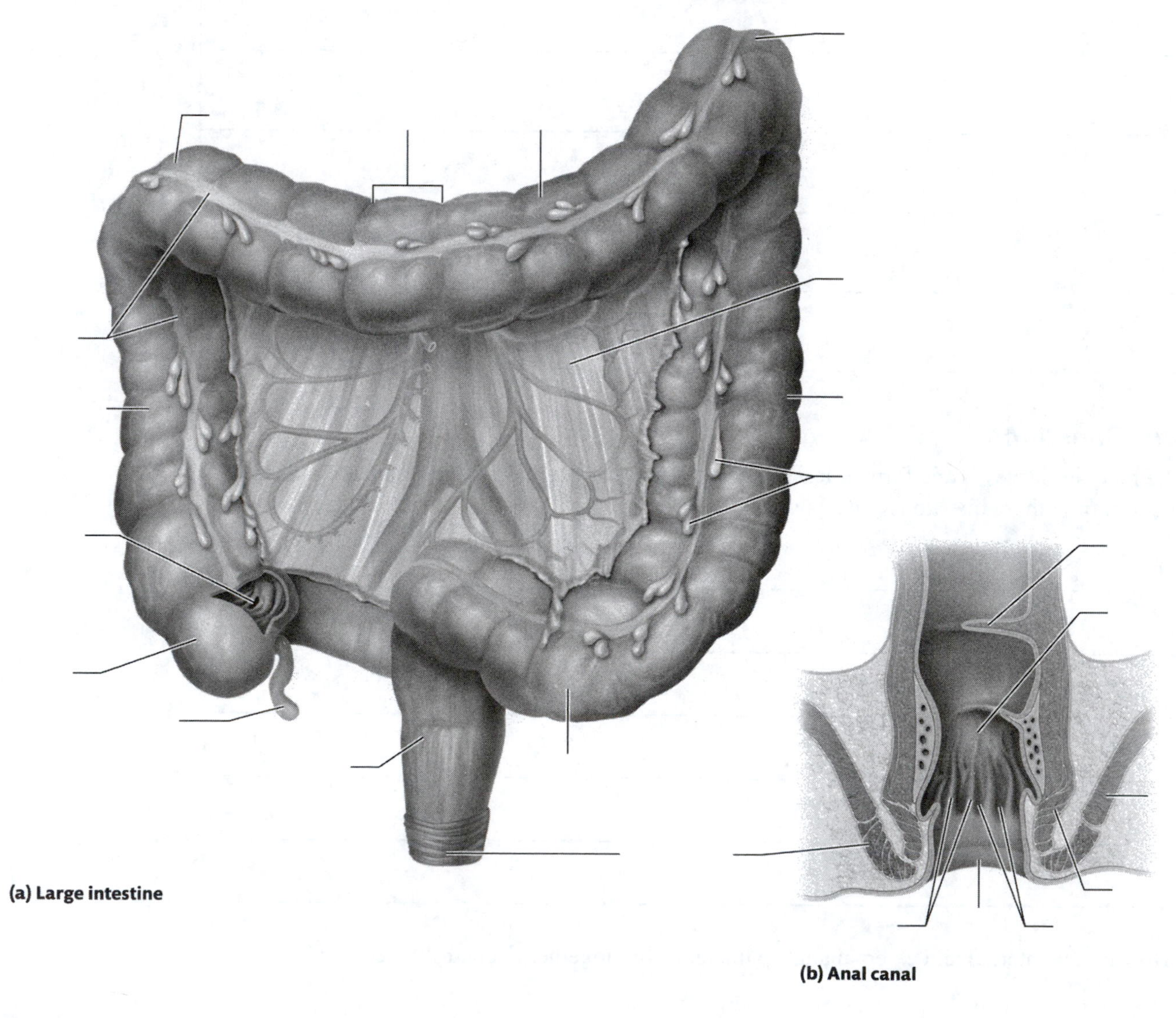

Figure 22.6 Gross anatomy of the large intestine.

Describe It: The Defecation Reflex

Write in the steps of the process of the defecation reflex in Figure 22.7, and label and color-code key components of the process. You may use text Figure 22.18 as a reference, but write the steps in your own words.

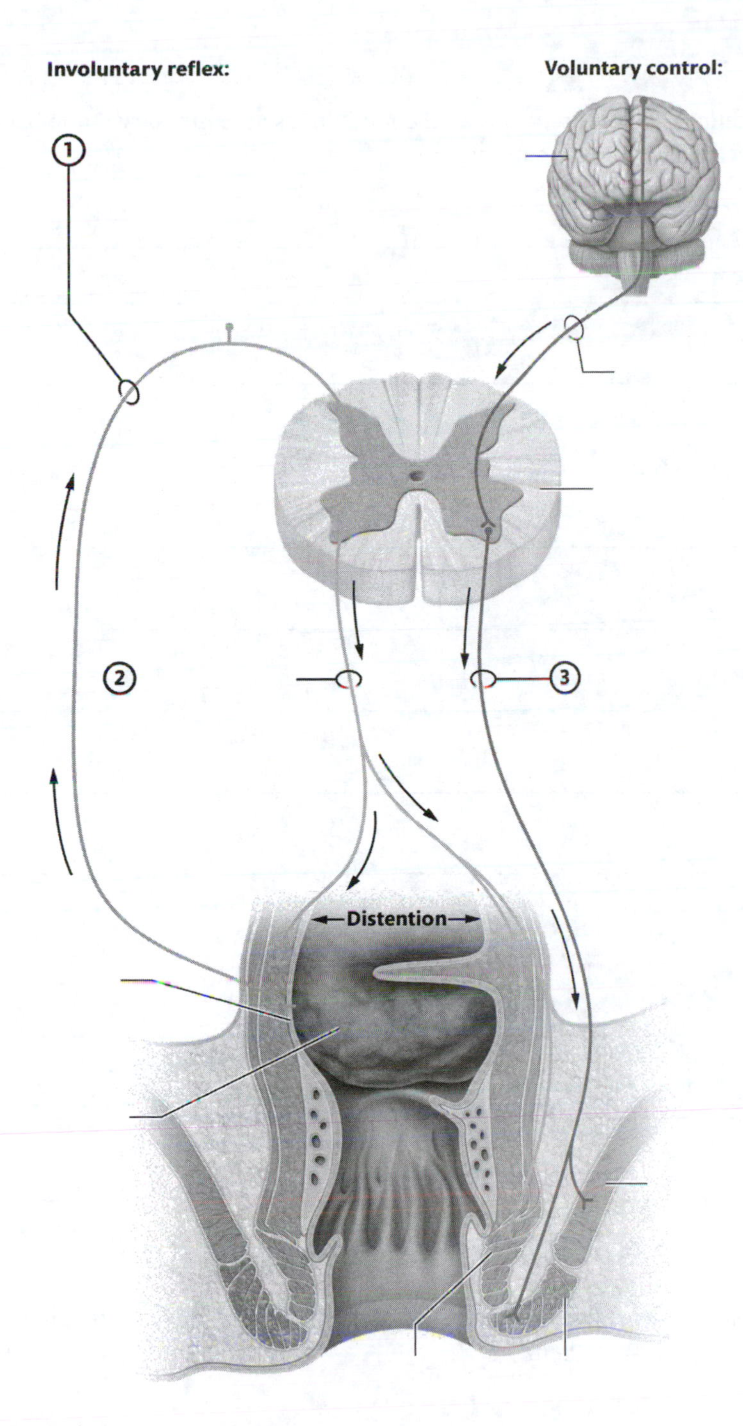

Figure 22.7 The defecation reflex.

Module 22.6: The Pancreas, Liver, and Gallbladder

Module 22.6 in your text explores the accessory organs of the digestive system. By the end of the module, you should be able to do the following:

1. Describe the gross and microscopic structure of the pancreas and its digestive functions.
2. Describe the gross and microscopic structures of the liver and gallbladder.
3. Describe the functions of the liver pertaining to digestion.
4. Explain the structural and functional relationship between the liver and the gallbladder.
5. Explain how pancreatic and biliary secretions are regulated.

Build Your Own Glossary

Below is a table listing key terms from Module 22.6. Before you read the module, use the glossary at the back of your book or look through the module to define the following terms.

Key Terms for Module 22.6

Term	Definition
Main pancreatic duct	
Acini (of the pancreas)	
Cholecystokinin	
Liver	
Gallbladder	
Hepatocytes	
Portal triad	
Bile	
Cystic duct	
Common bile duct	
Hepatopancreatic sphincter	

Survey It: Form Questions

Before you read the module, survey it and form at least two questions for yourself. When you have finished reading the module, return to these questions and answer them.

Question 1: ______________________________

Answer: ______________________________

Question 2: ______________________________

Answer: ______________________________

Key Concept: What are the main components of pancreatic juice, and what roles do they play in the duodenum?

____

____

Identify It: The Structure of a Liver Lobule

Label and color-code each component of the liver lobule in Figure 22.8. Then, list the main function(s) of each component. (*Note:* You may refer to Figure 22.23 in your textbook if you get stuck.)

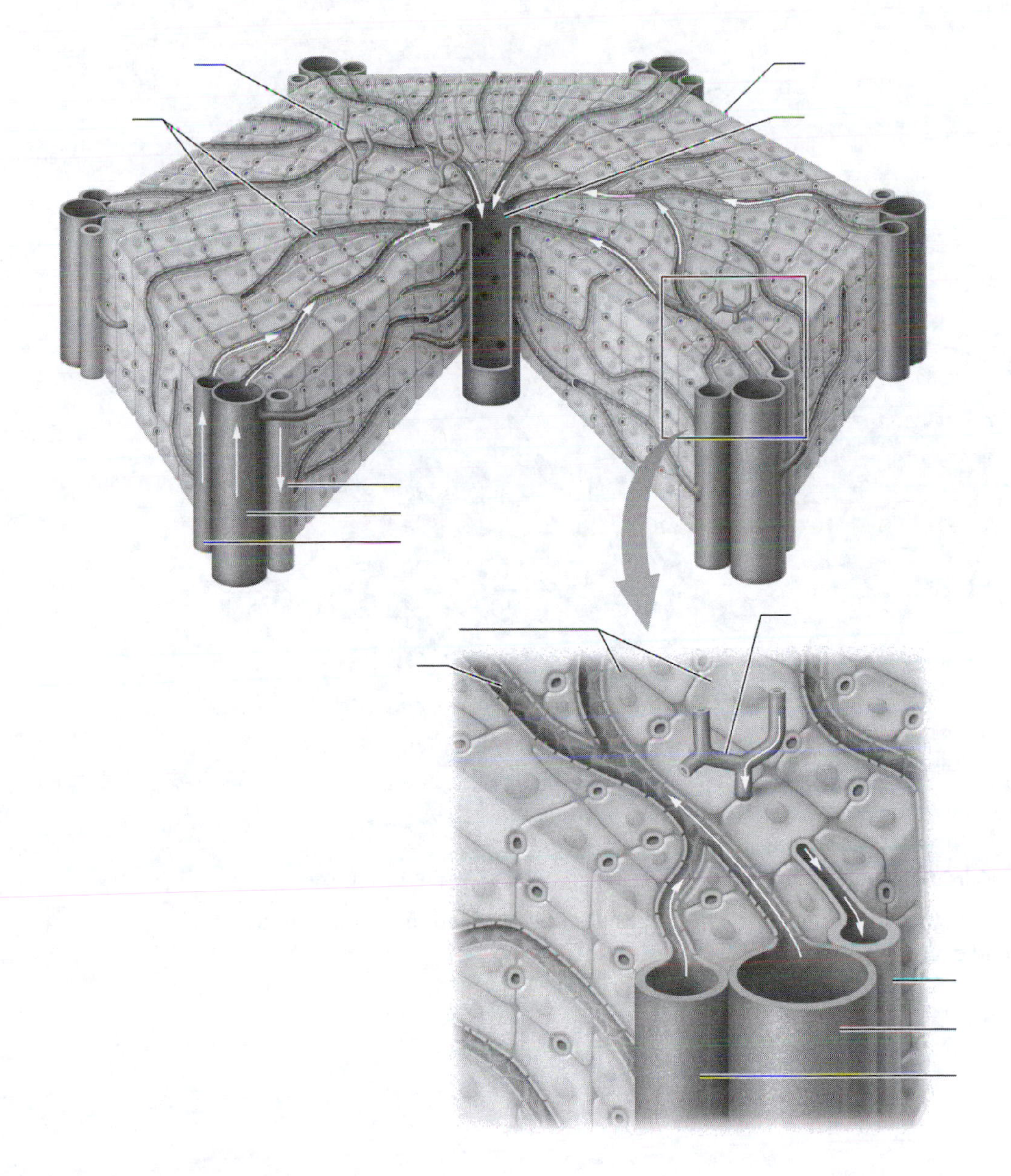

Figure 22.8 The structure of a liver lobule.

Key Concept: What is the relationship between the gallbladder and the liver?

____

____

Describe It: Secretion of Bile

Write in the steps of the process by which bile is released into the duodenum from the gallbladder and liver in Figure 22.9, and label and color-code key components of the process. You may use text Figure 22.26 as a reference, but write the steps in your own words.

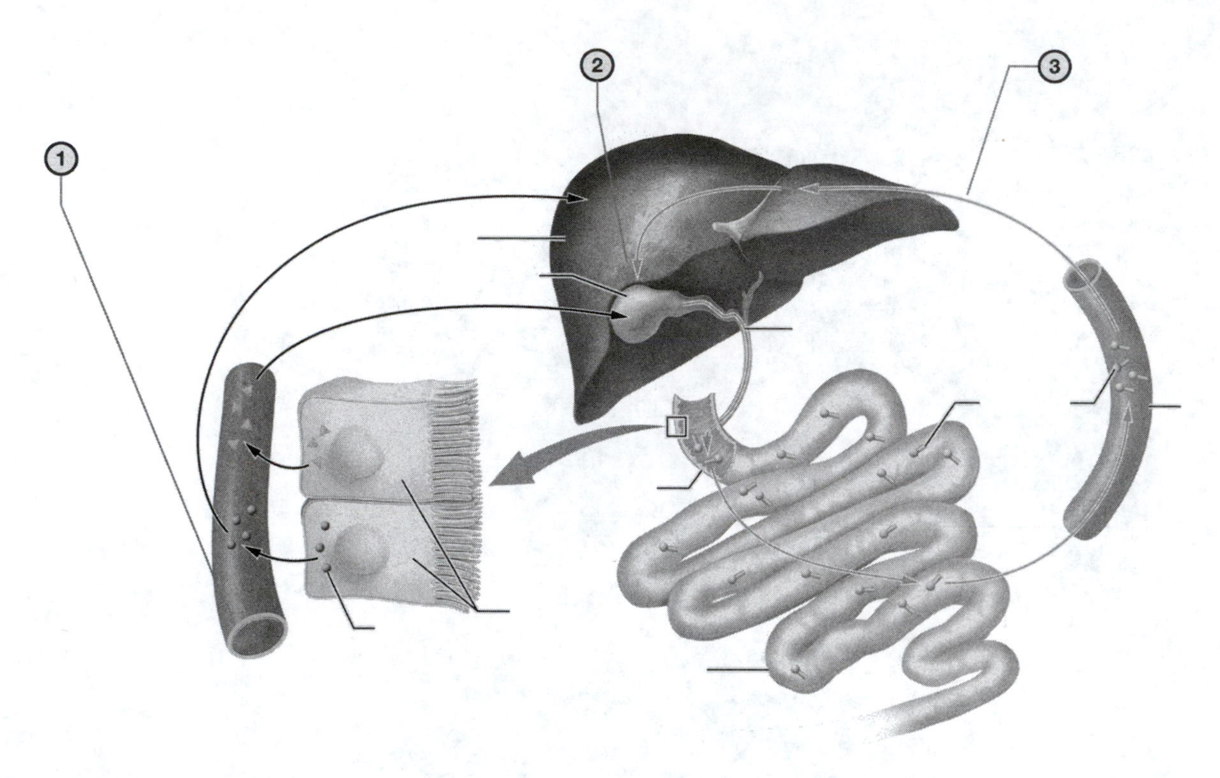

Figure 22.9 Secretion of bile.

Describe It: Other Functions of the Liver

Write a paragraph describing the other functions of the liver (besides bile production), such as nutrient metabolism, detoxification, and excretion.

Module 22.7: Nutrient Digestion and Absorption

Most of the content of the previous modules introduced you to the players in the game of digestion and absorption. Module 22.7 in your text details specifically how those players accomplish the work of digestion and absorption. By the end of the module, you should be able to do the following:

1. Explain the process of enzymatic hydrolysis reactions of nutrients.
2. Describe the enzymes involved in chemical digestion, including their activation, substrates, and end products.

3. Describe the process of emulsification, explain its importance, and discuss how bile salts are recycled.
4. Explain the processes involved in absorption of each type of nutrient, fat-soluble and water-soluble vitamins, and vitamin B12.

Build Your Own Glossary

Below is a table listing key terms from Module 22.7. Before you read the module, use the glossary at the back of your book or look through the module to define the following terms.

Key Terms for Module 22.7

Term	Definition
Pancreatic amylase	
Brush border enzymes	
Na^+/glucose cotransporter	
Trypsin	
Emulsification	
Pancreatic lipase	
Micelles	
Chylomicrons	
Vitamins	

Survey It: Form Questions

Before you read the module, survey it and form at least three questions for yourself. When you have finished reading the module, return to these questions and answer them.

Question 1: ______________________________

Answer: ______________________________

Question 2: ______________________________

Answer: ______________________________

Question 3: ______________________________

Answer: ______________________________

Key Concept: How are larger carbohydrates digested into monosaccharides for absorption?

____

____

Describe It: Amino Acid Digestion and Absorption in the Small Intestine

Write in the steps by which amino acids are digested and absorbed in the small intestine in Figure 22.10, and label and color-code key components of the process. You may use text Figure 22.30 as a reference, but write the steps in your own words.

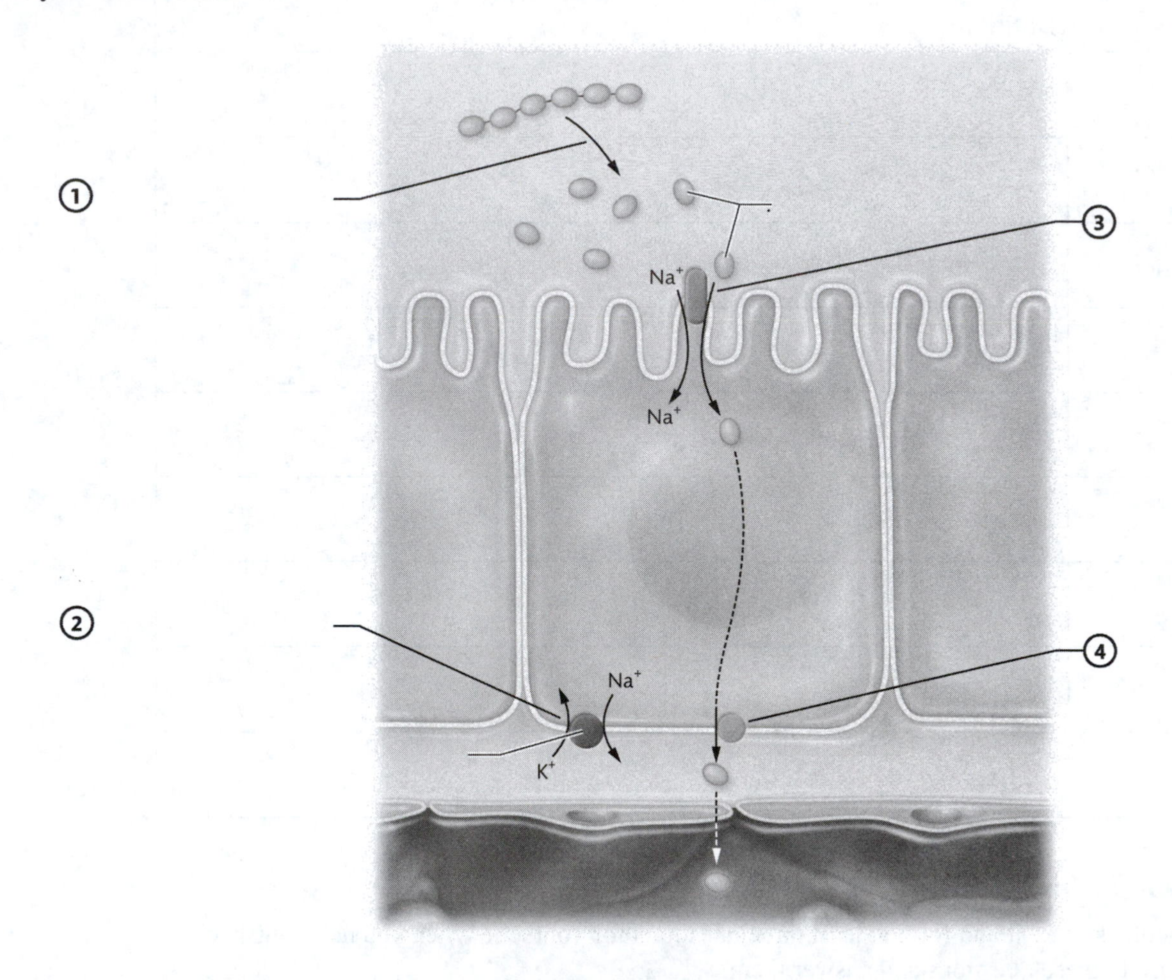

Figure 22.10 Amino acid digestion and absorption in the small intestine.

Key Concept: In what ways is lipid digestion different from carbohydrate and protein digestion?

____

____

Describe It: Lipid Absorption in the Small Intestine

Fill in the steps for lipid absorption in the small intestine in Figure 22.11, and label and color-code key components of the process. You may use Figure 22.33 in your text for reference, but try to write the steps in your own words.

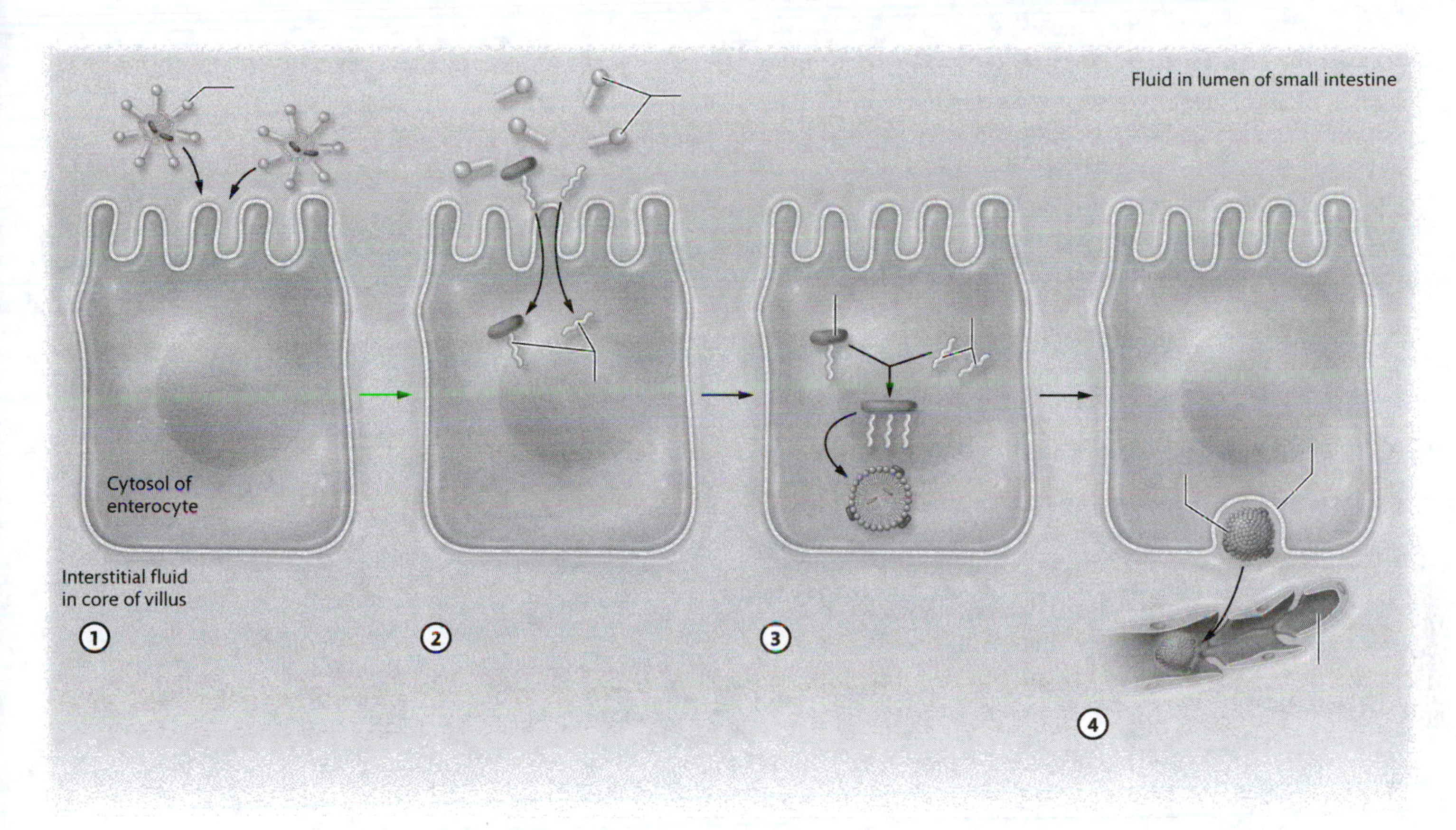

Figure 22.11 Lipid absorption in the small intestine.

Team Up

Make a handout to teach how each of the main classes of macromolecules is digested and absorbed. You can use Figures 22.31 and 22.32 in your text on pages 891–892 as a guide, but the handout should be in your own words and with your own diagrams. At the end of the handout, write a few quiz questions. Once you have completed your handout, team up with one or more study partners and trade handouts. Study your partners' diagrams, and when you have finished, take the quiz at the end. When you and your group have finished taking all the quizzes, discuss the answers to determine places where you need additional study. After you've finished, combine the best elements of each handout to make one "master" diagram for how each of the main classes of macromolecules is digested and absorbed.

Key Concept: How are water and electrolytes absorbed from the intestines?

____

____

Module 22.8: Putting It All Together: The Big Picture of Digestion

Module 22.8 takes in all that you have learned in the previous modules and ties it together by starting with the ingestion of food and ending with the defecation of the remaining, indigestible portion. By the end of the module, you should be able to do the following:

1. Describe the overall big picture of digestion and digestive processes.

Describe It: The Big Picture of Digestion

Write in the steps of the overview of digestion in Figure 22.12, and label and color-code key components of the process. You may use text Figure 22.34 as a reference, but write the steps in your own words.

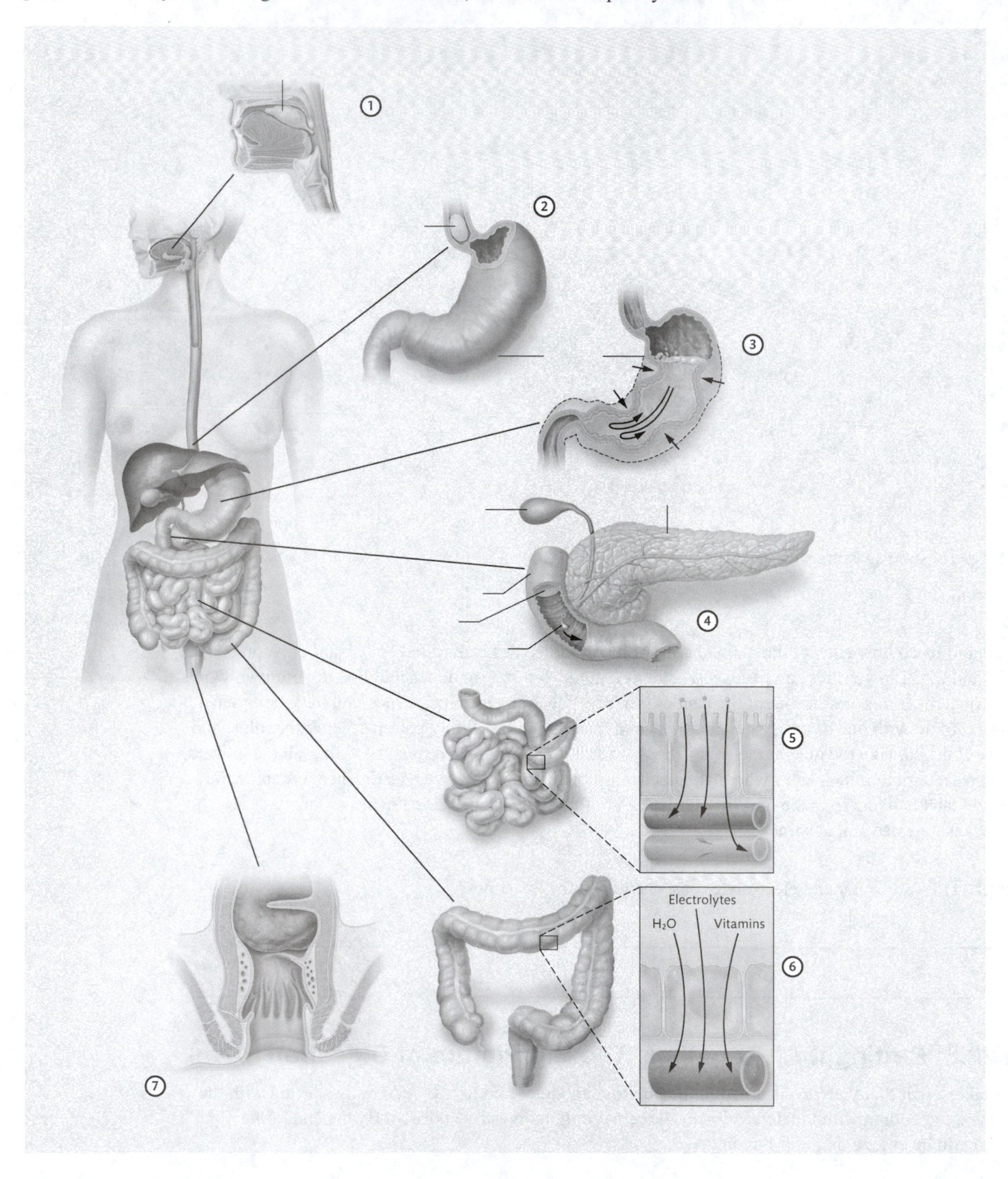

Figure 22.12 The big picture of digestion.

What Do You Know Now?

Let's now revisit the questions you answered in the beginning of this chapter. How have your answers changed now that you've worked through the material?

- What are the main organs of the digestive system?

  ____

- What are building-block molecules of the four classes of macromolecules (carbohydrates, lipids, proteins, and nucleic acids)?

  ____

- What are enzymes, and how do they make chemical reactions occur more easily?

  ____

23 Metabolism and Nutrition

We now turn to how the body manipulates the chemical energy in the molecules that ultimately came from the food we have eaten. This chapter introduces you to the story of metabolism, the body's chemical reactions, and its connection with nutrition.

What Do You Already Know?

Try to answer the following questions before proceeding to the next section. If you're unsure of the correct answers, give it your best attempt based on previous courses, previous chapters, or just your general knowledge.

- What is the most common energy exchange molecule in cells, and which organelle is responsible for most of its production?

  ____

- What are the four categories of macromolecules, and of those, which two are more commonly associated with energy storage?

  ____

- What mechanisms does the body have for adjusting body temperature? (*Hint:* You learned about some of these in some of the chapters in the first half of the textbook.)

  ____

Module 23.1: Overview of Metabolism and Nutrition

Module 23.1 in your text introduces you to the types of chemical reactions of metabolism occurring in cells, as well as the frequently encountered molecules. By the end of the module, you should be able to do the following:

1. Define the terms metabolism, catabolism, and anabolism, and identify the nutrients the body is able to use for fuel.
2. Compare and contrast endergonic and exergonic reactions.
3. Describe the process of phosphorylation.
4. Describe the hydrolysis of ATP, and explain why this reaction is exergonic.
5. Explain what happens in an oxidation-reduction reaction and how electrons are transferred between reactants, including NADH and $FADH_2$.

Build Your Own Glossary

Following is a table listing key terms from Module 23.1. Before you read the module, use the glossary at the back of your book or look through the module to define the following terms.

Key Terms for Module 23.1

Term	Definition
Metabolism	
ATP	
Catabolism	
Anabolism	
Exergonic reactions	
Endergonic reactions	
Phosphorylation	
Oxidation-reduction reactions	
Electron carriers	

Survey It: Form Questions

Before you read the module, survey it and form at least two questions for yourself. When you have finished reading the module, return to these questions and answer them.

Question 1: ____________________

Answer: ____________________

Question 2: ____________________

Answer: ____________________

Key Concept: What are the three main energy-storing molecules used by cells for catabolism? Which one is the preferred energy source for most cells?

____

____

Complete It: Endergonic and Exergonic Reactions

Fill in the blanks to complete the following paragraphs that describe the relationship of endergonic and exergonic reactions. (*Hint:* Since the exercise involves related metabolic reactions, you will use most of the same terms a second time in the second paragraph.)

Nutrients are broken down via ____________ reactions that are energy-releasing ____________ reactions. This energy is used to fuel the energy-storing ____________ reactions of ATP synthesis. Synthesis of molecules such as ATP is an ____________ process.

In another exergonic ____________ reaction, the third phosphate group from ATP is removed and energy is ____________. The energy from ATP breakdown fuels other endergonic ____________ reactions in the cell.

ATP breakdown is a highly ____________ process that provides the energy needed for many ____________ reactions in the cell.

Key Concept: What is the main difference between oxidation reactions and reduction reactions?

____

____

Module 23.2: Glucose Catabolism and ATP Synthesis

As glucose is the preferred energy source for most cells, next we will carefully examine how it is broken down so its energy can be used to synthesize ATP. When you finish this module, you should be able to do the following:

1. Explain the overall reaction for glucose catabolism.
2. Describe the processes of glycolysis, formation of acetyl CoA, and the citric acid cycle.
3. Describe the process of the electron transport chain.
4. Discuss the process of chemiosmosis and its role in ATP production.
5. Give the energy yield of each part of glucose catabolism and the overall energy yield of glucose catabolism.

Build Your Own Glossary

Following is a table listing key terms from Module 23.2. Before you read the module, use the glossary at the back of your book or look through the module to define the following terms.

Key Terms for Module 23.2

Term	Definition
Glycolytic (anaerobic) catabolism	
Oxidative (aerobic) catabolism	
Pyruvate	

Term	Definition
Lactate	
Coenzyme A	
Citric acid cycle	
Electron transport chain	
ATP synthase	

Survey It: Form Questions

Before you read the module, survey it and form at least three questions for yourself. When you have finished reading the module, return to these questions and answer them.

Question 1: ______________________________

Answer: ______________________________

Question 2: ______________________________

Answer: ______________________________

Question 3: ______________________________

Answer: ______________________________

Key Concept: How are substrate-level phosphorylation and oxidative phosphorylation different?

____

____

Key Concept: What happens to pyruvate after the end of glycolysis under aerobic conditions?

____

____

Describe It: The Reactions of Glycolysis

Describe the steps for the reactions of glycolysis in Figure 23.1. You may use Figure 23.4 in your text for reference, but try to write the steps in your own words.

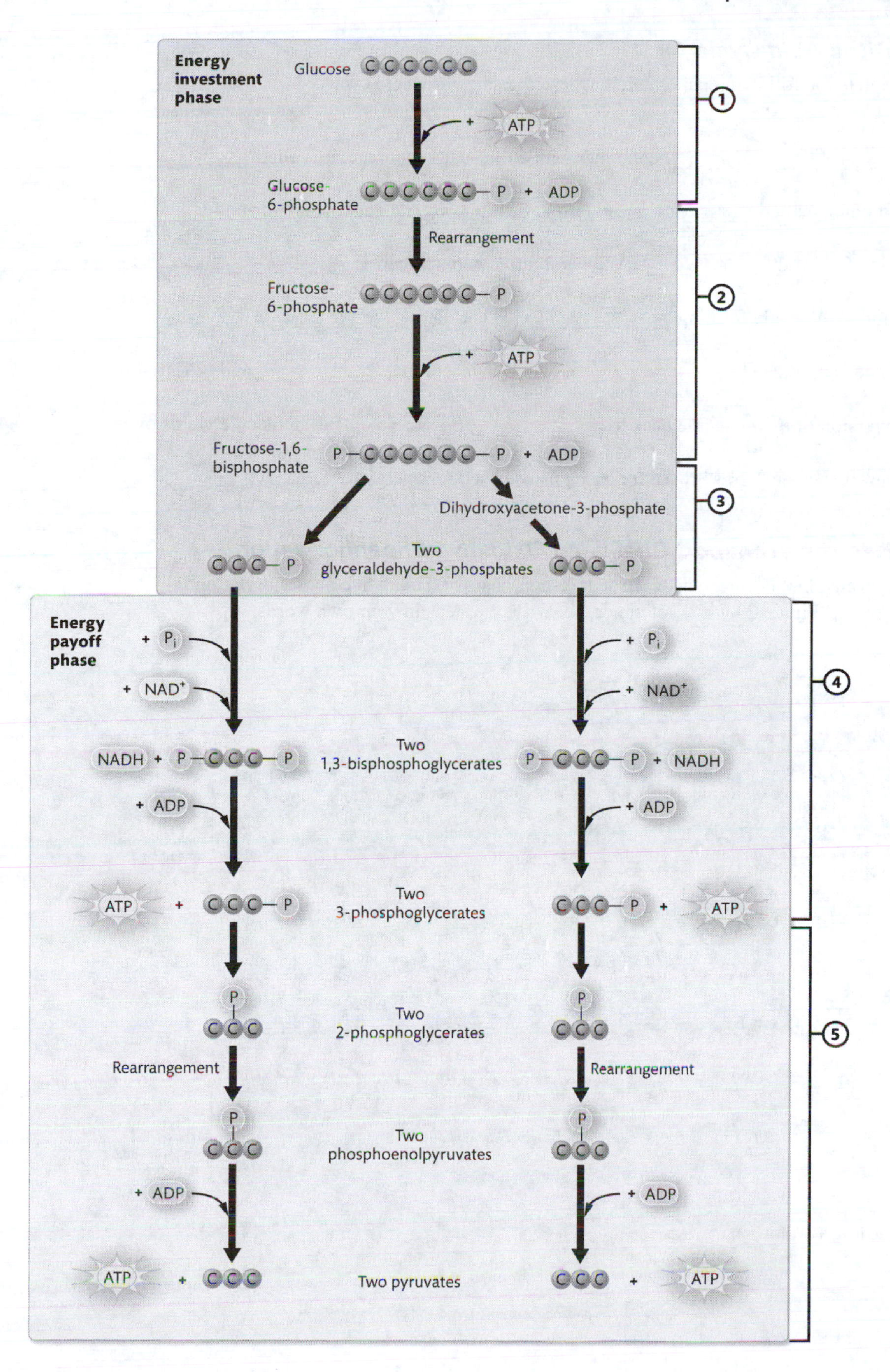

Figure 23.1 The reactions of glycolysis.

Complete It: The Citric Acid Cycle

Fill in the blanks to complete the following paragraph that describes the citric acid cycle.

Acetyl-CoA from ____________ ____________ combines with a four-carbon compound called ____________ to form a six-carbon compound called ____________ and CoA. Citrate is rearranged, then oxidized by ____________, generating CO_2 and ____________. Another intermediate molecule, ____________, is converted to succinate and CoA, while forming ____________. This is called ____________ ____________ phosphorylation. Succinate is oxidized by ____________ and NAD^+, generating ____________ and NADH, which results in the conversion of the molecule back to ____________. Because a glucose molecule yields two ____________ molecules, this cycle "turns" twice for each glucose used for energy.

Describe It: The Electron Transport Chain and Oxidative Phosphorylation

Write in the steps of the events that occur in the electron transport chain and oxidative phosphorylation in Figure 23.2. You may use text Figure 23.7 as a reference, but write the steps in your own words.

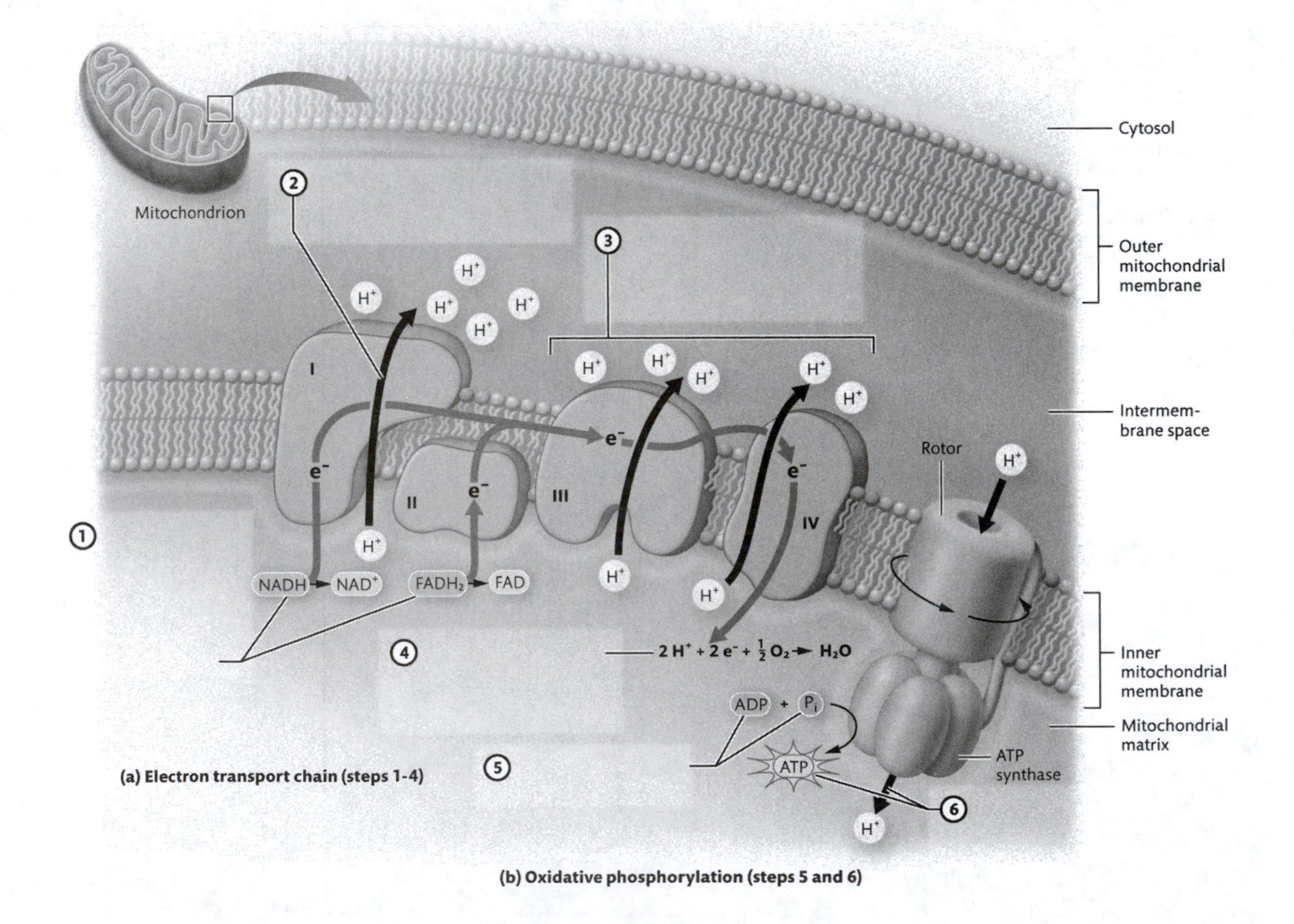

Figure 23.2 The electron transport chain and oxidative phosphorylation.

Key Concept: The core principle of *Gradients* figures into the final stages of mitochondrial respiration. How does this gradient facilitate ATP production?

____

____

Module 23.3: Fatty Acid and Amino Acid Catabolism

This next module will allow you to build on what you have just learned. Some of the very same metabolic pathways used for glucose catabolism are also used for fatty acid and amino acid catabolism. When you complete this module, you should be able to do the following:

1. Describe lipolysis, deamination, and transamination.
2. Summarize the β-oxidation of fatty acids, and explain how it leads to ATP production and relates to ketogenesis.
3. Explain how amino acid catabolism leads to ATP production.
4. Describe the effect of amino acid catabolism on ammonia and urea production.

Build Your Own Glossary

Below is a table listing key terms from Module 23.3. Before you read the module, use the glossary at the back of your book or look through the module to define the following terms.

Key Terms of Module 23.3

Term	Definition
β-oxidation	
Fatty acids	
Glycerol	
Ketone bodies	
Ketogenesis	
Transamination	
Oxidative deamination	
Urea cycle	
Urea	

Survey It: Form Questions

Before you read the module, survey it and form at least two questions for yourself. When you have finished reading the module, return to these questions and answer them.

Question 1: ______________________________

Answer: ______________________________

Question 2: ______________________________

Answer: ______________________________

Key Concept: What familiar molecules from glucose catabolism also play a role in fatty acid catabolism? What similar roles do these molecules play here?

____

____

Describe It: Fatty Acid Catabolism and β-oxidation

Write in the steps of fatty acid catabolism and β-oxidation in Figure 23.3. You may use text Figure 23.9 as a reference, but write the steps in your own words.

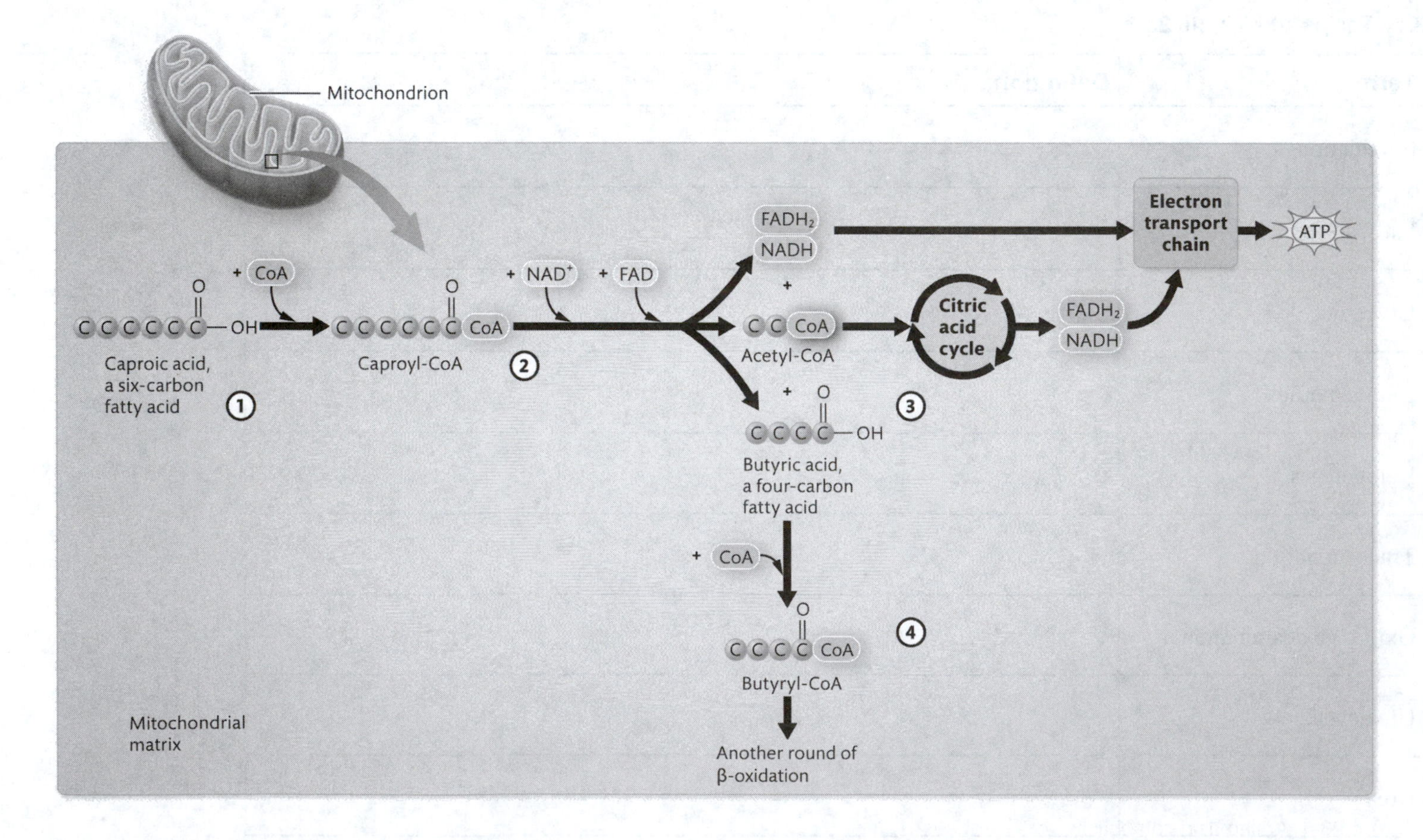

Figure 23.3 Fatty acid catabolism and β-oxidation.

Key Concept: How do the details of fatty acid catabolism explain the claim that fat has more than twice the calories of carbohydrate or protein?

___

___

Describe It: Ketogenesis

Write a paragraph describing the connection between ketogenesis and fat metabolism as if you were explaining to an audience of nonscientists.

Describe It: Amino Acid Catabolism

Write in the steps of amino acid catabolism in Figure 23.4. You may use text Figure 23.10 as a reference, but write the steps in your own words.

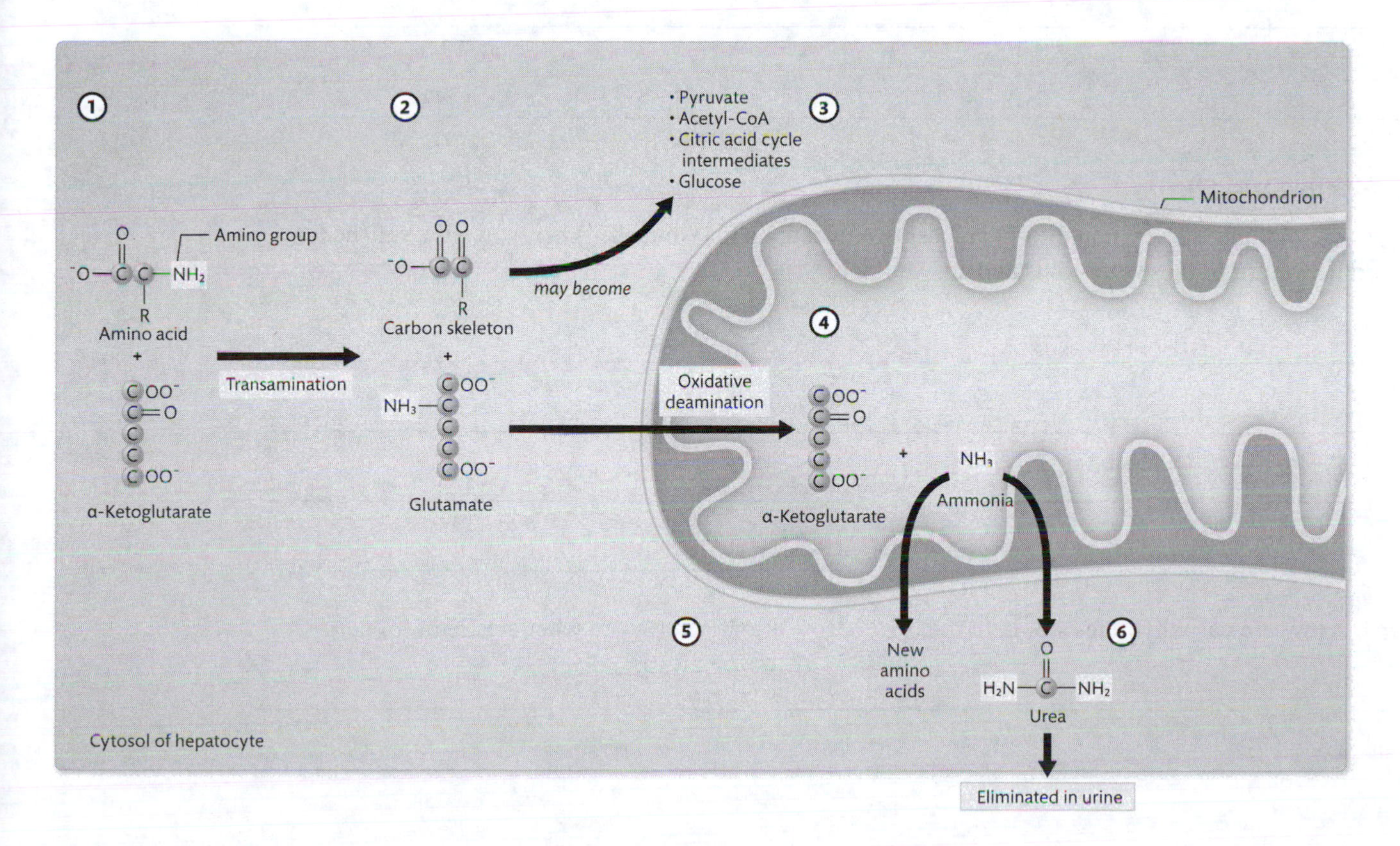

Figure 23.4 Amino acid catabolism and β-oxidation.

Module 23.4: Anabolic Pathways

Module 23.4 in your text explores the other side of metabolism—the *anabolic* reactions that store nutrients and build nutrient molecules. At the end of this module, you should be able to do the following:

1. Describe the processes of glycogenesis, glycogenolysis, and gluconeogenesis.
2. Describe the process by which fatty acids are synthesized and stored as triglycerides in adipose tissue.
3. Explain how nutrients may be converted to amino acids and lipids if needed.
4. Explain the fate of excess dietary proteins and carbohydrates.

Build Your Own Glossary

Below is a table listing key terms from Module 23.4. Before you read the module, use the glossary at the back of your book or look through the module to define the following terms.

Key Terms of Module 23.4

Term	Definition
Glycogenesis	
Glycogenolysis	
Gluconeogenesis	
Lipogenesis	

Survey It: Form Questions

Before you read the module, survey it and form at least two questions for yourself. When you have finished reading the module, return to these questions and answer them.

Question 1: ______________________________

Answer: ______________________________

Question 2: ______________________________

Answer: ______________________________

Key Concept: How are carbohydrates stored in the body? How are they retrieved when needed for energy?

____

____

Describe It: Nutrient Anabolism

Write a paragraph summarizing how the four main categories of nutrient building-block molecules (glucose, fatty acids, glycerol, and amino acids) are built up into macromolecules.

Key Concept: Can excess carbohydrate and protein in the diet be converted into fat? Explain your answer.

____

____

Module 23.5: Metabolic States and Regulation of Feeding

This module turns to metabolic states that ensure the body has the energy and materials it needs to carry out life processes. At the end of this module, you should be able to do the following:

1. Compare and contrast the processes that occur in the absorptive and postabsorptive states.
2. Explain the roles of insulin and glucagon in the absorptive and postabsorptive states.
3. Describe the role of the hormones insulin and glucagon in regulating glucose and amino acid catabolism and anabolism.
4. Explain the significance of glucose sparing for neural tissue in the postabsorptive state.
5. Describe how feeding behaviors are regulated.

Build Your Own Glossary

Following is a table listing key terms from Module 23.5. Before you read the module, use the glossary at the back of your book or look through the module to define the following terms.

Key Terms of Module 23.5

Term	Definition
Absorptive state	
Postabsorptive state	
Glucose sparing	
Leptin	
Ghrelin	

Survey It: Form Questions

Before you read the module, survey it and form at least three questions for yourself. When you have finished reading the module, return to these questions and answer them.

Question 1: ______________________________

Answer: ______________________________

Question 2: ______________________________

Answer: ______________________________

Question 3: ______________________________

Answer: ______________________________

Key Concept: In what ways are the metabolic changes of the absorptive state largely controlled by the hormone insulin?

____

____

Key Concept: In what ways are the metabolic changes of the postabsorptive state largely controlled by the hormone glucagon?

____

____

Describe It: Regulation of Feeding Behaviors

Write a paragraph describing the roles of the hormones leptin and ghrelin in regulating feeding behaviors as if you were explaining to an audience of nonscientists.

Module 23.6: The Metabolic Rate and Thermoregulation

Module 23.6 in your text explores the total amount of energy expended by the body to power all of its processes, especially maintenance of body temperature. By the end of the module, you should be able to do the following:

1. Define metabolic rate and basal metabolic rate.
2. Describe factors that affect metabolic rate.
3. Differentiate between radiation, conduction, convection, and evaporation.
4. Explain the importance of thermoregulation in the body.

Build Your Own Glossary

Below is a table listing key terms from Module 23.6. Before you read the module, use the glossary at the back of your book or look through the module to define the following terms.

Key Terms for Module 23.6

Term	Definition
Metabolic rate	
BMR	
Radiation	
Conduction	
Convection	
Evaporation	
Thermoregulation	
Hypothermia	
Hyperthermia	

Survey It: Form Questions

Before you read the module, survey it and form at least two questions for yourself. When you have finished reading the module, return to these questions and answer them.

Question 1: ______________________________

Answer: ______________________________

Question 2: ______________________________

Answer: ______________________________

Key Concept: What is the *basal metabolic rate*, and what are some factors that may influence it?

____

____

Key Concept: What are the main ways in which heat is exchanged between the body and the environment?

___

___

Team Up

Make a handout to teach the differences between the negative feedback loops that kick in to the main core body temperature in response to rising versus falling body temperatures. You can use Figure 23.18 in your text on page 931 as a guide, but the handout should be in your own words and with your own diagrams. At the end of the handout, write a few quiz questions. Once you have completed your handout, team up with one or more study partners and trade handouts. Study your partners' diagrams, and when you have finished, take the quiz at the end. When you and your group have finished taking all the quizzes, discuss the answers to determine places where you need additional study. After you've finished, combine the best elements of each handout to make one "master" diagram for the differences between the negative feedback loops that kick in to the main core body temperature.

Module 23.7: Nutrition and Body Mass

Module 23.7 in your text takes a look at the dietary sources of nutrients and how nutrient intake contributes to body mass. By the end of the module, you should be able to do the following:

1. Define the terms nutrient, essential nutrient, macronutrient, and micronutrient.
2. Describe the dietary sources, relative energy yields, and common uses in the body for carbohydrates, lipids, and proteins.
3. Discuss the major roles and sources of fat- and water-soluble vitamins and dietary minerals.
4. Describe the components of a balanced diet, including the concept of recommended dietary allowances and energy balance.
5. Explain how the liver processes cholesterol, and identify the roles of the different lipoproteins.
6. Describe how body mass is determined.

Build Your Own Glossary

Below is a table listing key terms from Module 23.7. Before you read the module, use the glossary at the back of your book or look through the module to define the following terms.

Key Terms for Module 23.7

Term	Definition
Macronutrients	
Micronutrients	
Essential nutrients	
Fiber	

Term	Definition
Complete protein	
Vitamin	
Mineral	
Cholesterol	
Body mass index	
Energy balance	

Survey It: Form Questions

Before you read the module, survey it and form at least two questions for yourself. When you have finished reading the module, return to these questions and answer them.

Question 1: ______________________________

Answer: ______________________________

Question 2: ______________________________

Answer: ______________________________

Key Concept: What are macronutrients, and how do they differ from micronutrients?

____

____

Key Concept: If we cannot fully digest the molecules of dietary fiber, why are they important to the human diet?

____

____

Key Concept: What is the difference between complete and incomplete proteins? Is it possible to get all the necessary amino acids from incomplete protein sources? Explain your answer.

____

____

Describe It: Micronutrients

Write a paragraph describing the importance of vitamins and minerals as if you were explaining to an audience of nonscientists.

Key Concept: What are HDLs and LDLs? Why is one considered "good" and the other "bad"?

____

____

Team Up

Make a handout to teach the concept of energy balance and its implications for a healthy diet and obesity. At the end of the handout, write a few quiz questions. Once you have completed your handout, team up with one or more study partners and trade handouts. Study your partners' diagrams, and when you have finished, take the quiz at the end. When you and your group have finished taking all the quizzes, discuss the answers to determine places where you need additional study. After you've finished, combine the best elements of each handout to make one "master" diagram for the concept of energy balance and its relationship to a healthy diet.

What Do You Know Now?

Let's now revisit the questions you answered in the beginning of this chapter. How have your answers changed now that you've worked through the material?

- What is the most common energy exchange molecule in cells, and which organelle is responsible for most of its production?

  ____

- What are the four categories of macromolecules, and of those, which two are more commonly associated with energy storage?

  ____

- What mechanisms does the body have for adjusting body temperature? (*Hint:* You learned about some of these in some of the chapters in the first half of the textbook.)

  ____

24 The Urinary System

The organs of the urinary system are organs of excretion—they "clean the blood" of *metabolic wastes*, which are substances produced by the body that it cannot use for any purpose. This chapter examines this process as well as the many different functions performed by the urinary system that are vital to maintaining homeostasis.

What Do You Already Know?

Try to answer the following questions before proceeding to the next section. If you're unsure of the correct answers, give it your best attempt based on previous courses, previous chapters, or just your general knowledge.

- What are the main functions of the urinary system?

  ____

- What do the kidneys do?

  ____

- What stimulates urination?

  ____

Module 24.1: Overview of the Urinary System

Module 24.1 in your text gives you an overview of the structures and functions of the urinary system. By the end of the module, you should be able to do the following:

1. List and describe the organs of the urinary system.
2. Describe the major functions of the kidneys.

Build Your Own Glossary

Following is a table listing key terms from Module 24.1. Before you read the module, use the glossary at the back of your book or look through the module to define the following terms.

Key Terms for Module 24.1

Term	Definition
Kidneys	
Urine	
Urinary tract	

Survey It: Form Questions

Before you read the module, survey it and form at least two questions for yourself. When you have finished reading the module, return to these questions and answer them.

Question 1: ______________________________

Answer: ______________________________

Question 2: ______________________________

Answer: ______________________________

Identify It: Structures of the Urinary System

Identify and color-code the structures of the urinary system shown in Figure 24.1.

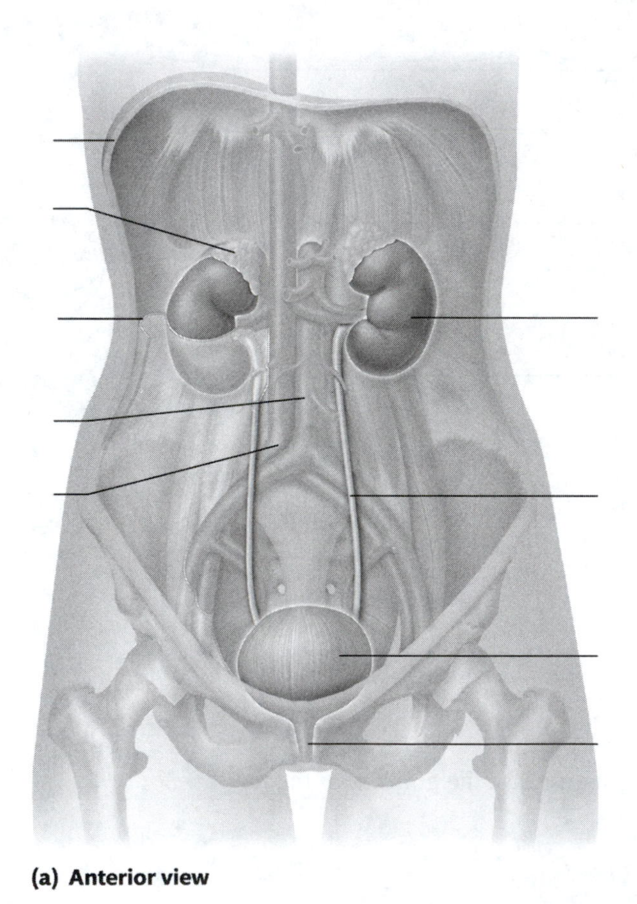

(a) Anterior view

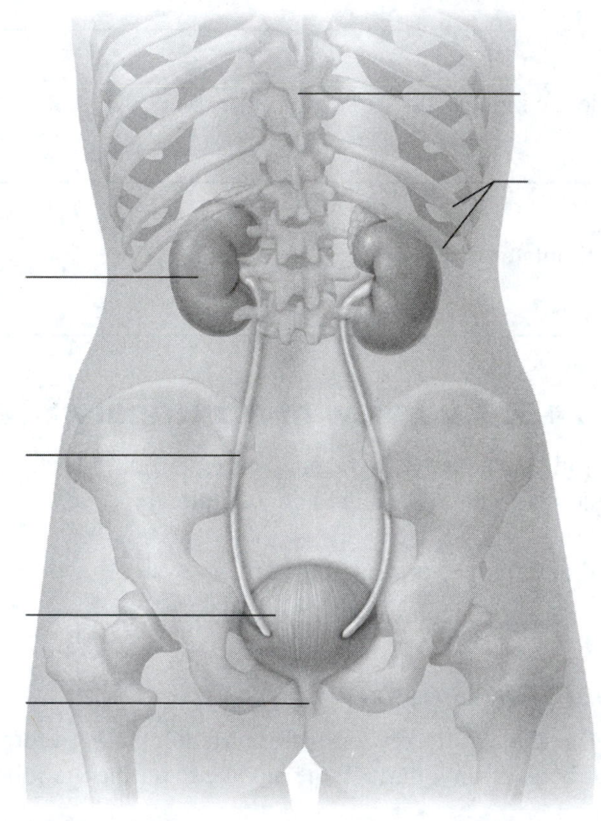

(b) Posterior view

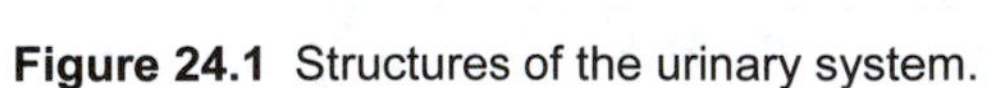

Figure 24.1 Structures of the urinary system.

Key Concept: How does the urinary system maintain overall homeostasis?

____

____

Module 24.2: Anatomy of the Kidneys

This module in your text examines the gross and microscopic anatomy of the kidneys. When you finish this module, you should be able to do the following:

1. Describe the external structure of the kidney, including its location, support structures, and coverings.
2. Trace the path of blood through the kidneys.
3. Describe the major structures, subdivisions, and histology of the renal corpuscle, renal tubule, and collecting system.
4. Trace the pathway of filtrate flow through the nephron and collecting system.
5. Compare and contrast cortical and juxtamedullary nephrons.

Build Your Own Glossary

Following is a table listing key terms from Module 24.2. Before you read the module, use the glossary at the back of your book or look through the module to define the following terms.

Key Terms for Module 24.2

Term	Definition
Renal cortex	
Renal medulla	
Renal pelvis	
Nephron	
Afferent arteriole	
Glomerulus	
Efferent arteriole	
Peritubular capillaries	
Renal corpuscle	
Proximal tubule	
Nephron loop	
Distal tubule	
Juxtaglomerular apparatus	

Term	Definition
Collecting system	
Cortical nephron	
Juxtamedullary nephron	

Survey It: Form Questions

Before you read the module, survey it and form at least two questions for yourself. When you have finished reading the module, return to these questions and answer them.

Question 1: ____________________

Answer: ____________________

Question 2: ____________________

Answer: ____________________

Identify It: The Gross and Microscopic Anatomy of the Kidney

Color and label Figure 24.2 with the structures of the kidney and nephron.

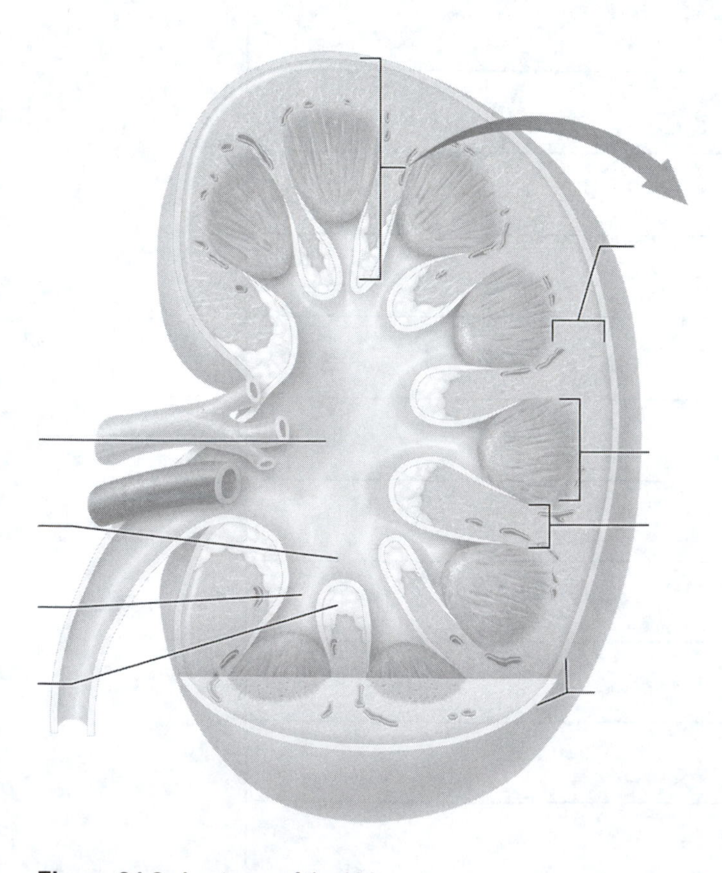
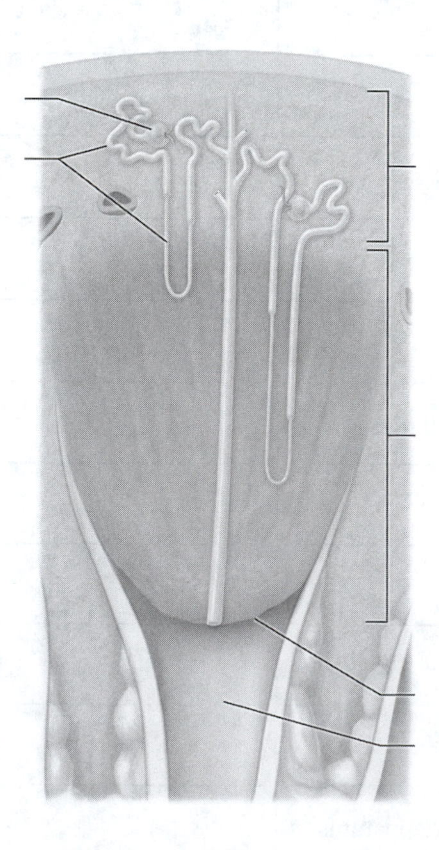

Figure 24.2 Anatomy of the kidney.

Describe It: Pathway of Blood Flow through the Kidney

Trace the pathway of blood as it flows from the renal artery through the renal vein. Draw arrows on Figure 24.3 to show the path that an erythrocyte would take through the kidney, and label the corresponding blood vessels.

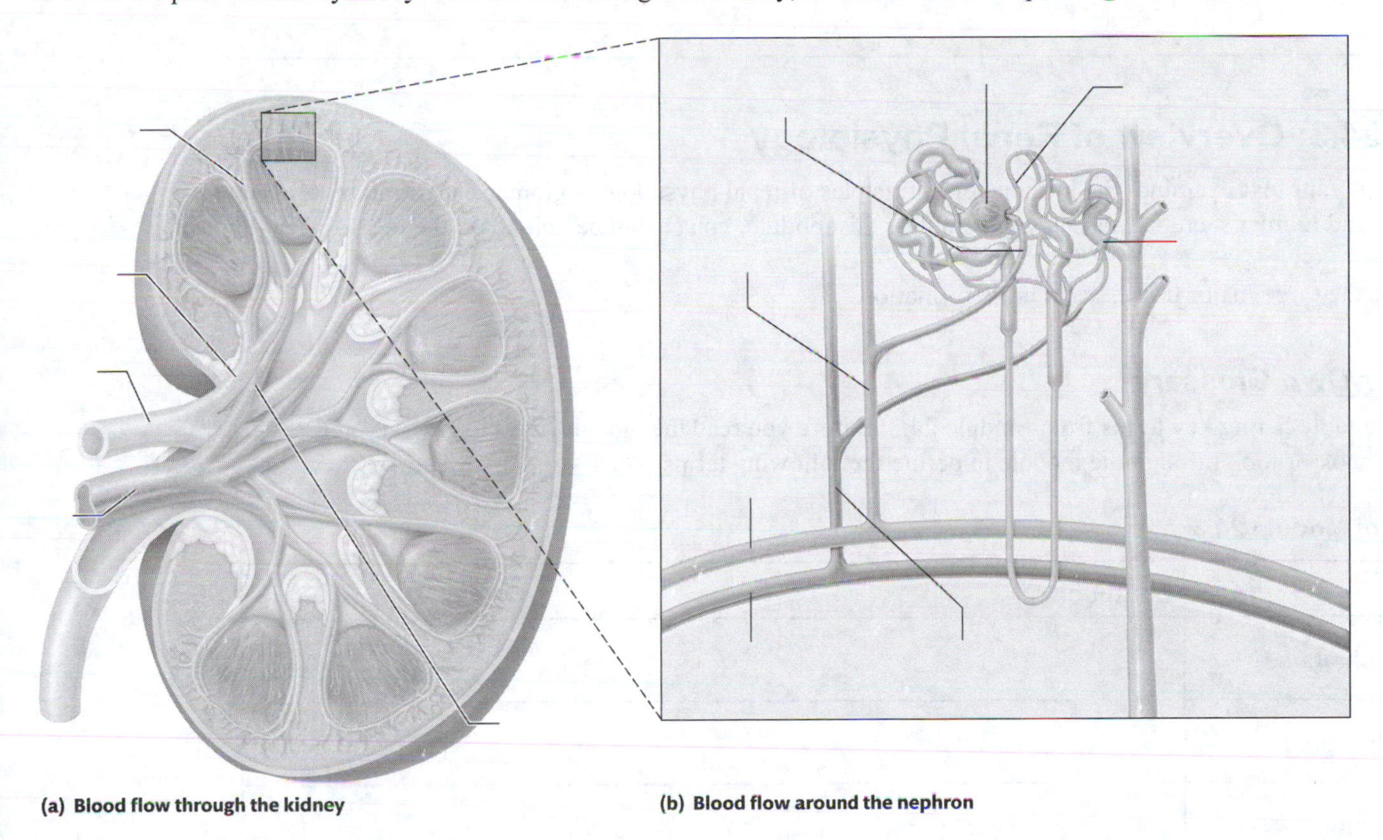

(a) Blood flow through the kidney

(b) Blood flow around the nephron

Figure 24.3 Blood flow through the kidney.

Key Concept: How is blood flow through the kidney different from other organs?

____

____

Draw It: The Nephron

Draw, color, and label a nephron using Figure 24.10 in your text for reference. Organize the labels in a way that makes sense to you, and include pertinent details about each segment in your labels.

Key Concept: What is a nephron? What takes place within a nephron?

____

____

Module 24.3: Overview of Renal Physiology

This module in your text examines the introductory principles of renal physiology—glomerular filtration, tubular reabsorption, and tubular secretion. When you complete this module, you should be able to do the following:

1. Describe the three major processes in urine formation.

Build Your Own Glossary

Following is a table listing key terms from Module 24.3. Before you read the module, use the glossary at the back of your book or look through the module to define the following terms.

Key Terms of Module 24.3

Term	Definition
Glomerular filtration	
Tubular reabsorption	
Tubular secretion	

Survey It: Form Questions

Before you read the module, survey it and form at least three questions for yourself. When you have finished reading the module, return to these questions and answer them.

Question 1: ____

Answer: ____

Question 2: ____

Answer: ____

Question 3: ____

Answer: ____

Key Concept: Where do the three basic processes of renal physiology take place? Be specific.

____

____

Module 24.4: Renal Physiology I: Glomerular Filtration

Module 24.4 in your text teaches you about the first process in renal physiology: glomerular filtration. At the end of this module, you should be able to do the following:

1. Describe the structure of the filtration membrane.
2. Define the glomerular filtration rate (GFR) and its average value.
3. Explain how the hydrostatic and colloid osmotic pressures combine to yield the net filtration pressure in the glomerulus.
4. Predict specific factors that will increase or decrease the GFR.
5. Explain how the myogenic and tubuloglomerular feedback mechanisms affect the GFR.
6. Describe the role of each of the following in the control of the GFR: renin-angiotensin-aldosterone system, atrial natriuretic peptide, and sympathetic nervous system activity.

Build Your Own Glossary

Following is a table listing key terms from Module 24.4. Before you read the module, use the glossary at the back of your book or look through the module to define the following terms.

Key Terms of Module 24.4

Term	Definition
Filtration membrane	
Podocyte	
Filtration slit	
Glomerular hydrostatic pressure	
Glomerular colloid osmotic pressure	
Capsular hydrostatic pressure	
Net filtration pressure	
Glomerular filtration rate	
Myogenic mechanism	
Tubuloglomerular feedback	
Angiotensin-II	
Atrial natriuretic peptide	

Survey It: Form Questions

Before you read the module, survey it and form at least two questions for yourself. When you have finished reading the module, return to these questions and answer them.

Question 1: ______________________________

Answer: ______________________________

Question 2: ______________________________

Answer: ______________________________

Identify It: The Filtration Membrane

Color and label the filtration membrane illustrated in Figure 24.4.

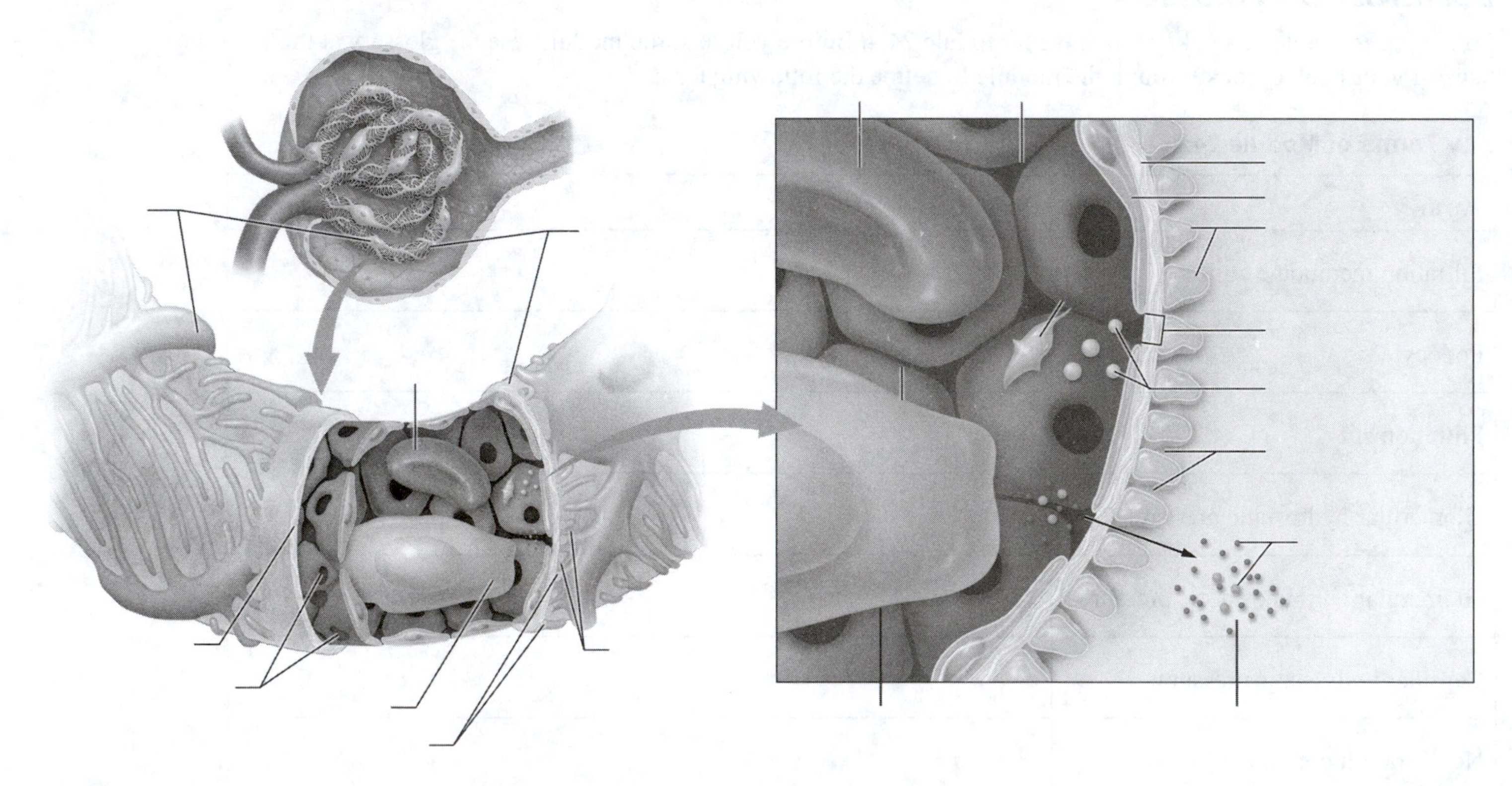

Figure 24.4 Structure of the filtration membrane.

Key Concept: How does the filtration membrane determine the composition of the filtrate?

____

____

Calculate It: The NFP

Calculate the net filtration pressure for the given pressure values. You may want to draw a renal corpuscle and arrows to help you envision the process.

Scenario 1: Systemic hypertension (high systemic blood pressure)

GHP: 62 mm Hg

GCOP: 30 mm Hg

CHP: 11 mm Hg

NFP = ____________________

In this case, would the amount of filtrate formed increase or decrease? ____________________

Scenario 2: Obstruction in the urinary tract

GHP: 51 mm Hg

GCOP: 31 mm Hg

CHP: 18 mm Hg

NFP = ____________________

In this case, would the amount of filtrate formed increase or decrease? ____________________

Key Concept: How do the three pressures in the renal corpuscle interact to determine the glomerular filtration rate?

____

____

Describe It: Tubuloglomerular Feedback

Figure 24.5 diagrams the process of tubuloglomerular feedback when the GFR increases abnormally. Describe the events that are occurring at each step of the feedback loop.

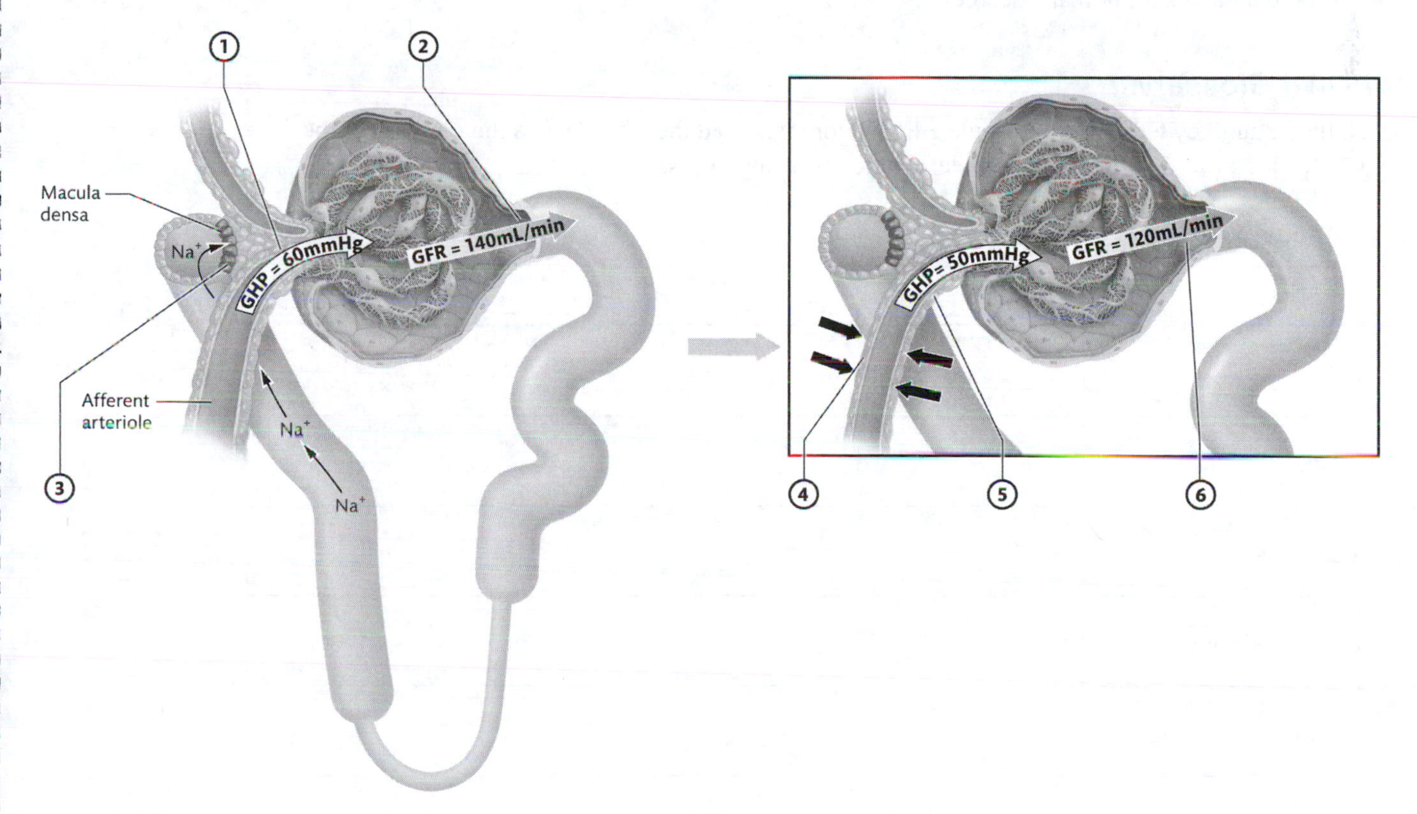

Figure 24.5 Tubuloglomerular feedback.

Team Up

Make a handout to teach the renin-angiotensin-aldosterone system. You can use text Figure 24.14 as a guide, but the handout should be in your own words and with your own diagram. At the end of the handout, write a few quiz questions. Once you have completed your handout, team up with one or more study partners and trade handouts. Study your partners' diagrams and, when you have finished, take the quiz at the end. When you and your group have finished taking all the quizzes, discuss the answers to determine places where you need additional study. After you've finished, combine the best elements of each handout to make one "master" diagram for the renin-angiotensin-aldosterone system.

Key Concept: How does the renin-angiotensin-aldosterone system affect systemic blood pressure and the GFR? What are other mechanisms that regulate the GFR?

____

____

Module 24.5: Renal Physiology II: Tubular Reabsorption and Secretion

Module 24.5 covers the details of tubular reabsorption and secretion. At the end of this module, you should be able to do the following:

1. Describe how and where water, organic compounds, and ions are reabsorbed in the nephron by both passive and active processes.
2. Describe the location(s) in the nephron where tubular secretion occurs.
3. Describe how the renin-angiotensin-aldosterone system, antidiuretic hormone, and atrial natriuretic peptide each work to regulate reabsorption and secretion.

Build Your Own Glossary

Following is a table listing key terms from Module 24.5. Before you read the module, use the glossary at the back of your book or look through the module to define the following terms.

Key Terms of Module 24.5

Term	Definition
Paracellular route	
Transcellular route	
Obligatory water reabsorption	
Facultative water reabsorption	
Antidiuretic hormone	

Survey It: Form Questions

Before you read the module, survey it and form at least two questions for yourself. When you have finished reading the module, return to these questions and answer them.

Question 1: ______________________________

Answer: ______________________________

Question 2: ______________________________

Answer: ______________________________

Draw It: Reabsorption and Secretion

Following are diagrams of cells of the nephron. Draw the transport proteins and other molecules required for the absorption of each of the following substances.

1. In Figure 24.6, draw the pumps and gradients necessary for the reabsorption of glucose in a proximal tubule cell.

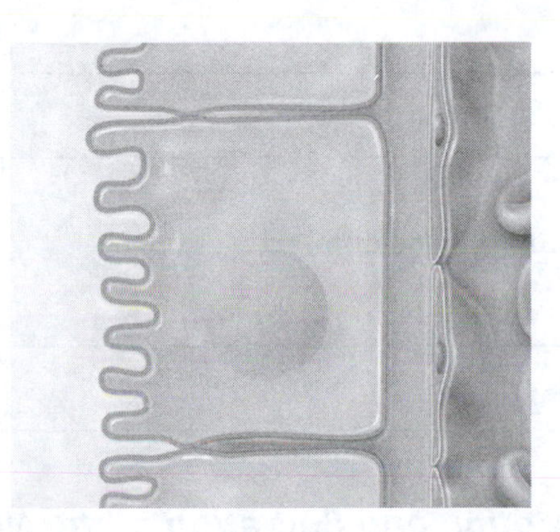

Figure 24.6 Cell of the proximal tubule.

2. In Figure 24.7, draw the pumps and gradients necessary for the reabsorption of water in a proximal tubule cell.

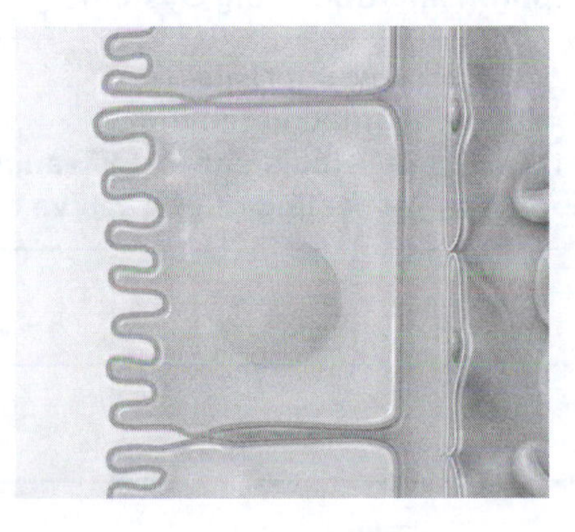

Figure 24.7 Cell of the proximal tubule.

3. In Figure 24.8, draw the pumps, gradients, and hormones necessary for the reabsorption of water in a late distal tubule cell.

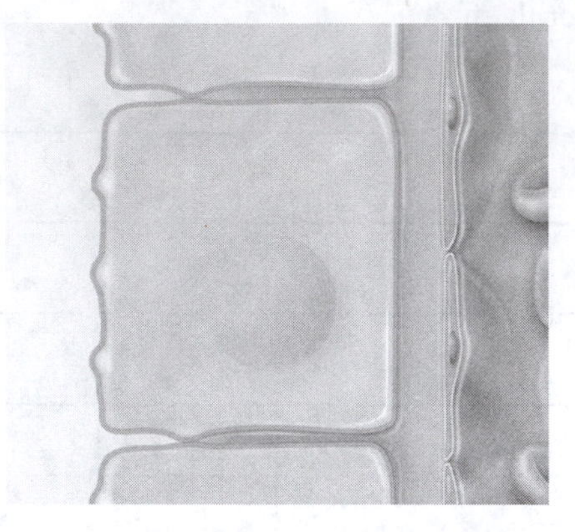

Figure 24.8 Cell of the distal tubule.

Key Concept: What role do sodium ions play in the reabsorption of water and other solutes?

___

___

Key Concept: How does reabsorption differ in the proximal and distal tubules?

___

___

Build Your Own Summary Table: Reabsorption and Secretion in the Nephron and Collecting System

Build your own summary table about the reabsorption and secretion of various substances by filling in the information in the following table. For each region of the renal tubule, be sure to include if the reabsorption or secretion is influenced by a hormone.

Summary of Reabsorption and Secretion in the Nephron and Collecting System

Substance	Amount reabsorbed in the proximal tubule?	Amount reabsorbed in the nephron loop?	Amount reabsorbed in the distal tubule/cortical collecting duct?	Transport passive or active?	Substance secreted? If yes, where?
Water					
Na^+					
K^+					

Glucose					
Ca^{+2}					
Bicarbonate ions					
Creatinine					
Urea					
H^+					

Module 24.6: Renal Physiology III: Regulation of Urine Concentration and Volume

This module examines how the final concentration of urine is controlled through facultative water reabsorption. The information about concentrated urine production and the countercurrent multiplier is, to be perfectly honest, fairly complex. Take your time, and break down each step into its component pieces to better understand the whole process. At the end of this module, you should be able to do the following:

1. Explain why the differential permeability of specific sections of the nephron tubules is necessary to produce concentrated urine.
2. Predict specific conditions that cause the kidneys to produce dilute versus concentrated urine.
3. Explain the role of the nephron loop, the vasa recta, and the countercurrent mechanism in the concentration of urine.

Build Your Own Glossary

Following is a table listing key terms from Module 24.6. Before you read the module, use the glossary at the back of your book or look through the module to define the following terms.

Key Terms for Module 24.6

Term	Definition
Medullary osmotic gradient	
Countercurrent mechanism	
Countercurrent multiplier	
Countercurrent exchange	

Survey It: Form Questions

Before you read the module, survey it and form at least two questions for yourself. When you have finished reading the module, return to these questions and answer them.

Question 1: ____________________

Answer: ____________________

Question 2: ____________________

Answer: ____________________

Key Concept: What hormone must be secreted in lower amounts to produce dilute urine? Why?

____

____

Identify It: The Countercurrent Multiplier

Label and describe, in your own words, the steps of the countercurrent multiplier, illustrated in Figure 24.9.

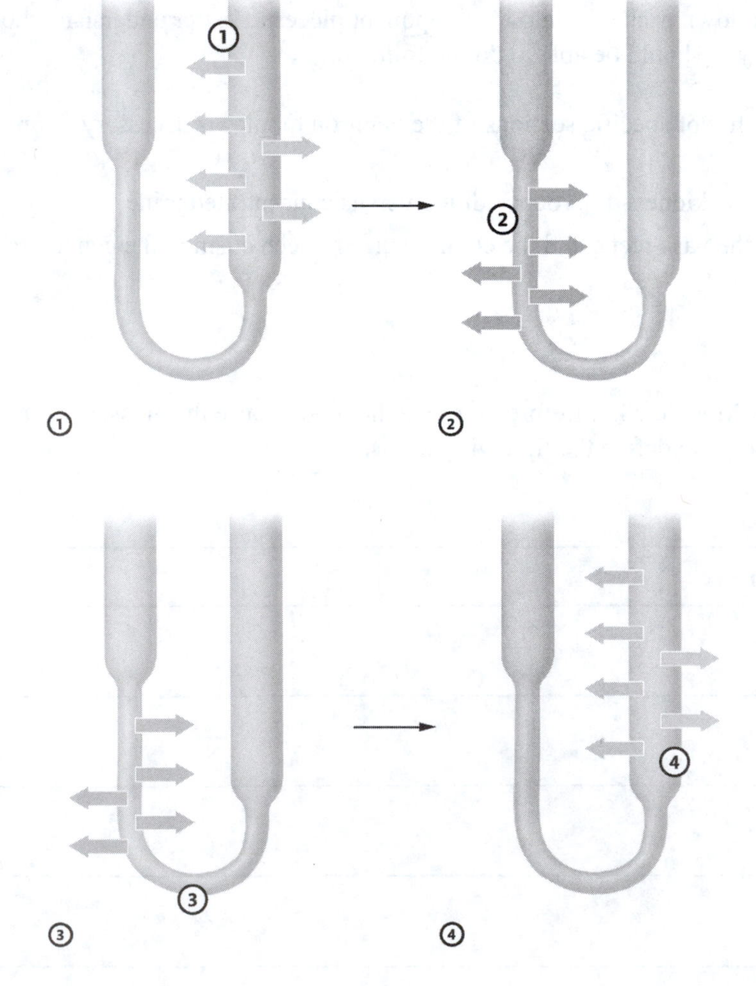

Figure 24.9 The countercurrent multiplier.

Draw It: The Countercurrent Multiplier and the Medullary Osmotic Gradient

Following are blank nephron loops and collecting system tubules. Draw what would happen to the medullary osmotic gradient and, therefore, water reabsorption in each scenario.

Scenario 1: Normal functioning of both limbs of the nephron loop with low secretion of ADH. In Figure 24.10, draw in the solute dots in the loop, collecting system, and the interstitial fluid, and determine if the urine produced will be dilute or concentrated.

Scenario 2: The pumps in the thick ascending limb are blocked. In Figure 24.11, draw in the solute dots in the loop, collecting system, and the interstitial fluid, and determine if the urine produced will be dilute or concentrated.

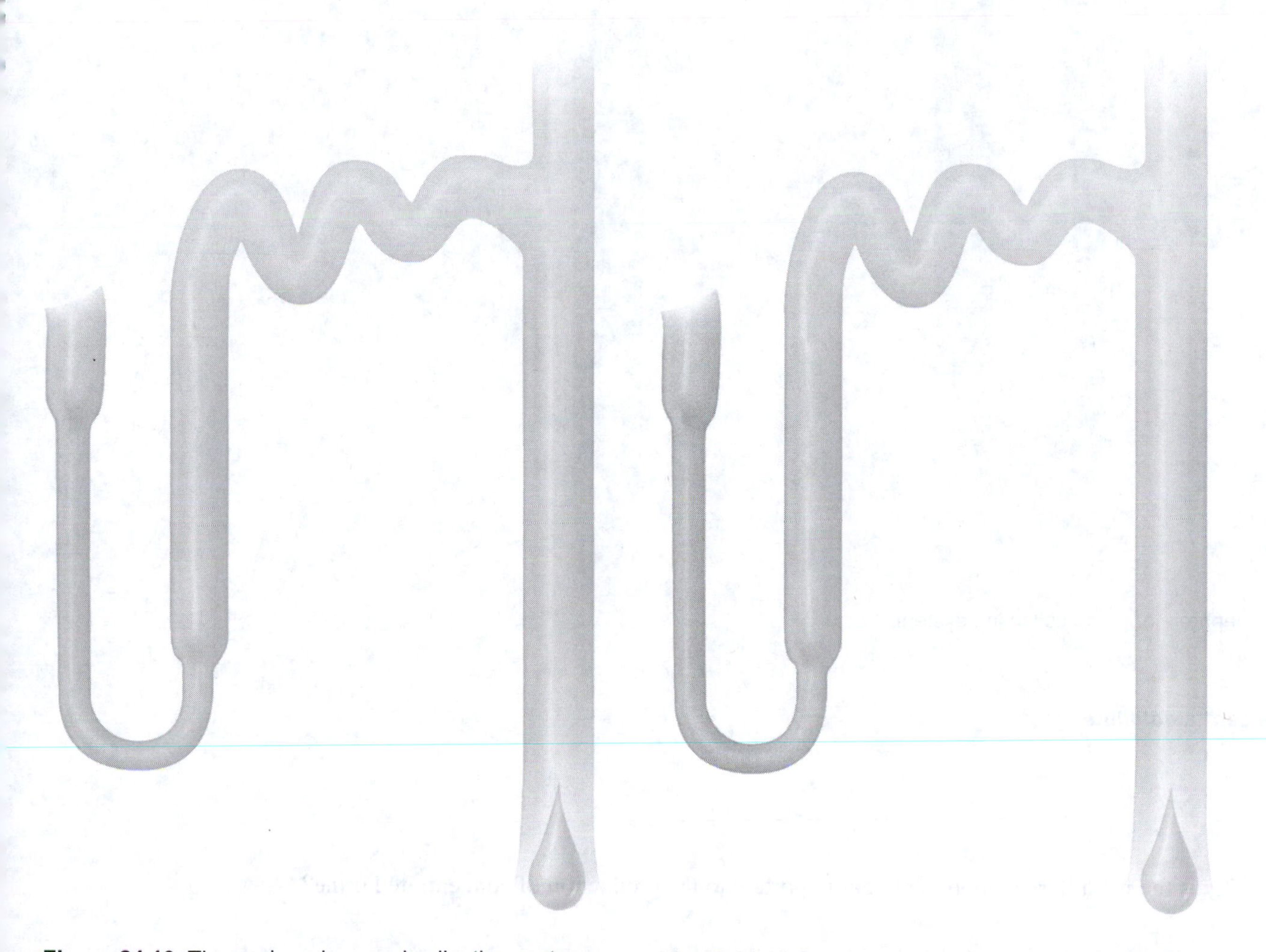

Figure 24.10 The nephron loop and collecting system.

Figure 24.11 The nephron loop and collecting system.

Urine will be: concentrated/dilute

Urine will be: concentrated/dilute

Scenario 3: Water reabsorption in the thin descending limb is blocked. In Figure 24.12, draw in the solute dots in the loop, collecting system, and the interstitial fluid, and determine if the urine produced will be dilute or concentrated.

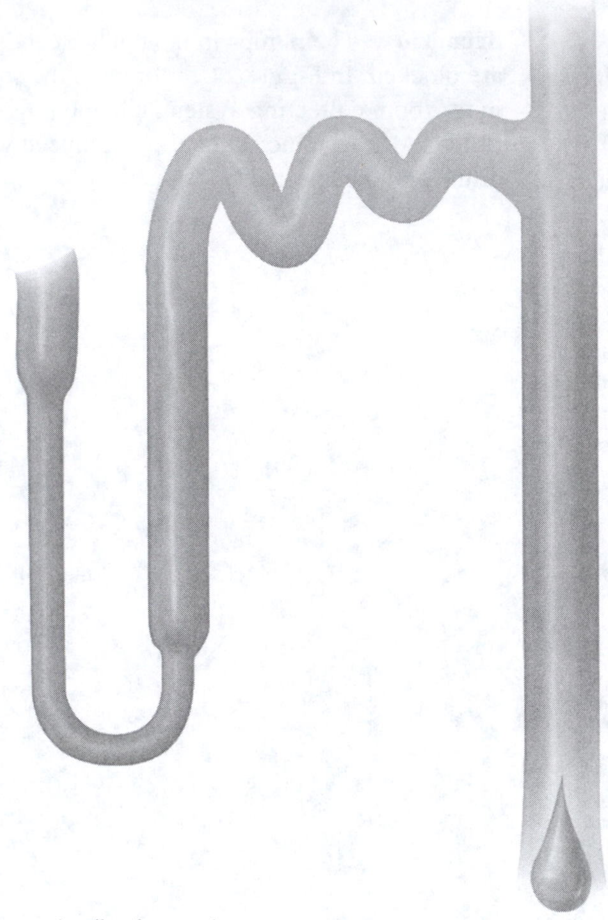

Figure 24.12 The nephron loop and collecting system.

Urine will be: concentrated/dilute

Key Concept: Why is the medullary osmotic gradient important to the production of concentrated urine?

____

____

Key Concept: How do the vasa recta and countercurrent exchange maintain the medullary osmotic gradient?

____

____

Module 24.7: Putting It All Together: The Big Picture of Renal Physiology

Now we will look at the big picture of the renal physiology presented in Modules 24.3 through 24.6. At the end of this module, you should be able to do the following:

1. Describe the overall process by which blood is filtered and filtrate is modified to produce urine.

Describe It: The Big Picture of Renal Physiology

Label and describe, in your own words, the steps of the big picture of renal physiology, illustrated in Figure 24.13. In addition, label and color-code key components of these processes.

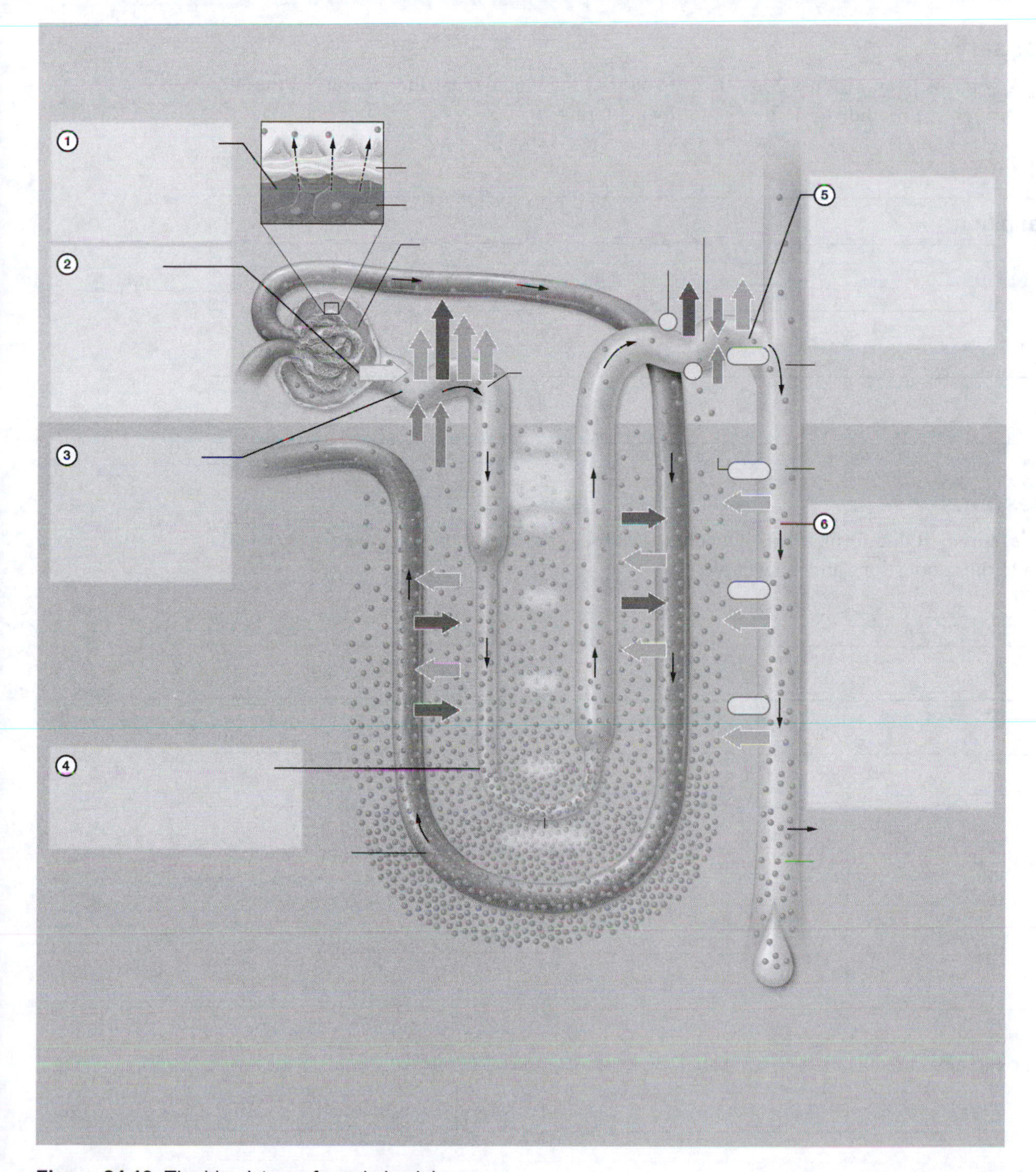

Figure 24.13 The big picture of renal physiology.

Module 24.8: Urine and Renal Clearance

Once the filtrate has been fully modified and leaves the papillary ducts, it becomes urine. This module examines the final composition of urine and how the components of urine can be used to assess an individual's overall health. At the end of this module, you should be able to do the following:

1. Explain how the physical and chemical properties of a urine sample are determined and relate these properties to normal urine composition.
2. Explain how filtration, reabsorption, and secretion determine the rate of excretion of any solute.
3. Explain how renal clearance rate can be used to measure the GFR.

Build Your Own Glossary

Following is a table listing key terms from Module 24.8. Before you read the module, use the glossary at the back of your book or look through the module to define the following terms.

Key Terms for Module 24.8

Term	Definition
Urinalysis	
Renal clearance	
Creatinine	

Survey It: Form Questions

Before you read the module, survey it and form at least two questions for yourself. When you have finished reading the module, return to these questions and answer them.

Question 1: ______________________________

Answer: ______________________________

Question 2: ______________________________

Answer: ______________________________

Key Concept: What are the normal components of urine? What should not be present in the urine under normal conditions?

____

____

Key Concept: How can renal clearance be used to estimate the glomerular filtration rate?

____

____

Module 24.9: Urine Transport, Storage, and Elimination

The final module of this chapter looks at what happens to urine after it leaves the kidneys and the organs that transport and store it. By the end of this module, you should be able to do the following:

1. Describe the structure and functions of the ureters, urinary bladder, and urethra.
2. Relate the anatomy and histology of the bladder to its function.
3. Compare and contrast the male and female urinary tracts.
4. Describe the micturition reflex.
5. Describe voluntary control of micturition.

Build Your Own Glossary

Following is a table listing key terms from Module 24.9. Before you read the module, use the glossary at the back of your book or look through the module to define the following terms.

Key Terms for Module 24.9

Term	Definition
Ureter	
Urinary bladder	
Detrusor muscle	
Trigone	
Urethra	
Micturition	
Micturition reflex	

Survey It: Form Questions

Before you read the module, survey it and form at least three questions for yourself. When you have finished reading the module, return to these questions and answer them.

Question 1: ______________________________

Answer: ______________________________

Question 2: ______________________________

Answer: ______________________________

Question 3: ______________________________

Answer: ______________________________

Identify It: Anatomy of the Urinary Tract

Figure 24.14 illustrates the anatomy of the male and female urinary tracts. Label and color each component.

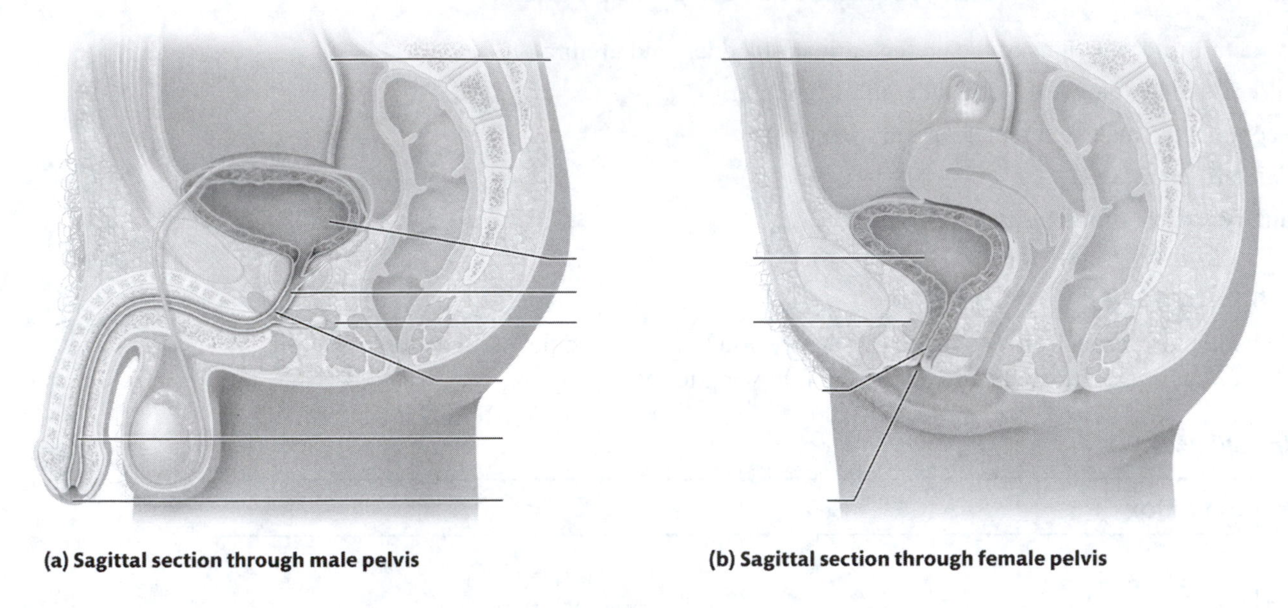

Figure 24.14 The male and female urinary tracts, midsagittal section.

Key Concept: How do the male and female urinary tracts differ?

____

____

Describe It: Micturition

Figure 24.15 illustrates the neural mechanisms that control micturition. Label, draw in the appropriate arrows, and fill in the text for each step in your own words.

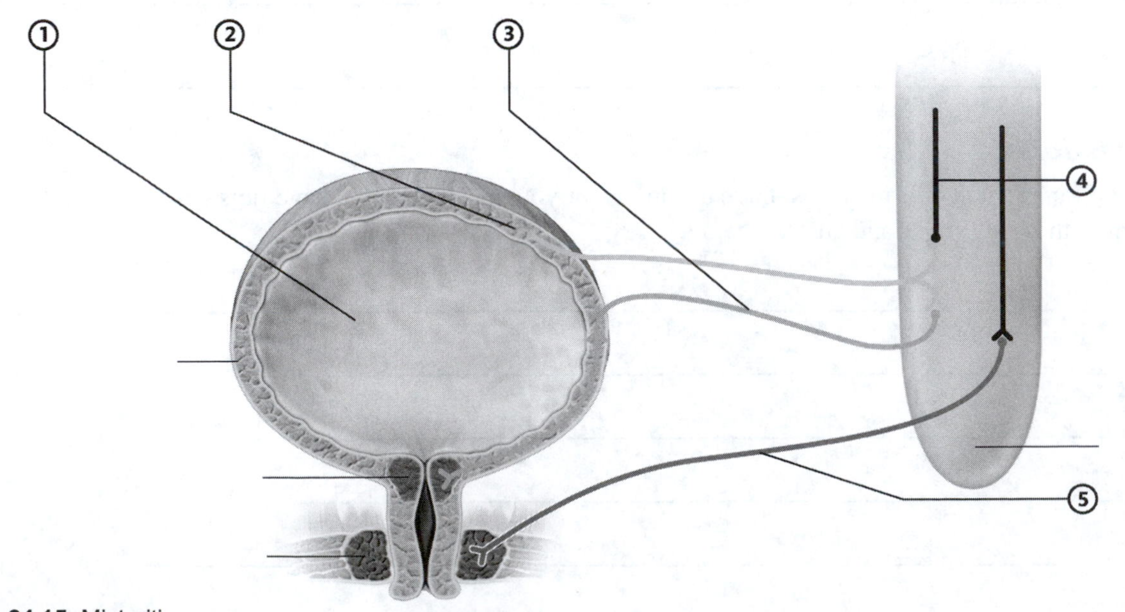

Figure 24.15 Micturition.

Key Concept: What part of the nervous system mediates the micturition reflex? How does the central nervous system exert voluntary control over micturition?

____

____

What Do You Know Now?

Let's now revisit the questions you answered in the beginning of this chapter. How have your answers changed now that you've worked through the material?

- What are the main functions of the urinary system?

  ____

- What do the kidneys do?

  ____

- What stimulates urination?

  ____

25 Fluid, Electrolyte, and Acid-Base Homeostasis

We now examine fluid, electrolyte, and acid-base homeostasis. Although you are already familiar with some of this information from the chemistry chapter (and other chapters, too), this chapter covers body fluids, electrolytes, and acid-base balance in detail, including their homeostasis and potential imbalances.

What Do You Already Know?

Try to answer the following questions before proceeding to the next section. If you're unsure of the correct answers, give it your best attempt based on previous courses, previous chapters, or just your general knowledge.

- What do the terms "hypotonic" and "hypertonic" mean?

  ____

- What are the most common electrolytes in the human body, and what does the term "electrolyte" mean?

  ____

- What are acids, bases, and buffers?

  ____

Module 25.1: Overview of Fluid, Electrolyte, and Acid-Base Homeostasis

Module 25.1 in your text introduces you to the general concepts of fluid, electrolyte, and pH homeostasis. By the end of the module, you should be able to do the following:

1. Explain the concept of balance with respect to fluids and electrolytes and acids and bases.
2. Define the terms body fluid, electrolyte, acid, base, pH scale, and buffer.

Build Your Own Glossary

Below is a table listing key terms from Module 25.1. Before you read the module, use the glossary at the back of your book or look through the module to define the following terms.

Key Terms for Module 25.1

Term	Definition
Fluid balance	
Electrolyte	

Term	Definition
Acid	
Base	

Survey It: Form Questions

Before you read the module, survey it and form at least two questions for yourself. When you have finished reading the module, return to these questions and answer them.

Question 1: ____________________

Answer: ____________________

Question 2: ____________________

Answer: ____________________

Key Concept: What is the principle of mass balance?

____

____

Draw It: Acids and Bases

In the boxes provided, draw what happens when an HCl and an NaOH are mixed together. Label your drawings with the following terms, where appropriate: base, acid, salt, and water.

	+	→		+

Module 25.2: Fluid Homeostasis

Now we look at water as a major component of the body and at regulation of its intake and output. When you finish this module, you should be able to do the following:

1. Describe the fluid compartments, and explain how each contributes to the total body water.
2. Compare and contrast the relative concentrations of major electrolytes in intracellular and extracellular fluids.
3. Explain how osmotic pressure is generated, and compare and contrast the roles that hydrostatic and osmotic pressures play in the movement of water between fluid compartments.
4. Describe the routes of water gain in and loss from the body.
5. Describe the mechanisms that regulate water intake and output, and explain how dehydration and overhydration develop.

Build Your Own Glossary

Following is a table listing key terms from Module 25.2. Before you read the module, use the glossary at the back of your book or look through the module to define the following terms.

Key Terms for Module 25.2

Term	Definition
Total body water	
Intracellular compartment	
Extracellular compartment	
ECF	
Obligatory water loss	
Sensible water loss	
Insensible water loss	
Osmoreceptors	
ADH	
Dehydration	
Overhydration	
Edema	

Survey It: Form Questions

Before you read the module, survey it and form at least three questions for yourself. When you have finished reading the module, return to these questions and answer them.

Question 1: ______________________________

Answer: ______________________________

Question 2: ______________________________

Answer: ______________________________

Question 3: ______________________________

Answer: ______________________________

Complete It: Movement of Water between Compartments

Fill in the blanks to complete the following paragraph that describes the ways in which water moves between intercellular fluid and extracellular fluid.

A hydrostatic pressure ____________ tends to push water away from an area of ____________ hydrostatic pressure to one with ____________ hydrostatic pressure. An osmotic pressure ____________ pulls water toward the solution with the more ____________ solution. Osmotic pressure is determined by its ____________ or the number of ____________ particles present. A hypotonic ECF causes a cell to ____________ water, since its osmotic pressure is ____________ than that of the cytosol. A hypertonic ECF causes a cell to ____________ water, since its osmotic pressure is ____________ than that of the cytosol.

Key Concept: What is obligatory water loss, and why is it necessary?

____

____

Build Your Own Summary Table: Factors That Influence Water Loss and Water Gain

As you read Module 25.2, build your own summary table about the different factors that influence water loss and water gain by filling in the information in the table below. Note that you should put a minus sign in front of the "output" amount.

Factors That Influence Water Loss and Water Gain

Factor	Description	Input/output	Total
Sensible water loss		(ml)	(ml)
Insensible water loss		(ml)	
Catabolism		(ml)	(ml)
Food		(ml)	
Water consumed		(ml)	

Key Concept: What is the most important hormone involved in fluid balance, and what does it do?

____

____

Describe It: Imbalances of Fluid Homeostasis

Write a paragraph describing the imbalances of fluid homeostasis in simple terms.

Module 25.3: Electrolyte Homeostasis

Next we look at the regulation of electrolyte levels in body fluids. This module explores how the body maintains homeostasis of crucial electrolytes such as sodium, potassium, and calcium ions. When you complete this module, you should be able to do the following:

1. Describe the function of the most prevalent electrolytes found in body fluids.
2. Describe the hormonal regulation of electrolyte levels in the plasma.
3. Explain how calcium ion regulation is related to phosphate ions.

Build Your Own Glossary

Below is a table listing key terms from Module 25.3. Before you read the module, use the glossary at the back of your book or look through the module to define the following terms.

Key Terms of Module 25.3

Term	Definition
Hypernatremia	
Hyponatremia	
Hyperkalemia	
Hypokalemia	
Hypercalcemia	
Hypocalcemia	

Survey It: Form Questions

Before you read the module, survey it and form at least three questions for yourself. When you have finished reading the module, return to these questions and answer them.

Question 1: ______________________________

Answer: ______________________________

Question 2: ______________________________

Answer: ______________________________

Question 3: ____________________

Answer: ____________________

Key Concept: What effect would sodium and potassium ion imbalances have on membrane potentials?

____

____

Describe It: The Renin-Angiotensin-Aldosterone System

Write in the steps of the renin-angiotensin-aldosterone system in Figure 25.1. You may use text Figure 25.7 as a reference, but write the steps in your own words.

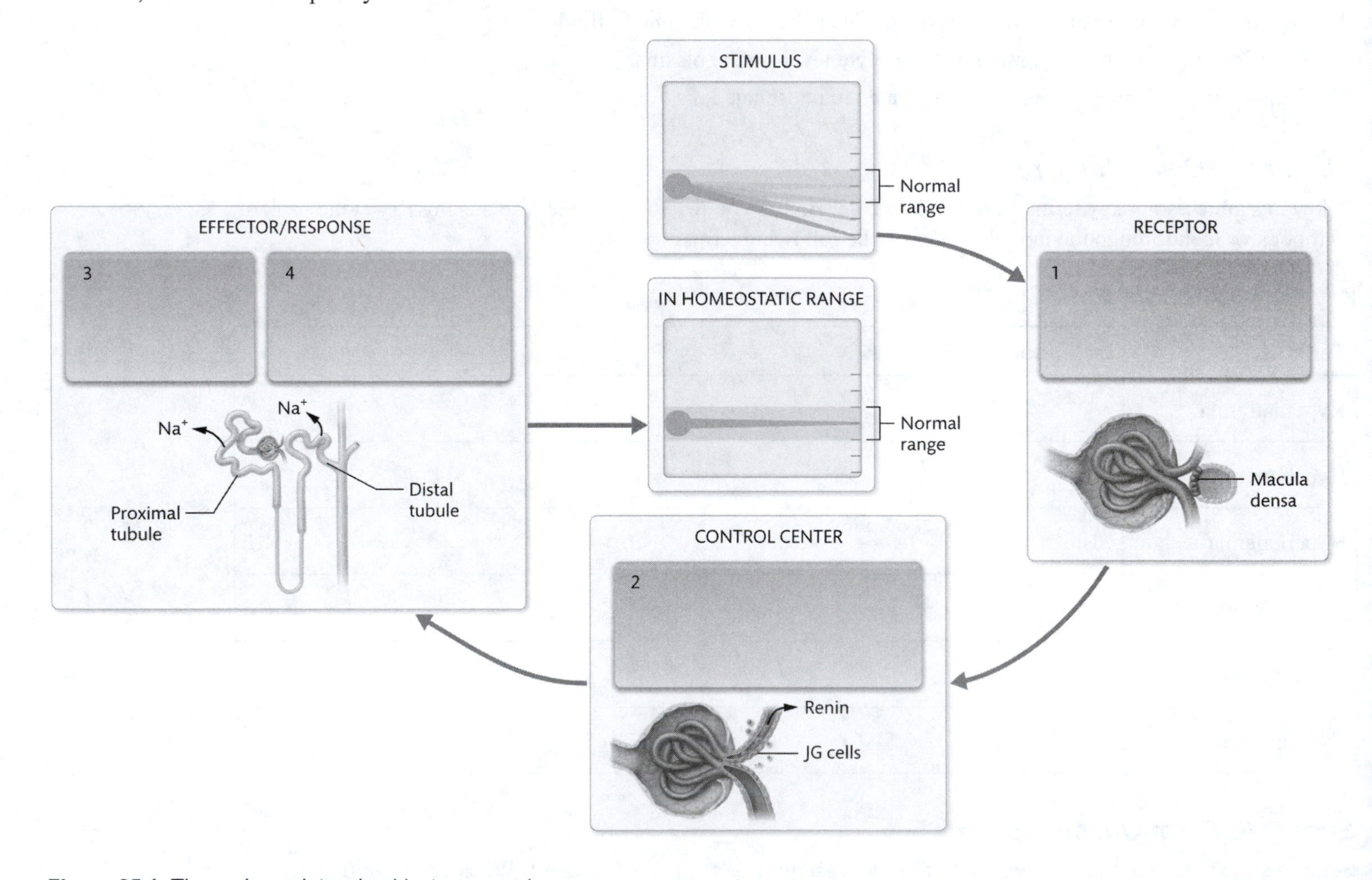

Figure 25.1 The renin-angiotensin-aldosterone system.

Key Concept: Why is hyperkalemia potentially one of the most dangerous electrolyte imbalances?

____

____

Describe It: Calcium Homeostasis

Write a paragraph describing the roles played by parathyroid hormone (PTH) and vitamin D3 (calcitriol) in calcium homeostasis.

Key Concept: What important roles do magnesium ions play in the human body?

____

____

Module 25.4: Acid-Base Homeostasis

The body must maintain a pH range of about 7.35–7.45. Module 25.4 in your text explores the mechanisms by which body fluids maintain this important, narrow pH range. At the end of this module, you should be able to do the following:

1. Explain the factors that determine the pH of blood, and describe how it is maintained within its normal range.
2. Describe the buffer systems that help to keep the pH of the body's fluids stable.
3. Describe the relationship of P_{CO_2} and bicarbonate ions to blood pH.
4. Describe the role of the respiratory system in regulating blood pH, and predict how hypo- and hyperventilation will affect blood pH.
5. Explain the mechanisms by which the kidneys secrete or retain hydrogen and bicarbonate ions, and describe how these processes affect blood pH.
6. Discuss the concept of compensation to correct respiratory and metabolic acidosis and alkalosis.

Build Your Own Glossary

Below is a table listing key terms from Module 25.4. Before you read the module, use the glossary at the back of your book or look through the module to define the following terms.

Key Terms of Module 25.4

Term	Definition
Buffer systems	
Volatile acid	
Fixed acids	

Term	Definition
Protein buffer system	
Acidosis	
Alkalosis	

Survey It: Form Questions

Before you read the module, survey it and form at least two questions for yourself. When you have finished reading the module, return to these questions and answer them.

Question 1: ______________________________

Answer: ______________________________

Question 2: ______________________________

Answer: ______________________________

Key Concept: What is the major source of metabolic acids in the body, and what is its origin?

____

____

Draw It: Carbonic Acid–Bicarbonate Ion Buffer System

In the boxes provided, draw the carbonic acid–bicarbonate ion buffer system by inserting these chemical formulas in the appropriate boxes: CO_2, HCO_3^-, H_2CO_3, H_2O, and H^+. Label your boxes with the following terms where appropriate: Carbon dioxide, Carbonic acid, Bicarbonate ion, Water, and Hydrogen ion.

The Carbonic Acid–Bicarbonate Ion Buffer System

	+	⟷		⟷		+

Key Concept: What happens when an acid is added to a solution buffered with bicarbonate ions?

____

____

Describe It: Secretion of Hydrogen Ions in the Kidney Tubules

Write in the steps in secretion of hydrogen ions in the kidney tubules in Figure 25.2. You may use text Figure 25.11 as a reference, but write the steps in your own words.

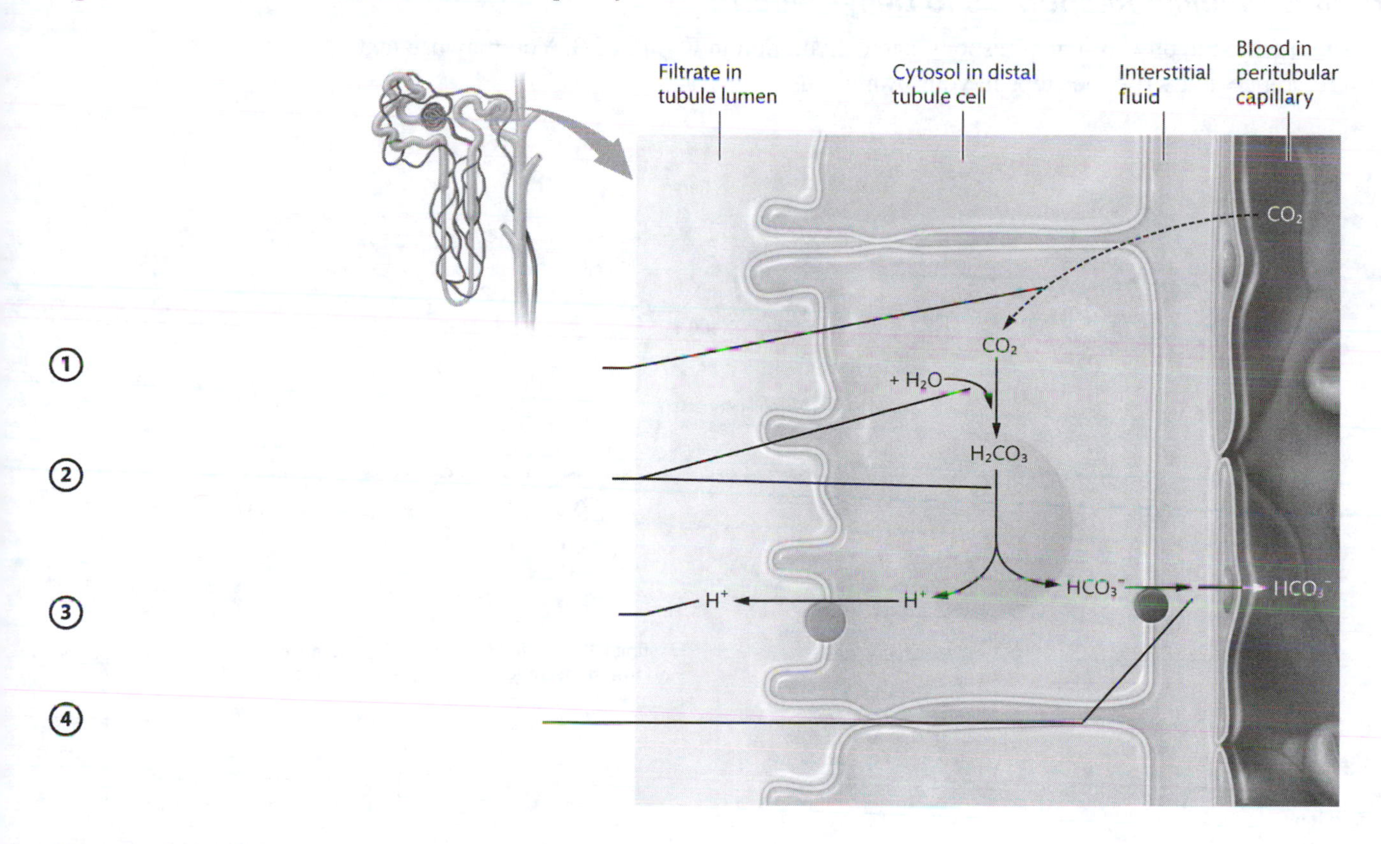

Figure 25.2 Secretion of hydrogen ions in the kidney tubules.

Key Concept: What are the main differences between respiratory acidosis and metabolic acidosis?

____

____

Describe It: Respiratory Alkalosis and Compensation

Write a paragraph describing respiratory alkalosis and compensation as if you were explaining to an audience of nonscientists.

Module 25.5: An Example of Fluid, Electrolyte, and Acid-Base Homeostasis

This module ties together the main concepts of fluid, electrolyte, and acid-base homeostasis with an example. At the end of this module, you should be able to do the following:

1. Provide specific examples to demonstrate how the cardiovascular, endocrine, and urinary systems respond to maintain homeostasis of fluid volume, electrolyte concentration, and pH in the body.

Describe It: Physiological Responses to Dehydration

Write in the steps related to the physiological responses to dehydration in Figure 25.3. You may use text Figure 25.12 as a reference, but write the steps in your own words.

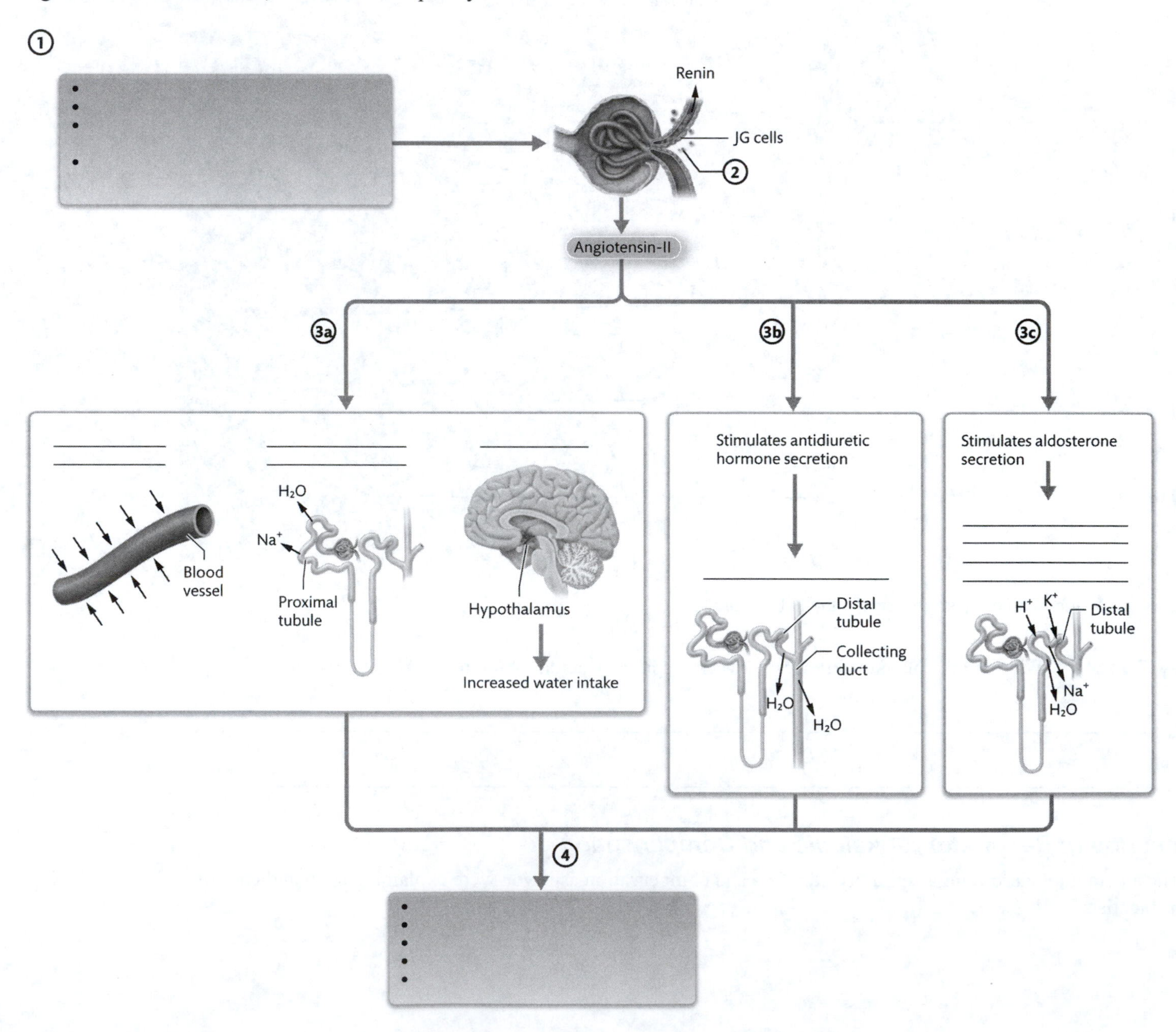

Figure 25.3 Physiological responses to dehydration.

What Do You Know Now?

Let's now revisit the questions you answered in the beginning of this chapter. How have your answers changed now that you've worked through the material?

- What do the terms "hypotonic" and "hypertonic" mean?

____

- What are the most common electrolytes in the human body, and what does the term "electrolyte" mean?

____

- What are acids, bases, and buffers?

____

26 The Reproductive System

We now turn to the reproductive system—a system more necessary for the survival of the species than the individual. This chapter covers the anatomy and physiology of the male and female reproductive systems, as well as related topics like birth control and sexually transmitted diseases.

What Do You Already Know?

Try to answer the following questions before proceeding to the next section. If you're unsure of the correct answers, give it your best attempt based on previous courses, previous chapters, or just your general knowledge.

- What are the main organs and glands of the male and female reproductive systems, and what are their basic functions?

  ____

- What hormones are produced by the glands of the male and female reproductive systems, and how do they affect men and women, respectively?

  ____

- What are the common methods of birth control and common types of sexually transmitted infections?

  ____

Module 26.1: Overview of the Reproductive System and Meiosis

Module 26.1 in your text introduces you to terminology and a basic overview of the reproductive system. It also covers meiosis, the type of cell division needed to produce sperm and eggs. By the end of the module, you should be able to do the following:

1. Compare and contrast the basic structure and function of the male and female reproductive systems.
2. Contrast the overall processes of mitosis and meiosis.
3. Describe the stages of meiosis.

Build Your Own Glossary

Following is a table listing key terms from Module 26.1. Before you read the module, use the glossary at the back of your book or look through the module to define the following terms.

Key Terms for Module 26.1

Term	Definition
Gonads	
Gametes	
Zygote	
Meiosis	
Haploid	
Diploid	
Alleles	
Sister chromatids	
Tetrads	
Crossing over	
Independent assortment	

Survey It: Form Questions

Before you read the module, survey it and form at least two questions for yourself. When you have finished reading the module, return to these questions and answer them.

Question 1: ______________________________

Answer: ______________________________

Question 2: ______________________________

Answer: ______________________________

Key Concept: Why is meiosis necessary for cells destined to become gametes?

____

____

Draw It: Mitosis and Meiosis

In the boxes provided, draw a side-by-side comparison of mitosis and meiosis. Label your drawings with the following terms, where appropriate: prophase, metaphase, anaphase, telophase, prophase 1, metaphase 1, anaphase 1, telophase 1, prophase 2, metaphase 2, anaphase 2, telophase 2, mother cells, and daughter cells. Note that some labels may be used twice. (*Hint:* If you get stuck, you may refer to Figure 26.2 in the text, but try to do this on your own as much as possible.)

Mitosis	**Meiosis**
	Meiosis 1
	Meiosis 2

Module 26.2: Anatomy of the Male Reproductive System

Now we look at the organs of the male reproductive system. When you finish this module, you should be able to do the following:

1. Describe the structure and functions of the male reproductive system.
2. Trace the pathway that sperm travel from the seminiferous tubules to the external urethral orifice of the penis.
3. Describe the organs involved in semen production.
4. Discuss the composition of semen and its role in sperm function.

Build Your Own Glossary

Following is a table listing key terms from Module 26.2. Before you read the module, use the glossary at the back of your book or look through the module to define the following terms.

Key Terms for Module 26.2

Term	Definition
Testes	
Scrotum	
Seminiferous tubules	
Interstitial cells	
Epididymis	
Ductus deferens	
Urethra	
Penis	
Seminal vesicles	
Prostate gland	
Semen	
Spermatic cord	

Survey It: Form Questions

Before you read the module, survey it and form at least two questions for yourself. When you have finished reading the module, return to these questions and answer them.

Question 1: ______________________________

Answer: ______________________________

Question 2: ______________________________

Answer: ______________________________

Key Concept: What are the two main functions of the testes? Which cells are responsible for performing each of those functions?

____

____

Identify It: Internal Structures of the Testis and Epididymis

Identify and color-code each component of the internal structures of the testis and epididymis in Figure 26.1. Then, list the main function(s) of each component.

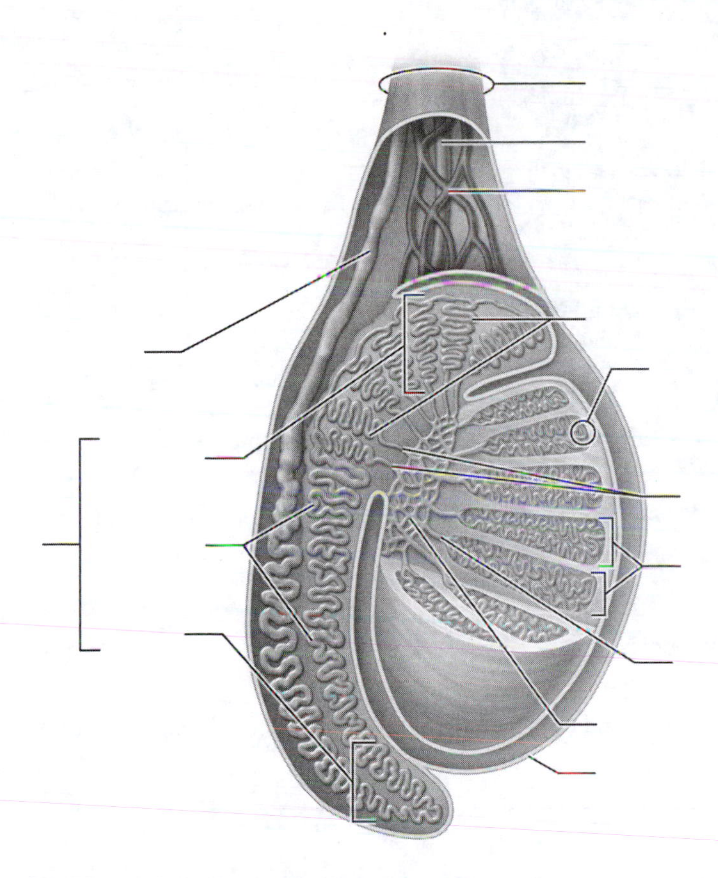

Figure 26.1 Internal structures of the testis and epididymis.

Key Concept: What are the main prostatic secretions, and what are their functions?

____

____

Identify It: Male Reproductive Duct System and Penis

Identify and color-code each component of the male reproductive duct system and penis in Figure 26.2. Then, list the main function(s) of each component.

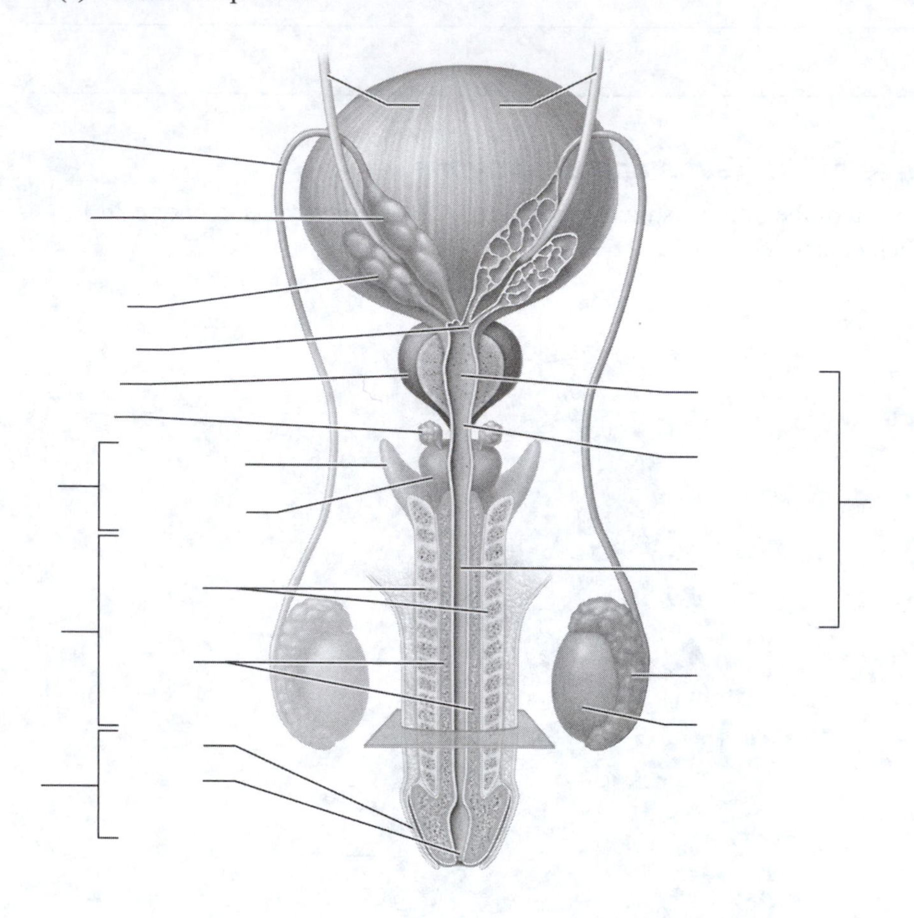

Figure 26.2 Male reproductive duct system and penis.

Key Concept: What are the coagulating proteins and enzymes found in semen, and why are they important?

____

____

Module 26.3: Physiology of the Male Reproductive System

Now that we have examined the anatomy of the male reproductive system, we turn to how those structures produce sperm and deliver them to the female reproductive tract. When you complete this module, you should be able to do the following:

1. Relate the general stages of meiosis to the specific process of spermatogenesis.
2. Discuss endocrine regulation of spermatogenesis.
3. Discuss the events and endocrine regulation of male puberty.
4. Describe the male sexual response.
5. Describe male secondary sex characteristics and their role in reproductive system function.
6. Explain the effects of aging on reproductive function in males.

Build Your Own Glossary

Following is a table listing key terms from Module 26.3. Before you read the module, use the glossary at the back of your book or look through the module to define the following terms.

Key Terms of Module 26.3

Term	Definition
Spermatogenesis	
Spermatocytes	
Sustentacular cells	
Spermiogenesis	
Acrosome	
Erection	
Ejaculation	
Puberty	

Survey It: Form Questions

Before you read the module, survey it and form at least two questions for yourself. When you have finished reading the module, return to these questions and answer them.

Question 1: ______________________

Answer: ______________________

Question 2: ______________________

Answer: ______________________

Describe It: Spermatogenesis in the Seminiferous Tubules

Describe the sequence of events of spermatogenesis in the seminiferous tubules in Figure 26.3. In addition, label and color-code key structures of the process. You may use text Figure 26.7 as a reference, but write the steps in your own words.

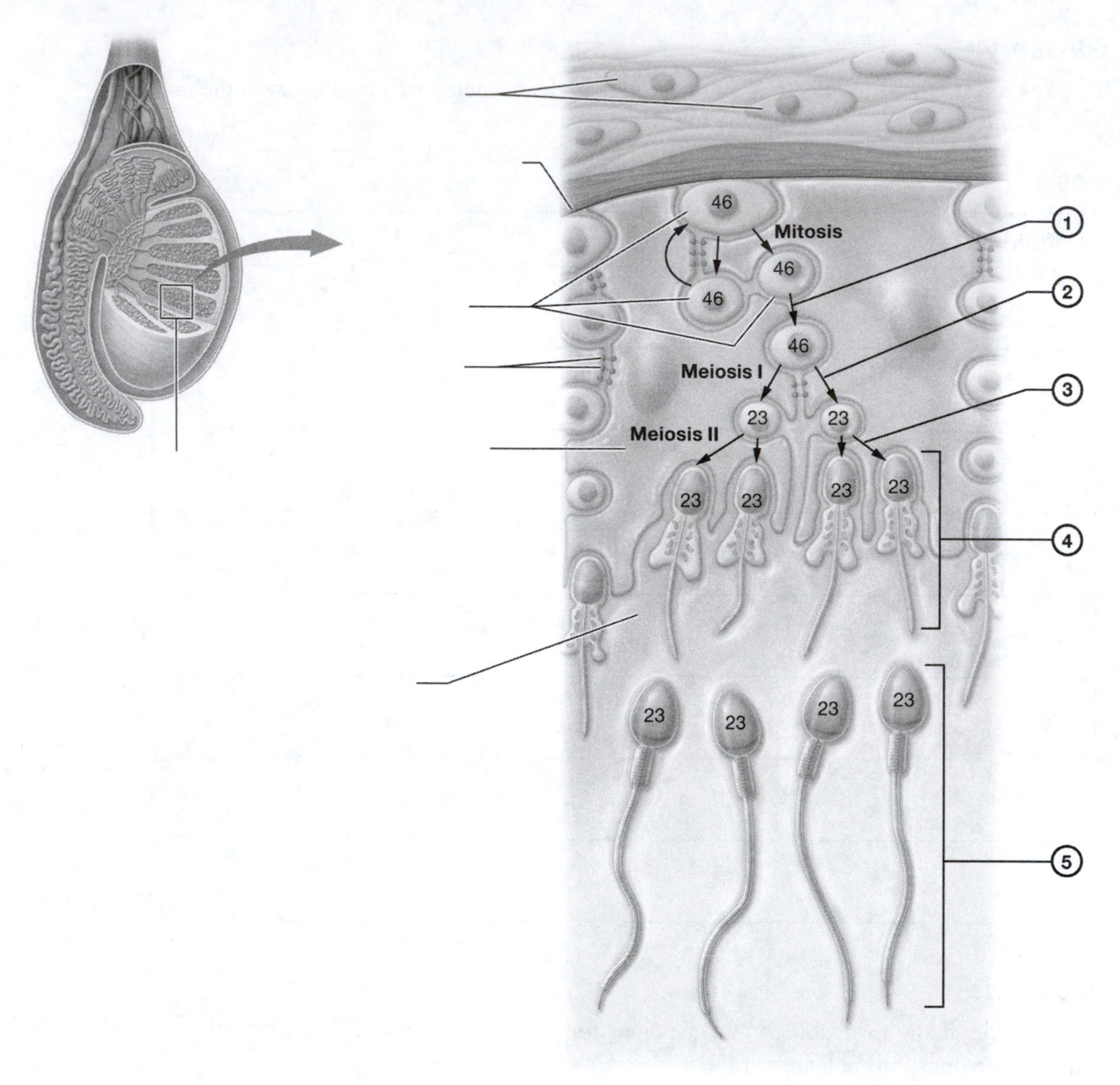

Figure 26.3 Spermatogenesis in the seminiferous tubules.

Key Concept: What are the main functions of sustentacular cells of the testis?

___

___

Describe It: Hormonal Control of Male Reproduction

Describe the steps involved in hormonal regulation of testicular function via the hypothalamic-pituitary-gonadal axis in Figure 26.4. In addition, label and color-code key structures of the process. You may use text Figure 26.9 as a reference, but write the steps in your own words.

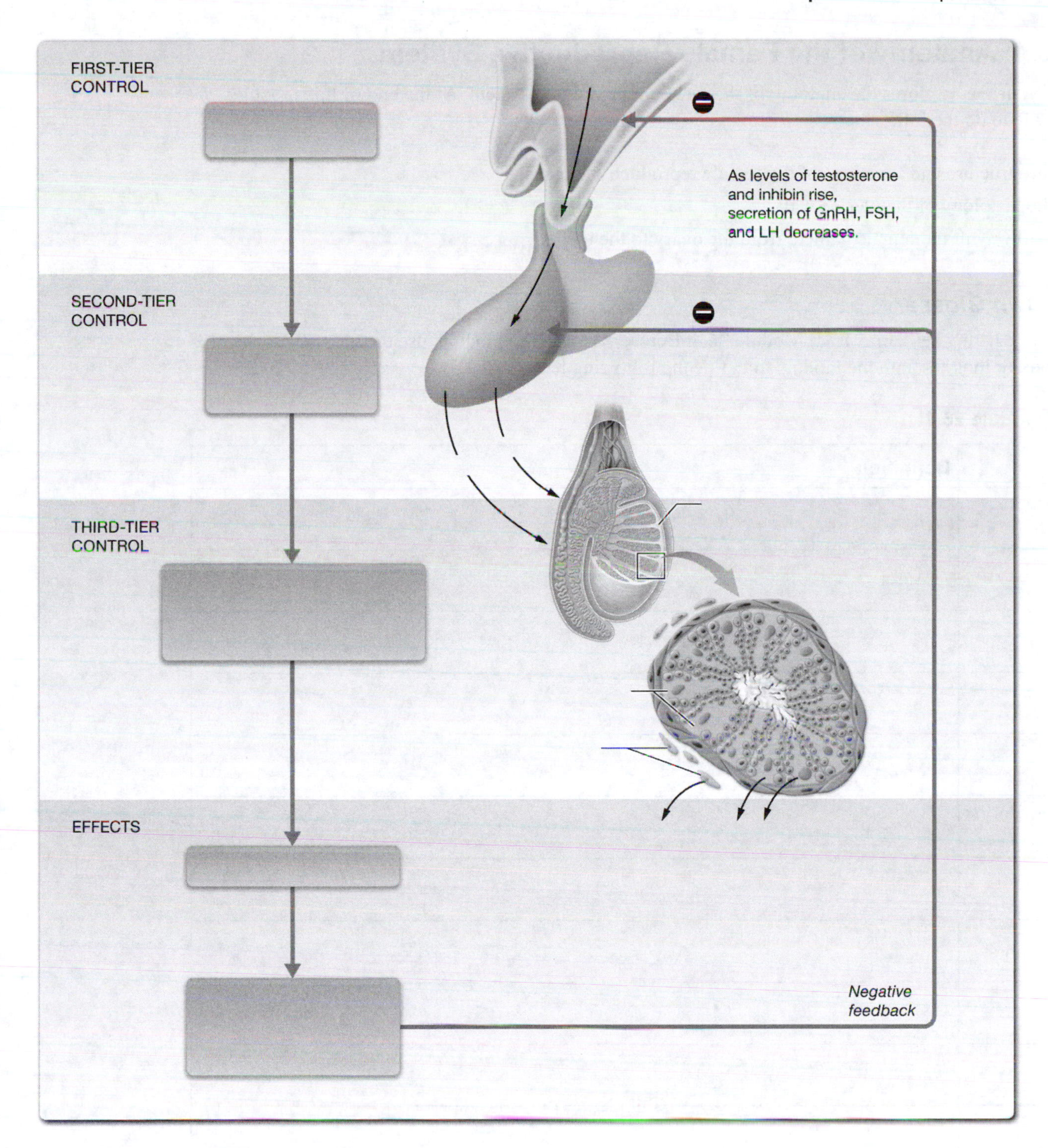

Figure 26.4 Hormonal regulation of testicular function.

Key Concept: What does the term "male secondary sex characteristics" mean? What are three examples of male secondary sex characteristics?

____

____

Module 26.4: Anatomy of the Female Reproductive System

Module 26.4 in your text explores the anatomy of the female reproductive system. At the end of this module, you should be able to do the following:

1. Describe the structure and functions of the female reproductive organs.
2. Describe the histology of the uterine wall.
3. Trace the pathway of the female gamete from the ovary to the uterus.

Build Your Own Glossary

Following is a table listing key terms from Module 26.4. Before you read the module, use the glossary at the back of your book or look through the module to define the following terms.

Key Terms of Module 26.4

Term	Definition
Ovary	
Uterine tubes	
Uterus	
Cervix	
Myometrium	
Endometrium	
Vagina	
Labia majora/minora	
Clitoris	
Perineum	
Mammary glands	

Survey It: Form Questions

Before you read the module, survey it and form at least three questions for yourself. When you have finished reading the module, return to these questions and answer them.

Question 1: ______________________________

Answer: ______________________________

Question 2: ______________________________

Answer: ______________________________

Question 3: ______________________________

Answer: ______________________________

Key Concept: What are the major ligaments that attach and hold the parts of the female reproductive system in the pelvic cavity?

____

____

Identify It: Internal Organs of the Female Reproductive System

Identify and color-code each component of the internal organs of the female reproductive system in Figure 26.5. Then, list the main function(s) of each component.

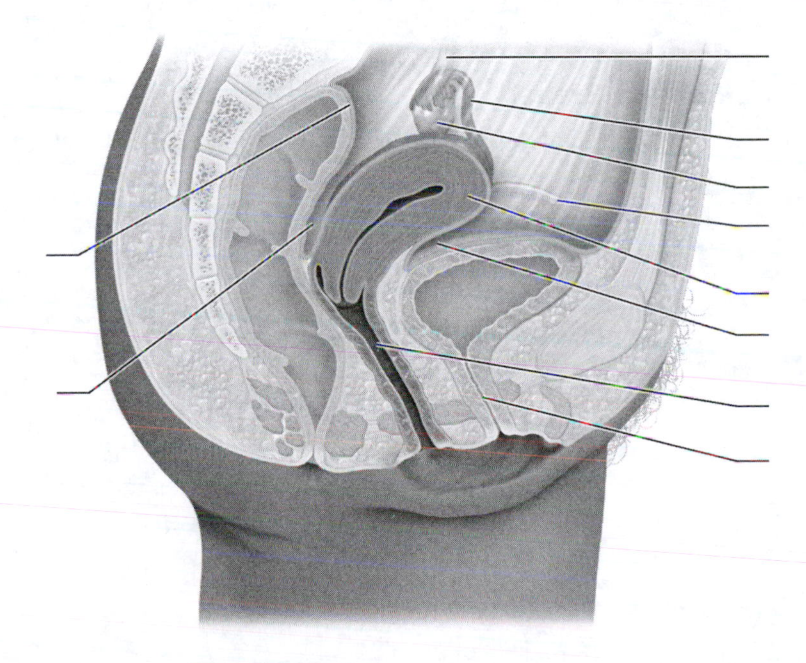

Figure 26.5 Internal organs of the female reproductive system.

Describe It: The Travels of the Oocyte

Write a paragraph describing how an ovulated oocyte is moved toward the uterus as if you were explaining to an audience of nonscientists.

Key Concept: How does the structure of the myometrium differ from that of the endometrium? How does this reflect their functions?

____

____

Describe It: External Female Genitalia

Write a paragraph describing the external female genitalia as if you were explaining to an audience of nonscientists.

Identify It: Internal Anatomy of the Female Breast

Identify and color-code each component of the internal anatomy of the female breast in Figure 26.6. Then, list the main function(s) of each component.

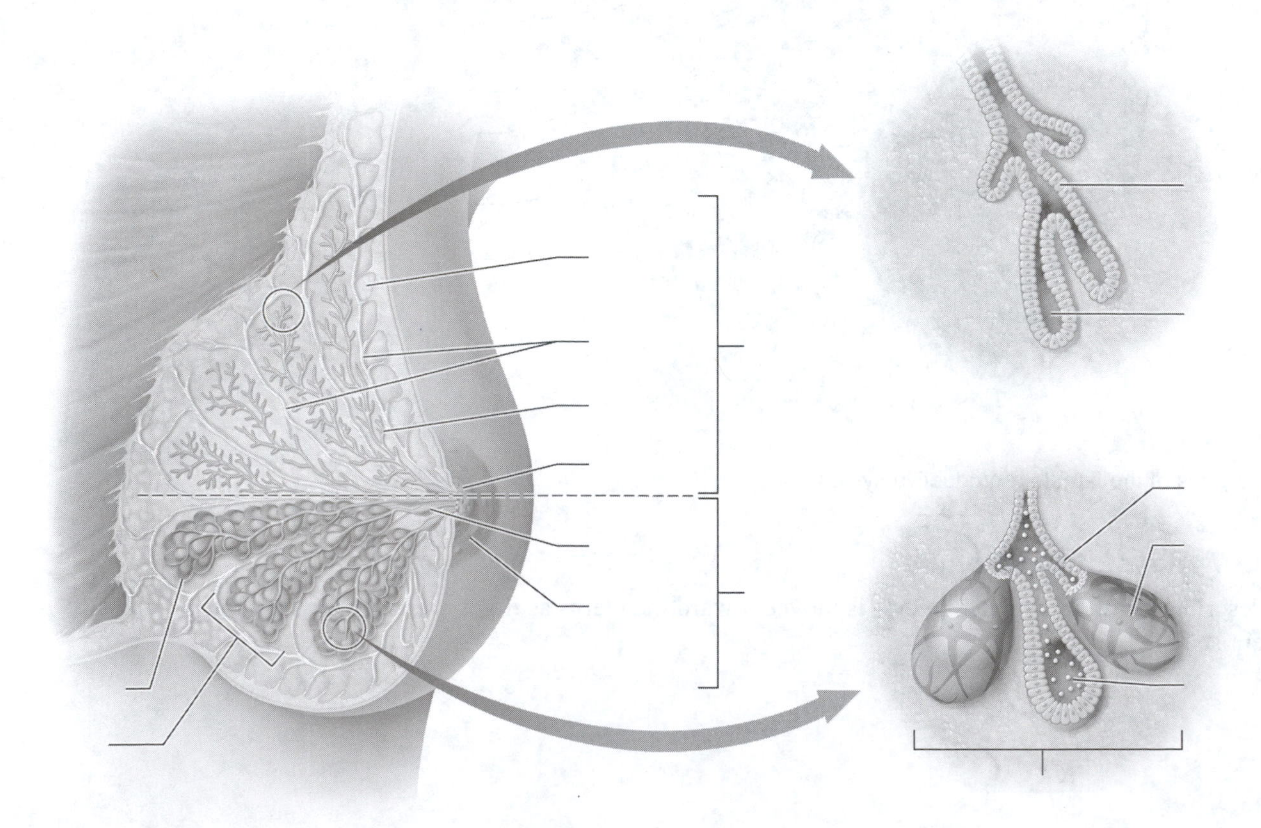

Figure 26.6 Structure of the non-lactating and lactating breast.

Module 26.5: Physiology of the Female Reproductive System

This module turns to female gamete production, or oogenesis, and then examines the female ovarian and uterine cycles (and the hormones that control them). At the end of this module, you should be able to do the following:

1. Relate the general stages of meiosis to the specific process of oogenesis.
2. Describe the events of the ovarian cycle and the uterine cycle.
3. Discuss endocrine regulation of oogenesis.
4. Analyze graphs depicting the typical female monthly sexual cycle and correlate ovarian activity, hormonal changes, and uterine events.
5. Describe female secondary sex characteristics and their role in reproductive system function.
6. Describe the physiological changes associated with menopause, and explain the fertility changes that precede menopause.

Build Your Own Glossary

Following is a table listing key terms from Module 26.5. Before you read the module, use the glossary at the back of your book or look through the module to define the following terms.

Key Terms of Module 26.5

Term	Definition
Oogenesis	
Polar body	
Ovarian follicle	
Corpus luteum	
Menstruation	
Human chorionic gonadotropin	
Menarche	
Menopause	

Survey It: Form Questions

Before you read the module, survey it and form at least two questions for yourself. When you have finished reading the module, return to these questions and answer them.

Question 1: ______________________________

Answer: ______________________________

Question 2: ______________________________

Answer: ______________________________

Key Concept: What are at least two ways that oogenesis is different from spermatogenesis?

____

____

Describe It: The Ovarian Cycle

Describe the sequence of events of the ovarian cycle in Figure 26.7. Please also note, in brackets, which steps are part of the follicular phase, ovulation phase, and luteal phase. In addition, label and color-code key structures of the process. You may use text Figure 26.15 as a reference, but write the steps in your own words.

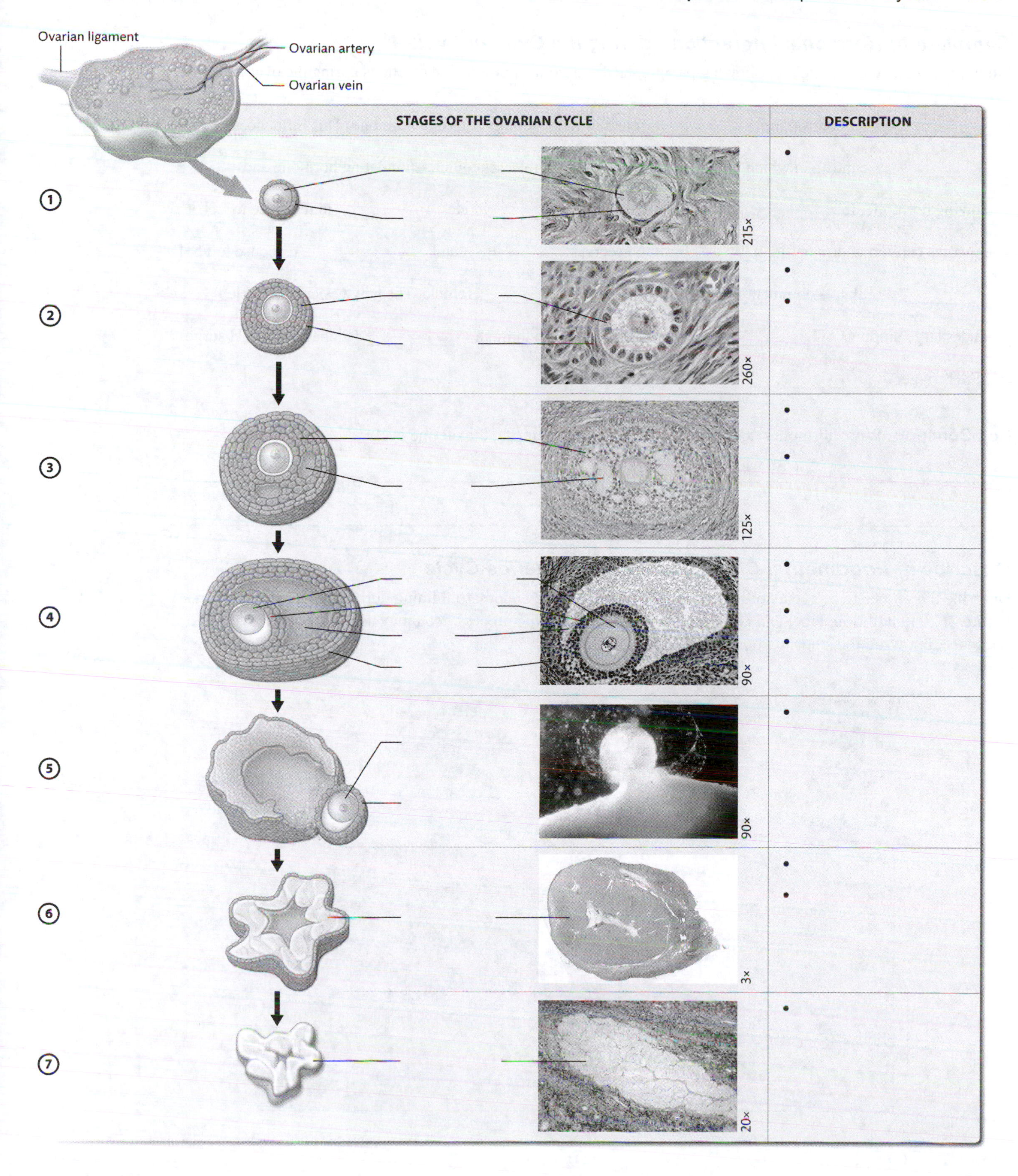

Figure 26.7 The ovarian cycle.

Complete It: Hormonal Interactions during the Ovarian Cycle

Fill in the blanks to complete the following paragraphs that describe hormonal interactions during the ovarian cycle.

In ____________ -tier control, the ____________ releases gonadotropin-releasing hormone. This influences ____________-tier control, in which the ____________ pituitary releases follicle-stimulating hormone and luteinizing hormone. In ____________-tier control, the ____________ secrete ____________ in response to LH. The ovaries convert androgens to ____________ and secrete ____________ and ____________ in response to FSH. ____________ stimulate a dominant follicle to mature to a ____________ follicle. The new vesicular follicle produces large amounts of ____________, triggering an LH surge through ____________ feedback. The LH surge and FSH trigger ____________.

Key Concept: What hormones control changes in the uterus during the uterine cycle?

____

____

Describe It: Endometrial Changes during the Uterine Cycle

Describe the sequence of events of the changes that occur to the endometrial lining during the uterine cycle in Figure 26.8. In addition, label and color-code key structures of the process. You may use text Figure 26.17 as a reference, but write the steps in your own words.

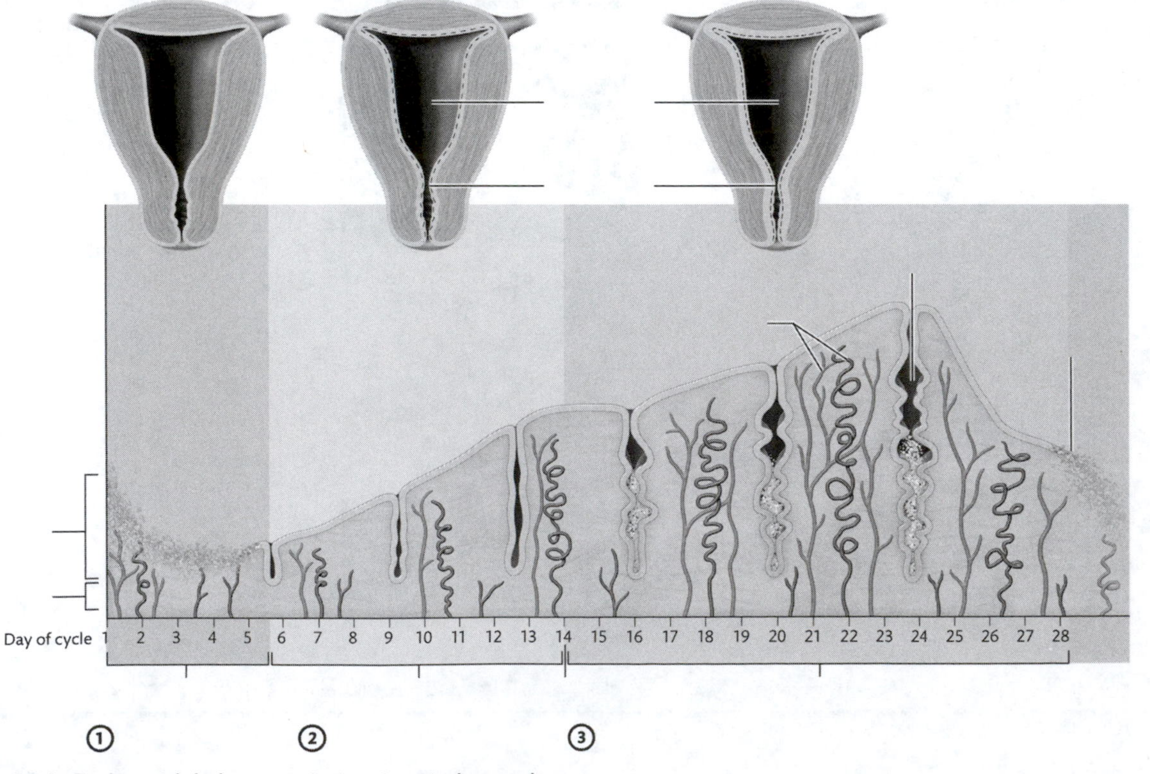

Figure 26.8 Endometrial changes during the uterine cycle.

Key Concept: What hormone causes the midcycle surge in LH? Does this result from positive or negative feedback? Explain your answer.

____

____

Draw It: Hormone Interactions during the Uterine Cycle

Draw a diagram of the two graphs that depict hormone interactions during the uterine cycle. You may use Figure 26.18 in your text as a reference, but make sure that your diagram and words are your own so that they make the most sense to you. (*Hint:* You are mainly describing what hormones go up and down, and why they do so.)

Key Concept: What are the female secondary sex characteristics, and which hormone is mainly responsible for producing them?

____

____

Module 26.6: Methods of Birth Control

Module 26.6 in your text explores the methods intended to prevent pregnancy, or methods of birth control. By the end of the module, you should be able to do the following:

1. Explain why changes in cervical mucus can predict a woman's monthly fertility.
2. Provide examples of how birth control methods relate to normal reproductive functions.

Build Your Own Glossary

Following is a table listing key terms from Module 26.6. Before you read the module, use the glossary at the back of your book or look through the module to define the following terms.

Key Terms for Module 26.6

Term	Definition
Barrier methods	
Spermicide	
Oral contraceptives	
Intrauterine device	
Vasectomy	
Tubal ligation	

Survey It: Form Questions

Before you read the module, survey it and form at least two questions for yourself. When you have finished reading the module, return to these questions and answer them.

Question 1: ____________________

Answer: ____________________

Question 2: ____________________

Answer: ____________________

Key Concept: What are the behavioral methods of birth control, and why are they generally less reliable?

____

____

Build Your Own Summary Table: Methods of Birth Control

As you read Module 26.6, build your own summary table about the different methods of birth control.

Category	**Name**	**Description**	**Theoretical failure rate**	**Typical failure rate**
Behavioral methods				
Permanent methods				
Hormonal methods				
Barrier methods				
Intrauterine devices and systems				

Module 26.7: Sexually Transmitted Infections (STIs)

Module 26.7 in your text introduces you to some of the most common sexually transmitted infections (STIs). By the end of the module, you should be able to do the following:

1. Predict factors or situations affecting the reproductive system that could disrupt homeostasis.
2. Describe the causes, symptoms, and potential complications of common sexually transmitted infections.

Build Your Own Glossary

Following is a table listing key terms from Module 26.7. Before you read the module, use the glossary at the back of your book or look through the module to define the following terms.

Key Terms for Module 26.7

Term	Definition
Chlamydia	
Gonorrhea	
Syphilis	
Trichomoniasis	
Human papillomavirus	
Genital herpes	

Survey It: Form Questions

Before you read the module, survey it and form at least two questions for yourself. When you have finished reading the module, return to these questions and answer them.

Question 1: ______________________________

Answer: ______________________________

Question 2: ______________________________

Answer: ______________________________

Key Concept: What are the similarities and differences between bacterial and parasitic STIs?

____

____

Team Up

Make a handout to teach the different kinds of sexually transmitted infections. There is more than one right way to do this, but suggestions may include: make a summary table that incorporates the information from this module about the type of organisms, signs, symptoms, treatments, and so forth; do an Internet search for images of organisms to either print or to use in producing your own drawings of them; search for the effects on untreated people—images and/or descriptions. At the end of the handout, write a few quiz questions. Once you have completed your handout, team up with one or more study partners and trade handouts. Study your partners' diagrams, and when you have finished, take the quiz at the end. When you and your group have finished taking all the quizzes, discuss the answers to determine places where you need additional study. After you've finished, combine the best elements of each handout to make one "master" handout covering STIs.

What Do You Know Now?

Let's now revisit the questions you answered in the beginning of this chapter. How have your answers changed now that you've worked through the material?

- What are the main organs and glands of the male and female reproductive systems, and what are their basic functions?

  ____

- What hormones are produced by the glands of the male and female reproductive systems, and how do they affect men and women, respectively?

  ____

- What are the common methods of birth control and common types of sexually transmitted infections?

  ____

27 Development and Heredity

We now examine what happens after a successful start to pregnancy, how a child develops during pregnancy, and how traits are inherited. This chapter covers embryonic and fetal development and the science of heredity.

What Do You Already Know?

Try to answer the following questions before proceeding to the next section. If you're unsure of the correct answers, give it your best attempt based on previous courses, previous chapters, or just your general knowledge.

- Where does a conceptus develop in the female reproductive system?

  ____

- What hormones are involved in the female cycle and pregnancy?

  ____

- What are genes, and how are they copied as part of a chromosome?

  ____

Module 27.1: Overview of Human Development

Module 27.1 in your text introduces you to human developmental biology, from embryology until immediately after birth. By the end of the module, you should be able to do the following:

1. Explain the difference between prenatal and postnatal development.
2. Describe the periods of prenatal development.

Build Your Own Glossary

Below is a table listing key terms from Module 27.1. Before you read the module, use the glossary at the back of your book or look through the module to define the following terms.

Key Terms for Module 27.1

Term	Definition
Prenatal period	
Postnatal period	
Pregnancy	

Term	Definition
Conception	
Gestation period	
Embryo	
Fetus	
Senescence	

Survey It: Form Questions

Before you read the module, survey it and form at least two questions for yourself. When you have finished reading the module, return to these questions and answer them.

Question 1: ______________________________

Answer: ______________________________

Question 2: ______________________________

Answer: ______________________________

Key Concept: What are the pre-embryonic, embryonic, and fetal periods?

____

____

Key Concept: What are the five stages of the postnatal period?

____

____

Module 27.2: Pre-embryonic Period: Fertilization through Implantation (Weeks 1 and 2)

Now we look at the period of development that occurs during the first 2 weeks after fertilization. When you finish this module, you should be able to do the following:

1. Describe the process of fertilization.
2. Describe the events of sperm capacitation, acrosomal reaction, sperm penetration, cortical reaction, and fusion of pronuclei.
3. Explain the formation and function of the extraembryonic membranes.

Build Your Own Glossary

Following is a table listing key terms from Module 27.2. Before you read the module, use the glossary at the back of your book or look through the module to define the following terms.

Key Terms for Module 27.2

Term	Definition
Fertilization	
Acrosomal reaction	
Cortical reaction	
Pronucleus	
Morula	
Trophoblast cells	
Inner cell mass	
Implantation	
Amnion	
Chorion	

Survey It: Form Questions

Before you read the module, survey it and form at least three questions for yourself. When you have finished reading the module, return to these questions and answer them.

Question 1: ____________________

Answer: ____________________

Question 2: ____________________

Answer: ____________________

Question 3: ____________________

Answer: ____________________

Key Concept: How does the cortical reaction prevent fertilization by more than one sperm?

____

____

Describe It: The Events Leading to Fertilization

Write in the sequence of events leading to fertilization in Figure 27.1, and label and color-code key components of this process. You may use text Figure 27.2 as a reference, but write the steps in your own words.

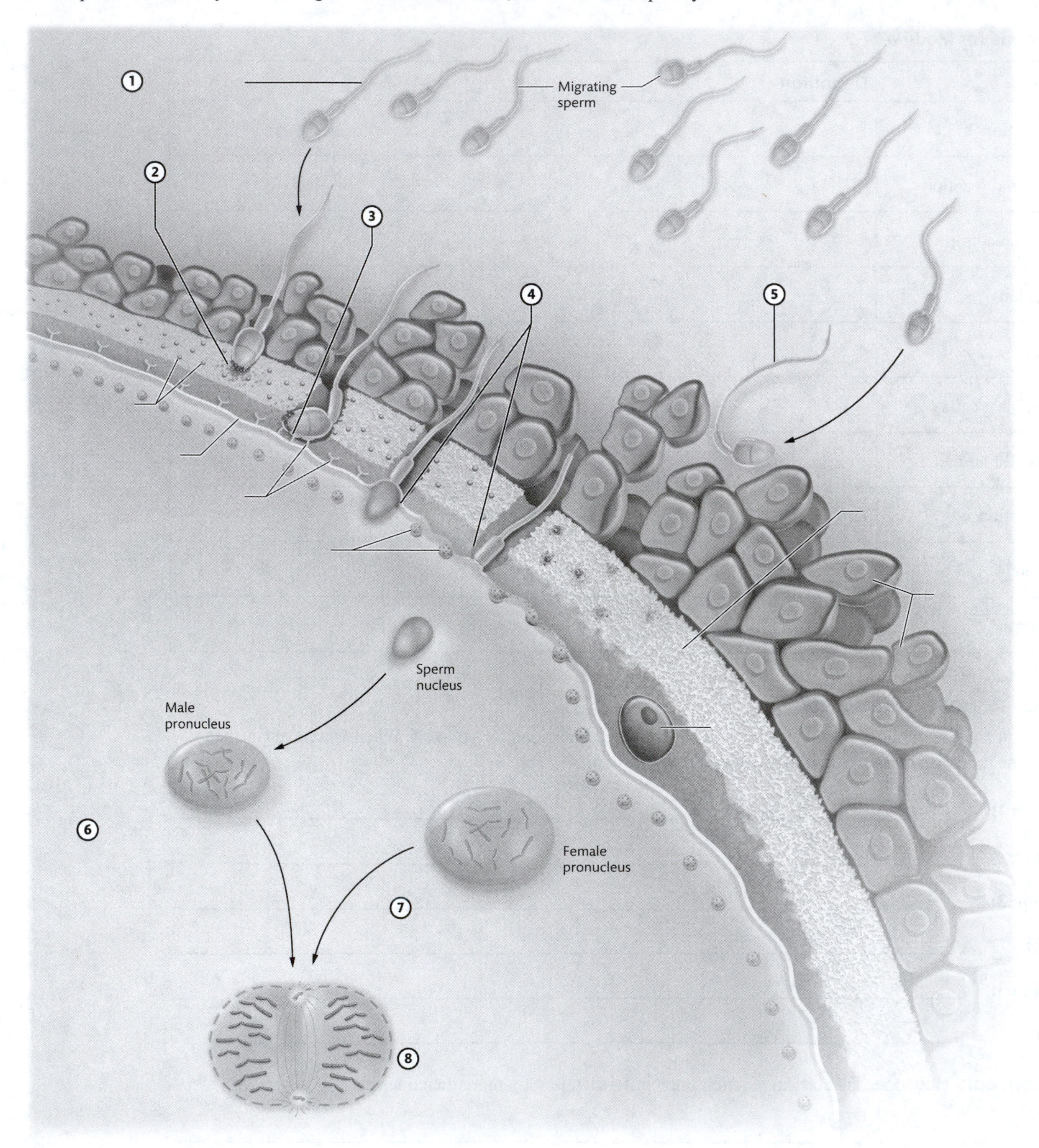

Figure 27.1 The events leading to fertilization.

Identify It: Fertilization through Implantation

Identify and color-code each stage of development that occurs from fertilization through implantation in Figure 27.2. Then, write a short description of each component.

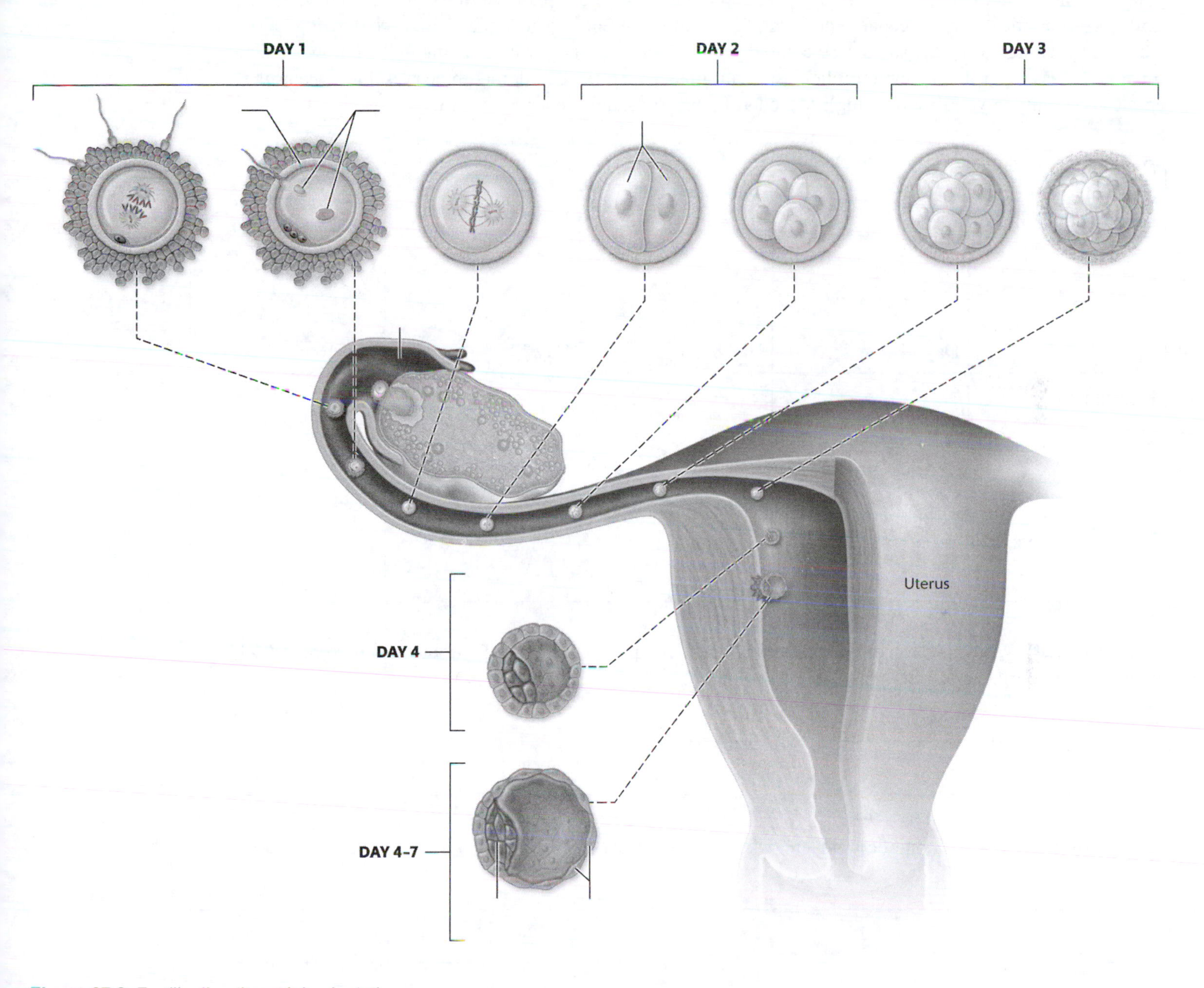

Figure 27.2 Fertilization through implantation.

Key Concept: What are the two distinct cell populations of the blastocyst? What does each cell population grow to form?

___

___

Draw It: Implantation of the Blastocyst in the Uterine Lining

Draw a sketch of the way in which the blastocyst implants and grows into the uterine lining. You may use text Figure 27.4 as a guide, but make sure it is your own drawing and the descriptions are in your own words. Label your drawings with the following terms, where applicable: Syncytiotrophoblast, Inner cell mass, Hypoblast, Epiblast, Amniotic cavity, Lacunae, Ectoderm, Mesoderm, and Endoderm. Underneath each drawing, describe what is happening during the stage. Use colored pencils to color parts of your drawing in the following shades: Inner cell mass = purple; Syncytiotrophoblast = tan; Epiblast = blue; Ectoderm = blue (yes, the same as the epiblast, but only in your last drawing); Mesoderm = brown; and Endoderm = yellow.

Drawing of growth of the blastocyst into the uterine lining	Days 4–7	Day 8	Day 12	Day 16
Description				

Key Concept: Which portion of the implanting blastocyst produces human chorionic gonadotropin (hCG)? What is the importance of hCG?

____

____

Key Concept: What is amniotic fluid, and why is it important?

____

____

Identify It: Formation of the Extraembryonic Membranes

Label each component of the formation of the extraembryonic membranes in Figure 27.3. Then, list the main function(s) of each component.

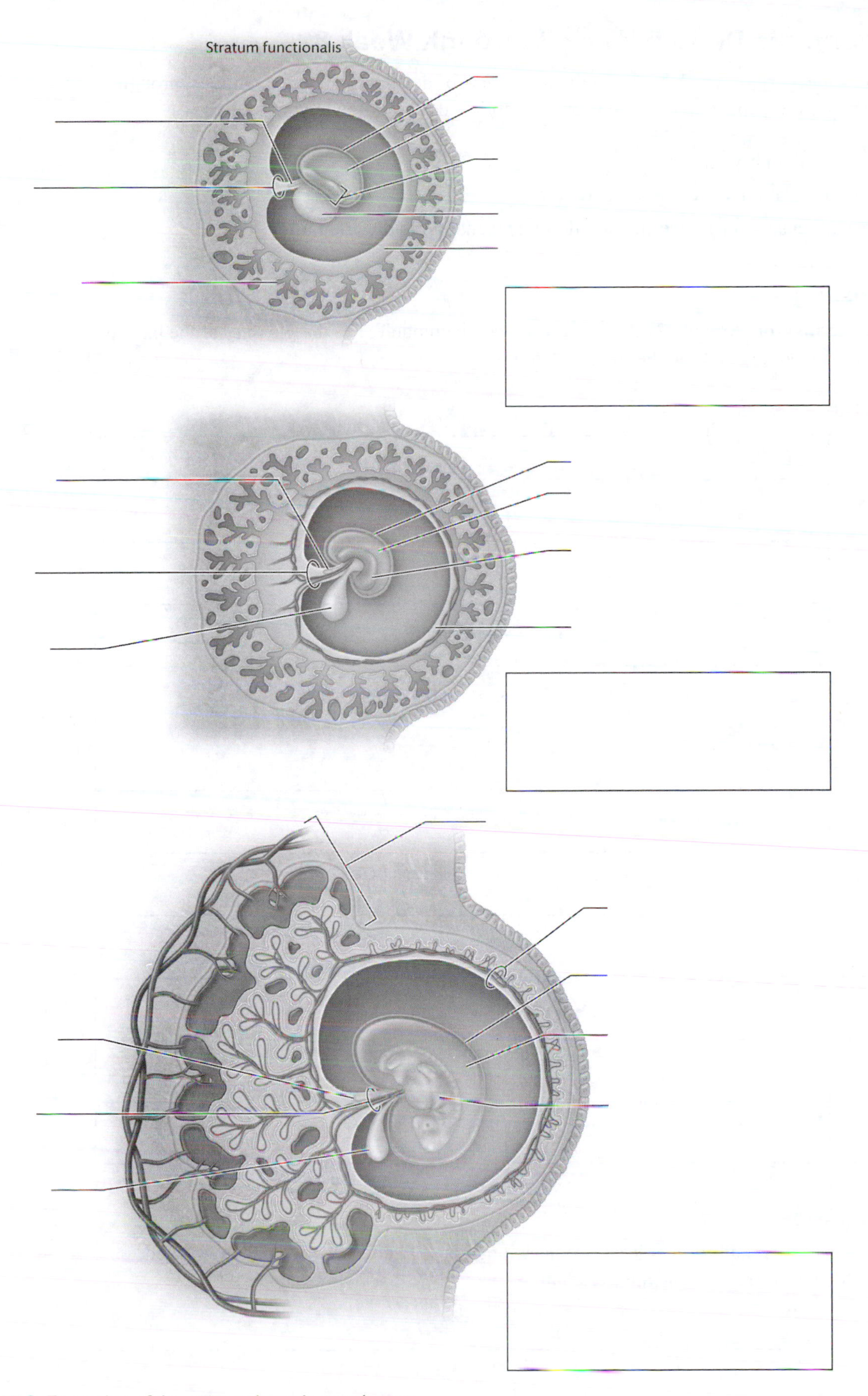

Figure 27.3 Formation of the extraembryonic membranes.

Module 27.3: Embryonic Period: Week 3 through Week 8

We now examine the embryonic period, which begins at the third week of development and continues through the eighth week. When you complete this module, you should be able to do the following:

1. Describe the major events of embryonic development.
2. Discuss the process of gastrulation and embryonic folding.
3. Describe organogenesis, focusing on ectoderm, mesoderm, and endoderm differentiation.

Build Your Own Glossary

Below is a table listing key terms from Module 27.3. Before you read the module, use the glossary at the back of your book or look through the module to define the following terms.

Key Terms of Module 27.3

Term	Definition
Gastrulation	
Primitive streak	
Cephalocaudal folding	
Transverse folding	
Organogenesis	
Neurulation	
Primary brain vesicles	
Neural crest cells	
Notochord	
Somites	
Teratogen	

Survey It: Form Questions

Before you read the module, survey it and form at least two questions for yourself. When you have finished reading the module, return to these questions and answer them.

Question 1: ______________________________

Answer: ______________________________

Question 2: ____

Answer: ____

Key Concept: What is gastrulation, and why is it important?

____

____

Complete It: Gastrulation and Formation of Germ Layers

Fill in the blanks to complete the following paragraph that describes gastrulation and formation of germ layers in the embryo.

Around week three, a thin groove on the dorsal surface of the epiblast develops, called the ____________ ____________. This establishes the ____________/____________ regions, the right and left sides, and the ____________/____________ (posterior/anterior) surfaces of the embryo. Some cells detach from the ____________ layer and move into and then underneath the primitive streak in a process called ____________. The first cells that migrate in this way become the inner germ layer, or ____________. More cells migrate to a position between the epiblast and endoderm, and become the middle layer, or ____________. The remaining cells of the epiblast form the outer layer, or ____________.

Key Concept: What is neurulation, and how does it set the stage for development of the nervous system?

____

____

Team Up

Make a handout to teach the differentiation of the three germ layers (ectoderm, mesoderm, and endoderm). You can use Table 27.2 in your text on page 1078 as a guide, but the handout should be in your own words and with your own diagrams. At the end of the handout, write a few quiz questions. Once you have completed your handout, team up with one or more study partners and trade handouts. Study your partners' diagrams, and when you have finished, take the quiz at the end. When you and your group have finished taking all the quizzes, discuss the answers to determine places where you need additional study. After you've finished, combine the best elements of each handout to make one "master" diagram for the differentiation of the three germ layers.

Module 27.4: Fetal Period: Week 9 until Birth (about Week 38)

Module 27.4 in your text turns to the fetal period, the longest period of development that extends from the beginning of week 9 until birth. At the end of this module, you should be able to do the following:

1. Describe the formation and function of the placenta.
2. Identify the major events of fetal development.
3. Describe the specific structures unique to fetal circulation.

Build Your Own Glossary

Below is a table listing key terms from Module 27.4. Before you read the module, use the glossary at the back of your book or look through the module to define the following terms.

Key Terms of Module 27.4

Term	Definition
Placentation	
Placental sinus	
Umbilical cord	
Decidua basalis	
Decidua capsularis	
Lanugo	
Vernix caseosa	
Vertex	
Ductus venosus	
Foramen ovale	
Ductus arteriosus	

Survey It: Form Questions

Before you read the module, survey it and form at least two questions for yourself. When you have finished reading the module, return to these questions and answer them.

Question 1: ____________________

Answer: ____________________

Question 2: ____________________

Answer: ____________________

Key Concept: What are the main vascular shunts, and what do they do?

____

____

Identify It: Development and Nutritive Functions of the Placenta

Identify and color-code each component of the placenta in Figure 27.4. Then, list the main function(s) of each component.

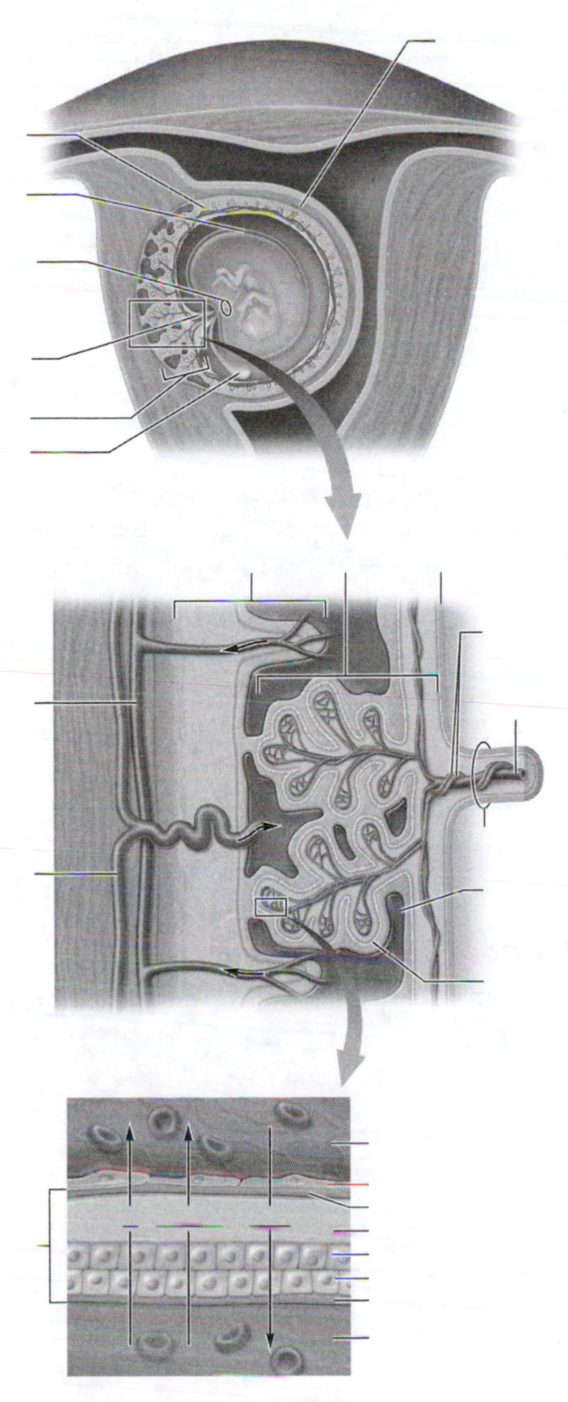

Figure 27.4 Development and nutritive functions of the placenta.

Build Your Own Summary Table: Major Events in Fetal Development

As you read Module 27.6, build your own summary table about the major events in fetal development by filling in the information in the table below.

Major Events in Fetal Development

Month(s)	**Major Changes Occurring in the Fetus**
Month 3	
Month 4	
Month 5	
Month 6	
Month 7	
Months 8 and 9	

Key Concept: What are the main vascular shunts, and what do they do?

____

____

Identify It: Comparison of Fetal and Newborn Cardiovascular Systems

Identify and color-code each item in the fetal and newborn cardiovascular systems in Figure 27.5. Then, list the main function of each component.

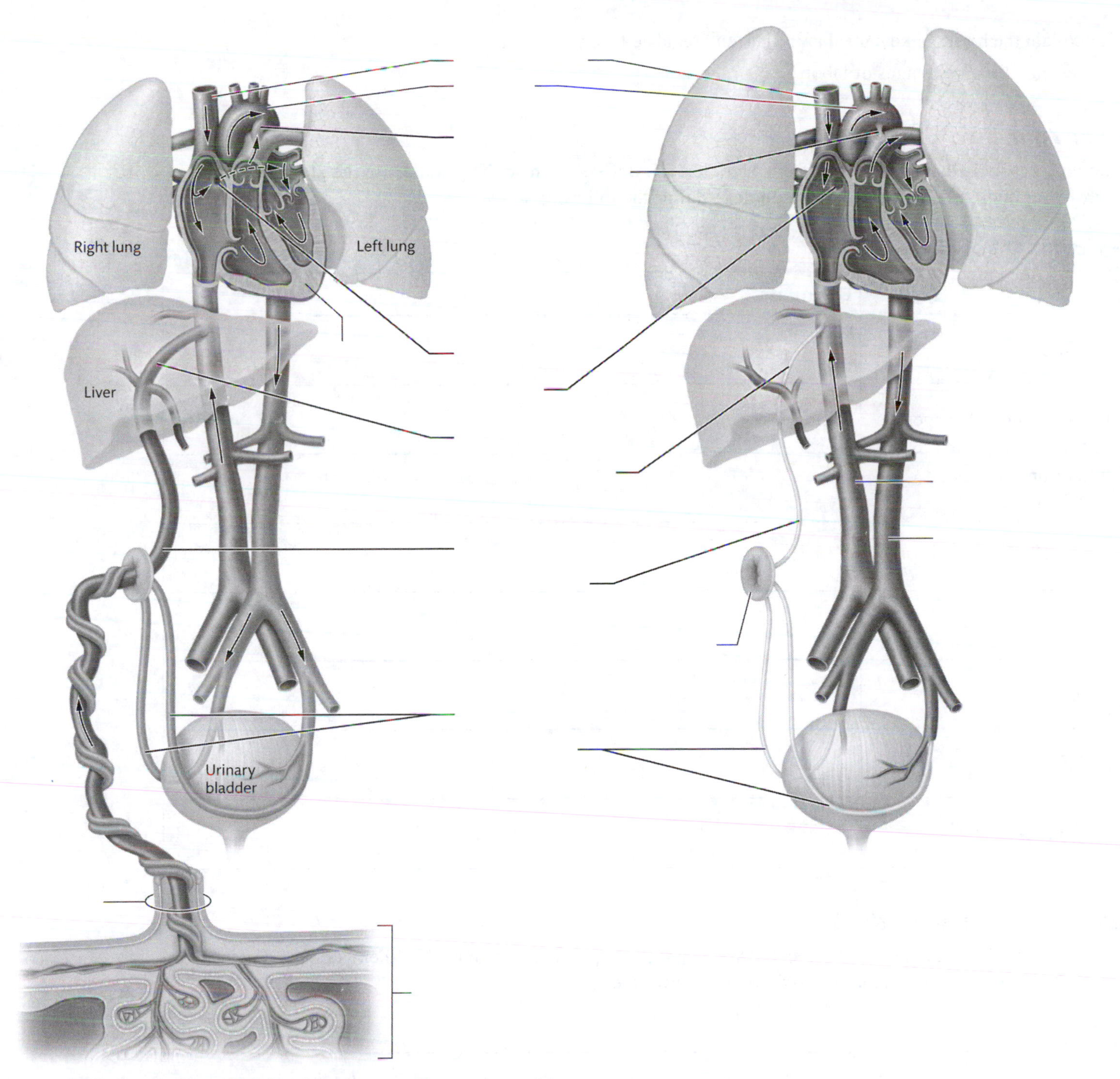

Figure 27.5 Comparison of fetal and newborn cardiovascular systems.

Module 27.5: Pregnancy and Childbirth

This module turns our attention to the anatomical and physiological changes that occur in the mother's body during this 9-month process of pregnancy, and then to the natural end result—childbirth. At the end of this module, you should be able to do the following:

1. Describe hormonal changes during pregnancy and the effects of the hormones.
2. Describe the functional changes in the maternal reproductive, endocrine, cardiovascular, respiratory, digestive, urinary, and integumentary systems during pregnancy.

3. Explain the hormonal events that initiate and regulate labor.
4. Describe the three stages of labor.

Build Your Own Glossary

Following is a table listing key terms from Module 27.5. Before you read the module, use the glossary at the back of your book or look through the module to define the following terms.

Key Terms of Module 27.5

Term	Definition
Relaxin	
Human placental lactogen	
Corticotropin-releasing hormone	
Parturition	
Dilation stage	
Expulsion stage	
Crowning	
Placental stage	

Survey It: Form Questions

Before you read the module, survey it and form at least three questions for yourself. When you have finished reading the module, return to these questions and answer them.

Question 1: ______________________________

Answer: ______________________________

Question 2: ______________________________

Answer: ______________________________

Question 3: ______________________________

Answer: ______________________________

Key Concept: What are the main hormonal changes that occur during pregnancy?

Build Your Own Summary Table: Maternal Anatomical and Physiological Changes

As you read Module 27.5, build your own summary table about the anatomical and physiological changes that occur in a woman's body during pregnancy by filling in the information in the table below.

Maternal Anatomical and Physiological Changes

System	Major Changes Occurring in the Mother
Reproductive	
Cardiovascular	
Respiratory	
Digestive	
Urinary	
Integumentary	

Key Concept: What pregnancy side effects are directly related to the enlargement of the uterus into the abdominal cavity in the last trimester?

Describe It: Initiation and Regulation of Childbirth

Write in the sequence of events of the initiation and regulation of childbirth in Figure 27.6. You may use text Figure 27.14 as a reference, but write the steps in your own words.

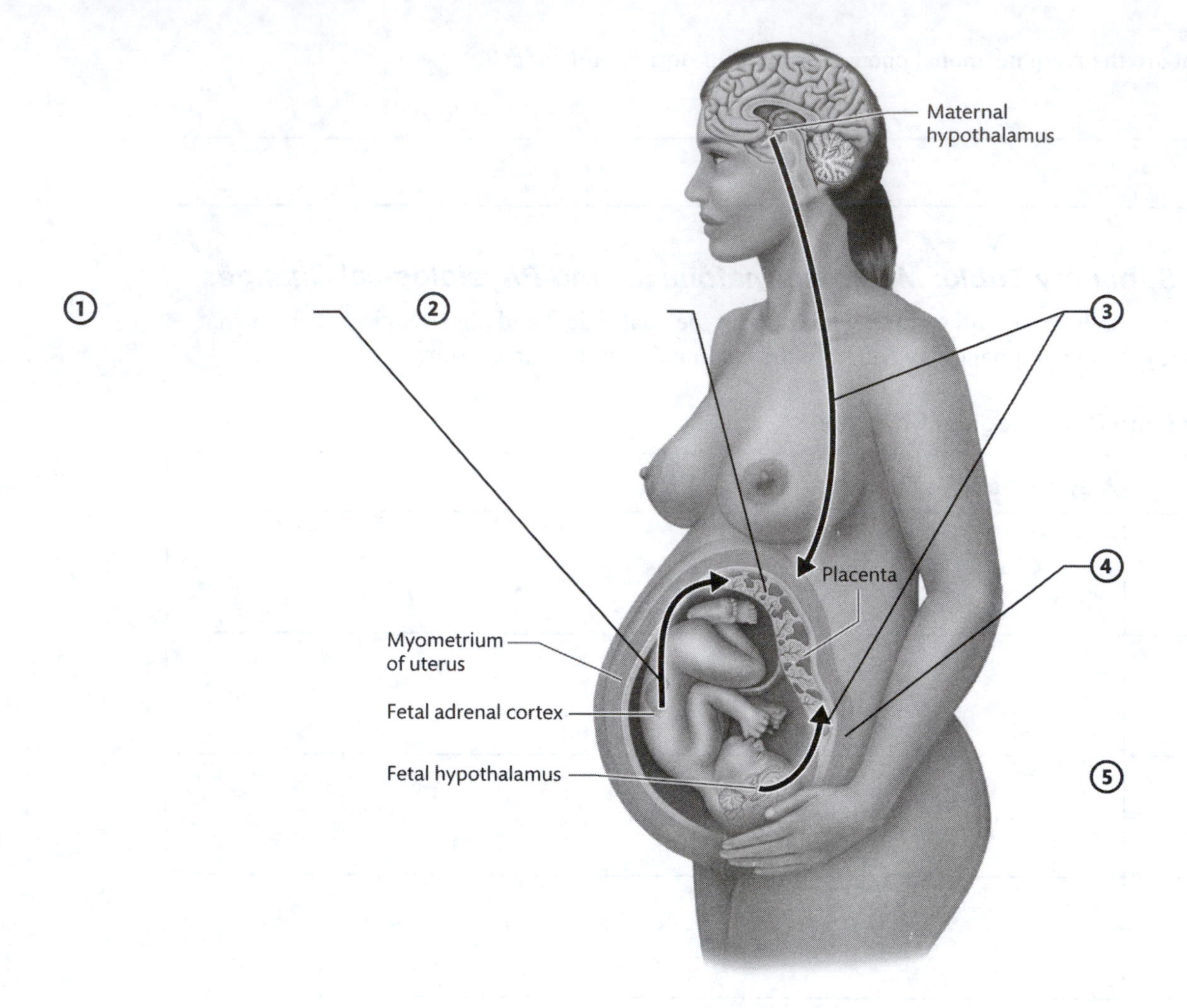

Figure 27.6 Initiation and regulation of childbirth.

Describe It: The Three Stages of Labor

Write a paragraph describing the three stages of labor as if you were explaining it to an audience of nonscientists.

Module 27.6: Postnatal Changes in the Newborn and Mother

Module 27.6 in your text examines the postnatal period, but mostly focuses on the neonatal period: immediately after and the first few weeks after birth. By the end of the module, you should be able to do the following:

1. Describe the major changes that occur to the newborn during the neonatal period.
2. Describe the major anatomical and physiological changes to the mother after parturition.
3. Describe the structure and function of the mammary glands.
4. Explain the hormonal regulation of lactation.

Build Your Own Glossary

Below is a table listing key terms from Module 27.6. Before you read the module, use the glossary at the back of your book or look through the module to define the following terms.

Key Terms for Module 27.6

Term	Definition
Neonate	
Apgar score	
Postpartum period	
Lochia	
Lactation	
Colostrum	
Meconium	
Let-down reflex	

Survey It: Form Questions

Before you read the module, survey it and form at least two questions for yourself. When you have finished reading the module, return to these questions and answer them.

Question 1: ______________________________

Answer: ______________________________

Question 2: ______________________________

Answer: ______________________________

Key Concept: Why is "the first breath" considered the most dramatic change that occurs in the newborn after birth?

____

____

Describe It: Hormonal Regulation of Lactation

Write in the sequence of events in the hormonal regulation of lactation in Figure 27.7. You may use text Figure 27.16 as a reference, but write the steps in your own words.

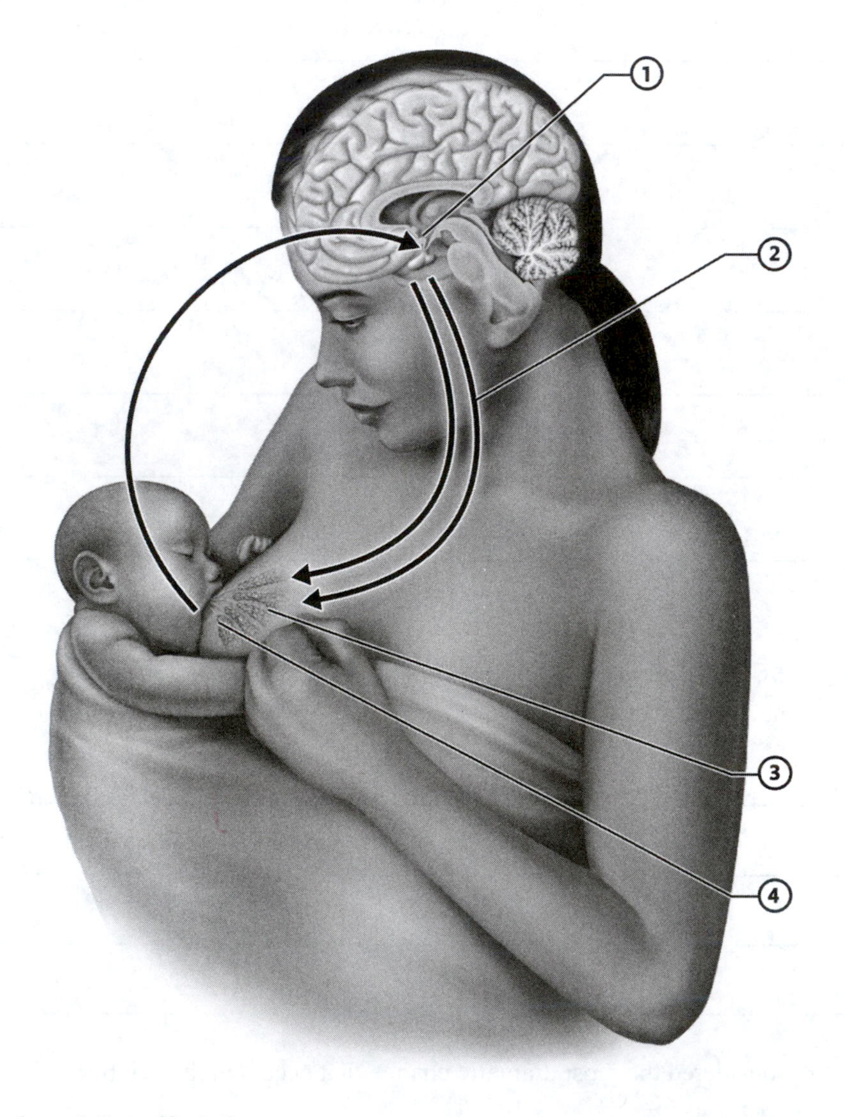

Figure 27.7 Hormonal regulation of lactation.

Key Concept: What is the connection between postpartum depression and the changes in maternal hormone levels after birth?

____

____

Module 27.7: Heredity

Module 27.7 in your text introduces you to the science of heredity or how genes transmit characteristics from parent to offspring. By the end of the module, you should be able to do the following:

1. Define chromosome, gene, allele, homologous, homozygous, heterozygous, genotype, and phenotype.
2. Analyze genetics problems involving dominant and recessive alleles, incomplete dominance, codominance, and multiple alleles.
3. Explain how polygenic inheritance differs from inheritance that is controlled by only one gene.
4. Explain how environmental factors can modify gene expression.
5. Discuss the role of sex chromosomes in sex determination and sex-linked inheritance.

Build Your Own Glossary

Below is a table listing key terms from Module 27.7. Before you read the module, use the glossary at the back of your book or look through the module to define the following terms.

Key Terms for Module 27.7

Term	Definition
Heredity	
Genome	
Autosomes	
Alleles	
Homozygous	
Heterozygous	
Genotype	
Phenotype	
Punnett square	
Incomplete dominance	
Sex-linked traits	
Polygenic inheritance	

Survey It: Form Questions

Before you read the module, survey it and form at least two questions for yourself. When you have finished reading the module, return to these questions and answer them.

Question 1: ______________________________

Answer: ______________________________

Question 2: ______________________________

Answer: ______________________________

Key Concept: What are alleles, and how do they contribute to the variety of characteristics in individuals?

____

____

Practice It: The Punnett Square

Having significant hair growth on the back of the hands is a dominant trait (compared to having relatively little hair on the hands). Fill in the boxes of the Punnett square to determine the possible offspring from the parental alleles shown.

Mother ↓ Father →	**H**	**h**
H	Genotype: ____________	Genotype: ____________
h	Genotype: ____________	Genotype: ____________

Key Concept: How does a blood type such as "AB" demonstrate codominance?

____

____

Key Concept: Why are X-linked disorders expressed far more often in males?

____

____

Team Up

Make a handout to teach how polygenic inheritance can produce the wide range of variation we find in traits such as height and skin color. You can use Figure 27.20 in your text on page 1097 as a guide, but the handout should be in your own words and with your own diagrams. At the end of the handout, write a few quiz questions. Once you have completed your handout, team up with one or more study partners and trade handouts. Study your partners' diagrams, and when you have finished, take the quiz at the end. When you and your group have finished taking all the quizzes, discuss the answers to determine places where you need additional study. After you've finished, combine the best elements of each handout to make one "master" diagram for polygenic inheritance.

What Do You Know Now?

Let's now revisit the questions you answered in the beginning of this chapter. How have your answers changed now that you've worked through the material?

- Where does a conceptus develop in the female reproductive system?

  ____

- What hormones are involved in the female cycle and pregnancy?

  ____

- What are genes, and how are they copied as part of a chromosome?

  ____

Credits

Chapter 4 4.5: William Karkow, Pearson Education, Inc.

Chapter 5 5.3: Lisa Lee, Pearson Education, Inc.; 5.4: William Karkow, Pearson Education, Inc.

Chapter 8 8.1a: Larry DeLay, Pearson Education, Inc.

Chapter 10 10.1: Don W. Fawcett/Science Source

Chapter 16 16.9b: Stephen W. Downing, Pearson Education, Inc.

Chapter 19 19.1a: William Karkow, Pearson Education, Inc.

Chapter 21 21.2b: Jupiterimages/Stockbyte/Getty Images; 21.2d: Living Art Enterprises, LLC/Science Source; 21.4d: Nina Zanetti, Pearson Education, Inc.

Chapter 26 26.7 top: Stephen W. Downing, Pearson Education, Inc.; 26.7 second and third from top: William Karkow, Pearson Education, Inc.; 26.7 middle: Ed Reschke/Photolibrary/Getty Images; 26.7 third from bottom: Claude Edelmann/Science Source; 26.7 second from bottom: William Karkow, Pearson Education, Inc.; 26.7 bottom: Stephen W. Downing, Pearson Education, Inc.